Interpreting the Earth

Interpreting the Earth

Robert R. Compton

STANFORD UNIVERSITY

Harcourt Brace Jovanovich, Inc.

NEW YORK   CHICAGO   SAN FRANCISCO   ATLANTA

To my Mother and Father

ISBN: 0-15-541547-6
Library of Congress Catalog Card Number: 76-62596
Printed in the United States of America

Cover: Jasper specimen (approximately 3″ × 6″) from the original Bigg's claim in Oregon. The color bands and dendrites are iron and manganese oxides deposited by water in the rock's minute openings. Collection of Mr. and Mrs. D. R. Noel, Noel's Earthen Artifacts, San Jose, California.

Base relief map: Richard Edes Harrison, copyright by A. J. Nystrom & Co.

Design: Geri Davis

Illustration: John Waller with Jack Foster, Jean Foster, Mark Schroeder, and Judy Waller

Preface

his book stemmed from discussions with my beginning students, who have been much interested in the classical content of geology but have also wanted a more intimate look at places and scientific studies, a more relevant approach to the real world. These students have been particularly intrigued with things in their natural settings, evidently because one can there see interrelations—the wholeness of things. When I set out to write a book, however, I faced the fact that many students could not take field trips or work with laboratory materials. I therefore tried to describe areas and studies as though the students were there, following the scene and the data as well as the rationale of an investigation. To be able to use that approach, I traveled widely and examined most of the areas described in the book. In addition to collecting samples and taking photographs, I talked (or corresponded) with most of the original investigators and compared their papers and diagrams with what I could see myself. Most of the studies are thus recent ones, many of them made by persons with whom I was already acquainted.

It seemed desirable to start the book with things for which we all have a natural interest and "feel," and the topic of water and water-related processes seemed ideal. Chapter 1 thus opens with some water-related problems and then gets directly to the nature of water and how it interacts with sediment. Chapter 2 presents a view of a total river system, and Chapter 3 relates how some remarkable studies of a dry-land river, the Rio Grande, help us understand erosion as well as how all rivers work. Chapters 4, 5, and 6 then carry surface water systems to their natural completion—from rivers to estuaries to the sea. The importance of glaciers and related climatic variations become apparent through their connections with the sea, and studies of the great ice sheets (Chapter 7) serve to return the student to land processes.

Minerals are only mentioned in the first seven chapters, but my strong hope is that the student will read Supplementary Chapter 1 (Minerals) at some stage before reading Chapter 8, on soils. The color plates and other photographs in the supplementary chapters should help in visualizing minerals and rocks as real things. Supplementary Chapter 2 (Rocks) might also be read at an early stage; however, rocks are also introduced broadly in the main chapters, first as sediments (in the first nine chapters) and then as sedimentary and metamorphic rocks (Chapter 10), and finally as igneous rocks (Chapter 12). Earthquakes (Chapter 11) are placed within this sequence because water in rocks evidently sets off fault movements, and water-charged sediments are the chief cause of strong surface motions and earthquake damages.

The concept of geologic time and the stratigraphic basis of geology are presented in Chapters 13 and 14, and studies of layered rocks are then expanded to the interpretation of folds, faults, and rock flow in Chapter 15. The student now has the tools with which to examine the earth's deeper structure and movements. The earth's main layers, their state of inner balance, their origin, and their systematic movements (plate tectonics) are covered in Chapters 16, 17, 18, and 19 and are brought together in a review of North America's recent geologic history in Chapter 20. Chapter 21 closes the book with the question of how to use this knowledge in supplying ourselves adequately but conservatively with natural resources.

I have tried to make the chapters stand sufficiently on their own so that they can be used in other sequences if desirable. The book could be started with the supplementary chapters on rocks and minerals, for example, if supportive lectures or laboratory exercises were provided. I have tried to suggest additional readings and projects that would permit any chapter to be expanded into a week or so of studies. This may be an important possibility in places where the local geology is highly specialized.

Many persons have helped me during my studies and writing. The book would have ended at an early stage without the supportive advice of my beginning students and of Victoria R. Todd, who was a graduate student at Stanford at that time. James C. Ingle, Jr., and Caroline M. Isaacs gave enthusiastic support from the start, and Robert H. Meade introduced me to a new world of literature on rivers, estuaries, and the sea and helped plan visits to oceanographic institutions on the Altantic coast. I wish also to thank persons and agencies who went considerably out of their way to help during my investigations: for topics in Chapter 2, the U.S. Geological Survey and the Virginia Division of Water Resources in Richmond; for Chapter 3, the Bureau of Reclamation staff in Albuquerque and John P. Wilson and Helene Warren, Museum of New Mexico, Santa Fe; for Chapter 4, Robert N. Ginsburg, University of Miami and James D. Howard, University of Georgia; for Chapter 5, Donald J. P. Swift, NOAA–AOML, John C. Ludwick, Old Dominion University, Norfolk, and David B. Duane and E. M. Meisburger, U.S. Army Coastal Engineering Research Center; for Chapter 6, Charles D. Hollister, Frank T. Manheim, John D. Milliman, David A. Ross, and Gilbert T. Rowe, Woods Hole Oceanographic Institution, and Monty A. Hampton, U.S. Geological Survey; for Chapter 7, Herbert E. Wright, Jr., Paul J. Conlon, and Roger LeB. Hooke, University of Minnesota, and the staff of the U.S. Army Cold Regions

Research and Engineering Laboratory; for Chapter 8, J. Hatten Howard III, University of Georgia; for Chapter 9, Leland H. Gile and John W. Hawley, Soil Conservation Service; for Chapter 10, Allan B. Griggs, U.S. Geological Survey; for Chapter 11, George Plafker, U.S. Geological Survey; for Chapter 12, Clifford A. Hopson, University of California at Santa Barbara, Aaron C. Waters, University of California at Santa Cruz, and Wendell A. Duffield, U.S. Geological Survey; for Chapter 13, James C. Ingle, Jr., Caroline M. Isaacs, and Gerta Keller, Stanford University; for Chapter 14, Robert G. Young, then at Mesa College, Grand Junction, and James D. Howard, University of Georgia; for Chapter 15, Victoria R. Todd, U.S. Geological Survey; for Chapter 16, Hugh L. Davies, Geological Survey of Papua New Guinea, and George A. Thompson and Tjeerd H. Van Andel, Stanford University; for Chapter 17, George M. Stanley, Fresno State University; for Chapter 18, Odette B. James and Don E. Wilhelms, U.S. Geological Survey, and Carleton B. Moore, Arizona State University; for Chapter 20, Clarence A. Hall, University of California at Los Angeles, John T. Hack, U.S. Geological Survey, and Arthur D. Howard, Stanford University; for Chapter 21, J. David Lowell, Pillar, Lowell, and Associates.

The final manuscript was read in its entirety by Russel R. Dutcher, University of Southern Illinois at Carbondale; Warren Huff, University of Cincinnati; Harry Lawrence, Pasadena City College; Richard Paull, University of Wisconsin at Milwaukee; Virginia R. Gilson, and Christine Bucey Lockwood. In addition, specific chapters were reviewed by the following persons: Robert H. Meade, U.S. Geological Survey (Chapters 1, 2, 3, 4, and 5); Carl F. Nordin, Jr., and William W. Emmett, U.S. Geological Survey (Chapter 3); Herbert E. Wright, Jr., and Roger LeB. Hooke, University of Minnesota (Chapter 7); Leland H. Gile and John W. Hawley, Soil Conservation Service (Chapter 9); George Plafker and C. Barry Raleigh, U.S. Geological Survey (Chapter 11); Clifford A. Hopson, University of California at Santa Barbara (Chapter 12); Caroline M. Isaacs, Stanford University (Chapters 13, 14, 15, 16, and 17); Robert G. Young, Bendix Field Engineering Corporation (Chapter 14); Victoria R. Todd, U.S. Geological Survey (Chapter 15); and Odette B. James and Don E. Wilhelms, U.S. Geological Survey (Chapter 18).

I am grateful to the many additional persons who helped with specific items. Those who supplied samples or photographs, or gave permission to copy drawings, are credited in the figure captions. I wish also to thank all who helped prepare the manuscript and the book. Bea Sanders and Christine Bucey Lockwood typed early versions of the chapters, and Vicki L. LaBrie prepared the entire final typescript. Candace B. Compton printed my photographs (all the photographs without credits). Ruperto Laniz and Perfecto Mari helped with rock preparations and photography. The expert work and help of Geri Davis, who designed the book, and John Waller, the principal artist, are appreciated greatly, as is the enthusiastic support of the Harcourt Brace Jovanovich staff.

Robert R. Compton

Contents

21. Materials and Energy from the Earth   451

Supplementary
Chapter

1. Minerals   475

CONTENTS

Interpreting the Earth

One of our shrinking estuaries—houses constructed on a filled part of San Francisco Bay, which formerly extended from the foreground to the shore in the far distance. Such fills not only reduce the bay's potential for supporting living things but are likely to be hazardous during major earthquakes. U.S. Geological Survey photograph by Norman Prime.

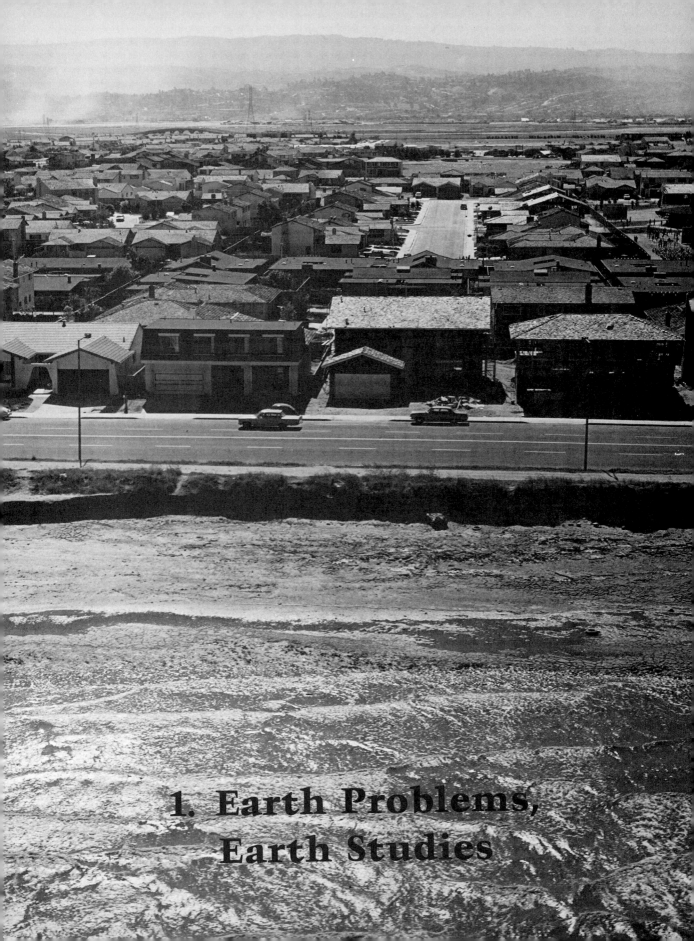

1. Earth Problems, Earth Studies

Earth Problems, Earth Studies

The San Francisco Bay System
A Plan for the Bay
Causes and Remedies
Beginning a Specific Study
    The structure of water
    Particles in water
Water and Sediment in Motion
    Turbulence
    Formation of waves
    Some findings overall

We, the total society, do not understand the earth well. We tend to use it unwisely and this is causing increasing problems. Fuels, metals, water, and food are in short supply, and destruction of our surroundings, especially of living things, has reduced our spirit. Shortages of materials and losses of spirit lead to human conflict and thereby to additional losses. One can think of the waste and anguish of recent warfare or of the increasing destruction of clean water (and its life) that was finally publicized in the 1960s.

Those crises resulted in many promises but few remedies. Problems were treated piecemeal, as though totally unrelated, a practice that often led to further problems. A new consciousness, however, arose in that decade. It was expressed largely as a grassroots movement among young people, but it was supported by the press, by literary writings, and by scientific studies. This major step toward understanding was actually a rediscovery of the one-ness or wholeness of the earth and its living things.

It is very important that we understand that this consciousness is not the "increased awareness of the environment" that one still hears of occasionally, or the popular concept called "the family of man." Consider this carefully: "the environment" is the *human* environment. It is the world seen through human eyes and concocted partly from the human ego. If we want to truly understand the earth, including ourselves, we must move out of that position.

It is necessary to see, too, that I am not simply restating the science called ecology. Ecology treats the relations among all organisms (life forms) and their inorganic surroundings. In actual fact, though, it is a biologic science and deals rather lightly with inorganic surroundings. This is not to put ecology down in the least. Broader understanding of the earth would be impossible without the studies that discovered the great food chains and other intricate webs that connect living forms.

But an intricate web of interaction also exists among nonliving things. It reaches back through billions of years to the primitive earth and presumably effected the origin of life itself. Like the processes of the living world, some processes of the nonliving world are in a delicate balance, and some are re-markably resilient. All are related and all have come to be intertwined with living things.

This total linkage of living and nonliving things constitutes what I will call an *earth system*. In Figure 1-1, for example, the lines represent link-ages, each an interdependency that may be simple or highly complex. Remove one of the lines and imagine the changes this action would impose on the others. Some effects would be large, some small, but I think you will agree that nothing can be changed without affecting the whole. Expanded gigantically, the diagram might illustrate the oneness of the earth and why some of its crises cannot be resolved by dealing with parts or processes separately.

Our need is to know how the total system works, but we quickly find that we cannot explore it all at once. We must start with certain parts and discover their connections. To do this precisely we have to define or delimit each part and such a part may also be called a system. The Mississippi River can be treated as an earth system if we draw a geographic line around the entire area it drains and concentrate on all interactions within that perimeter.

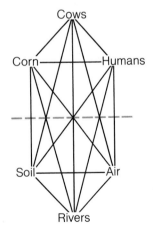

FIGURE 1-1

Linkages among some living and nonliving parts of an earth system. Each line is supportive to some degree of the entire array.

FIGURE 1-2
San Francisco Bay and its surroundings. The map shows the original extent of the bay (the outer perimeter) as compared to the parts filled with earth and rock or closed off by dikes. The medium gray areas in the bay are shallows susceptible to being filled. Heavy lines indicate faults, steeply inclined fractures (breaks) along which the earth is offset. The photograph is an image relayed from Earth Resources Satellite I on May 23, 1973.

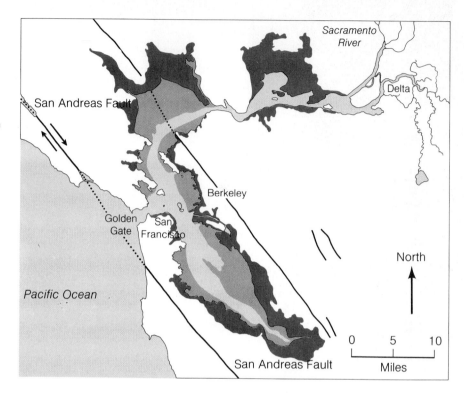

Connections between systems will be discovered by working within them, especially near their boundaries. This method is no more or less scientific than living: each of us was born into a small world that we got to know well and then gradually enlarged. By tracing the flow of the Mississippi, tidal currents from the Gulf of Mexico become obvious and may eventually lead to understanding the river's total attachment to the sea.

A more complete example should help explain these ideas, and I have chosen San Francisco Bay because it has been afflicted with problems that have led to recent human conflicts. In addition, I have lived near it for many years and can sense it with some heart.

The San Francisco Bay System

Originally almost 700 square miles in extent, San Francisco Bay has been shrunk by human-emplaced dikes and earth-fills to about 400 square miles of open water (Figure 1-2). As the bay was diminished, the human population around it grew, from 1.7 million in 1940 to about 4 million in 1965, and perhaps to 7.5 million in 1990. From one of the hillside communities surrounding it the bay seems a quiet and simple body of water, but this view is misleading. Ocean tides move a gigantic volume of water in and out of it twice daily, and inflow from the Sacramento and San Joaquin rivers affects its dynamics (movements) greatly (Figure 1-2). The bay is an *estuary*, which

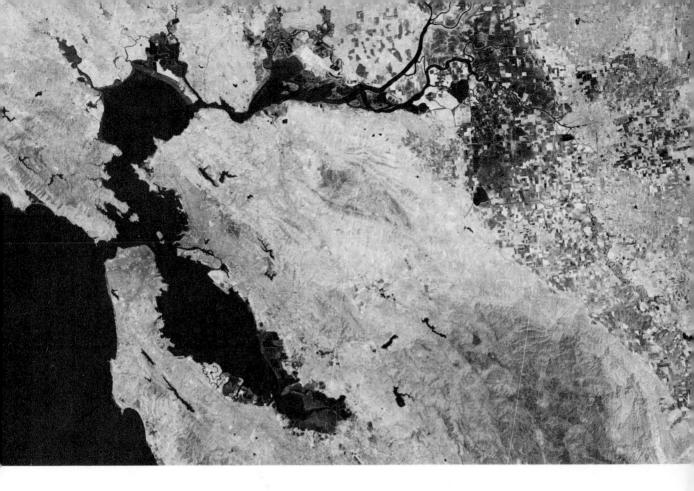

means that at the same time it is an extension of a river and the sea. It shares the water, the life, and the processes from each source, and in the mixture something unique is created. The resulting system is greater than the sum of its parts.

Most of the earth's surface under the bay and immediately around it has very little relief (is almost flat) and is therefore affected greatly by the tides, which raise and lower the bay's surface by 4 to 12 feet. Peripheral flats called *marshes* lie just underwater at high tide and are covered by plants. Water is carried to them and away from them through a myriad of *tidal channels* (Figure 1-3). The gently sloping bottom of the bay is partly exposed at low tide when it is called a *mudflat*. The principal basins and channels of the bay are hidden at all tidal stages by turbid (muddy) water, but depth soundings show them to slope gradually through each lobe of the bay in such a way that they eventually connect to the ocean through the Golden Gate (Figure 1-2).

From an ecological viewpoint, the bay has become human-dominated. In the modern world such systems are studied by *planners*, specialists who attempt to integrate all uses as effectively as possible. To these persons the bay is a "multiuse system," and we can appreciate the planner's task by examining a list of principal human uses:

1. Refuse and sewage disposal
2. Industries needing water access
3. Shipping and port facilities

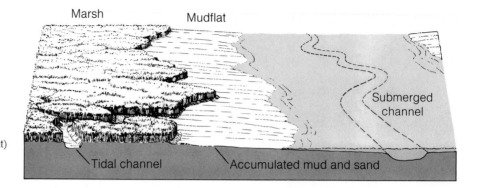

Marsh　　　　Mudflat

Submerged channel

FIGURE 1-3

View and cutaway section (front) showing an arm of the bay at low tide.

Tidal channel　　　Accumulated mud and sand

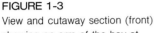

FIGURE 1-3

View and cutaway section (front) showing an arm of the bay at low tide.

4. Airports on fill (earth and rock used to raise the surface above sea level)
5. Highways on fill; bridges
6. Recreation (boating, fishing, hunting, swimming)
7. Housing and public facilities on fill (see the chapter opening page)
8. Commercial fishing
9. Salt production
10. Dredging of shells (for cement), sand and gravel

The greatest single human use is not listed and is difficult to define, though it might be called living-in-the-vicinity. The bay improves the environment by moderating the climate in both summer and winter. Far more than that, it presents a sense of space and variations of color, form, and movement that are unquestionably beautiful and therefore attractive to humans.

Uses by other animals have decreased greatly as a result of human uses. This change is important because the bay is the only major estuary on 1,200 miles of coast. It provides the broad gradation from salt to fresh water necessary to the life cycle of many sea animals. Fish and shellfish use it for a living and breeding ground. It is also a major wintering ground for migratory wildfowl and a year-round feeding ground for hundreds of thousands of water-based birds.

The nonliving parts of the system became reasonably well understood only after human uses had put them under great pressure. A fundamental attribute of the bay turned out to be as simple a thing as the amount of oxygen dissolved from the air. Oxygen is required by all animals living in the bay. It also promotes the chemical decomposition of sewage and industrial wastes as well as of the natural organic wastes of the bay. The oxygen consumed in these processes limits the amount of wastes that the bay can contain before becoming oxygen-depleted and foul. The bay gains oxygen in several ways, chiefly: (1) directly from the air, especially when the wind makes waves curl and froth; (2) from the inflow of fresh water, which carries approximately 25 percent or so more oxygen than sea water; (3) from the tidal inflow of sea water; and (4) from marsh plants, which produce oxygen. Each means of gaining oxygen is a connection within the system and each has been inadvertently affected by humans. By reducing the bay's surface area we have reduced its capacity to gain oxygen from the air. By damming and diverting rivers and streams we have reduced the inflow of oxygen-rich fresh water. By reducing the volume of the bay we have reduced the amount of water flowing in and out with the tides. By diking and filling the bay's shallows we have destroyed or locked off 80 percent of its marshes.

From the relations of oxygen to the system, the linkages expand in large numbers. In addition to generating oxygen, the marsh plants are the basic source of food for the life chains that pass up through small invertebrates (chiefly crustaceans) to fishes, birds, humans, and other forms. Also, the plants grow so close together that they catch the fine mud and organic particles that get stirred up by wind waves and by tidal currents that well into the marshes at rising tide. When the tide level falls, a thin accumulation of ooze is left high and dry for several hours, and sun and air combine to oxidize its fine organic particles and thus help to purify it. The accumulating mud, in turn, becomes the base of growth for the plants, which use the nutrients produced by the breakdown of the organic wastes. Nutrients not used by the plants are freed to the bay where they nourish minute green algae and diatoms (unicellular plants having a shell of silica). These minute plants are a useful food source in moderate numbers but tend to die out and befoul the water when generated massively. The marsh plants thus have many important connections to both the living and nonliving parts of the system. The gain realized by their destruction was small and went to a comparatively few persons; the losses generated are already large, affect all life in and near the bay, and will continue to increase into the distant future.

The mud accumulating around the base of the marsh plants also provides some connections to the geological past as well as to the future. Slow accumulation of similar mud over tens of thousands of years has resulted in a mud layer, ten to several hundred feet thick, under almost all the bay. This material must bear the weight of bay-fills (chiefly stony soil) and whatever people build on them. The surface of the fill may subside somewhat irregularly, although this is not a major engineering problem. The serious problem is that the region has experienced several major earthquakes in the past 100 years, and in the most violent (that of 1906) damage was by far greatest on filled parts of the bay. Evidently the porous earth-fills contained abundant water and the jiggling of the earthquake made them flow like a thick liquid. How the more recently loaded bay mud will respond to major tremors is not really known.

We do know that the earthquake of 1906 and others since were generated when the earth slipped suddenly along fractures that extend deep underground, and that such fractures, called *faults*, are widespread in the bay area (Figure 1-2). The faults link the bay area to powerful deep-earth processes. The San Andreas fault divides the western part of California into two huge masses pressing horizontally past one another in the sense shown by the two arrows. The masses stick along the fault surface until pressures build to the point that they suddenly slip—thus generating an earthquake. The mountains around the bay and the valley partly filled by the bay owe their geographic shapes and their vertical relief to other, similar movements in the past. All these movements will continue to affect the bay and our uses of it, as far into the future as we can imagine.

A Plan for the Bay

Degradation of the bay by filling and pollution had become so noticeable by the early 1960s that many citizens were aware of it and in a mood to act. Leaders finally emerged to organize and campaign for corrections, but as is often the case, they were not members of local or state governments. Three women who were aware of the bay's problems in general reacted when they learned that their city, Berkeley, was proposing to fill in 2,000 acres of the bay along its waterfront. This particular fill was not large, but on researching the situation, the women found that about half the remaining bay was susceptible to similar filling (Figure 1-2). They interested friends in forming the Save San Francisco Bay Association, which grew rapidly and carried arguments systematically to government officials and the public. Some have called the ensuing conflict the Battle of San Francisco Bay. Its effectiveness was due partly to the large number of people who presented their views to their elected representatives and partly to vital help from a few persons in positions to act directly, for example, a state senator, a leading planner, and a radio announcer. After three years of hard work by concerned citizens, a bill was passed by the state legislature in 1965 forming a commission that was to look into all aspects of the bay and compose a regional plan for its future.

The San Francisco Bay Conservation and Development Commission was thereby empowered to study and plan across boundaries and interests of 32 cities and 9 counties! It could do this effectively only because it included representation from many of these local governments as well as persons representing state and federal government agencies and private interests, 27 members in all. Using an additional array of specialists and advisors, but probing very actively themselves, the commissioners studied the bay system for three years and presented a firm plan to the state legislature in 1969. Another tremendous effort was mounted by local citizens, and the plan was finally voted to acceptance by the legislature.

It thereby became unlawful to fill, dike, or dredge any part of the bay without review by the commission. The water pollution problems were placed, by a separate action, under a new regional Water Quality Control Board empowered to monitor and regulate sewage outfalls in the bay and other natural waters. Other aspects of bay usage are spelled out in a series of recommendations to local governments and include a strong appeal for more public access along the shoreline.

An especially significant action was the retention of the commission as a reviewing and overseeing group. New evidence can thus be considered at once by persons acquainted with the system and in a position to act. The plan, in fact, spelled out several important topics that required further research.

One topic, the cause of water circulation in the southern part of the bay, was at that time under study by oceanographers of the U.S. Geological Survey, who presented a preliminary report in 1970 and have enlarged the study since (Reference 1, in list at the end of the chapter). Their principal data are of three kinds: (1) the amounts of fresh water entering the bay through the Sacramento and San Joaquin rivers (Figure 1-2); (2) the amount of salt dis-

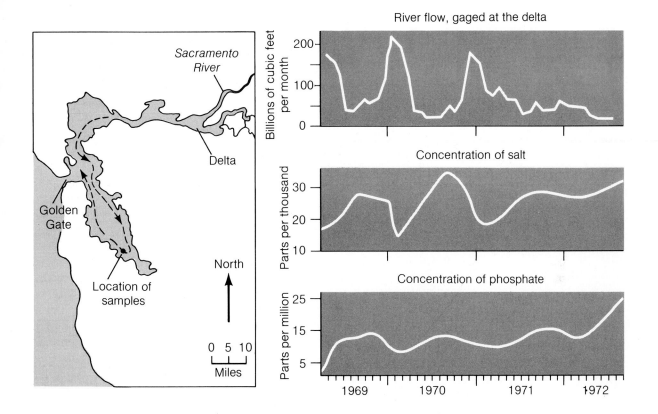

River flow, gaged at the delta

Billions of cubic feet per month

200
100
0

Concentration of salt

Parts per thousand

30
20
10

Concentration of phosphate

Parts per million

25
15
5

1969   1970   1971   1972

Sacramento River

Delta

Golden Gate

Location of samples

North

0 5 10
Miles

FIGURE 1-4

Graphs showing how amounts of water flowing into San Francisco Bay from the Sacramento and San Joaquin rivers compared with concentrations of dissolved salt and phosphate at the station labeled on the map. Data from David S. McCulloch, "Seasonal flushing of south San Francisco Bay: 1969–1972" (pp. 39–46 in *Progress report on the USGS Quaternary Studies in the San Francisco Bay area*, an informal collection of preliminary papers. Friends of the Pleistocene guidebook, 164 pp., 1972).

solved in the bay water, which indicates the proportions of river water and sea water at any one place; and (3) the amount of phosphate in the water, which is a measure of pollution (phosphate is introduced mainly in the garbage, excrement, and detergents in sewage and in the wastes from food-processing industries). As Figure 1-4 shows, the relations between the data are strikingly consistent. During the winter and spring, when river flows were high, the concentrations of salt and phosphate were low. The concentrations then increased gradually when the dry season began. This cycle indicates that comparatively fresh water passing seaward from the delta circulates in a broad loop through the southern lobe of the bay, flushing out the polluted and relatively saline water (see map in Figure 1-4). Especially important was the discovery that after a winter of unusually low river flows, the phosphate concentration rose nearly to a level indicating total befoulment.

The southern part of the bay and its million human neighbors were thus linked quite suddenly to a distant river system and all that affected it. And nothing promised to affect it so profoundly as a series of projects sponsored by the state of California and the federal government that may divert most of the Sacramento River's water to the more southerly, drier parts of the state. The consequent array of dams, aqueducts, and pumping stations is essentially built and operative. Because they may reduce river flow to the bay by as much as 75 percent, it appears that they may cause befoulment of the bay during years of low river flows.

Causes and Remedies

Who is to blame for such oversights in planning, for the already broad and intense degradation? In the case just described, fingers have been pointed in various directions. The question has been answered more broadly, however, by the Federal Water Pollution Control Administration, after extensive investigations and hearings concerning the deterioration of *all* our coastal waterways (Reference 2). I quote their entire General Summary (page 480), which is concentrated and worth reading carefully.

The Nation's estuarine and coastal resources today are seriously impaired and, in some cases, have suffered impairment which is irreversible. Fundamentally, this loss is the result of unwillingness or inability of the governments sharing responsibility and authority for their management to do the things necessary to protect these resources for all beneficial uses today and to conserve or preserve their maximum future usefulness.

The reasons for this unwillingness or inability are various and highly complex. Most basic, perhaps, are four reasons:

1. Shortsighted, imbalanced, or otherwise inadequate public policies governing the use of these resources up to now necessarily have reflected the dominant values of the American people. These traditionally have given a high priority to economic growth and technological development without adequately considering the adverse effects upon the estuarine and coastal environment.
2. Another reason, undoubtedly, is ignorance concerning the sometimes fragile and always interdependent nature of the complex of resources found in the estuarine and coastal zone.
3. Fragmentation and conflicts among governmental programs charged with the management of these resources have handicapped sound management. Closely related are, on the one hand, the limited use of plans which in fact coordinate the fragmented activities of the numerous agencies and governments involved in the management of these resources; and, on the other hand, the limited effectiveness of institutional arrangements now in being which were intended to overcome this fragmentation through interagency and/or governmental review and consultation or through joint or cooperative action. Also a contributing factor is the absence in these programs of policies and organization focusing specifically on the resources of the estuarine and coastal zone.
4. Although governments in the more recent period have moved to establish essential programs to conserve or preserve these resources, inadequate funding has prevented these programs from adequately accomplishing their mission.

These points may be summarized still more briefly: (1) we are all responsible; (2) our ignorance requires that we all acquire greater understanding; (3) governments are not likely to do the job; and (4) priorities of funding should be changed. Data bearing on the last point are presented in Figure 1-5. If there is any hope from open scientific studies, we appear to be headed toward serious trouble.

We have seen that local citizens, and very few of them at the outset, created the pressures leading to corrections in the San Francisco Bay case. From reason 3, especially, we see that this is likely to be the need elsewhere. Regardless of where action begins, local people must be involved in studies and corrections. Earth systems are local, particular things, requiring the feel that comes with long residence and many touches. Though well trained, a specialist raised in Iowa, schooled in New York, and living in Sacramento may well have trouble understanding San Francisco Bay in the time a state or federal agency typically allots for its study.

It should also be realized that citizen action will typically be precipitated just short of disaster. Even if he or she looks at a bay every morning and evening, an untrained person cannot easily recognize the signs of gradual worsening that precede rapid deterioration and collapse. It is essential to actually measure rates of change at an early or "background" stage if one is to offset real problems.

In the studies I made for this book I was repeatedly impressed by the scarcity of such information. Apparently, government agencies have not been able to provide the coverage needed, either in making measurements, in storing data, or in planning studies. Industry is generally constrained or secretive. Possibly, citizens could make systematic measurements, even broad studies, that might be reported to suitable agencies. Specialists could then move toward areas of trouble or interest. During the San Francisco Bay studies, 12 university freshmen, guided by a professor of geophysics, studied streams in a hilly part of the bay area before and after a major logging operation. Their measurements of soil erosion helped in understanding the total bay system

FIGURE 1-5

Distributions of government expenditures, during 1969, for research and development in three nations with highly advanced industries. From *Science indicators 1972* (Report of the National Science Board, 1973, U.S. Government Printing Office, 145 pp.).

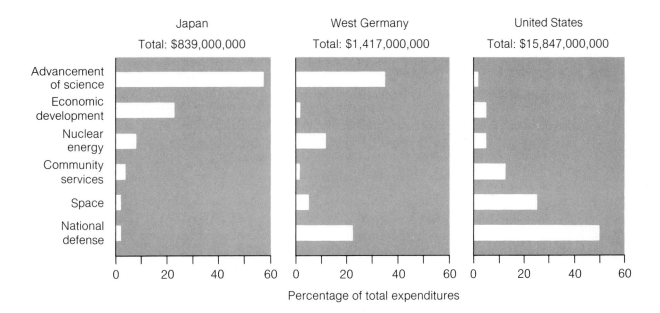

	Japan	West Germany	United States
	Total: $839,000,000	Total: $1,417,000,000	Total: $15,847,000,000

Percentage of total expenditures

and led directly to new laws governing logging in San Mateo County specifically as well as providing more broad regulation for the entire state. The important point is that legislation had *not* been pressed theretofore, even though California's logging industry had been active on a massive scale for decades.

Beginning a Specific Study

Our broad look at San Francisco Bay and its problems led us to the results of specific studies—the ultimate sources of data and new ideas. Such studies are the personal efforts of certain individuals (which is why I will often cite names in my descriptions). Like the parts of the system, the studies are also interrelated and intersupportive. An examination of the velocities of currents in one part of the bay is helpful to persons interested in another part. One can thus look very particularly at one part of a system, as we will do in this section, and feel confident that any discoveries will be broadly useful.

A principal purpose of this book is to show how ideas may be derived from observations of the real world. By this I mean observations of actual systems rather than of written descriptions and interpretations of them. Many of the examinations of water systems described in the first several chapters are of this nature. Many could, in fact, be made by anyone, and I hope you will become interested in trying one yourself. If you do, try to find a place that is small enough and handy enough for you to get to know well. The crucial thing is the thoroughness with which you get to know it. You should start in whatever way seems natural, but perhaps the information on the nature of water that follows will help. We must, however, switch scale from a thing as big as San Francisco Bay to a small number of the water molecules in it!

The structure of water   Water molecules seem quite prosaic at first: Each consists of one oxygen and two hydrogen atoms held together strongly by shared pairs of electrons, as suggested diagrammatically in Figure 1-6A. The oxygen carries two additional pairs of electrons, and because like-charged particles tend to repel one another, the four pairs of electrons are forced apart to give a symmetrical array like that indicated in Figure 1-6B. As a result, the two hydrogen atoms must lie on one side of the molecule, and the free electron pairs on the other side. This configuration causes an imbalance in electrical charges, for the hydrogen nuclei give that part of the molecule a positive charge, and the electron pairs give the opposite part a negative charge. Water molecules are thus said to be *polar*—each has a positively charged end lying opposite a negatively charged end.

The polarity of water molecules is the physical basis of water systems. Because oppositely charged particles attract one another, water molecules tend to arrange themselves so that a hydrogen nucleus of one molecule lies next to a free electron pair of an adjoining molecule. This arrangement results in a three-dimensional structuring among all molecules, one joined by electrical forces (Figure 1-6C). Water thus has a "body" or resistance to flow, an important property called *viscosity*. One can think of liquids such as syrup that are more viscous than water, and those such as ether that are less viscous.

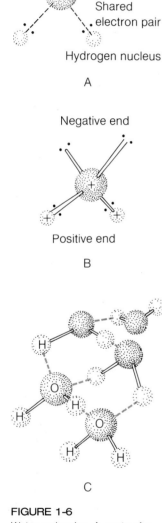

FIGURE 1-6
Water molecules. A, parts of a single molecule; B, lines indicating the positions of the electron pairs around an oxygen atom (weighted to give a sense of position in and out of the page); C, aggregate of water molecules.

The polar nature of the water molecule also causes interactions with solid bodies. Almost all mineral substances in rocks, sand, and mud have abundant electrons at their outer surfaces (largely because of the abundant oxygen atoms there). These electrons attract the positively charged ends of water molecules. We say water "wets" these substances; in fact it truly sticks to them. If you wet a pebble and place it on a table, the water film does not run off and leave the pebble dry; it stays there until it evaporates.

Particles in water Some important interactions of grains (particles) and water can be understood on this basis. Try releasing a sand grain at the top of a large jar of water. It will accelerate as it falls the first few inches but then fall at a constant rate, called its *settling velocity*. The force that accelerated the grain downward (the grain's weight in water) must have been balanced by the water's resistance. Water resists the falling grain because it sticks to the grain and because it is viscous. Thus, small grains may take hours to sink to the bottom of San Francisco Bay, even in places where the water is quiet.

For grains smaller than $\frac{1}{2}$ millimeter ($\frac{1}{50}$ of an inch), the arresting force is the drag created by the viscous flow of water around the grain. This is suggested by Figure 1-7A, which we must imagine in three dimensions because the total drag depends on the total surface area of the grain. With increasing grain size, the weights of grains (a function of their volumes) increase faster than their surface areas, so that large grains settle more quickly than small ones (Figure 1-8). This relation can be verified easily by experiment.

A second kind of arresting action becomes dominant for grains larger than 1 millimeter ($\frac{1}{25}$ inch), as shown by the change in slope of the curve in Figure 1-8. These grains fall so fast that the fluid cannot stream in a simple, connected way behind them. It separates to form a rotating body called an *eddy* and a turbulent train called a *wake* (Figure 1-7B). Wakes can be seen experimentally by letting small pebbles fall through a water suspension of very fine mica or other lustrous particles. The separation and eddy create an appreciable upward pressure behind a grain. The falling grain also generates an upward reaction or pressure in front of it. These forces act together to slow the grain, an effect proportional to the cross-sectional area imposed on the fluid by the falling grain. You can test this concept by releasing two identical, disc-shaped pebbles in water, one flat-side down and the other edge-side down. The pebbles have the same surface area yet the latter descends dramatically faster.

The categories of *sediment size* shown in Figure 1-8 are used widely elsewhere in this book. One reason they are important is that sizes of grains on the bed of an estuary or stream are indicators of the transporting power of the water currents. In San Francisco Bay, for example, the swiftest moving waters in the main channels are underlain by sand, the quieter parts of the bay by silt, and the marsh flats by fine silt and clay. The sand sizes illustrated on the right may thus be valuable in classifying sediments in the beds of other estuaries or streams, and if still larger grains are present, a scale divided in millimeters can be used to measure them directly.

Silt and clay-sized grains, which are tremendously important in many water systems, are too fine to be seen without a microscope, but their sizes

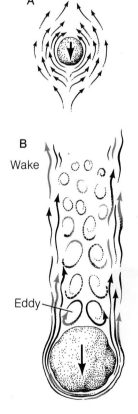

FIGURE 1-7
Grains falling in water. A, stream lines of water flowing around small, slowly falling grain; B, separation of flow and eddy behind large, rapidly falling grain.

13

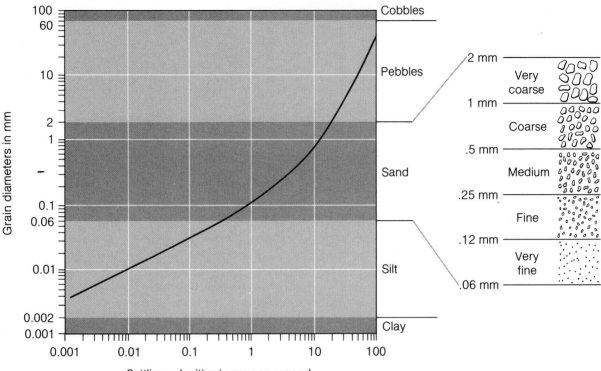

FIGURE 1-8

Graph showing the settling velocities of roughly spherical grains of typical minerals and rocks. The shaded and unshaded bands indicate the size limits for each of the sediment classes listed on the right. Sand sizes are generally classified further because they are easily visible and handy to work with.

can be determined approximately by a procedure based on their settling velocities. Suppose that you have a sample consisting of silt and clay and you want to determine the proportion of coarse silt in it, for example, grains larger than 0.01 millimeter in diameter. Figure 1-8 shows that grains of that size settle at a velocity of 0.009 centimeters per second. If you mix the sediment into water in a large jar and measure the depth of the water, you can calculate the time it will take for all grains larger than 0.01 millimeter to settle to the bottom. For example, if the water is 8 inches (20 centimeters) deep, you should let the grains settle for about 37 minutes. After that period, the finer sediment can be accumulated by pouring the muddy water into a second jar and letting it stand until clear. You have thus separated the sample into two size categories and can estimate or weigh the amounts of each.

An important result of this procedure is to see that sediment is *sorted* (separated) by size when it settles in water. Big grains settle ahead of little ones. If sediment of mixed sizes is introduced all at once and allowed to settle completely, the resulting layer will be coarse grained at the base and grade upward into finer sizes. We will meet such layers in later chapters.

Water and Sediment in Motion

Interactions of water and sediment in still water are basic to understanding water systems, but those of sediment in flowing water are even more basic. Visit a stream if you can. One in a gutter is fine, or make one with a hose, if necessary. Test some of the relations that will be described in this section.

Transport of sediment by moving water is one of the more important earth processes, and one can generally observe two ways in which it takes place. Silt, clay-size, and organic (lightweight) particles are generally carried throughout the body of water, making it turbid (muddy). These grains settle so slowly that once they are mixed into a stream they are washed along for great distances. The larger grains such as sand and perhaps gravel, are rolled or bounced along the bed of the stream. Their transport is typically spasmodic, starting when an extra swift thread of current drags against their projecting ends and starts them rolling. They eventually stop when they pass into a sluggish thread of current or lodge against large grains on the bed.

Turbulence The fact that streams consist of swift and sluggish threads of current introduces an extremely important aspect of water motion: different parts of streams move at different rates and in various directions. This phenomenon is part of what is called *turbulence* (note that turbulence is not the same as turbid). The easiest way to see turbulence is to release a handful of sandy dirt in a fairly swift stream (one moving 2 to 4 feet per second). The resulting muddy cloud will billow downstream like a smoke plume on a windy day. By watching closely you may see that some of the sand grains settle quickly through the cloud and to the bottom, but that most stay suspended beyond the time they would normally settle, and some even ride along in the stream until out of sight. The weaving, spiraling courses of the grains define the patterns of current threads (Figure 1-9). The threads often change very rhythmically, commonly forming rotating eddies near the base of the stream that tend to rise toward the surface.

Turbulence is caused in part by irregularities in the stream bed, but is produced more generally by the bed dragging against the stream. As shown in Figure 1-9, the sediment cloud traveled the same distance in all parts of the stream except the lowest few inches. This regularity must mean that the main body of the stream is riding along like a slab on the lower part, where the velocity falls off rapidly due to the water's adhering to the bed. Motion in the lower part is thus unsteady and tends to generate pulsating wobbles and eddies (turbulence) that rise into the stream, carrying sediment with them. Since the turbulence becomes more and more energetic as the overall velocity of the stream increases, swift streams transport much more sediment than sluggish ones, other factors being equal.

Formation of waves Another intriguing result of stream motion is the formation of waves, as shown beautifully in a stream studied by Robert Anderson and Steven Hug, students in an introductory geology class. The stream is 40 miles south of San Francisco and flows across a sandy beach about 150 feet wide. As shown in Figure 1-10, water flowing out of the nearly still pond in the foreground begins to develop ripples where its velocity (at the surface) was found to be about 1.5 feet per second. The slope of the stream toward the sea increases gradually as does its velocity—up to about 4.5 feet per second in the narrow part of the channel. The velocity also varies with the stream's depth; in the shallow parts along the banks velocity is only about 1 to 2 feet per second.

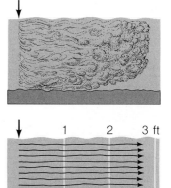

FIGURE 1-9
Vertical sections through a stream into which sandy mud was dropped at the arrow. Typical directions of sediment movement are depicted at top; distances of transport during one second are shown at bottom.

These variations in depth and velocity are reflected by the sizes of the waves, which are larger in the deeper, swifter parts of the stream (Figure 1-10). Note how the shallows along the banks are ·marked by a smooth surface or by small ripples. The waves are not like those caused by the wind, for they stay more or less in one place. They are essentially wrinkles produced in the stream as it either quickens its downstream motion, or is retarded. You can produce a local set of these waves by placing an obstruction in a stream. The waves show an intriguing thing that could not be observed otherwise: the stream is divided into a series of bands parallel to the direction of flow. They narrow and converge downstream and are finally lost in a very turbulent stretch of water.

Figure 1-11, left, shows the fan-shaped part of the stream several hours later, when the stream had divided into three subchannels separated by low islands. The sinuously lined patterns on the surface of the islands were made by currents, proving that the islands were once sand bars underwater. Anderson and Hug observed that the islands emerged when the stream eroded (deepened) the three subchannels. The subchannels probably evolved from the swiftest of the bands shown in Figure 1-10.

The waves in Figure 1-11 show by their sizes where the subchannels are deep and swift and where they are shallow and sluggish. When Anderson and Hug examined the sand on the bed, they found that it was sorted to coarse grain sizes (1–2 millimeters) in the swift parts of the subchannels, and to finer and more poorly sorted (more mixed) sizes in the less swift parts. Evidently the swifter, more turbulent currents could carry all but the largest sand grains in suspension, so that the large grains are the chief particles on the bed. This sorting process is called *winnowing*—the separation of small grains from large ones by a moving current. The slower currents cannot suspend the fine sand, so their bed consists of both large and small grains.

Sand transport associated with the largest of the waves was another intriguing and important interaction in this stream. Figure 1-11, right, shows the narrow part of the channel that slopes steeply toward the ocean. So much sand was being carried along the bed here that it formed a dense carpet of moving grains. The large waves mark the deepest and swiftest part of the

FIGURE 1-10
Stream flowing across a sand beach, slowly deepening its channel as sea level falls during ebbing tide (note, for example, the eroded forms just left of center that show that the stream was once wider and shallower). Photograph by Steven Hug.

entire stream, and each was seen to slowly peak (in a matter of a minute or so) and then break in a froth of turbulent water that discharged a large amount of sand from the bottom and carried it downstream. By comparing the two parts of Figure 1-11, you can get a feel for the greater power of the narrower, deeper stream. Realize that the same amount of water must be flowing through each channel!

Some findings overall Because these findings have come out all together (as they do when you examine real things), it may be helpful to list some of the major ones:

1. Stream velocities depend on steepness of slope but also (greatly) on local depth of water.
2. Waves are largest in the deepest and swiftest parts of streams.
3. Sediment is carried partly along the bed and partly suspended in the main body of the stream.
4. Because turbulence increases with velocity, so does the amount of sediment being carried.
5. Streams sort sediment by size, partly by the winnowing action of currents along the bed and partly by the different rates at which the grains settle.

Perhaps the main thing to note is that all these things are interrelated. Now look back at Figure 1-11. How differently the channels appear—yet they are parts of one stream. Think, too, about this stream being one of many within the workings of the San Francisco Bay region. Have we not looked at specific, remarkable things that help us understand what is happening in the broader system because they apply to all water systems? Do you see some advantages in studying systems that can be encompassed completely and visited often?

FIGURE 1-11

Left. Upper part of the stream in a view opposite that of Figure 1-10 (flow here is from left to right). Deepening of the channel, overall, is indicated by the relics of earlier channel edges in the right foreground. Right. Waves (one breaking) in the narrowest and deepest part of the channel. Note the smaller scale roughness (turbulence). Flow is from left to right. Photographs by Steven Hug.

REFERENCES CITED

1. U.S. Geological Survey. *A preliminary study of the effects of water circulation in the San Francisco Bay estuary*, Circular 637-A,B, 1970.
2. *The National Estuarine Pollution Study.* Report of the Secretary of the Interior to the United States Congress, Senate Document No. 91–58, 633 pp., 1970.

ADDITIONAL IDEAS AND SOURCES

1. Surface velocities of streams can be determined by measuring a convenient course (10 to 100 feet) and clocking the passages of floating objects through it. For velocities near the bottom, stir up bits of water-logged leaves, etc., with a stick. When you've determined the surface and bottom velocities, you can estimate an average velocity by assuming a distribution like that in Figure 1-9, bottom.

2. In examining sand deposits you may notice that mica and other flaky grains (such as fragments of sea shells) tend to be found with smaller grains of spherical shape. This mixing occurs because the flaky ones have larger surface areas for their weights. Compare the settling velocities of grains of various shapes in a jar of water.

3. Before making a study of sands or gravel, you might find it useful to read about the common mineral varieties (quartz, mica, feldspar) that make up sand grains. Supplementary Chapter 1 describes minerals and Supplementary Chapter 2 rocks.

4. The plan for San Francisco Bay has been published in two parts, and although it is out of print, it has become such an important model for other planning studies that it is worth borrowing from a library and examining. The references are *San Francisco Bay Plan* and (a separate volume) *San Francisco Bay Plan Supplement,* 1969. Both are publications of the San Francisco Bay Conservation and Development Commission, Sacramento, Calif.

5. If you would like to read some specifics of the involvement of law and politics with earth systems, I recommend the *Ecology Law Quarterly* (if unavailable locally, you can buy separate copies from School of Law, University of California, Berkeley, Calif. 94720). Vol. 3, No. 1 (Winter 1973) has several excellent articles, one of which deals with overall interactions of planning (or lack of it) for a river system in Illinois.

6. If you are interested in the sort of view one gets by living close to a system all one's life, I recommend the essays of Wendell Berry collected in *The Long-legged House* (New York: Harcourt Brace Jovanovich, 1969; also available in paperback). "A Native Hill" is especially pertinent.

Slope in hardwood forest on the Blue Ridge in Virginia—the part of the James River system that regulates its water supply.

2. The James River System

The James River System

irginia's James River, shown in Figure 2-1 at midcourse, is one of the dozen or so principal rivers draining the eastern seaboard of the United States. A number of attributes make it especially interesting. It is large enough to show how major rivers work yet small enough to be observed easily. Unlike most major rivers, it is not impounded by dams and unlike many it is only locally polluted. Because many of its tributaries rise in unspoiled, beautifully wooded country, its stream and valley forms can be viewed in a near-natural state. It is nonetheless modern in that its lowermost valley is undergoing rapid urban and industrial development, and pollution has become critical in some of these lower reaches (stretches). Of special interest is the fact that the Commonwealth of Virginia has studied the entire river system in light of these problems and is preparing a plan for its uses, the first step in planning for all the river systems of the state.

But what is a *river system*—this one in particular? Its most tangible parts, shown in Figure 2-2, are the main river, its many tributaries, and all the land they drain. The map emphasizes the wholeness of the system (it is not easy to trace the main river). The even spacing of lines links with another definitive aspect: rain and snow distribute a roughly equal amount of precipitation across the entire area, which annually averages about 45 inches in the wettest parts and 40 inches in the driest. Most of this water soaks into the ground and only eventually joins surface streams, so we must include in the system

FIGURE 2-1

The James River at the village of Wingina, about 70 miles upstream from Richmond.

the soils and rocks through which it flows. All plants and animals that affect the water must also be included. So must the direct source of the water, the atmosphere, although in this regard the system is very much open-ended —subject to weather variations generated far away.

The system is thus many things, but so defined it is terribly abstract. Anyone who has lived near a river can probably sense a fallacy. The lines on the map are not the James River! Their pattern is intriguing and informative but the river is something else. The real river is what one looks at, listens to, and wades in. It turns out to be movement as much as water. Pulled by gravity, the river is flowing down a valley slope to the sea. Its sounds and movement patterns express the system's essence—interaction between the freedom of water to move and the constraints imposed by the land, down to the last pebble.

Thus, we see why the system must include land as well as water. Each works on the other, as we have already noted. Water adheres to mineral substances. It sticks to the sand and solid rock forming the James River channel. Water also impedes its own flow because of its viscosity (internal bonding). The channel thus retards the river and the river drags against the channel. A large amount of mechanical energy is converted into heat, which warms the water to a small degree and is dissipated continuously into the air.

The interaction ultimately results in sculpturing of the land by the river. The entire James River system has become ordered into broad terrains that grade into one another yet are characterized by certain land forms, river forms, and life forms. The terrains are so basic to human uses that they were named long ago (Figure 2-3). The Appalachian Mountains and the mountains

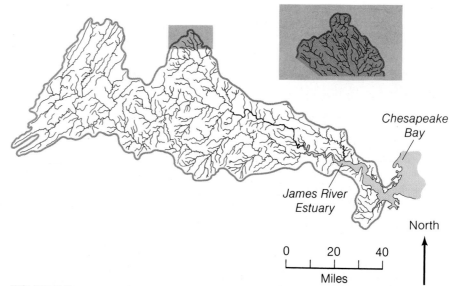

FIGURE 2-2

The James River and its principal tributaries, with one segment enlarged to show the typical density of smaller tributary streams. The gray line joins all ridges and low divides that separate water flowing to the James River from water flowing to adjoining river systems. This perimeter is thus the boundary of the James River system.

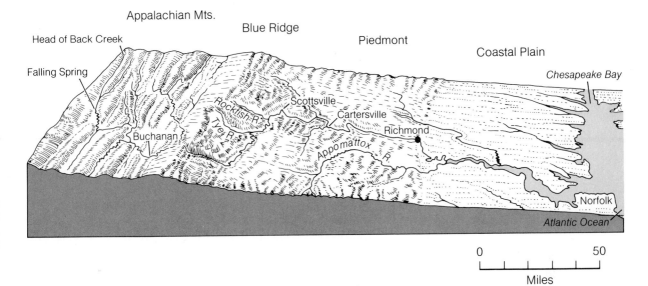

Appalachian Mts.

Blue Ridge

Piedmont

Coastal Plain

Head of Back Creek

Falling Spring

Chesapeake Bay

Rockfish R.

Scottsville

Tye R.

Cartersville

Buchanan

Richmond

Appomattox R.

Norfolk

Atlantic Ocean

0   50

Miles

FIGURE 2-3

Diagrammatic view of the terrains drained by the James River, as named above the diagram. The other place names are river gaging stations and other localities mentioned in the chapter. The relief is exaggerated.

called the Blue Ridge are characterized by long straight ridges that rise as much as 4,000 feet above sea level and are being eroded by short but relatively steep streams. Between the ridges are equally straight valleys down which major tributaries flow in winding courses. The Piedmont is a broad tract of hills, few rising more than 800 feet above sea level, divided by intricately branching tributary valleys. As shown in Figure 2-1, the main river flows through this terrain in a broad valley. The valley has always provided the chief access to the region. The Coastal Plain varies from low, broad hills on the west to nearly flat plains and swampy areas near Chesapeake Bay. Being nearly flat and handy to the sea, this part of the region has always been used extensively by people.

Beginnings of the River

Living things have such a tremendous influence on the way rain joins the river system that the river's beginnings vary markedly from one terrain to another. The mountain ridges to the west are especially interesting because they are nearly as they were when Europeans first settled Virginia. In fact, the entire Commonwealth was originally covered by hardwood and conifer forests much like those that grow now on the Blue Ridge.

Under these forests is a layer of leaves, branches, and decaying mulch an inch to a foot thick. The stuff seems ordinary enough, but it is as vital to the river as it is to the life of the forest. Even when large amounts of rain reach the ground, which typically slopes at 10 to 30 degrees, little if any runs off on the surface. The water wets the leaves and twigs, which constitute a rough, intricate surface that greatly retards surface flow. The water thus sinks into the soil before it can accumulate into rivulets and small streams.

The photograph on the opening page of the chapter shows a typical view of this part of the system. One can walk for miles along many of these wooded slopes without seeing a single gully or stream channel. Here and there, however, the slopes are interrupted by swales (subdued valleys) that suggest some kind of erosive action. By digging away leaves and mulch in the central parts of the swales, one can see trains of angular stones in crude channelways, and among the stones flowing water can be seen or heard for hours after a heavy rain. The hidden channels grade downslope into partly exposed lines and patches of stones, and these grade in turn into bouldery brooks that flow most of the year.

The streams slowly enlarge down valley even though they have no obvious tributaries. One reason is that they catch enough direct rainfall to rise during each downpour. Test pits in other forests have shown, additionally, that rain falling on the surrounding slopes moves readily in the upper part of the soil—in animal burrows, in cracks, and in tubes left by decayed roots. These narrow underground passages are so well protected by roots that they are eroded little, if at all. Nevertheless they transmit water fast enough that the adjoining stream rises steadily soon after a heavy rainfall. This intricate plumbing system evidently protects the slopes from erosion, for the streams carry little sediment (though the water may be stained brownish by organic substances dissolved from decaying plant litter). That the forest soil is tender and susceptible to erosion is shown by gullies and muddy streams where construction or logging has removed the forest litter.

In the Piedmont, forested lands soak up and transmit rain much as in the Blue Ridge and Appalachian terrains. The forest litter tends to be more packed and sparse here but this firmness is offset by the fact that the slopes are typically less steep. Croplands, however, make up a fourth of the Piedmont and they interact with rain very differently. The streams in these areas are always muddy after rains, and erosion can be seen in action on almost any plowed field during a rainfall of 2 inches or more.

Erosion by impact of rain Careful studies, chiefly by the Soil Conservation Service, have shown that field erosion is caused mainly by the impact of raindrops, and anyone willing to observe a patch of bare ground during a heavy rain will find the effects remarkable. Several actions take place. The upper inch of soil, packed by the momentum of the falling drops, soaks up the rain less quickly. This condition increases the amount of water moving across the surface and commonly generates a nearly continuous sheet one-eighth to one-fourth inch deep. In addition, each drop explosively stirs the uppermost grains of soil into the moving sheet. As a result, a large amount of sediment is transported even though the water is shallow and thus slow moving (because water sticks to the ground). I saw many large sand grains being moved in this way across a gently sloping dirt road near McDowell, Virginia, during a downpour. When the rainfall suddenly stopped, the water sheet cleared of all but the finest sediment within a few seconds although the water continued to move at the same velocity (about 1 inch per second). Raindrop impact was a far more powerful erosive agent than the force of the shallow stream.

Row crops such as corn, tobacco, and cotton protect the surface only moderately. Even slightly sloping furrows concentrate runoff into streams that

FIGURE 2-4
Roadcut along Interstate 64 east of Charlottesville, a surface sloping 30 degrees toward the viewer. It was covered with a mulch of straw and planted with grass eight months before the photograph was taken. The mulch was accidentally removed from the left part of the surface, and the gullies were eroded in that short interval of time.

erode small gullies or wash away the finer soil materials. Grasses impede downhill runoff and also provide cover from raindrop impact. Figure 2-4 shows how effectively a steep surface was protected by straw mulch and grass and how rapidly gullies formed where the mulch was accidentally removed.

Most (roughly 90 percent) of the eroded materials are redeposited before reaching the streams of the system. Their resting places can be found easily enough after a heavy rain: pastures at the base of steep slopes, patches of brush and grass along roads, low-lying parts of plowed fields, wherever currents were slowed by plants or a decrease in slope. These deposits, however, are typically sandy because clay and silt have been winnowed out and carried preferentially to streams. Moreover, the finer sediments include the light-weight organic materials that were the principal nutrients in the eroded soils.

These processes may seem to be trivial and piecemeal, but the history of the region shows otherwise. Most of the upper (richer) soil was stripped from the Piedmont of the eastern United States during the 1700s and 1800s, a truly incalculable loss that has continued to affect us in many ways. The damage was caused by clearing forests and burning their litter, by plowing furrows directly downslope, and by planting row crops almost to the exclusion of cover crops. When it became unsuited for agriculture, most of the land was returned to woodland. Improved farming practices have reduced erosion greatly, but it is important to realize that much of Virginia's impressive woodlands cover ground that was once used so destructively that it was abandoned.

The Coastal Plain presents a contrasting story. Some plantations have been farmed successfully there since the 1600s. Original soils are still present in many places. The main reason is that the land slopes very gently, a factor that results in sparse runoff and therefore little erosion. Even the streams flow sluggishly, commonly winding across bottomlands grown with trees and bushes.

Flow Underground

In the James River region most precipitation soaks into the ground because the countryside is mainly woodland and grassland. It is crucial that we understand what happens to this water. In an average year, something like 10 to 20 percent evaporates directly back into the atmosphere. A greater amount, nearly half, is drawn back up by plants, which incorporate a small percentage in new tissues and return (*transpire*) the rest to the atmosphere. The transpired water is not a total loss, for it cools and moistens the air and thereby reduces the rate of direct evaporation. In any case, the circuiting of rain through soil and plants to the atmosphere is a sort of subsystem that supports all land-based life.

The remaining water that enters the soil, something like a third of the total precipitation, seeps downward to become part of what is called *groundwater*. It flows for long distances through soil and rocks and eventually surfaces in springs and channels, although groundwater under the Coastal Plain seeps directly into the bottom of Chesapeake Bay.

Movement of groundwater Exactly where and how does the groundwater move? I have already described observations made after heavy rains on forested slopes—how the water soaks in rapidly and then flows through openings in the soil to adjacent streams. This phenomenon is an example of very shallow, rapid groundwater flow. Somewhat deeper flow can best be visualized by examining a deep excavation in rocks typical of the Blue Ridge and Piedmont (Figure 2-5). The rock is *granite*, formed deep in the earth and now exposed because of long continued erosion (Appendix B). Note that the cracks and breaks are most abundant just under the dark forest soil but also occur here and there in the lower part of the excavation. All these cracks seep water for days after a soaking rain, long after the ground surface is dry or nearly so. As seepage takes place, the height to which the rocks in the excavation are wet slowly drops. In the photograph, taken after a long dry period, water is still flowing only in the deepest cracks. In mines, similar cracks occur at the deepest levels penetrated, and they, too, transmit groundwater. Apparently, cracks and other small openings are filled with groundwater up to a comparatively shallow level under the entire region.

The rock between visible cracks appears solid, but in fact contains countless microscopic openings in and among the mineral grains. These openings may also contain water, as can be proven by weighing a piece of rock fresh from a mine or quarry and then drying it in an oven for a week or so, after which it will weigh appreciably less than when it was damp. The microscopic pores and cracks make up only 1 percent or so of tightly knit rocks like those underlying the Blue Ridge and Piedmont, but as much as 25 percent of porous *sandstone* (a rock made up of sand grains cemented together) that occurs under the Coastal Plain.

Because water adheres to rocks and to itself, the sizes of these various openings govern the rates at which groundwater flows through them. The microscopic pores in the more tightly knit rocks transmit water exceedingly slowly. If a fist-sized piece of granite typical of much of the Blue Ridge is dried and then placed in a bucket of water, it will take hours to fill most of its

FIGURE 2-5

Typical array of cracks in rock exposed by a recent excavation, 60 feet deep, near Massies Mill in the Tye River valley. Note the forest and soil at the top, the abundance of cracks at shallow depths, and the water (dark) seeping from the deepest of the cracks.

minute openings. Appreciably faster flow takes place in cracks like those in Figure 2-5. At this excavation, the rate at which water was oozing out of the cracks indicated it was flowing through them at a rate of approximately 1 inch per minute. Groundwater may move at rates of inches per second where rocks have been greatly crushed or large openings have been dissolved in them. These channels form major springs where they intersect the surface.

The water table Our understanding of groundwater circulation comes largely from observing water levels in wells, which are used widely in the James River country. As shown in Figure 2-6, each well fills with water just to the level at which the surrounding rocks are saturated with water. By connecting the levels between the wells we can locate the top of the groundwater reservoir, a surface called the *water table*. Note in Figure 2-6 that the water tables lies farthest beneath the surface at hilltops and closest to the surface in valleys. This is because groundwater moves under the force of gravity from high positions to low, finally emerging in springs and streams.

explain this better

The position of the water table also varies with time. It rises slowly to the surface during prolonged rains and rises correspondingly in wells. During dry periods it falls slowly as groundwater flows out through springs and into streams and is pumped out of wells.

This connection is proven by the variations in flow that can be observed in streams. They flow most voluminously immediately after rains but continue to flow at gradually decreasing rates for weeks afterward. Apparently, rains send large amounts of water through soil and rocks to streams and at the same time refill the groundwater reservoir. Between rains, the reservoir continues to feed the surface streams but does so less and less voluminously because of reduction of the vertical distance (the hydraulic head) between the highest parts of the water table and the adjacent streams.

Solution of rocks by groundwater The chemical substances dissolved in groundwater show that groundwater also reacts with rocks. Chemical analyses of various waters in the James River system show that small streams contain about the same proportions of dissolved substances as wells nearby. The dissolved substances vary from one place to the next because the underlying rocks vary. The Appalachian Mountains are underlain partly by *limestone*, a rock consisting largely of the mineral substance calcium carbonate ($CaCO_3$), which is somewhat soluble in rainwater. Springs and wells in these areas, contain appreciable amounts of dissolved calcium, and streams and the

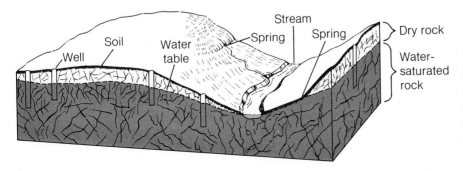

FIGURE 2-6
Diagram showing how wells in a typical part of the Piedmont have filled with water to the level of the water table, the surface below which all openings in the rocks are filled with water.

main river in that part of the system average around 25 milligrams per liter of dissolved calcium. Springs and wells in the nearly insoluble granites of the Blue Ridge and Piedmont contain only a few milligrams per liter of calcium and the tributaries draining these areas carry similarly small amounts. This close correspondence between rock and water compositions indicates that the surface streams are fed largely from underground.

In the Appalachian terrain, the slow dissolving of limestone by groundwater has formed large openings that are an important part of the water system. The upper photograph of Figure 2-7 shows solution-enlarged cracks in limestone that may be compared with the cracks typical of granite, in Figure 2-5. The lower photograph of Figure 2-7 shows major vertical fractures that have been enlarged by groundwater into channelways and caverns many feet across. Major springs are formed where such channelways emerge in the main Appalachian valleys, many flowing at rates of 1,000 gallons or more per minute. In contrast, springs flowing from granites of the Piedmont and Blue Ridge, which are fed mainly through cracks like those of Figure 2-5, flow at rates of only a few gallons per minute.

He stresses this repeatedly; I would place more stress on balance, adjustment.

Order Within the System

As groundwater seeps into the river from beneath and tributaries join it at the surface, the river's width and depth gradually increase. At the same time the slope of the river toward the sea gradually decreases. The latter effect is so gradual that it cannot be detected at the river itself; however, it can be seen by constructing *a longitudinal profile*, a diagram showing the altitudes of the water surface along the river's course (Figure 2-8). These altitudes and distances were determined directly from topographic maps, which are described generally in Appendix C. The distances represent miles *on the river*, determined by laying a string along the river's course shown on a map. The diagram's vertical dimension was expanded greatly to show the changes in slope clearly (note the two numbered scales). This has no effect on the basic result: the river's slope changes in a remarkably orderly way, decreasing downstream in a beautifully smooth, concave-upward curve.

Increase in discharge downstream Also shown in the figure are the amounts of water, in cubic feet per second, that typically pass through the channel at the places indicated. Each amount, called the *discharge*, was measured at a gage of the U.S. Geological Survey or the Commonwealth of Virginia and represents an average of measurements taken over many years. These data express another orderly aspect of the system: the average discharge increases regularly downstream. In the main, this increase results from the more or less even distribution of precipitation and tributaries over the region (Figure 2-2). The area drained becomes greater and greater as one progresses downstream.

Increase in velocity downstream An additional generalization, however, does not seem to fit. Because the river is flowing down its valley due

FIGURE 2-7
Limestone in a deep quarry at Linville partly dissolved by groundwater flowing along cracks in the rock. Above, enlarged cracks and rounded surfaces resulting from solution of the white limestone. Left, caverns formed along cracks extending from the surface to a depth of 120 feet. Both were formed before quarrying exposed them.

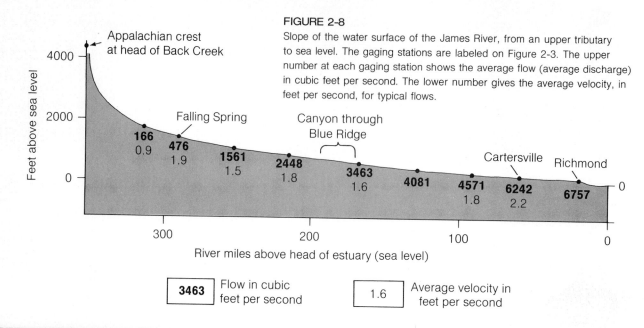

FIGURE 2-8

Slope of the water surface of the James River, from an upper tributary to sea level. The gaging stations are labeled on Figure 2-3. The upper number at each gaging station shows the average flow (average discharge) in cubic feet per second. The lower number gives the average velocity, in feet per second, for typical flows.

| 3463 | Flow in cubic feet per second | 1.6 | Average velocity in feet per second |

FIGURE 2-9

Spring flow in Piney River, a tributary of the Tye River in an upland valley of the Blue Ridge.

Quantitative explanation – Strahler

p. 324, 326

chezy eq

$\bar{V} = C \sqrt{R \cdot S}$

to the force of gravity, it would seem that it should flow swiftest where the slope is steepest. It does not; in fact, its velocity more generally increases downstream (Figure 2-8). This paradox is caused by the stream's tendency to stick to its bed, which causes further interactions that we will examine in detail.

The smaller tributaries show the most obvious interactions. In the steepest streams, in the Blue Ridge and Appalachians, the water flows between boulders and tree roots that make the channels extremely rough (Figure 2-9). The large amount of contact area compared to the cross-sectional area of the stream causes a great deal of drag on the moving water. Moreover, the irregularities in the bed send eddies and subcurrents in all directions, with some pools totally arresting the down-valley momentum of the stream. The sounds produced and the abrasion of rock surfaces are evidence of the stream's work on its bed as well as on itself.

A few miles downstream the stones in the bed are much smaller and the increased depth and width of the channel place a larger proportion of the stream well out of touch with bed and banks (Figures 2-10 and 2-11). This part of the stream thus flows faster than the brook even though its slope is much less.

Still farther downstream the bed materials are mainly sand mounded by current flow much as wind forms small dunes on a beach, but the mounds offer less resistance to flow than does a bouldery bed. The main change, however, is the increased depth and width of the channel, which permit more of the water to flow more freely (Figure 2-11C). This effect can be judged approximately by comparing velocities at one gaging station, measured at different times. Table 2-1 shows the increase in average velocity as the river rose from a low flow, to a typical flow, and to a high flow (a moderate flood).

Together, these relations indicate an extremely important kind of order within the system: impedance to flow (drag) decreases regularly downstream.

FIGURE 2-10

Rockfish River a few miles east of the Blue Ridge, where it heads. This tributary of the James River lies just north of the Tye River and is similar to it (see Figure 2-3).

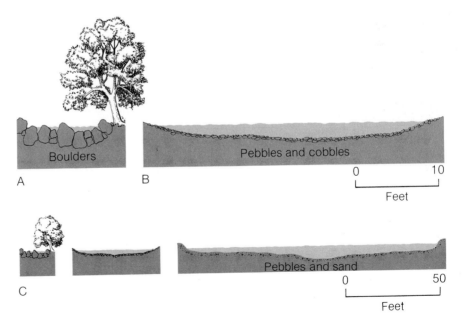

FIGURE 2-11

Cross sections of stream channels. (A) A mountain stream and (B) a medium-sized tributary, at the same scale; (C) the main river with the other two channels at the same scale.

This is why the river can flow on a decreasing slope without a consequent decrease in velocity.

Relations among erosion, transport, and deposition Impedance of the stream by the bed implies that the bed is being worked on by the stream. These are the two equal and opposite parts of the interaction mentioned heretofore as the essence of the river system. Theoretically, the river should be eroding its bed and deepening its valleys. Is this actually taking place, and if so, where?

By observing the river almost anywhere, one can see that it is moving sand grains and finer sediment, some suspended in it and some rolling along the bed. Whether or not more grains are arriving at a given place than are leaving, however, can be determined only by precise measurements made over a period of years. No such studies have been made on the James River, but some other relationships give a general idea of what must be happening.

The ancient rocks (granites and the like) that underlie the surrounding countryside are exposed here and there in the stream bed *throughout the river system*. These stream-bed outcrops are rounded and scratched, presumably by grains thrown against them by the stream.

Between the rock exposures, and covering most of the river bed, is a layer of sand and gravel, typically a few inches to a few feet thick. Because most of these particles are abraded to roundness and consist of rocks other than those exposed in the bed and in the nearby valley, they must have been transported from somewhere upstream and deposited on the bed.

Judged from these relations, the stream *in every reach* is (1) slowly eroding its bed (the abraded exposures of solid rock), (2) depositing materials carried from upstream (the layer of sediment on the bed), and (3) carrying some sediment toward the estuary (the stuff we can see in transit). Combined with the smoothly curving slope of Figure 2-8, these things must mean that erosion and deposition are closely in balance within each reach of the system. Thus, we see another aspect of order that must be based in some way on the regular changes in discharge, channel size, slope, and impedance we have already noted. In a long-term view, the system must slowly be reducing the land and lowering its overall slope to the sea. As it does so, we can predict that all its attributes will also be changed but so slowly that balance will tend to be maintained.

TABLE 2-1.

Channel Dimensions and Flow Data Measured by the U.S. Geological Survey at the Cartersville Gaging Station

DATE	WIDTH (ft)	DEPTH* (ft)	VELOCITY* (ft/sec)	DISCHARGE (ft³/sec)
September 1, 1969	614	3.3	1.88	3,830
January 2, 1968	648	4.6	2.29	6,830
March 1, 1966	764	13.0	3.83	37,900

*Depth and velocity are averages for the entire river at the gaged crossing. The depth measurements were made at 20-foot intervals across the river, and velocities were measured at several depths at each 20-foot-interval station.

Because this river system, like most, shows such remarkable order and balance, it might seem that the various interrelations could be expressed in some simple mathematical form, but if so it has not been discovered. The variables within the system are many and some of their precise interrelations are not well known. Like a living thing, the system is based in part on internal complexities, such as the details of current flow, and in part on outside effects, such as storms and droughts, that are not readily predicted. Several of these attributes tell us so much about how the system works that we should examine them thoroughly.

no simple mathematical relationship

Low Flows and Flood Flows

Extreme high and low flows of the river must be well understood because they have much to do with the working of the system and limit many human uses of it. Floods result from specific storms. The river rises quickly and then almost as quickly subsides. The great storm of August 19th and 20th, 1969, generated a peak flow of 248,000 cubic feet per second at Cartersville, only 30 hours after the river started to rise from an average flow of 6,500 cubic feet per second. The flow then dropped to 50,000 cubic feet per second 48 hours after the peak and continued dropping slowly until it returned to an average flow 7 days after the peak.

Low flows result from protracted periods of scarce rain. The river falls slowly as the groundwater reservoir is depleted and flows at minimal rates for the duration of the drought. The drought of 1966 resulted in flows at Cartersville of less than 900 cubic feet per second for the periods June 28th to August 12th and August 22nd to September 14th. During 23 days the flows were less than 500 cubic feet per second, the lowest being 330 cubic feet per second.

cf. Calif. drought

In order to plan for this tremendous range in flows, hydrologists (water resources scientists) have examined all river flow records, some as far back as 1898. Using statistical methods, they have been able to calculate the probabilities of high and low flows in the future. These data do not tell what the river will do at a specific time, but they are useful to the industrial planner whose factory will need a certain input of water, or to the health officer who must contend with concentrations of pollutants in a river flowing one-tenth of normal.

Damage by lowest and highest flows cannot be anticipated accurately, however, because our record may be too short to predict the true potential of the system. A case supporting this point is the flood caused in 1969 by Hurricane Camille, which produced flows in some parts of the James River system that were eight times greater than any measured previously. This flood is worth looking at specifically because it showed how a river system works when running at full strength, as well as some other intriguing things.

Hurricane Camille moved in from the Gulf of Mexico on August 17th, raking the eastern side of the Mississippi delta with winds of over 150 miles per hour, and passed up the Mississippi Valley as a weakening tropical storm. Shifting toward the east, it intensified when it crossed the Appalachians and Blue Ridge because it met a moist air mass moving in from the Atlantic. The resulting torrents of rain centered over the eastern James River system; 4 inches fell in 36 hours over much of this area and the maximum, on the Blue Ridge, was estimated at over 30 inches.

Landslides and sediment flows The resulting stream flows were catastrophic, particularly in the tributaries that drain the Blue Ridge. Soils on the forested slopes apparently soaked up rainfall as usual, but in some locations they became so water-charged as to slide and flow in segments many tens of feet across and hundreds of feet long (Figure 2-12). In many cases, the slides churned into avalanches and then became completely mixed with water at the base of the steep slopes, moving as mud-and-debris flows down the steep tributary canyons to the more gentle valley slopes. When these viscous masses came to rest they were partly reworked by subsequent stream flows, which carried the finer materials downstream but spread sand and pebbles in fan-shaped arrays over fields and around dwellings (Figure 2-13). Where the landslides and avalanches moved directly into torrential tributaries, most of their materials were transported downstream, but the larger fragments remained in the mountain canyons, typically entangled with tree trunks in a series of mounds, ridges, and debris dams. Other parts of the load were deposited next to the streams, on valley flats that were covered at peak flow.

To get a feel for the water power generated, consider the flow measured on the Tye River, a principal tributary draining about 90 square miles of the Blue Ridge (Figure 2-3). This river is gaged near the eastern front of the Blue Ridge, where it was flowing $2\frac{1}{2}$ feet deep on August 19th, just before the storm. Just after the peak of the storm, at 4:00 A.M. on the 20th, it was 29 feet deep! The amount of flow at the gaging station averaged only 142 cubic feet per second for the period 1936–1967. At the peak of the flood it was 80,000 cubic feet per second (the gage was washed away and the figure is a conservative estimate based on high-water marks along the valley).

These figures are not so remarkable in themselves as they are in the context of their effects on the river valley. Channels and valley floors in the mountains were eroded appreciably but the more general effect downstream was deposition of sediment on valley flats. Apparently, most of the gigantic load of sediment fed in by landslides and bank erosion in the mountain canyons was moved through the system quite efficiently. The channel and valley must be about the right size and shape to transmit floods of this magnitude. These relations indicate that the flood's specific interactions are significant for the entire system and should be examined more closely.

Flow in the channel and on the valley floor The Tye River, like the James River, flows in a distinct channel in a valley flat appropriately called a *floodplain* (Figure 2-14). By the time the Tye River rose to fill its banks, its velocity must have at least doubled and its turbulence increased manyfold. It did not, however, erode its bed and banks much at this stage, presumably because it was carrying so much *detritus* (fragmental material) from upstream that erosion and deposition were approximately balanced. When the river overflowed its banks, the sediment suspended in the roiling channel tended to be deposited from the shallow sheet that spread out slowly across the flood-

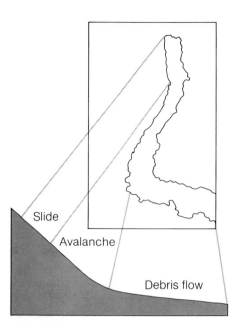

FIGURE 2-12
Scar and deposits left by a combined slide, avalanche, and debris flow in a valley of the Blue Ridge. The approximate slopes of each are shown in the diagram, which is oriented at right angles to the photograph. Photograph by Virginia Department of Mineral Resources.

FIGURE 2-13
Debris-flow deposits below mountain canyons reworked by streams into a distinct fan (below the trees near center) and a valley train (lower right). Much of the finer sediment was washed downstream. Photograph by Virginia Department of Mineral Resources.

FIGURE 2-14

Channel and floodplain of the Tye River, with the James River in the background flowing from right to left. Taken six days after the flood peak, the photograph shows the eroded abutment of a highway bridge washed away by the flood and repairs in progress on the abutment of a railway bridge. Below is a diagrammatic vertical section through the channel and adjacent parts of the floodplain. Photograph by Joe Hirn, Virginia Department of Highways.

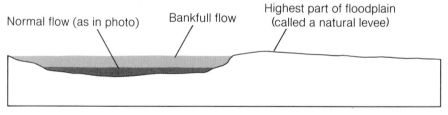

Normal flow (as in photo)   Bankfull flow   Highest part of floodplain (called a natural levee)

plain. A foot or so of muddy sand was deposited near the channel and lesser amounts away from it, as shown by the light area in Figure 2-14. That this sort of thing has happened many times in the past is shown by the low slope of the floodplain *away* from the river bank (Figure 2-14). The finest deposits accumulated in low areas away from the river, especially where the floodwater formed temporary ponds (as on the right side of Figure 2-14 just beyond the railroad).

The flood on the main river was larger in scale but even more benign. The banks are protected by large trees and the river moved so slowly when it spread over the floodplain (for widths up to a mile) that it left barns and other structures more or less intact. A typical case is illustrated by Figure 2-1, where the water rose almost to the bridge deck, covering everything from the lowest left corner of the view to the slopes in the distance. The floodplain was eroded locally where water was funneled along roads or back into the main channel, but the typical effect was the widespread deposit of a thin layer of mud and sand.

Richmond is at the Fall line

Within the main channel, however, flow was turbulent and powerful, especially so in the steep reach at Richmond shown in Figure 2-15. The circular wakes of violent eddies show clearly in the left side of the photograph, as does the phenomenon of giant, rhythmically spaced waves, forming a train that extends from the foreground into the distance, about one-third of the way from the right edge of the picture. Like the bands of waves in the stream on the beach (Chapter 1), the train marks the deepest and swiftest part of the river, and experiments in flumes (channels built in laboratories) have shown that each wave lies over a mound of sediment on the bed. Such mounds migrate and disperse as the waves slowly shift and break, and tremendous amounts of sediment are transported in this way.

Long-term work by floods In addition to showing how energetic a river system can become at peak flow, these various flood data give insights into the overall working of the system, one of which is the disclosure of how erosion takes place in the forested mountains. Recall that trees and litter protect the slopes from erosion by raindrop impact and runoff, yet paradoxically the slopes are steep and the stream beds peppered with large boulders. Garnett P. Williams and Harold P. Guy, who studied the Camille storm slides and eroded canyons in detail, determined that the total erosion averaged over large areas was as great in a few days as that caused by normal-type erosion in several thousand years (Reference 1).

Have such events occurred before? Definitely so, for one can easily spot groups of giant boulders that mark the sites of former debris dams. These groups now form lovely boulder cascades from which the tree trunks have long since rotted and the finer materials have been winnowed by stream flow (Figure 2-9). The remaining boulders are too large or are set too tightly to be

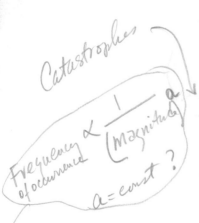

Catastrophes

$$\text{Frequency of occurrence} \propto \frac{1}{[\text{magnitude}]^a}$$

a = const ?

FIGURE 2-15
Main channel of the James River in flood at Richmond, viewed downstream on August 22, 1969. Photograph by Ken Soper, Virginia Department of Highways.

eg: B value for earthquakes magnitude closely measured

moved, but sand carried in the water has scoured them into rounded and fluted forms.

Another major discovery was that the large load of sediment acquired in the Blue Ridge headwaters by the Camille storm was largely balanced by the increased transporting power of the river. The many kinds of order within the river and the characteristics of the valley were maintained. It can only be concluded that the channel and valley forms are well adjusted to massive floods and, therefore, were probably generated by such floods in the past. The system is thus based on events that seem unusual from the viewpoint of a human lifetime but must have happened repeatedly. This observation has great practical significance: *human modifications of the river and floodplain must be thought out with care because the river will always tend to reconstruct its most suitable channel during floods.*

Origin of the Floodplain

The floodplain's flatness, its rich soils, and its easily excavated surface will continue to attract human uses in spite of risks. The 1969 flood showed what the floodplain is and how it works. It would also be valuable to know how it formed.

Small tributary valleys tell most about its origin because they can be examined easily in countless places. As shown in Figure 2-16, the typical stream winds from one side of the floodplain to the other. Judged from the underlying materials, which are depicted somewhat diagrammatically in the figure,

FIGURE 2-16

Small floodplain typical of upper tributaries in the Blue Ridge with detail below showing the channel and the materials deposited in and near it. The forest has been omitted and the thickness of the floodplain deposits approximately doubled for clarity.

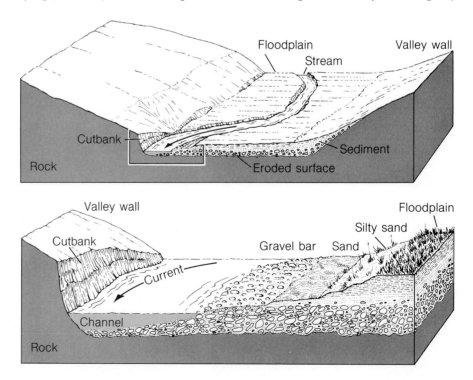

FIGURE 2-17
Stages in the widening of a
floodplain by lateral erosion at
the outside of bends.

the stream has evidently eroded the solid rock down to a nearly horizontal though rough surface. On this base it has deposited sediments, first a layer of coarse gravel, and then sand and silt. The floodplain itself is the upper surface of these deposits. Gravel like that in the lower layer covers most of the stream's bed and thickens toward the inside of the bend to form a gravel bar. The gravel bar is partly overlain by a sand bar and this, in turn, by deposits of silty sand. The development of these features must spell out the origin of the floodplain.

How the features develop is indicated by the relations at the bend. The stream is deepest and most powerful at the outside of the bend and has eroded laterally (to the left) to cut a steep bank in solid rock. This erosion is caused by the momentum of the current as it moves into the bend. The current on the inside of the bend (on the right in Figure 2-16) is correspondingly slack, and as a result gravel moving downstream tends to be deposited there. As the stream cuts on the left it thereby fills on the right, gradually widening the valley. Figure 2-17 suggests stages in such a process projected over thousands of years. Erosion of the floodplain is thus balanced by deposition on it.

But we still need to see how the deposits on the inside of the bend are built up to floodplain level. The deposition of the first sand is easily observed during moderate flows after any heavy rain. The stream rises a foot or so and drops sand in the shallows near the bank because the currents there are weakest. This deposit provides a footing for plants, which further arrest the stream and therefore catch sediment from every succeeding flow high enough to reach them. The deposit becomes finer in grain size as it grows upward because the finer materials are stirred to the top of the stream and become entrapped by the plants. Upward growth slows as the deposit builds higher because each new addition requires floods of greater magnitude. Flows that can overtop the initial gravel bar occur many times a year, but only a few floods a year will reach halfway up to the floodplain, and perhaps one flood in 10 years will spill onto the floodplain itself. The top of the floodplain thus grows upward so much more slowly than the deposits at lower levels that the latter sooner or later reach floodplain level.

Once the features of small floodplains are seen, it is easier to spot similar features in larger ones. Figure 2-10 shows a medium-sized tributary, flowing from right to left, that has undercut a rock bank and deposited a gravel bar

FIGURE 2-18

Upper tributary valleys of the Tye River, near the eastern border of the Blue Ridge, in early spring. The light material covering the floor of the valley to the lower right of the large tree is sandy silt deposited when the stream by the tree flooded.

on the opposite bank. Note especially how the finer, grass-covered deposits rise to the right of the bar toward the nearly flat floodplain surface. Similar relationships on the main river are difficult to photograph because many of the steep bluffs at bends are covered by trees or have been modified by human works. Figure 2-1 was taken from a high bluff at the outside of a large bend, and Figure 2-15 shows the river directing its full power against the bend in the distance.

Variation in the floodplain's width These interpretations of the origin of the floodplain are partly deductive, however, and therefore we should seek additional evidence. One generalization that we can make after touring the system or examining maps of it is that the average width of the floodplain increases downstream. The average width along small brooks is ten feet or so, along medium-size tributaries a few hundred feet, along main tributaries a thousand feet, and along the main river two or three thousand feet. The size of bends in the stream also increases downstream, and this appears to be the more basic relationship because it is true of other rivers that are not confined between valley walls. The size of the bends therefore appears to be a major factor in determining the width of the floodplain, which supports the idea that the floodplain is cut by the migration of bends.

Another kind of evidence supporting erosional growth of the floodplain is the correspondence between its width and the kinds of rocks that lie beneath it. Many of the thick rock layers that make up the Appalachians are easily eroded sedimentary rocks (mudstone, limestone, and sandstone), and where the river flows across them the floodplain is typically a half mile wide. It then narrows abruptly to nothing where the river crosses the resistant rocks that make up the Blue Ridge (Figure 2-3). East of the Blue Ridge, where the river once more flows on less-resistant rocks, the floodplain widens to a half mile again.

The upper tributaries show the same sort of correlation in detail. Mountain canyons have narrow floodplains, if any, where the rocks are resistant. The valley walls are steep and show many bare outcrops of rock. But where rocks are easily eroded, floodplains may extend to the very headward end of the tributary networks. Figure 2-18 is a view down a tributary valley that ends in a hillslope only 200 yards behind the point where the camera was held. The valley ends downstream just beyond the tobacco shed, where it joins the valley of a larger tributary. The tiny stream (under the oak) flows only after heavy rains. Yet the white patches near the center of the photograph are sand and silt deposited during a recent high flow. This small valley floor is thus a floodplain that acts basically like that of the larger tributaries. Significantly, this part of the Blue Ridge gave rise to most of the landslides during the 1969 flood. Perhaps the slides and the unusually extensive floodplains are both due to the nonresistant nature of the underlying rocks.

The width of the floodplain and the erodability of the rocks are connected in another important way. When the river is at flood, its depth and therefore its erosive power are far greater where there is no floodplain (Figure 2-19). We have just noted that the narrow parts of valleys are underlain by the more resistant rocks. Evidently it is these rocks that receive by far the strongest erosive attack during floods. This effect compensates for their resistance to erosion and may be why the river slope in Figure 2-8 extends quite smoothly across a terrain that consists alternatively of more and less resistant rocks. The river canyon across the Blue Ridge, for example (labeled on Figure 2-8), is only a little steeper than the valley in adjoining reaches.

River valley with broad floodplain

River valley with
no flood plain

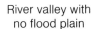
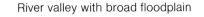

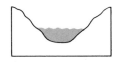

FIGURE 2-19
Channel shapes and water depths during average flow (top) and major floods, showing dependence on shapes of the valley.

In conclusion, the floodplain expresses the interaction between solid land and flowing water more than any other feature of the system. It is related in an orderly way to downstream changes in the river, especially to increasing flow and increasing size of bends. The floodplain is generated by floods yet it acts to regulate their erosive force, which is doubtless why the system's valleys are well adjusted to major floods. By catching and conserving soils eroded upstream, the floodplain has formed a porous, periodically enriched deposit uniquely suited to life processes (the largest trees formerly grew there). Its importance within the system suggests it be used with care.

Users and Uses

Some links between the natural system and human users have probably become apparent by now, but who are the chief users and what are the principal uses? According to the Virginia Department of Conservation and Development, total daily withdrawals from the river by individuals (mainly farmers) average 5 million gallons; those by public water systems, 192 million gallons; those by industry, 495 million gallons, and those by steam plants, 1,559 million gallons. Most users return the greater share of this water to the system, though it is polluted to various degrees. The causes of pollution are identifiable and can doubtless be corrected when a will develops. Develop it must if needs for water continue to grow as they have, for water will have to be reused extensively. The human population in the James River system is now 2 million and is expected to exceed 4 million in the year 2,000.

The water shortage that is likely to ensue will not result simply from the increased numbers of persons. It will be linked appreciably to the kind of industrial growth likely to occur. Average per capita income in the James River area for 1967 ranged from a low of $1,632 in poorer agricultural areas to a high of $2,954 in cities with industry; United States per capita income that year averaged $3,158. These figures imply a comparatively cheap labor force that makes the area attractive to industry. The presence of abundant good water will also tend to attract industries that are heavy water users and possibly above-average water polluters. Moreover, as incomes increase in municipal areas, water uses will also increase. In more industrialized parts of the United States, for example, the increased numbers of home garbage disposals have placed an enormous load of wastes on sewage plants and typically on river systems.

Increasing a water supply   What can be done to increase the amount of water available? Wells on the Coastal Plain gain water that would otherwise be lost to Chesapeake Bay. This source might be developed to a greater degree. Additional wells in the Piedmont and mountains would not add water to the amount basically available, but would provide widely distributed, modest sources for more upland users. Such sources would create a means of using more of the system's water twice, for wastes from upland users are returned via groundwater flow to streams and the river. When waste water emerges from underground its quality is generally improved, and some hydrologists have suggested that this would be the least costly way to remove pollutants. An argument against this solution is the possibility of polluting entire underground reservoirs, so this idea must be tested precisely.

Dams could be constructed to impound the extra water flowing through the system during floods, but it is questionable whether the water that could be stored would be valuable enough to offset the costs of moving towns, railroads, etc. The loss of major parts of the valley itself would have to be weighed carefully. We can predict other possible factors. Because water flow is now well balanced with respect to the sediment load from upstream, would not the sediment-free water leaving the dam erode the valley just below it? And would not this eroded sediment tend to be deposited in still lower reaches because of the reduction in flow caused by the dam? Such things have happened where other rivers have been dammed, in some cases with considerable dollar losses and with disastrous effects on life forms and life webs. What of the time when the dam is so filled with sediment that it cannot supply the extra water needed for the new developments downstream?

Another possible way of creating more water is to cut down forests in the Appalachian and Blue Ridge terrains in order to conserve the water lost by their life processes. Some tests have indicated that stream flow increases appreciably when forest cover is replaced by grasses. Aside from the grimness of this measure, problems are likely to arise because the soils and slopes are adjusted to forest growth. With the loss of the large tree roots that bind soil and loose rock to the underlying solid rock, landsliding would tend to increase (recall the Hurricane Camille slides). With the disappearance of the forest litter, surface runoff and hillslope erosion would increase. Increased runoff would force the entire downstream channel system to readjust. Increases in flood peaks and sediment loads would insure greater flood losses. Will the possible damages to the system be worth the water gained?

Erosion due to urbanization Whether or not the water supply can be increased, what stream-related problems other than pollution can be anticipated for urban areas? M. Gordon Wolman and Asher P. Schick studied the effects of urbanization on streams in parts of Maryland similar to the lower Piedmont and upper Coastal Plain of Virginia (Reference 2). They found that sediment loads of streams rose sharply whenever the ground was stirred or bared by construction. Erosion rates of more than 150 tons per acre per year were measured in some cases, which may be compared to the Virginia Piedmont average (for all types of uses) of 3.2 tons per acre per year. The heavy loads clogged the channels with gravel and sand and often forced the streams to flow against their banks and erode them rapidly. When construction was completed, runoff increased sharply because rain could not soak through the new rooftops and asphalt pavements into the ground. Moreover, the water carried little sediment and therefore tended to erode its channels—especially their banks—at rates of more than 1 foot per year in several cases. This lateral cutting was due in large part to currents deflected around bars of gravel deposited during the preceding period of deposition, augmented by artifacts of the city—bent supermarket carts and the like.

How do we plan for a water system that rides on such a variety of circumstances as cast-off junk, expanding industry, and natural order? Formal planning generally takes place in two stages. In the first, the professional planner studies the system and devises various possible projects and tentative plans. He or she then casts the various costs of the projects and the attributes of the system into numbers, which permits using an electronic computer to make the laborious comparisons between alternative plans. At the next stage, the results of the planner's work are considered by the government that wants to adopt a plan. Many value judgments made at this stage go beyond what the planner could set in firm numbers. A dam site may have been recommended on a cost-return basis but the government must also consider the dam's possible *presence*. Human needs include human feelings. Anyone who has lived in the valley and lovingly preserved some part of it for most of a lifetime is very much a part of the system!

Does our examination of the river system suggest any priorities for setting costs or making broad value judgments? Both the order within the system and the mechanics that appear to generate that order seem to say that the system should be left as unchanged as possible. Any changes should certainly be studied with care, not just with respect to immediate costs and effects but also with respect to the entire system for the future.

This seems to say that there is a scientific basis for opposing progress. More correctly, I think, it says there is a basis for opposing growth whose effects are at best unknown. All of us, however, must study and judge for ourselves.

[handwritten margin note: An alternative conclusion: The river will simply adjust itself to any changes we make to a new order]

[handwritten margin note: WEAK]

Summary

The James River system is based on many parts and actions, the more important of which apply to other river systems in regions with temperate, moist climates.

1. The regular spacing of tributaries results from the even distribution of precipitation across the area and from the previously forested nature of the entire terrain.
2. Almost all precipitation sank into the soil of that forested terrain, as does a large proportion today. This water joins the pervasive reservoir of groundwater, from which it flows slowly to the surface and thus prolongs river flow between storms.
3. Removal of the forest cover resulted in increased surface runoff during storms and thus in the erosion of much of the original topsoil.
4. Confluence of tributaries and upwelling of groundwater gradually increase the river's discharge downstream; as a result, the river gradually deepens and widens and its slope gradually decreases downstream.
5. The average velocity increases downstream in spite of the decrease in slope, partly because of the increase in channel size and partly because of decrease in the size of particles on the bed, which impede flow.

6. The system's capacity for erosion and transportation of sediment increases enormously during major floods, which have evidently generated the channel and valley forms.
7. The floodplain is produced by lateral cutting at bends, an action controlled by the local resistance of the rocks. The floodplain generally becomes wider (and river bends larger) downstream.
8. Because the system is entirely integrated, we must be cautious in using it and especially in modifying it. We should particularly respect the river's use of the floodplain.

REFERENCES CITED

1. Garnett P. Williams and Harold P. Guy. Erosional and depositional aspects of Hurricane Camille in Virginia, 1969. U.S. Geological Survey Professional Paper 804, 80 pp., 1973.
2. M. Gordon Wolman and Asher P. Schick. Effects of construction on fluvial sediment, urban and suburban areas of Maryland. *Water Resources Research*, vol. 3, pp. 451–64, 1967.

ADDITIONAL IDEAS AND SOURCES

1. The development of sinuous rivers is an important topic discussed in an excellent article by Luna B. Leopold and W. B. Langbein, "River meanders." Published in the June 1966 issue of *Scientific American* (Vol. 214, No. 6); it is also available as a reprint (No. 869) from W. H. Freeman and Co., 660 Market St., San Francisco, Calif. 94104.
2. Further evidence of ordering of the river system with its landscape is suggested by Figures 2-2 and 2-3. Note how many of the tributaries and ridges are aligned northeast-southwest. This orientation is the trend of the bands of rock that make up the western part of the terrain. The river has thus "settled in" to the underlying rock pattern by eroding its principal tributary valleys in the bands of less-resistant rocks.
3. The sudden steepening of the river profile at Richmond (Figure 2-8) is also an expression of fit to rock varieties. At this point easily eroded rocks of the Coastal Plain (sandstone and the like) give way to the more resistant rocks of the Piedmont, creating a rapid (the Fall Line of earlier American history).
4. Of great value to me in examining the James River system was a series of planning bulletins prepared by the Virginia Department of Conservation and Development, Division of Water Resources, as part of their comprehensive water resources plan. I mention this not so much to suggest you read them as to emphasize that similar plans and materials on other systems may be available to you. Contact your county and state planning commissions (or whatever). Besides their own documents, planning groups often keep libraries of pertinent materials that can be examined at their offices.

5. If you are studying a stream or small river (one you can wade safely), you can determine the discharge by the following procedure. (1) Stretch a tape or a string knotted at intervals of a foot across the stream just above its surface. (2) Use a yardstick to measure the depth at enough points so that you can construct a fairly accurate profile (vertical section) of the bottom of the stream (like those in Figure 2-11). (3) Measure the area of this cross section in square feet (for example, by plotting the profile on cross-ruled paper and counting the squares). (4) Measure the approximate average velocity of the stream as described in the Additional Ideas and Sources section of Chapter 1. (5) Calculate the discharge by multiplying the velocity by the area of the cross section.

Thunderstorm over the Rio Grande valley 25 miles north of Santa Fe. The river winds toward the foreground in White Rock Canyon, which it has cut from horizontal layers of volcanic rocks.

3. Sediment Problems on the Rio Grande

Sediment Problems on the Rio Grande

Sizing Up the Accumulation of Sediment
    Effects on irrigation systems
    Surveying the accumulation of sediment
Sediment Transport in the Main River
    Roughness due to grains
    Roughness due to bed forms
    Erodable beds and banks
    Suspended clay and silt
    Effects of temperature
    The results overall
The High Mountain Streams
    Orderly nature of the streams
Erosion in the Arroyo Country
    Measuring rates of erosion
    Amounts of sediment eroded
Causes and Remedies
    Measuring effects of grazing
Summary

Linkages between earth systems and climate are truly basic, as we shall see by moving from the well-watered eastern seaboard to the semiarid Southwest. The Rio Grande epitomizes water systems in semiarid regions, and the view on the chapter opening page illustrates some of its climate-related problems. The gigantic sweep of country is largely in sun yet a thunderstorm is bringing heavy rain to several square miles of the river valley in the distance. Hillsides will come alive with sheets and rills of muddy water and some gullies will briefly carry torrents, but most of the water will soak into sandy channels before reaching the river. By noon the next day the ground will be dry, and although some other part of the vast system will be receiving rain, that storm, too, will generate only local flow. Snow in the Rocky Mountains (in the right distance) yields considerable water in the spring but otherwise the river is fed first here, then there, and as a result flows modestly indeed. At the point shown in the view (1,000 miles from its mouth) the Rio Grande drains an area twice as large as the entire James River system, yet its annual flow here is one-sixth as much. That is, its *water* flow is one-sixth as much; the concentration of sediment in its average flows is fifty times that of the James!

The resulting problems are not unique to the Rio Grande. They develop wherever easily eroded, semiarid country is pressed by human uses. Erodability and low water flows result from low rainfall and sparse vegetative cover. Rapid evaporation is another contributing factor. The combined effects can be judged from a simple comparison. The James River system receives an average of 43 inches of precipitation annually, and about one-third of it escapes evaporation and transpiration to flow out through the river. The Rio Grande system receives about 16 inches a year, and something like one-thirtieth of it flowed out through the prehistoric system. Human uses have so increased evaporation and transpiration that less than one one-hundredth of the 16 inches now reaches the Gulf of Mexico.

The dryland nature of the river system is emphasized by a map of its continuously flowing parts (Figure 3-1). Compared to the James River (Figure 2-2), it is a long but spindly vine. Nonetheless, the Rio Grande can support large human uses because its spring flows are stored behind dams. The resulting reservoirs feed irrigation works that are part of an intensive agriculture on the floodplain. The surrounding country is used for grazing, but in spite of its gigantic size, it yields only a fraction of the values derived from the crops raised on the valley floor and the animal herds that feed on them. Most cities and towns have arisen on the floodplain, but they include almost no water-based industry for the simple reason that agriculture has preempted all the available water.

All uses of the floodplain, however, have been in jeopardy for more than 100 years. The river has flooded repeatedly and has deposited considerable sediment in the Middle Valley, New Mexico's most populated area (Figure 3-1). This chapter will explore the basic question that is simple yet axiomatic to understanding all river systems: why is this river filling its valley rather than eroding it?

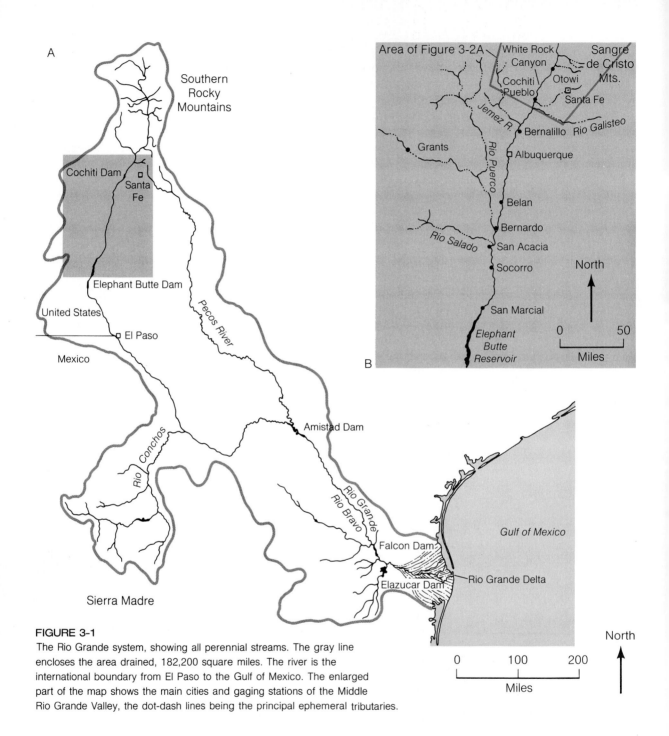

FIGURE 3-1

The Rio Grande system, showing all perennial streams. The gray line encloses the area drained, 182,200 square miles. The river is the international boundary from El Paso to the Gulf of Mexico. The enlarged part of the map shows the main cities and gaging stations of the Middle Rio Grande Valley, the dot-dash lines being the principal ephemeral tributaries.

Sizing Up the Accumulation of Sediment

It is important at the outset to see that we are dealing with part of a larger system. History has determined our choice—the Middle Valley is where the problems have occurred and where intensive studies have been made. The subsystem is limited at the lower end by Elephant Butte Reservoir, which is thought by some to be involved in the problems (Figure 3-1). At the upstream end is White Rock Canyon, the upper part of which is shown in the chapter opening photograph. The river flows through the canyon in a narrow course typified by shallow, turbulent reaches called *riffles* that alternate with more tranquil deep stretches called *pools*. It then changes abruptly where it emerges from the canyon near Cochiti Pueblo. Its valley opens out rapidly and its floodplain widens from near 0 to 2 miles (Figure 3-2A). The valley just below Cochiti Pueblo is the upper end of the Middle Valley proper, and its various parts are typical of the rest of the valley.

The floodplain that lies near the center of the broad valley contains an open, sandy wash or *floodway*, typically 1,000 to 2,000 feet wide (Figure 3-2B). During high flows the wash is largely covered by water, but at other

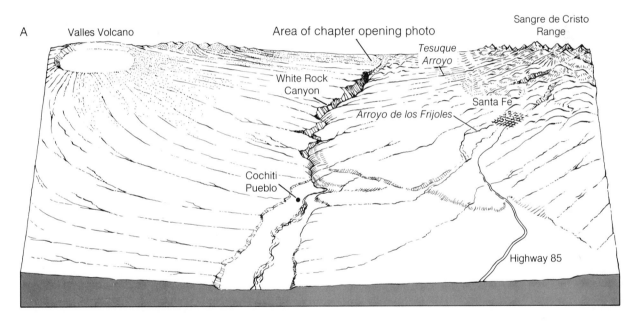

A, Valles Volcano, Area of chapter opening photo, Sangre de Cristo Range, Tesuque Arroyo, White Rock Canyon, Arroyo de los Frijoles, Santa Fe, Cochiti Pueblo, Highway 85

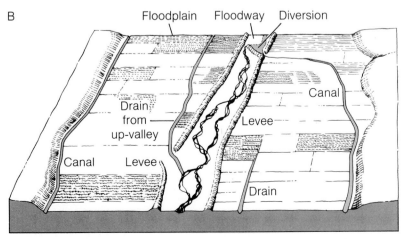

B, Floodplain, Floodway, Diversion, Drain from up-valley, Canal, Levee, Canal, Levee, Drain

FIGURE 3-2

A, upper end of the Middle Valley and vicinity, showing some of the places mentioned elsewhere in the chapter. B, an enlarged view of the floodplain, showing the levee-bounded floodway and the various canals of an irrigation system.

times it has distinct, sinuous channels that part and rejoin around huge sand bars. The floodway is bordered locally by low bluffs and is constrained in many places by levees and other engineered structures. Broadly, though, the floodway retains two general aspects crucial to the problems we are considering: (1) it lies at about the same level as the rest of the floodplain, and (2) the banks of the channel in the sandy wash are low and easily eroded by the river.

Effects on irrigation systems In order to see how a sediment-laden river in such a situation can damage floodplain installations, it is necessary to examine an irrigation system. As shown in Figure 3-2B, a small dam diverts water into a canal that carries it to the floodplain and down the valley. Water is drawn through lateral ditches to fields, and excess water must be drained off through another canal. To avoid the cost of pumping, the main canal is constructed with a very gradual slope, one less than that of the river. Thus, water in the canal comes to lie above the river and will flow by gravity to the entire floodplain.

Such works, however, may be set off balance by sediment from the river. River velocities are reduced at the diversion dam, and thus sediment tends to fill behind the dam and in the upper part of the main canal. Fine sediment remaining in suspension is deposited in the lesser ditches and on the fields, slowly elevating their surfaces and thereby reducing the rates of flow from the canal. The diversion also reduces flow in the river, which leads to sediment being deposited in the floodway below the dam. The river is held between levees, so the filling actually elevates its bed above the adjacent floodplain. Elevation of the river stops drainage from the fields, which then become water-logged, and plants suffocate as water rises around their roots. Finally, the elevated river becomes more and more likely to rise above the levees and flood massively, which is disastrous to the irrigation systems.

Although irrigation systems can be repaired, the overall history of irrigation in the Middle Valley shows that the damages accrued as long-term losses. Pueblo Indians first leveled ditches to water their fields more than 2,000 years ago, and by 1590 25,000 acres were being irrigated. Spanish colonization increased the total area irrigated to 100,000 acres by 1800, and further development under Mexican rule brought the total to 123,000 acres in 1850. The wave of U.S. settlers arriving after 1850 pressed further on the river system, but somehow the total area irrigated only increased to 125,000 acres in 1880 and then declined rapidly. In 1896 roughly 50,000 acres of the Middle Valley remained under irrigation and by 1924 the number had been reduced to 40,000 acres, possibly less.

Surveying the accumulation of sediment Spanish and Mexican records are too sparse to show exactly when serious troubles began, but surveys indicate the river was building up its bed measurably in the late 1800s and early 1900s. Flooding in the 1920s led to more and more organized action and study. Ninety sections across the river and floodplain were surveyed in 1936 and again in 1941–42; the differences in elevation showed that the floodway and floodplain had indeed built up slowly, especially in the lower part of the valley. Starting in 1948, the Bureau of Reclamation and the Army Corps of

TABLE 3-1.

Rates of Filling by the Rio Grande, in Feet Per Year, During each of Three Survey Periods

Miles above Elephant Butte Dam	206	183	143	91	62	42
1936–1944	.05	.041	−.047	.038	.40	
1944–1954	.00	.043	.044	.050	.27	
1954–1962	.00	.042	.058	.049	.10	

Data from U.S. Bureau of Reclamation summary report. *Rio Grande, aggradation or degradation,* 1936–1962, by Hydrology Division, Albuquerque, New Mexico, December 1967.

Engineers cleared vegetation from the floodway and straightened and narrowed it with levees. Dams were built across several major tributaries to catch sediment that would otherwise be carried to the Middle Valley.

To help in planning these projects and to determine their early successes, if any, floodway surveys were made in 1944, in 1952-54, and in 1962, using the ninety sections surveyed in 1936 and 1941–42 and seventy-four additional ones. The sections are about a mile apart, from Cochiti Pueblo to Elephant Butte Reservoir, and thus give an unusually complete measure of sediment filling along the entire Middle Valley. Table 3-1 shows the results of the surveys, as alloted to five divisions set up by the Bureau of Reclamation. Evidently, there has been little or no filling on the upper end of the valley during the twenty-six years surveyed, whereas filling at the lower end has been considerable. The average rates of filling in between are about the same from one division to the next and have remained fairly constant during the three survey periods. The rates of filling in the lowermost division have decreased considerably, and although this is probably the result of the channel works, many other factors must be judged. In order to interpret these measurements, we must see just how the river transports sediment and why it deposits it.

Sediment Transport in the Main River

In spite of the foregoing record, the Rio Grande transports gigantic sediment loads through the Middle Valley. To get a feel for this incredible capacity, the river's slopes and amounts of flow (discharges) may be compared to those of the James River, which carries far less sediment (Figure 3-3). Compare the discharges carefully. The two rivers flatten downstream almost identically, yet the Rio Grande's flows decrease rather than increase downstream! How can a dwindling river transport abundant sediment on a decreasing slope? The answers come from detailed studies made by hydrologists of the U.S. Geological Survey. Because each result is tremendously important to understanding all water systems, we will take them up one at a time.

Roughness due to grains The decrease in the average grain size of the Rio Grande's bed sediment downstream (Figure 3-3) produces a large decrease in the roughness of the bed downstream (Figure 3-4). This, in turn,

FIGURE 3-3

Slope of the Rio Grande in the
Middle Valley compared with an
equal length of the James River.
The numbers below the slopes
are average discharges, in cubic
feet per second, during 1950–65.
The numbers at the base are
average sizes, in millimeters, of
the bed sediment. Data from
U.S. Geological Survey Water
Supply Papers 1723, 1732, 1904,
1923 and, 1498–F.

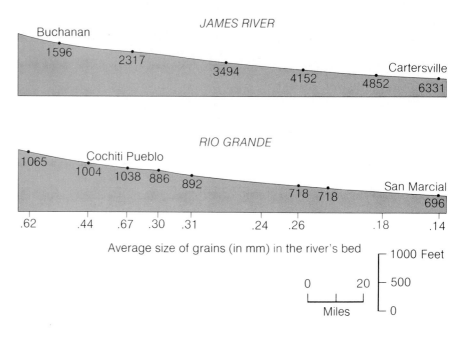

results in a decreasing drag on the river, which helps it flow efficiently on a lessening slope.

Roughness due to bed forms Probably the greatest factor in the river's efficiency is its ability to smooth its bed during rising floods. This discovery was made partly by exploring the river and partly by extensive studies using flumes (Figure 3-5). When the flume in the figure was started with a smooth sand bed, a stream flowing at around 1 to 1.5 feet per second eroded and redeposited the sand into small, rhythmically spaced *ripples* (Figure 3-6A). Similar bed forms are found in the Rio Grande, where it is shallower than about half a foot and moves at similar velocities (Figure 3-7). Sand grains roll and skip up the backsides of ripples and fall down the steeper frontsides, so that the ripples slowly migrate downstream.

As the velocity of the flume was increased to 1.5 feet per second, the ripples became fewer and larger and then coalesced into similarly shaped but much larger forms called *dunes* (Figures 3-5 and 3-6B). Dunes in the Rio Grande are concealed by muddy water but can be detected during low flows by wading and probing with a stick. They form when river velocities are less than about 3 feet per second. Dunes impede flow considerably by causing a separation of the main current at the summit and a rotating eddy that catches sand on the downstream face (Figure 3-6B). As the separated current strikes the next dune, it erodes sand and moves it up and over to the next downstream face. Every half minute or so this system becomes un-balanced and a swirling eddy rises through the river to erupt as a short-lived boil on the surface. The boils become more and more violent and sand-laden as the velocity of the river increases.

At river velocities around 3.5 feet per second the dunes in the laboratory flume flattened noticeably, and at somewhat higher velocities the bed was

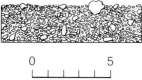

FIGURE 3-4

Enlarged sections through the
Rio Grande's bed showing the
effect of grain sizes on bed
roughness. Top, near Cochiti
Pueblo and below, near
Socorro.

FIGURE 3-5
Up-stream view of flume at Colorado State University, after a run that cast the sand bed into dunes. The ripples formed later, during reduced flow at the end of the run. From U.S. Geological Survey Professional Paper 462–1 ("Summary of alluvial channel data from flume experiments, 1956–1961" by H. P. Guy, D. B. Simons, and E. V. Richardson, 1966).

FIGURE 3-6
Sections along the flume, showing typical streams and bed forms during specific runs. The arrows show velocities, in distance traveled per second. Amounts of sediment transported per second are listed on the right. Data from Tables 4 and 14 in the source listed in Figure 3-5.

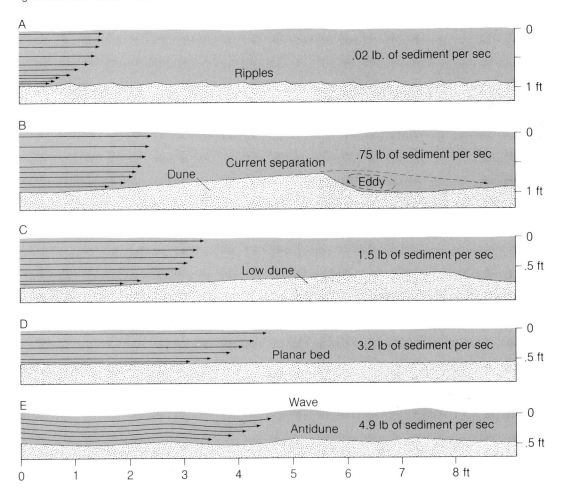

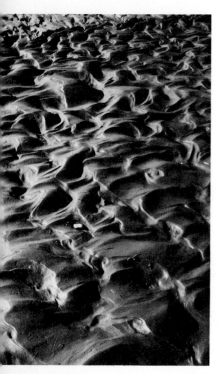

FIGURE 3-7

Ripples on a large sand bar near Bernalillo, formed when the water was about 6 inches deep. Flow was somewhat oblique from right to left, parallel to the streamlined ridges on the ripples. The shiny film of mud was deposited on the sand as the river fell. The ripples are 4 to 12 inches across.

wiped into a smooth plane across which sand was carried at greatly increased rates (Figures 3-6C and 3-6D). The same bed changes were detected in the river by probing from overhead cables and boats. The increased sand transport is apparently due to the greatly increased velocity near the bed, where small but closely spaced eddies spiral downstream parallel to the direction of flow.

At still higher velocities the plane bed of the flume was molded into low sand mounds over which the entire stream swept smoothly and continuously, with a water wave over each sand mound (Figure 3-6E). Surface waves on the Rio Grande are generally larger than flume waves and form in the swiftest part of the river during floods. The sand mounds are called *antidunes* because they often migrate slowly upstream. This migration takes place because some of the large amount of sand streaming over the mounds accumulates on the upstream side of them, whereas sand tends to be removed from the downstream side. As with the stream on the beach, the waves periodically crest and break, sending a great pulse of suspended sand downstream.

Erodable beds and banks The changes expressed by Figure 3-6 are possible because the bed consists of loose sand that has been size-sorted by the river and can therefore be eroded easily by increasing flows. Because the banks and entire floodway consist of similar sand, the river can widen its channel rapidly during a flood. By doing so it retains its flow in a steep-sided channel rather than spilling out over the floodplain as do rivers with less erodable banks, such as the James River. As Rio Grande floods wane, sediment is deposited near the banks first (where velocities are least), so that the river narrows its channel and thus conserves its depth. The sandy, open floodway thus permits the river to expand and contract, and thereby to flow efficiently during low as well as high flows.

Suspended clay and silt Also contributing to the river's efficiency is the large concentration of fine sediment in many flood flows. The suspended fine materials tend to buoy up sand grains. This phenomenon can be determined experimentally by casting a handful of muddy sand into a tall jar and watching the fine materials, which settle very slowly, arrest the fall of the larger grains. The fine materials, in effect, make the water more viscous. They also make it heavier, which increases its tendency to erode the river bed. The combined effects can be remarkable, as shown by two flows measured at Socorro and Bernalillo (Figure 3-8). Because river slopes, widths, depths, and temperatures were almost the same at these two places, the huge difference in the amounts of sand being transported must have been due largely to the concentration of suspended fine sediment at Socorro.

The effects of bed roughness are also emphasized by this case. The coarser grained bed at Bernalillo had dune forms, whereas the finer grained bed at Socorro was planar. As a result the sediment-laden Socorro flow moved at twice the velocity of the less laden flow!

Effects of temperature Finally, the Rio Grande's capacity to carry sediment is based partly on seasonal flows that are cold as well as rich in sediment. Melting snow and cold rains in April and early May result typically in river temperatures of 45° to 50°F. Typical summer flows, on the other hand,

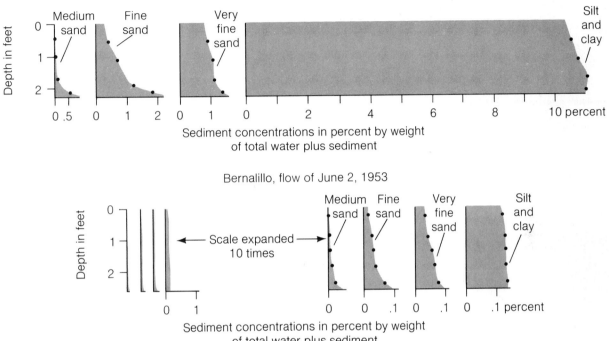

Socorro, flow of Aug. 17, 1954

Sediment concentrations in percent by weight
of total water plus sediment

Bernalillo, flow of June 2, 1953

Sediment concentrations in percent by weight
of total water plus sediment

FIGURE 3-8

Concentrations of various sizes of sediment in two Rio Grande flows, with black dots showing the depths of the samples that were analyzed. The flow measured at Socorro was produced mainly by a heavy summer storm in the Rio Puerco drainage. The Bernalillo data on the left are plotted at the same scale as the Socorro data and are shown at an expanded scale on the right. Data from U.S. Geological Survey Professional Paper 462–B ("Vertical distribution of velocity and suspended sediment, Middle Rio Grande, New Mexico," by C. F. Nordin, Jr., and G. R. Dempster, Jr., 1963).

have temperatures of 70° to 80°F. Since the viscosity of water increases with falling temperature, grains fall more slowly in the cold flows and are also rolled more forcefully along the river bed. Other factors being equal, cold flows also change the bed from dunes to planar form more readily than warm ones.

The results overall The combined effects of these factors may be judged from records of overall water flow and sediment transport in the Middle Valley. Measurements of the principal tributaries and of several stations on the main river are shown for two extreme years in Figure 3-9. In 1952 the flow in the main river was large because of a wet winter. Water arriving at the upper end of White Rock Canyon, gaged at Otowi, came mainly from snow-melts in the high Rockies. Because it was cold it moved sediment through most of the valley very effectively. Summer storms within the valley were unusually light that year, so local tributaries moved comparatively little sediment into the river.

In contrast, 1955 was a year of low winter precipitation and therefore of low river flows generally. Major summer storms, however, were frequent within the drainage area of the Middle Valley, and the data for the principal tributaries show that they introduced very large loads of sediment. The Rio Puerco drainage experienced several summer storms that each resulted in flows of over 1,000 cubic feet per second for a day or so and moved prodigious amounts of mud into the river. The river could not move such large loads during a low water year and deposited sediment massively below each of the

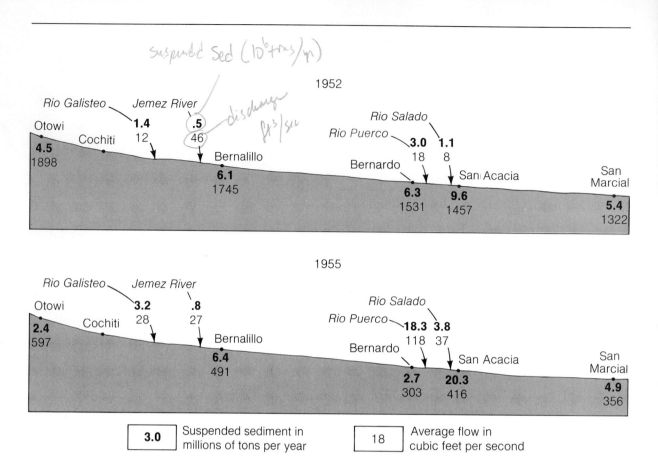

suspended Sed (10^6 tons/yr)

discharge ft³/sec

FIGURE 3-9

Amounts of suspended sediment in millions of tons per year (the heavy numbers) and average flows (discharges) in cubic feet per second for gaged stations on the Rio Grande and the four principal tributaries of the Middle Valley. Data from U.S. Geological Survey Water Supply Papers 1252 and 1402.

tributaries. Including amounts estimated for other tributaries, the river deposited approximately 4 million tons between Bernalillo and Bernardo, nearly 5 million tons between Bernardo and San Acacia, and 16 million tons between San Acacia and San Marcial.

In conclusion, low water years and high sediment yields of local summer storms have been causing the river to fill its bed in spite of its remarkable capacity to transport sediment. Cold spring floods and the continuing snowmelt and groundwater flow from wet winters are crucial to moving sediment through the valley. These flows should be dammed or diverted as little as possible. The Bureau of Reclamation is presently constructing a diversion from the upper San Juan River to increase the Rio Grande's flow, which should be helpful to the Rio Grande.

But the river will still carry sediment into Elephant Butte Reservoir, the source of irrigation water for large areas downstream. The more useful remedy would be to reduce the sediment loads introduced by tributaries. To judge that possibility we must discover the sources of the sediment in the outlying parts of the system.

The High Mountain Streams

The highest and most outlying parts of the Rio Grande system are the Rocky Mountains, fountainhead of the river's spring flow. The southernmost of these is Sangre de Cristo Range, which lies along the northeast side of the

FIGURE 3-10

Northwest side of Lake Peak (12,600 feet above sea level) showing the talus slope and the craggy, broken granite of the summit ridges. The entire end of the valley was formerly occupied by a glacier.

Middle Valley and about 20 miles from the main river (Figure 3-1). This rugged tract is inordinately important to the system because it receives about 30 inches of snow water and rain a year—three times as much as the main river valley. To sense the erosive potential of this water, one must climb to a summit ridge and look back toward the river. From that point to the foot of the range, water falls 6,000 feet in just 7 miles! One might easily conclude that erosion by this water was the source of the sediment plaguing the Middle Valley, but mountain streams should not be judged so quickly. As this section will relate, the mountain slopes are naturally protected against rapid erosion, and detailed measurements indicate the streams are well ordered and slowly evolved.

Two major factors in the mountains' resilience come from their geologic past. The range is made up of resistant rocks, mainly granite that crystallized over a billion years ago and was uplifted when the range was formed. When the climate cooled about 2 million years ago, glaciers developed in the highest parts of the mountains and scoured the soil and loose rock from most areas above an altitude of 11,000 feet. When the last glaciers melted about 10,000 years ago, they left an array of bare rock peaks and cliffed valley heads called *cirques* at the highest elevations, and low ridges and mounds of debris called *moraines* in the valleys just below.

The glaciers continue to influence the water system. Snow water that seeps into open cracks in the glaciated crags freezes at night, and because water expands when it freezes, it pries loose rock fragments that eventually bound down the steep valley heads to form a steeply sloping deposit called a *talus* (Figure 3-10). Surface water and whatever sediment it carries are captured by the labyrinth of the talus, and when the water emerges in springs far down-valley it is clear or nearly so. When sediment has filled the chinks and holes,

FIGURE 3-11
Downstream view of the valley of Tesuque Creek at 8,000 feet altitude, 2 miles from the mountain front. The north-facing (left) slope is grown with fir and long-leafed pines, the south-facing slope with pinyon pine and brush. The floodplain, hidden by the trees, is 40 feet wide in the left corner and 90 feet wide near the center of the photograph.

alpine plants root into the talus and protect it. The developing soil is occasionally swept down by slides, but it is then trapped in lakes or by the grasses of alpine meadows. Vegetation also stabilizes the stream channels in these upland valleys, aided by blocks of rock too large to be moved by the streams.

Below this alpine area, from about 11,000 to 8,000 feet altitude, plentiful precipitation (20 to 30 inches) gives rise to dense conifer and aspen forests that protect the mountain slopes much as the Virginia woods protect the Blue Ridge. Below 8,000 feet altitude, however, the precipitation falls off to about 15 inches and temperatures increase. Because evaporation and oxidation destroy plant litter rapidly on slopes facing the sun, plants on these slopes are sparse and the soil relatively unprotected (Figure 3-11). But once erosion cuts through the stony soil, it is slowed down greatly by the solid rock beneath, so that the slopes are not a continuing supply of abundant mud and sand. Moreover, the main stream in Figure 3-11 has not flooded strongly for many decades, as I learned from long-time residents and from examining old beaver dams and ponds that have trapped nothing coarser than sand and small pebbles.

Orderly nature of the streams   One can determine from topographic maps that the slopes of the streams gradually decrease downstream in areas underlain by rocks of roughly equal resistance to erosion (Figure 3-12). The smoothly concave curves of the slopes and the close similarity among the three streams can hardly be coincidence. These small streams have evidently interacted with the steep terrain in just as orderly a way as large rivers interact with low-lying terrains.

John P. Miller, who studied the mountain streams of the Sangre de Cristo Range over a period of several years, found considerable evidence to support this idea (Reference 1). He discovered that tributaries are distributed evenly, as in the James River drainage, and that channel widths and depths increase downstream as flow increases. By studying fragments of specific kinds of rocks, he found that the streams erode solid rock slowly and occasionally transport fragments as large as boulders. Generally the streams are nearly clear, flowing over beds of clean gravel and sand. His findings substantiate what one suspects after a brief examination: the mountain streams are not the source of the muddy sediment that plagues the Middle Valley.

Erosion in the Arroyo Country

Where they emerge from the mountain front, however, the valleys of the Sangre de Cristo Range change dramatically. In the space of a thousand feet the valley bottoms widen from around 150 to 400 feet or more and the stream channels from around 10 to 30 feet. The stream beds lose their boulderly, overgrown aspect to become open sandy washes with neatly cut banks (compare Figure 3-13 with Figure 3-11). Called *arroyos*, these flat-bottomed channels characterize the broad sweep of country that slopes gently from the base of the mountains to the river (Figure 3-2). Their freshly cut forms imply

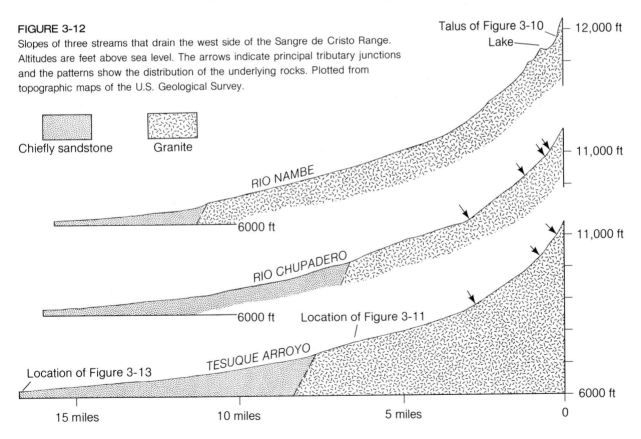

FIGURE 3-12

Slopes of three streams that drain the west side of the Sangre de Cristo Range. Altitudes are feet above sea level. The arrows indicate principal tributary junctions and the patterns show the distribution of the underlying rocks. Plotted from topographic maps of the U.S. Geological Survey.

Chiefly sandstone   Granite

RIO NAMBE

Talus of Figure 3-10 — 12,000 ft
Lake — 11,000 ft
6000 ft

RIO CHUPADERO

11,000 ft
6000 ft   Location of Figure 3-11

TESUQUE ARROYO

Location of Figure 3-13

6000 ft

15 miles   10 miles   5 miles   0

active erosion—by streams that may flow only a few days (or hours) out of the year! Moreover, their smooth sandy floors suggest that water moves down them as swiftly as in the laboratory streams flowing on planar beds.

That the arroyo country produces the sediment damaging the Middle Valley is proven by the fact that streams leaving the mountains carry little sediment whereas those flowing from arroyos into the river are heavily laden. The puzzle is why this should be, for the arroyos are not as steep as the mountain streams and the surrounding terrain is not nearly so rugged. John P. Miller's study gave a basic clue: the mountain streams become arroyos at precisely the place where the underlying rocks change from granite and other resistant rocks to less-resistant rocks that extend under most of the arroyo country (Figure 3-12). These rocks are former sedimentary deposits, consisting of sand, silt, and clay.

But many other parts of the world expose easily eroded rocks that have not caused major sediment problems, so that the large sediment loads of the Rio Grande arroyos must have additional causes. In the hope of discovering the causes, Luna B. Leopold, John P. Miller, William W. Emmett, and Robert M. Myrick studied arroyos south and west of Santa Fe extensively during 1951–53 and 1958–64 (Reference 2). Their first need was to observe and measure flows in washes that are almost always dry. They watched systematically for heavy thunderstorms during the summers of 1951–53 and chased many like that shown in the chapter opening photograph, finding that rainfalls of at least an inch in a half hour are required to bring the arroyos fully to life. The resulting flash floods arrived as low, wavelike fronts called *bores*, as much as 2 feet high. The bores typically filled only the deepest subchannel of the arroyo floor, but additional surges often flooded the entire floor in a matter of 5 to 10 minutes, after which the flows subsided rapidly, with the dwindling flow using (or constructing) a subchannel within the floor (as in

FIGURE 3-13

Arroyo of Tesuque Creek, 7 miles from the mountain front. The view is downstream and toward the Rio Grande, with typical arroyo country in the middle distance.

Figure 3-13). The greatest flow they observed, in Galisteo Arroyo, moved at an average velocity of 6.5 feet per second and averaged 1.3 feet deep and 110 feet wide. Sediment-laden water moves that fast over a broad wash because of the smoothness of the bed and because of the channel afforded by the last stages of the antecedent flow.

Caving of arroyo banks was studied systematically and was found to take place piecemeal. In Figure 3-13, for example, the piles of broken material fell after a recent flow; beyond them are older piles, partly washed over by subsequent floods. In the right foreground is a partly separated slab that caved the year after the picture was taken. Leopold and Miller found that such caving takes place *after* a flood passes.

At their upstream ends, some of the arroyos have steep *headcuts* eroded into nearly flat parts of the valley (Figure 3-14). Water falling over the headcut creates such turbulence as to undercut and cave the vertical face, slowly extending the trench up-valley. The trench is also widened slowly by piecemeal caving along its sides (Figure 3-15). As the figure shows, the arroyos become wider and shallower downstream and eventually pass into filled stretches—valley flats with no obvious stream channel. In Figure 3-14, for example, sand and silt had been accumulating recently among the grasses and bushes in the flat above the cut, and the cut exposes crudely layered, older deposits of the same materials.

FIGURE 3-14
Vertical cut, 10 feet high, at the head of an arroyo tributary to Tesuque Creek, 2.5 miles southeast of Tesuque Pueblo. Note that there is no obvious stream channel above the headcut.

Measuring rates of erosion To detect the total movements of sediment, Leopold and coworkers measured rates of caving of headcuts and sidewalls of the arroyos. They also measured rates of sediment accumulation in various reaches of some arroyos and also the rates of erosion of sediment from the surrounding hillslopes. One method of measuring erosion on hillslopes was to set nails into the slopes, flush with the surface, and repeatedly measure the parts exposed by raindrop and sheetwash erosion. The nails were exposed over 5 years at an average rate of 0.2 inch per year, which means that approximately 27,000 cubic feet of sediment were eroded annually from each square mile.

Another set of surveys measured the *downhill creep* (a very slow movement) of the soil layer itself. Such movement was suspected because clays in the soils swell when wetted and shrink when dried, as shown by crack-

$$\frac{0.2 \text{ in}}{\text{yr}} \times \frac{1 \text{ mi}^2 \text{ ft}}{12 \text{ in}} \times \frac{5280^2 \text{ ft}^2}{\text{unit}}$$

$$= 464640$$

$$\approx 465,000 \text{ ft}^3$$

? because not all of area sloping ?

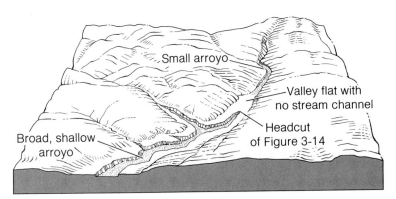

Small arroyo

Valley flat with no stream channel

Broad, shallow arroyo

Headcut of Figure 3-14

FIGURE 3-15
Land forms around the headcut of Figure 3-14, showing the typical segments of a valley in arroyo country. The distance from the headcut to the hill-top divide is 0.3 mile. The abundant pinyon and juniper trees have been left out for clarity.

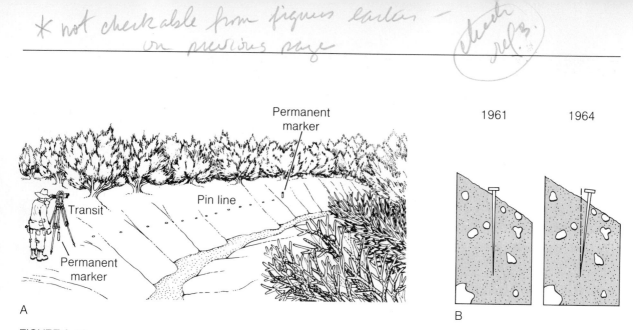

FIGURE 3-16
A, sighting positions of slope-creep pins along a tributary of Arroyo de los Frijoles. B, downhill rotation of 6-inch pin.

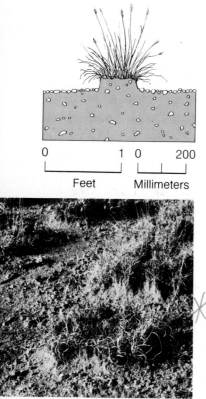

FIGURE 3-17
Soil relics (pedestals) sustained by grass bunches on a slope of 5 degrees, six miles north of Santa Fe, with cross section showing the distribution of pebbles.

ing and crumbling on dried surfaces. It seemed likely that gravity would give these movements a slight downhill component. To measure it, a line of pins was set in a 45° slope next to a typical tributary of the arroyos (Figure 3-16A). A transit (a precise sighting instrument) was then set up over a solidly based point at one end of the line and another solid point was established by driving a steel pipe deeply into the slope at the far end of the line. Sightings were then made repeatedly over 5 years, first on the solid marker, to align the instrument, and then on each pin. The results showed a downhill movement of the tops of almost all the pins, generally by a downhill rotation (Figure 3-16B). The ground surface had thus moved downhill at a rate of about 0.2 inch per year, almost all the movement taking place in the upper 6 inches of soil. As a result, about 55 cubic feet of soil were moving into each linear mile of stream channel each year.

Amounts of sediment eroded   When the various data were averaged across the area studied, it was found that each square mile was yielding these amounts of sediment: 13,500 tons per year by raindrop and sheetwash erosion, 200 tons per year by headcutting and sidecutting of arroyo walls, and 98 tons per year by soil creep. The large amount of raindrop and sheetwash erosion may seem remarkable but is supported by other evidence. Figure 3-17 shows a typical hillside several miles from the area that Leopold and others studied. Each bunch of grass grows from a prominence of soil called a *pedestal*, which stands 2 to 4 inches above the surrounding gravelly surface. By trenching the plot, I found that the soil in the pedestal consists of silt and sand, with about 5 percent of pebbles like those on the gravelly surface. As the figure depicts, the materials under the gravelly surface are the same as those forming the pedestals. These relations imply that fine sediment unprotected by grass was washed away, leaving pebbles that were too large to be moved. In my experience, this photograph could have been taken at countless places over the arroyo country.

Most of the eroded sediment, however, is redeposited before reaching the channel system. This was shown in the area studied by Leopold and coworkers by measuring the sediment trapped in a small reservoir. Allowing for some overflow of muddy water from the reservoir, one finds that about 15 percent of the eroded materials were moved into it. This figure would result in around 2,000 tons of sediment per year for each square mile being eroded, a figure supported by similar measurements in other parts of the western United States.

Is this enough to yield the high sediment loads of the Rio Grande? Geological maps show that rocks as erodable as those in the area studied underlie about 4,700 square miles of the Middle Valley's arroyo country. Assuming the studied area is reasonably average, one finds that the 4,700 square miles would have delivered about 9 million tons of sediment to the river annually, as averaged for the period of study, 1960–64. Data on suspended sediment in the main Middle Valley tributaries are incomplete for that period, but estimating from data for the Rio Puerco and the main river, it appears that the 9 million tons is more than all the sediment delivered to the river annually from the Middle Valley for that period. Raindrop and sheetwash erosion in the arroyo country therefore appear rapid enough to account for the valley's sediment problems. Careful measurements showed that the dry, low appearance of the terrain is misleading!

Causes and Remedies

Slope erosion by heavy summer rains evidently produces most of the sediment moving to the Rio Grande floodway. A basic cause of the erosion is the loosely cemented nature of the rocks underlying much of the arroyo country, and that cannot be remedied. Another basic cause is sparseness of vegetative cover in the arroyo country, which some persons attribute mainly to overgrazing and others to the dry climate. Because the first alternative can be remedied and the second cannot, the two must be judged with care.

The evidence most commonly cited for overgrazing is a supposed drastic reduction in grass cover after large animal herds were brought into the region. There is no question that grass has been thinned in grazed areas, for it comes back considerably when protected. The "luxuriance" described by some early travelers, however, may well have been localized, perhaps on broad valley flats like those at the foot of arroyo stretches. A careful historical study by Kirk Bryan showed that there were definitely broad flats in the Rio Puerco drainage in 1950, and that they alternated with arroyo reaches, the arroyos as much as 30 feet deep (Reference 3). The flats were irrigated and farmed extensively in 1880, but in about 1885 the stream cut arroyos through the flats, leaving fields so high and dry that they were abandoned.

The arroyo cutting brought a large quantity of sediment to the Rio Grande, leading Bryan to conclude that this was the cause of the sediment problems. We have seen, however, that slope erosion rather than arroyo cutting is now the main source of sediment. Might the Rio Puerco have received a large load of sediment from its surrounding slopes and *also* have cut the arroyos indicated by Bryan's study?

TABLE 3-2.

Annual Runoff and Resultant Erosion from Each of Three
Small Drainage Areas in the North-Central Part of the Rio Puerco Drainage*

Grazing Use	Period of Measurement	Average Runoff Per Year (in millions of cubic feet)			Average Erosion Per Year (in tons per square mile drained)		
		AREA 1	AREA 2	AREA 3	AREA 1	AREA 2	AREA 3
Year-around	1953–1958	13.8	14.0	12.9	1700	2200	1500
Winter	1958–1961	8.2	9.4	7.8	200	600	630

*Based on data from Earl F. Aldon (Reference 4).

Measuring effects of grazing   A firm answer comes from three small drainage areas on Rio Puerco tributaries gaged carefully by the U.S. Forest Service to test range management practices (Reference 4). The three areas were grazed year-round from 1952–58 and then only in the winter from 1958–63. Herds were adjusted to use about half of each year's growth of grass, and measurements showed that total utilizations during the two periods were almost identical. Average rainfalls for the two periods were also the same.

It is therefore striking that when water and eroded sediment were measured in catchment ponds, more runoff and much more erosion occurred during the period of year-round grazing (Table 3-2). The causes of the two different erosion rates are suggested by Table 3-3. Grass cover and plant litter increased during years of winter-only grazing and reduced the amount of bare soil by about 15 percent. Table 3-2 shows, however, that runoff was reduced by 35 percent and erosion by 75 percent of the earlier figures. It seems inescapable that it was the traffic of animals during the summer months—the working and turning of ground by hoofs—that caused the high erosion rates during the first test period. The discovery is a striking example of a linkage that would probably have been underestimated prior to the study.

The figures certainly show that grazing could have caused the enormous slope and arroyo erosion, but was the timing suitable? Prehistoric grazing was probably not damaging because wild animals grazed the arroyo country almost entirely during the winter. The numbers of animals during Spanish and Mexican occupation are not well known. Most historians agree that herds were increased throughout the Southwest starting in about 1850 or 1860 and that the increases were large between 1870 and 1880. Bryan's study of the Rio Puerco area disclosed large increases of herds just before the start of rapid arroyo cutting in about 1885, and early surveys showed that the river was filling rapidly in the Middle Valley before 1895.

Another kind of evidence comes from early maps of the Rio Grande in the Mesilla Valley, the next major valley downstream from the Middle Valley (Figure 3-18). The large changes in the river channel between 1844 and 1890 suggest that the river became so overloaded with sediment that it filled its meandering channel and finally overran it, constructing a broad floodway down the valley.

TABLE 3-3.

Materials at the Surface of Three Drainage Areas,
Measured in the Years Noted and Listed as Percentage of the Total Area*

Surface Materials	AREA 1			AREA 2			AREA 3		
	1952	1958	1961	1952	1958	1961	1952	1958	1961
Grass	7.2	5.2	12.0	4.9	3.0	7.9	4.6	3.0	5.9
Plant litter	6.1	2.9	12.3	5.9	5.3	14.6	3.8	4.4	11.3
Bare soil	70.2	82.0	68.0	65.5	77.9	63.7	79.7	81.0	74.2
Bare rock	4.4	2.3	4.3	6.8	6.0	8.5	2.4	3.5	5.6

*Based on data from Earl F. Aldon (Reference 4).

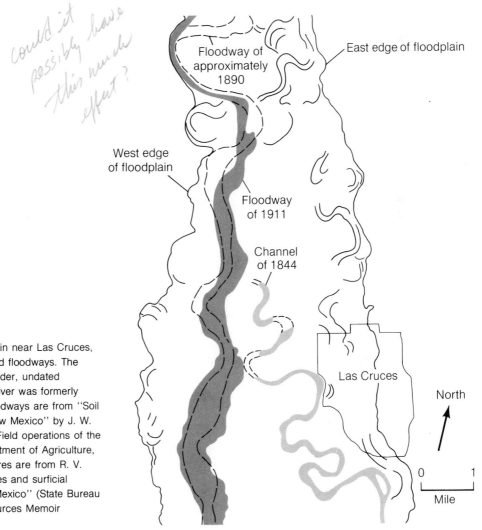

could it possibly have this much effect?

FIGURE 3-18

Map of Rio Grande floodplain near Las Cruces, showing dated channels and floodways. The curving lines are relics of older, undated channels that indicate the river was formerly like it was in 1844. The floodways are from "Soil survey of Mesilla Valley, New Mexico" by J. W. Nelson and L. C. Holmes (Field operations of the Bureau of Soils, U.S. Department of Agriculture, 1914). The remaining features are from R. V. Ruhe, "Geomorphic surfaces and surficial deposits in southern New Mexico" (State Bureau of Mines and Mineral Resources Memoir 18, 1967).

Did climatic changes contribute to the excessive erosion? The record of total precipitation at Santa Fe, which goes back to 1850, shows many ups and downs but no major long-term changes in total rainfall. The record, however, includes all rainfalls for each day, and, when Leopold analyzed these data for summer *vs* winter and large rains *vs* small, he found that the earliest part of the record shows relatively more precipitation from summer rains and relatively less from light winter rains (Figure 3-19). We have already noted the tremendously different erosional effects of wet summers compared to those of wet winters (Figure 3-9). It is also known that numerous light winter rains are especially nurturing to grass, partly because of frequency and partly because mosisture is conserved by cool weather. Dry, cold winters are detrimental to grass. Leopold's analysis therefore points to an early period of summer-dominated rainfall that probably contributed to excessive erosion. We have no record of the effects of grazing animals at that time.

In conclusion, we have determined neither the complete causes nor the exact timing of the Middle Valley's sediment problems but can set down some useful points:

1. Prime but irremedial causes are the easily eroded rocks and the fragile nature of grasses in the dry, variable climate of the valley.
2. The river can be modified toward carrying its sediment but only to the detriment of uses downstream.

FIGURE 3-19

Two aspects of the rainfall record at Santa Fe derived by Luna B. Leopold (Reference 2, 1966).

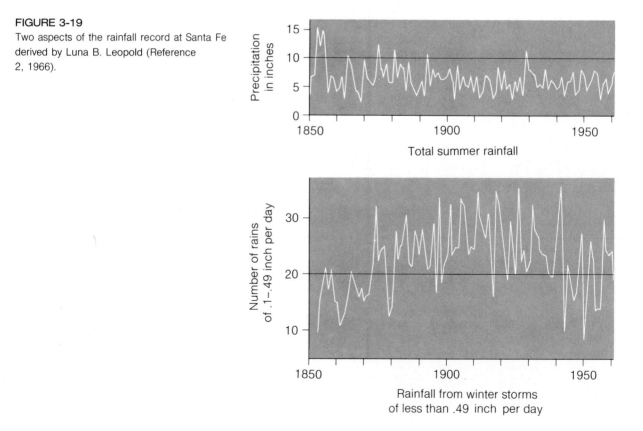

3. Summer grazing and traffic by herds caused much of the slope and arroyo erosion that led to damages on the floodplain.
4. Remedial measures such as restriction of grazing to winter months should be tried more extensively.

These specifics lead to a general conclusion: everything in the Middle Valley relates to everything else. Each of the studies that led to this conclusion produced one or more surprises. Evidently, when problems force us to look closely, especially to make real measurements, we discover important linkages in systems. I hope this does not glorify the problems as much as it says that much remains to be discovered. No part or aspect of a system is so small or obscure that it might not add greatly to our understanding of the whole.

Summary

The Rio Grande's sediment problems have led to some important discoveries about river systems:

1. In contrast to river systems in temperate, moist regions, those in arid and semiarid regions are largely ephemeral, that is, even the main river may have highly variable, localized flows.
2. Where such rivers flow on beds of loose sand or silt, they can transport enormous quantities of sediment because they can enlarge their channels rapidly during floods and can remold their hummocky (duned) beds to smooth planar surfaces.
3. Studies of the Rio Grande showed that two factors increasing the capacity of rivers to transport sediment are low temperatures of the water and large quantities of suspended clay, both of which increase the viscosity of the river and therefore its ability to move particles.
4. Sparseness of vegetation makes arid and semiarid lands especially susceptible to the erosive actions of raindrops and sheetwash and to the headward cutting of gullies and arroyos.
5. Measurements in the Rio Grande country show that large quantities of sediment are introduced to streams by downhill creep of soil and cutting of arroyos, but that hillslope erosion during rainstorms erodes the largest quantities of soil.
6. Most of the eroded sediment is redeposited on gently sloping ground or in arroyos, but the fine materials, especially, tend to be washed on to the main river.
7. Erosion is by far most rapid in areas of loosely cemented rocks (former sediments) and is intensified by grazing in summer months and by periods of sparse winter precipitation and frequent summer storms.
8. Where these factors cannot be corrected, the additions of eroded sediment to the river channel and floodplain may result in costly damages to the irrigation systems generally in use there. Such a river is filling its valley rather than eroding it.

REFERENCES CITED

1. John P. Miller. High mountain streams: effects of geology on channel characteristics and bed material. State Bureau of Mines and Mineral Resources (New Mexico) Memoir 4, 53 pp., 1958.
2. Luna B. Leopold and John P. Miller. Ephemeral streams—hydraulic factors and their relation to the drainage net. U.S. Geological Survey Professional Paper 282–A, 37 pp., 1956.

   Luna B. Leopold, William W. Emmett, and Robert M. Myrick. Channel and hillslope processes in a semiarid area, New Mexico. U.S. Geological Survey Professional Paper 352–G, 253 pp., 1966.
3. Kirk Bryan. Historic evidence on changes in the channel of Rio Puerco, a tributary of the Rio Grande in New Mexico. *Journal of Geology*, vol. 36, pp., 265–82, 1928.
4. Earl F. Aldon. Ground-cover changes in relation to runoff and erosion in west-central New Mexico. U.S. Forest Service Research Note RM-34, 4 pp., 1964.
5. Wayne C. Hickey, Jr., and George Garcia. Changes in perennial grass cover following conversion from yearlong to summer-deferred grazing in west central New Mexico. U.S. Forest Service Research Note RM-33, 3 pp., 1964.

ADDITIONAL IDEAS AND SOURCES

1. An important study of one of the main arroyos, by Stanley A. Schumm ("Effect of sediment characteristics on erosion and deposition in ephemeral-stream channels", U.S. Geological Survey Professional Paper 352-C, 1961), showed the sequence of headcut to arroyo to valley flat (as in Figure 3-15), repeating rhythmically through a distance of several miles along one of the valleys west of Santa Fe. If you live near a place where gullies can be studied (even very small ones), it would be interesting to see if they have rhythmically stepped courses, with lengths of steps more or less proportional to stream size.
2. If you are interested in determining the amounts of suspended sediment in various parts of a stream, you can use the following procedure (if you have access to a fairly precise balance). (1) Collect samples of water by opening jars at various positions and depths. (2) Filter the samples through pre-weighed filter papers. (3) Dry and weigh the papers and sediment. This will give you the concentration of sediment in the stream. If you also determine the discharge, as described in Additional Ideas and Sources for Chapter 2, you can calculate the approximate amount of sediment the stream is transporting in any given amount of time.
3. If you want to determine the amounts of sediment transported down hillslopes during rainstorms, plant cans approximately vertical and exactly flush to the surface to serve as traps. You can then remove and weigh the sediment after each storm. For heavy downpours, however, you'd be wise to observe the cans to see if turbid water is flowing back out of them.

Cord grass (*Spartina alterniflora*) in a marsh adjacent to Sapelo Island, Georgia. The horizontal mat of grass was floated in by the high tide; as it is disintegrated by bacteria and fungi, it will become the basis of the marsh's food chain.

4. Estuaries

Estuaries

Some kinds of earth systems are difficult to understand when studied individually, even if the studies are intense. Estuaries—the final attachments of rivers to the sea—are a case in point. Their essence is difficult to judge because of what seem to be major differences among them. Truly basic attributes are obvious in a few but not in others. One must thus move broadly among many estuaries to recognize the basic similarities, using one kind of study here, another kind there.

One important general discovery has come from dating, in years, the sediments that have accumulated in estuaries. This dating has shown that all estuaries were formed at about the same time—when huge glaciers far to the north largely melted away. The melting resulted in a rise of sea level and consequent flooding of river valleys and other low areas along coasts. This is how the estuaries were formed. Figure 4-1 shows part of the evidence: the shapes of the estuaries closely follow the shapes of stream valleys, out to their smallest tributaries. The figure also points to differences among the estuaries. Note the broad bay enclosed by narrow sand islands along the coast north of Cape Charles. As we shall see, these islands were formerly chains of sand dunes along the coast and were isolated when the sea rose and flooded the low ground behind them. The shape of the resulting estuary is markedly different from the filled river valleys, yet the two have the same origin.

Another attribute they share is that estuaries are places where fresh water from land mixes with salt water from the sea. To the hydrologist and oceanographer, this mixing is their most definitive characteristic. Mixing is obvious

FIGURE 4-1

Southern half of the Chesapeake Bay system, showing the intricately branching shapes of submerged stream valleys and the contrasting shape of the estuary along the coast north of Cape Charles. The photograph, supplied by the U.S. Geological Survey, is an image transmitted from Earth Resources Technology Satellite (ERTS-1).

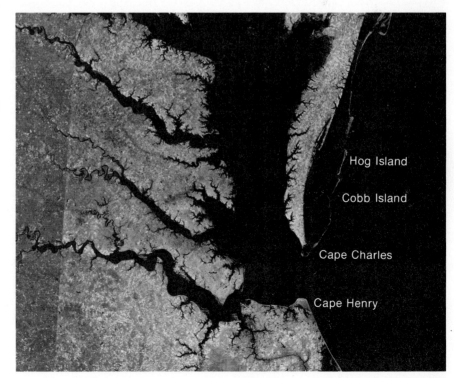

in the flooded river valleys, for the water is fresh (nonsaline) at the upper tip of each estuary and gradually becomes more saline toward the sea. The gradations are more complicated in estuaries like those north of Cape Charles because fresh water is less abundant and enters at a number of places, including the bottom of the estuary, where it seeps in from groundwater reservoirs such as those described in Chapter 2. Nonetheless, the water is generally brackish (mixed fresh and salt water); these estuaries are mixing grounds like the others.

Our studies must go beyond these two basic characteristics because considerable problems have arisen from our uses of estuaries. Situations such as that affecting San Francisco Bay are common and will shortly become more so unless we take corrective measures. Perhaps the general problem is that estuaries are located next to the flat ground that humans have used so long and so hard. Cities have grown there because estuaries provide waterways for shipping and for the disposal of sewage and other wastes. Where land has become scarce, humans have filled the shallow estuaries and used them as dry land. Such use leads to many kinds of problems, but the most fundamental is the loss of vital living systems. Acre for acre, estuaries produce more food than any other kind of earth system. Healthy stands of cord grass (*Spartina*) like that shown on the chapter opening page can produce as much as 10 tons of food per acre per year—six times the world-average productivity of cultivated wheat fields (Reference 1). Moreover, studies of the minute particles produced by the breakdown of the cord grass have shown them to contain as much as 25 percent of protein (Reference 2). The countless small animals that feed on the rich particles can also feed on abundant diatoms and other microscopic algae. These various minute grains are the basis of food chains that support all estuarine animals, many migratory birds, and most of the finfish and shellfish of the adjoining parts of the sea.

To conserve this vital marine resource and still use estuaries in other ways (which we will doubtless do), we must understand more of their internal workings. In this chapter we will examine three kinds of estuaries along the Atlantic coast: first, those of southern Florida, to see some exceptionally complete connections among living forms and the physical estuary; second, those of the Georgia coast, to fully view the effect of the ocean, including its tides; and third, the open bays of Virginia, to see the effects of river-driven circulation—both of water and sediment.

Florida's Life-dominated Estuaries

The waterways of southern Florida are an ideal example of a system dominated by life-related processes and substances. The fresh-water parts of the system contribute almost no ordinary (inorganic) sediment to the estuary. An annual rainfall of about 60 inches sends surface water creeping down the nearly flat but ever-so-lightly channeled plains of the Everglades (Figure 4-2). Humans have ditched and otherwise interrupted this amazing swampy river but much of it still functions in its original, unique ways. The entire terrain is a stream, miles wide but typically only a few inches deep. Densely grown saw grass (a *sedge*) so constrains the flow of water that sediment collects where it forms, making a layer of black mud a few inches to 10 feet thick. This material consists mainly of saw grass particles and calcium carbonate secreted by algae in relatively open pools.

Mangroves grow densely where the fresh water mixes with salt water at the northern edge of Florida Bay (Figure 4-2). The fallen leaves of these trees function much like the cord grass just mentioned, disintegrating into particles that become the basic food for small estuarine animals and thence support the rest of the estuary's fauna (Reference 3). As Figure 4-3 shows, the plants have downward-branching aerial roots that prop the plant against storm winds and serve as a living base for shellfish. The roots and attached animals reduce the turbulence of waves and thereby trap sediment, which slowly builds a mound under each plant. But the mangrove does not initiate this process. The white patches in the right foreground are algae-encrusted grasses that also trap sediment—until the water is shallow enough for floating mangrove seeds to root in the mud. It is from these rootings that the mangrove forest spreads (note the many young trees in the photograph) and eventually becomes an intricate complex of islands between open channels.

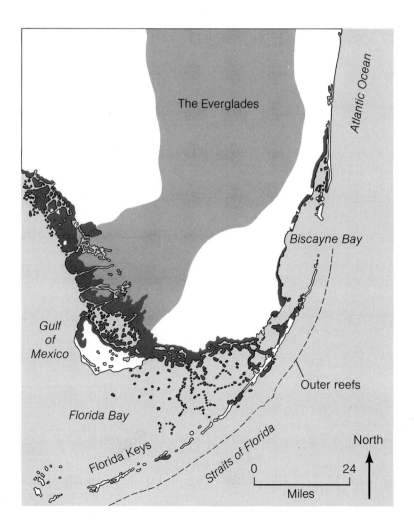

FIGURE 4-2

Map of southeastern Florida and the adjoining sea, showing the Everglades system (gray), the mangrove forests (dark pattern), and the reef-protected waters of Floriday Bay and the Keys. Almost all the Everglades system is less than 15 feet above sea level.

FIGURE 4-3

Forest of red mangroves from which young plants are spreading into the shallow lagoon. The white mud bottom can be seen under the prop roots of the two plants in the foreground. The white patches in the lower right are algae-encrusted sea grass that extends over much of the bottom.

Origin of calcium carbonate sediments   The white muds visible in Figure 4-3 consist almost entirely of calcium carbonate and are spread as a layer over the vast expanse of Florida Bay. The sediment consists partly of whole shells of mollusks and foraminifers (single celled animals that secrete shells of calcium carbonate), and partly of broken bits of shells and minute crystals of calcium carbonate released when certain algae decompose. These materials have been cast into linear banks and islands between 5 to 8 foot deep basins (Figure 4-2). Turbulence generated by waves shifts sediment from basins to banks, where it tends to be caught in closely grown turtle grass. The distribution of coarse and fine sediment in Figure 4-4 suggests that the banks shift downwind—at least until they capture mangrove seedlings, after which they commonly accumulate sediment and become stabilized as islands.

Figure 4-2 indicates that banks and islands first started forming in the northeastern end of the bay and that filling is progressing toward the southwest. The fact that filling has not taken place at the mouth of the Everglades river system supports the point that little sediment has been added to the estuary from land. The organic particles in the bay have thus either formed there or have been washed in from the sea.

FIGURE 4-4

Vertical section through mudbank and the limestone (rock made of calcium carbonate) that underlies it (the latter was pitted by erosion before the bay formed). The depths to which the waders have sunk indicates the firmness of sediments on the two sides of the bank. The slopes of the bank are exaggerated.

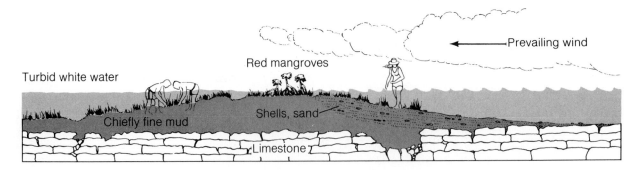

Prevailing wind

Red mangroves

Turbid white water

Chiefly fine mud

Shells, sand

Limestone

We can judge the latter possibility by examining the connections of the estuary to the sea—the gaps between the Florida Keys (Figure 4-2). Figure 4-5 shows that these areas are scored by distinct channels, much like stream channels. The channels pass water back and forth as the tides rise and fall and also transmit larger amounts of water back to sea after hurricanes. The small fan-shaped deltas at the seaward ends of the channels indicate that only small amounts of sediment are transported from estuary to sea and that even less is moved from sea to estuary. Transport of sediment is limited because the difference between the levels of high and low tide is small, and tidal currents are therefore weak.

The delicate little submarine deltas emphasize the role of the living reefs, which protect the Keys and their surroundings from storm waves. The great strength of the reefs rests on the intersupportive nature of their many animals and plants. The large corals are mainly moosehorn corals, which can grow upward at rates of about an inch a year and are streamlined toward the surge of waves. The living forms have maximal contact with nutrient-laden water because the reef has long spurs oriented parallel to the approach of waves. When hurricanes beat the corals down, the fragments are cemented to the reef by encrusting algae and bryozoa. Skeletons of mollusks and foraminifers are added to the mass in the same way.

The shallow sea bottom behind the reefs also has its particular array of inhabitants and sediments. Bare calcium carbonate rock (limestone) and small coral reefs occur here and there, but turtle grass and sand dunes cover most of the bottom (Figure 4-5). Although this tract is characterized by many plants and animals that produce fine calcium carbonate grains, few become incorporated in the reefs or get washed into Florida Bay. This relation was proven thoroughly by Robert N. Ginsburg, who learned to distinguish between the various kinds of calcium carbonate particles in the Florida system by comparing them with parts of living plants and animals (Reference 4). He collected bottom sediments at many localities across the reef-and-estuary complex and identified thousands of the grains in these materials. The results, shown in Figure 4-6, correspond closely with the proportions of living plants and animals in each part of the system.

This close correlation indicates that sediment grains are moved around moderately at most. Some reasons have already been mentioned: the weak tidal currents, the protection afforded by the outer reefs, the containment by the Keys, the cementing effects of encrusting animals, and the impeding effects of mangroves and turtle grass. Ginsburg discovered that another factor was the pelletizing actions of animals; about 80 percent of the fine mud of Florida Bay has been consumed by small animals and excreted as fecal pellets. Another discovery was that some living algae send soft filaments around the upper half inch or so of sediment, thus holding it against currents.

Interpreting an ancient environment from limestones Ginsburg's studies thus emphasize the many attachments and dependencies between the physical system and its great variety of inhabitants. His main interest, though, was to use these data in interpreting the origins of ancient limestones, especially the environments in which they formed. He identified the various grains in the limestones and then compared them to the propor-

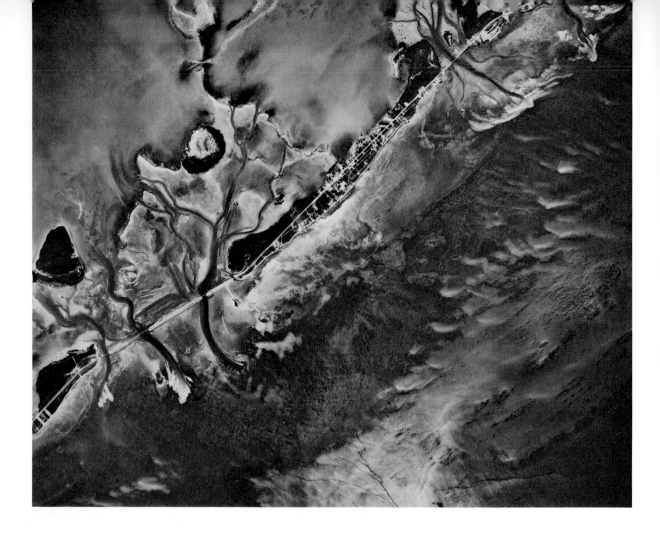

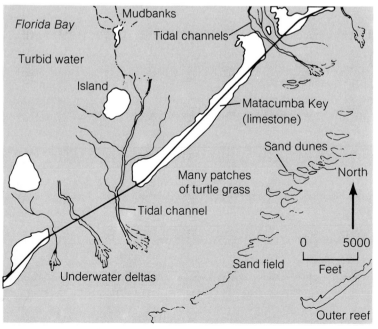

Florida Bay

Mudbanks

Tidal channels

Turbid water

Island

Matacumba Key
(limestone)

Sand dunes

North

Many patches
of turtle grass

Tidal channel

0   5000

Sand field

Feet

Underwater deltas

Outer reef

FIGURE 4-5

An aerial photograph of part of the Florida Keys
and vicinity taken on a day when the bay water
was turbid but the sea so clear that all bottom
features as far as the outer reefs could be seen.
Photograph by U.S. Army Map Service.

FIGURE 4-6

Distribution of the various kinds of sediment particles of the Florida Bay reef system. The vertical dimensions of each band are proportional to the amounts of the particles at the localities directly beneath. Generalized from several sets of data across the area. From Robert N. Ginsburg (Reference 4).

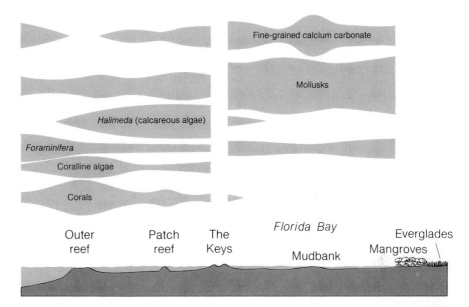

southern Florida. The method paid off handsomely with respect to understanding the history of the Florida region. The limestone that underlies Florida Bay and the Everglades and forms much of the outer Keys is composed of grains much like those of the current reef-bay system. The grains reveal that the outer reefs were about where the living ones are now and that a nearly flat platform, covered with shallow water like Florida Bay, extended across what is now southern Florida. Many of the Keys were small coral reefs on this platform.

This system was upset when glaciers began forming far to the north, as mentioned in the introduction. The sea fell and the accumulated sediments were eventually hardened to limestone and left high and dry—as shown by pits dissolved in the limestone by rainwater (Figure 4-4). When the glaciers melted and the sea rose again to cover the eroded limestone, the system was repopulated with animals and plants that began forming calcium carbonate particles anew. The changes in sea level that were the cause of all estuaries can thus be recognized in southern Florida—by seeing the many attachments of the living forms within the system.

Sand Barriers of the Southern Atlantic States

Estuaries constructed of river-borne sediment are shown especially clearly along the Atlantic coast north of Florida. Ocean waves and currents have sorted the sand from these sediments and have molded it into a chain of islands parallel to the coast (Figure 4-7). The islands are often called *sand barriers* because they contain and protect the marshy estuaries, thus serving the same function as the Keys and living reefs of Florida. The contrast in the

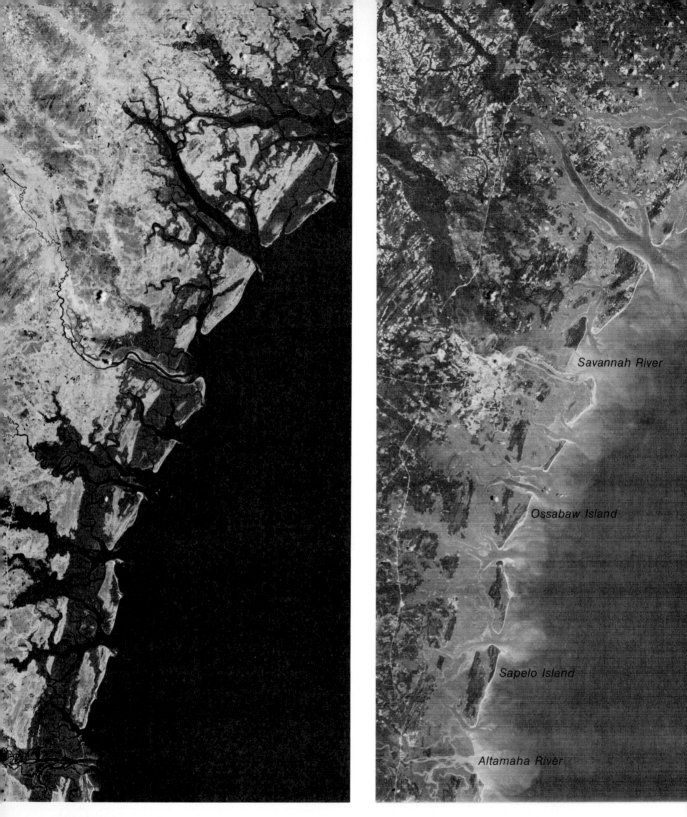

FIGURE 4-7

Sand islands (light areas in the left view) and marshes (dark areas) of the Georgia–Carolina coast. The photograph on the left was taken mainly with the infrared spectrum and thus shows water bodies as black, marshes as dark gray, and land as light gray. The photograph on the right was taken at the same time but with a full light spectrum, so it shows the turbid waters (light areas) of rivers and the sea near shore. ERTS-1 images supplied by the U.S. Geological Survey.

FIGURE 4-8

Southeastern end of Ossabaw
Island, with the Atlantic in the
background and a large tidal
channel on the right. Each
forested band is an old
sand-dune ridge, and the gray
areas are grassy marsh.

living parts of the two systems is due to the cooler waters to the north and
to the large amounts of mud and organic particles in the waters off the river
systems. Note in the right-hand photograph of Figure 4-7 the plumes of turbid
water extending seaward from the mouths of channels, especially at the
mouth of the Altamaha River near the lower edge of the view. If you were
to compare the sands on beaches in southern Florida with those along the
Georgia coast, you would see a dramatic change—from almost all calcium
carbonate fragments of organic origin to almost all quartz grains eroded from
the Piedmont and Appalachian Mountains.

The nature and history of the sand barriers and associated estuaries have
been worked out largely by John H. Hoyt and others at the Georgia Marine
Institute on Sapelo Island (Reference 5). The islands consist of parallel rows
and bands of large sand dunes, most of them overgrown by forest. The construction of an island can be imagined by watching the beach on a day of
strong onshore winds. Sand blown up the beach is generally caught in clumps
of beach grass. As a mound accumulates, sand can be seen to roll and bounce
up its windward side and fall and accumulate on its leeward (downwind)
side, much as on the river dunes described in Chapter 3. When the dunes
cover the grass, some begin to shift slowly downwind, and, because of local
differences in wind currents and supply of sand, some overtake and merge
with others. The dunes thus become larger and fewer in number, eventually
building to heights of 10 to 30 feet. Sand deposits of that thickness retain
enough rainwater to support bushes and small trees, which stabilize the dunes
and slowly add plant detritus to them, forming a soil. After thousands of years,
large forest trees cover the dunes completely. The episodic growth of each
island is thus recorded by its parallel bands of forested dunes (Figure 4-8).

Cycling of sand by waves   Most of the white band around the island in Figure 4-8 is normally sea bottom, for the photograph was taken during a very low tide. The fine sand is sorted to essentially one grain size as it is worked and reworked into delicate ripples by the waves. Shells of mollusks and crustaceans can be found everywhere, and the sand is commonly marked and burrowed by trails of various marine animals. One of particular interest is the "ghost" shrimp (*Callianassa major* Say), which lives permanently enclosed in a burrow about an inch in diameter, located just below sea level at midtide.

Larger features in Figure 4-8 exposed by the low tide are the hooked extension of sand in the distance and the faint bands that parallel the shore elsewhere. The bands are sand bars that normally form just offshore and migrate slowly toward the beach, much as sand dunes migrate on land (Figure 4-9). This migration is one stage of a cycling of sand that is probably typical of many beaches. As Figure 4-9 indicates, storm waves erode the beach and dunes, moving the sand offshore to form a low mound outside the normal surf zone. Fair-weather waves then move the sand back toward the beach in the form of slowly migrating bars. As the bars reach shore and rebuild the eroded beach, sand is blown landward and the dunes are replenished. The dunes thus function as a sand reserve as well as a barrier to the waves.

Lateral migration of the islands   In spite of this near-equilibrium, however, the islands are slowly shifting southward due to erosion of their northern ends and growth of their southern ends. On Sapelo Island, dune ridges deposited 1,000 or more years ago have been cut off along the north

FIGURE 4-9

The two main stages in the cycling of sand from shore to sea and back. The enlarged sections below show the patterns of small layers resulting from landward transport of sand across migrating dunes and bars.

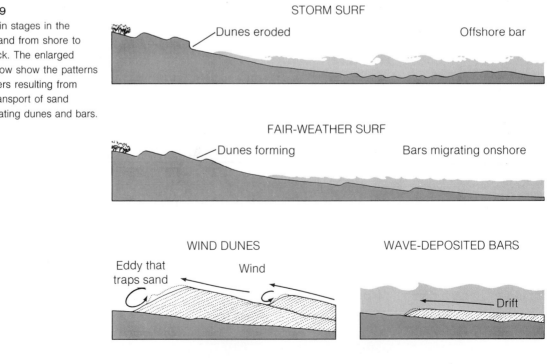

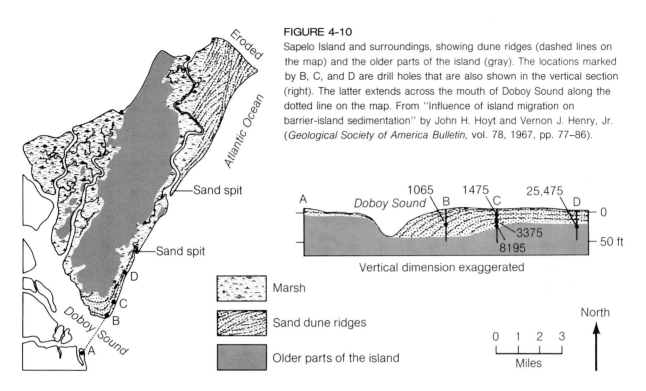

FIGURE 4-10

Sapelo Island and surroundings, showing dune ridges (dashed lines on the map) and the older parts of the island (gray). The locations marked by B, C, and D are drill holes that are also shown in the vertical section (right). The latter extends across the mouth of Doboy Sound along the dotted line on the map. From "Influence of island migration on barrier-island sedimentation" by John H. Hoyt and Vernon J. Henry, Jr. (*Geological Society of America Bulletin*, vol. 78, 1967, pp. 77–86).

end of the island (Figure 4-10). The southeast end of the island has grown outward some 1,100 feet since about 1750, largely during 1959–67. A continuing southward feed of sand is suggested by the two peninsulas called sand spits shown in Figure 4-10; they were probably built across the two channel mouths as sand was shifted southward along the beach and in the wave zone.

Southward shift of the large tidal channel, Doboy Sound, has been proven by drilling vertical holes into the beach and older sands at the locations marked B, C, and D in Figure 4-10. Plant remains brought up by the drill were dated by the radiocarbon method described in Appendix B. Their ages in years are shown in the vertical cross section of the figure, with a short line indicating the position of each sample. In hole C, for example, the plant remains dated the sand near the base of the more recent accumulation at 8,195 years, with younger layers being deposited at the dates shown. The dashed lines connect sand layers of equal age and show that the sand was filled southward (from right to left in the cross section) during the past 8,000 years or so. The historic shifts described in the preceding paragraph are thus only the most recent stages of a long-continuing migration.

A record of changes in sea level The cross section also shows that the sand layers were built one on top of another in the right side of the diagram, forming a sequence nearly 25 feet thick. Because each layer is a beach sand, the sequence indicates that sea level rose gradually since the lowest layer accumulated about 8,000 years ago. This sequence evidently records the period when the last of the glaciers were melting and the sea was filling to its present level.

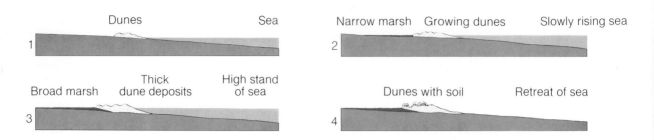

FIGURE 4-11

Stages in the formation of dune-beach barriers and marsh deposits.

The older sediments, those lying beneath the layered sands in Figure 4-10, are also significant with regard to changes in sea level and origin of estuaries. These sediments are beach and dune sands from about 25,000 to more than 35,000 years old and are about the same age as the dune sands in the main, forested part of the island. The upper surface of these deposits was evidently formed by erosion; note, in the cross section, for example, how the channel of Doboy Sound has been eroded into them. These relations are much like those in southern Florida, where an older accumulation of sediment was exposed and eroded and later covered by more deposits when the sea rose most recently.

John Hoyt used these relations to explain how episodic rise and fall of sea level led to the formation of the sand islands. His scheme is shown by the numbered steps in Figure 4-11. In stage 1 the sea is at a lower level than at present because of the growth of glaciers. Stage 2 shows the sea slowly rising as the glaciers melted, building up the sand barrier and flooding behind it to form a narrow estuary. In stage 3, the sea has reached its highest stand—the island has grown vertically and laterally and the estuary is larger. This stage brings us to the end of the deposition of the older deposits in Figure 4-10—to about 25,000 years ago. Glaciers then formed another time and the sea retreated again (stage 4); deep soil formed on the sand mound and erosion modified it. Now imagine the sea rising once more and the first three stages repeating. These stages added the layers of younger sand over the seaward parts of the eroded island, resulting in the relations shown in Figure 4-10.

The fit of this scheme to the data could, of course, be happenstance for one island; however, all the barrier islands have similar accretions of younger deposits on the seaward side of older, forested deposits. Moreover, Hoyt and coworkers have found still older barrier deposits landward of the estuaries and even older ones landward of those! Each deposit shows the same general design and details, down to burrows of *Callianassa* along the ancient shorelines. The sea has thus advanced and retreated several times during what must have been repeated glacial episodes. The record in the estuaries and their older deposits thus establishes an intimate connection among river systems, the ocean, and periods of glaciation.

Effects of Ocean Tides on the Marshes

As the sea rose after the last glacial period, the present estuaries formed behind the sand barriers and slowly filled with muddy sediment. Indeed, most parts are so nearly filled that they are often shown on regional maps as land areas; nonetheless, they are true estuaries—submerged at high tide and connected to rivers and to the sea by an intricate network of channels (Figure 4-7). Vast expanses of cord grass and forested islands make them among the most beautiful of wildlands.

The muddy sediment was originally introduced by rivers but has been distributed into most parts of the estuaries by the tides. It is essential to have an idea of how this tidal machine works, for it is a basic attribute of all estuaries and is shown especially clearly along the Georgia coast. The ocean tides are generated by the gravitational pull of the moon and sun, coupled with the rotation of the earth. The resultant forces press the oceans into broad wavelike bulges (*high tides*) and troughs (*low tides*). If the earth were covered by a continuous ocean, the tides would appear to sweep westward in a continuous progression with two high tides and two low tides per day (Figure 4-12). The real oceans, however, are separated largely by land masses and therefore generate their own tidal systems. Perhaps the best way to see how such systems work is to partly fill a glass tray with water and rock it gently back and forth (Figure 4-13A). The water will rise and fall on the right and left, as do the tides, but will remain at its original level at the center of the tray. We can now superimpose the effect of the earth's rotation, which causes the tidal bulges and troughs to sweep around ocean basins like the spokes on a wheel. In Figure 4-13B, for example, the tray has been rocked with a circular motion, thus generating bulges and troughs that are higher at the edges of the tray and diminish gradually to zero at the center. As the curved arrows indicate, the waves (tides) sweep around the tray as though pivoted at the center, the point of no tidal motion.

The North Atlantic Ocean has such a rotating tidal system, but the effects on the United States coast are simplified by the fact that the tides approach the coast nearly straight-on (Figure 4-14). Note how the *tidal range* (the vertical difference between high and low tide) increases markedly where the coast is broadly indented, as between Cape Hatteras and Florida. The tidal effects on the Georgia estuaries are thus pronounced. Note, in contrast, the minimal tidal ranges at the southern tip of Florida—one cause of the relative stability of the sediments there. The various Atlantic estuaries also have internal tidal effects due to the shapes of the water bodies. The shapes and orientations of some bays, as the Bay of Fundy, may cause enormously exaggerated tides (Figure 4-14). Local heights and times of estuarine tides are, in fact, unpredictable by theory but can be read from tide tables compiled from years of local measurements. Most United States estuaries have two high tides and two low tides during each *lunar day*, a period of 24 hours and 50 minutes—one rotation of the earth increased by the eastward movement of the moon in its orbit.

Estuarine flow caused by tides   In addition to raising and lowering the water level, tides cause water to flow in and out of estuaries. An approaching oceanic bulge (high tide) creates a surge up the estuary and also makes the water surface slope in that direction. Therefore, water flows up the estuary throughout the period of the rising tide. The currents slacken as high tide

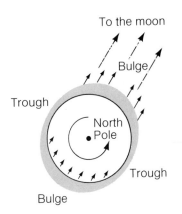

FIGURE 4-12
Tides in an idealized world-circling ocean, appearing to sweep westward because of the earth's rotation. The moon's gravitational attraction pulls the water away from the earth on the nearer side and, in effect, pulls the earth away from the water on the farther side—resulting in two bulges.

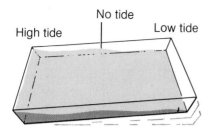

No tide

High tide

Low tide

A

Water surface high (a wave)

High tide

No tide

Low tide
Water surface low

B

FIGURE 4-13

A, water put in motion analogous to
tides by rocking a tray of water back
and forth. B, rotating tidal system
produced by keeping the tray aligned
as shown but moving it with a circular
motion. The resulting waves rotate
around the tray as indicated by
the arrows.

FIGURE 4-14

The approach of the tide on the Atlantic coast (arrows)
and its magnitude, as shown by lines connecting all points
with the tidal ranges indicated (in feet). Note that the
range increases toward the Georgia coast and that it is
minimal in southern Florida. From *The National Atlas of
the United States of America* (U.S. Geological
Survey, 1970).

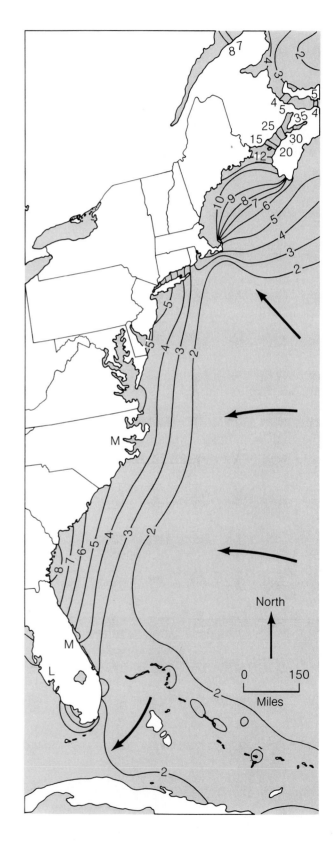

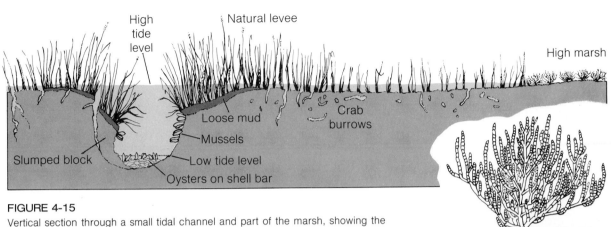

FIGURE 4-15
Vertical section through a small tidal channel and part of the marsh, showing the positions of high and low tides relative to forms and life of the marsh. The enlarged insert is of pickle weed (*Salicornia*), which is the typical plant on the higher parts of the marsh. The photo shows cord grass and marsh mud next to a small tidal channel with a group of fiddler crabs (escaping). Water level is about a foot below high tide and dropping. It contains about 25 parts per million of suspended sediment, whereas the turbid cloud stirred up by the crabs contains about 200 parts per million.

is attained, and, after about a half hour of near stillness, the water begins to move toward the sea in conjunction with the approach of the oceanic low tide. A lunar day in an estuary is thus divided between two periods when tidal currents move seaward (*ebb tides*) and two periods when they move landward (*flood tides*). The velocities of tidal currents in Georgia are typically 2 to 8 feet per second in constricted channels, rates that are comparable to those of swift rivers on land. The velocities, however, are only a few inches per second over broad shallows in the estuaries, where they compare with flow over floodplains of rivers on land.

Erosion and deposition by tidal currents   The actual transport of marsh mud during a tidal cycle can be detected easily by walking on the marsh and swimming in the channels. As the tide rises, moderately muddy water flows up the tidal channels, stirring up more mud as it goes. When the water finally spreads onto the marsh at high tide, much of the mud falls out among the cord grass near the channel. It becomes the base for the luxuriant grasses growing there and forms a slight rise (a *natural levee*) like that at the edge of a river floodplain (Figure 4-15). Cord grass behind the levee is markedly smaller because it has less contact with nutrient-bearing mud and water.

The turbid water clears noticeably during the half hour or so of slack before it begins moving off the marsh. Because ebb flow through the marsh grass is slow, rarely swifter than 2 inches per second, little of the freshly deposited mud is re-eroded and carried back to the tidal channels (Figure 4-15). Lightweight organic detritus, however, typically remains suspended and is carried off.

Because ebb flow is impeded by the natural levees and by the grasses and animal burrows, water continues to drain off the marsh for several hours after tidal channels have been emptied seaward. The upper tributaries thus act as though they were conventional (one-way) streams, eroding headward and developing branching patterns and rhythmically stepped courses much like arroyos on land. Evidence of their eroding action appears as mud fragments deposited where upper tributaries flow into larger, more gently sloping channels (Figure 4-16).

The channels do not generally fill with mud because the turbulence of their flow is great enough to keep silt and clay-size grains suspended (Chapter 1). The smooth mudbanks and the great depths of the channels compared to their widths make them highly efficient streams—they flow with a minimum of impedance. Nonetheless, they migrate slowly like streams on land, cutting and filling at bends and evolving meandering (sinuous) courses in which the sizes of bends are closely adjusted to the sizes of channels (Figure 4-17). Channels occasionally migrate so as to meet, as just above the center of Figure 4-17. If you examine the left photograph of Figure 4-7 closely, you can see that this has happened in many places over the marsh, producing an intersecting array of channels unlike those of streams on land. These intersections are important attributes of estuaries because they provide a means of distributing mud far more widely than otherwise. The orange floodwaters of the Altamaha River, for example, often appear in tidal channels many miles north of the normal outlets of the river.

Effects of tidal flow on marsh life   Interchanges of mud and plant detritus by the river-and-tide machine make the marshes exceptionally nutritive. Detrital food produced from the cord grass and other rooted plants, augmented by diatoms and other algae, is moved off the marshes on each ebb tide and thereby circulated to sessile (attached) animals of the tidal channels. Eugene P. Odum and Armando A. de la Cruz measured quantities of the

FIGURE 4-16

Part of gravel bar deposited where a small tidal tributary enters a larger one on the west side of Sapelo Island. The oyster shells (white) are about 5 inches long. All the other fragments, including the sand, are of marsh mud—typically abraded to roundish forms even though they are so soft a finger can easily be pressed through them.

FIGURE 4-17

Marsh along the southwestern edge of Ossabaw Island, the bands of forest on the left being old sand-dune ridges that once lay along the coast. Tide is at ebb, exposing the muddy bottom of the large tidal channel in the background. The white strip of beach sand (upper left) has been transported a half mile mile up the estuary from the Atlantic.

suspended detrital food at various tidal stages and found that somewhat more than half of what goes out on the ebb tide does not return on flood tide (Reference 2). The marshes must, therefore, be feeding a host of ocean creatures as well as those in tidal channels. The plant detritus also fertilizes the marshes, as do nutrients in both river and sea water. Burrows of fiddler crabs and other animals circulate oxygen deep into the organic-rich mud, keeping it from turning foul (otherwise the marsh plants would die).

An intriguing long-term aspect of balance within the marshy estuaries is the continuous addition of mud to their surface. The cord grass and many of the animals can live only as long as the marsh surface remains just below the level of high tide. Remains of these same plants and animals can be found at all levels in mud excavated from depths as great as 20 feet. These findings strongly support Hoyt's view of the origin of the sand barriers and the estuaries: the sea has slowly risen as mud was added to the surface of the marshes. Note that this relation implies a dramatic and drastic change in the living parts of the system were sea level to start falling. Estuaries with marshes are thus delicately in tune with growth and melting of glaciers far away.

Estuarine Circulation in Open Bays

A comparatively recent discovery is that the rivers flowing into estuaries generate a remarkable circulation that is all but invisible yet affects many estuarine processes greatly. This finding was made in the open bays north of Cape Hatteras. Let us first see why sand barriers typical of the southern estuaries do not extend across these bays. Figure 4-18 shows the basic evidence: the northern rivers transport far less sediment to the estuaries than do the southern rivers. This contrast in sediment loads, like the changes in sea level, is linked to recent glaciation. Glaciers eroded much of the northeastern United States, leaving a stony soil and extensive bare rock. River systems carry little mud from this terrain. The southeastern United States, on the other hand, was not glaciated, and the thick, clay-rich soils there have always yielded much fine sediment to rivers.

The sediment loads shown in Figure 4-18 cannot be projected back beyond the 1600s because they may be due largely to human uses; however, the shape of the coast indicates that the southern rivers have always carried more sediment than the northern ones. Note in the figure that the open bays not only disappear to the south but the coast projects seaward as capes at or near the mouths of most rivers. The projections were probably once deltas, formed mainly when the sea was lower during glacial periods. Waves and currents have subsequently eroded and sorted the delta sediments into the sweeping curves of the capes.

Discovery of a two-layer flow The open bays to the north have well-known values to commerce, but they also have a value in showing how estuarine systems work. Because they are open to both rivers and the sea, they provide a view of slow but important circulation caused by river flow seaward. Tidal currents are so obvious that this circulation was not well known until 1952, when D. W. Pritchard described a study he had made on the James estuary (Reference 6).

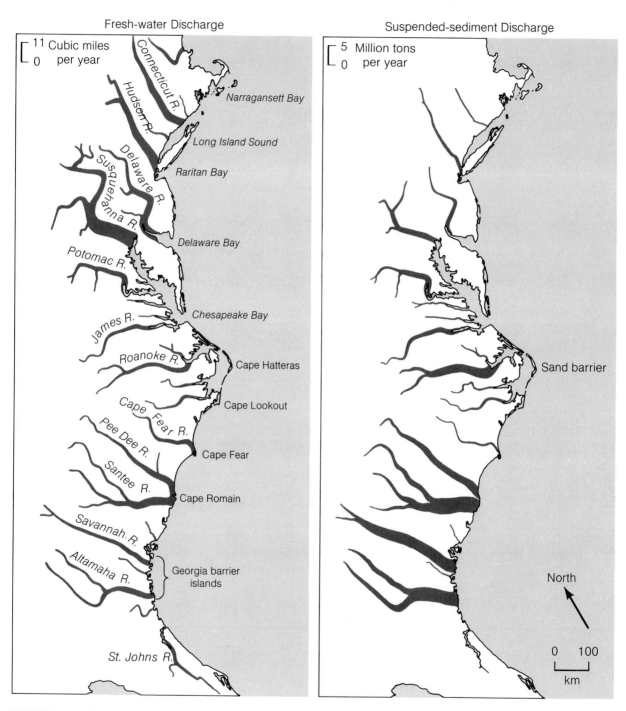

Fresh-water Discharge

⌈ 11 Cubic miles
⌊ 0   per year

Connecticut R.
Hudson R.
Narragansett Bay
Long Island Sound
Delaware R.
Susquehanna R.
Raritan Bay
Delaware Bay
Potomac R.
Chesapeake Bay
James R.
Roanoke R.
Cape Hatteras
Cape Fear R.
Cape Lookout
Pee Dee R.
Cape Fear
Santee R.
Cape Romain
Savannah R.
Altamaha R.
Georgia barrier islands
St. Johns R.

Suspended-sediment Discharge

⌈ 5 Million tons
⌊ 0 per year

Sand barrier

North

0   100
km

FIGURE 4-18

Estuaries and major rivers of the Atlantic seaboard, showing by the widths of the shaded forms the annual water and sediment discharges to the ocean. The sediment discharges are for the year 1908, before most of the rivers were dammed. From "Landward transport of bottom sediments in estuaries of the Atlantic coastal plain" by Robert H. Meade (*Journal of Sedimentary Petrology,* vol. 39, 1969, pp. 222–34).

The study was based mainly on measurements of tidal currents and water *salinities* (amounts of dissolved salt). Pritchard found that salinities increase gradually seaward because of the mixing of river water and sea water, and that water of a given salinity shifts up and down the estuary with the tides. He also found that the salinities increase with depth; a less saline layer of water grades rapidly downward into a more saline layer (Figure 4-19). On measuring velocities of tidal currents at various depths and averaging the velocities over many tidal cycles, he discovered that the upper layer is moving slowly down the estuary. This movement seems reasonable because the river is flowing in that direction. But he also found that the lower layer is moving up the estuary!

To visualize this two-layer flow, one must see it as a background to the more obvious tidal flows. Imagine placing two bottles in the estuary, weighted so that one floats in the upper layer, one in the lower. As the tide ebbs, both bottles will move down the estuary; as the tide floods, both will move up the estuary. As Figure 4-20 shows, however, flood currents are swifter in the lower layer than in the upper layer, and the opposite holds for ebb currents. The bottles will thus become increasingly separated with time, the upper one moving farther and farther down the estuary, and the lower one farther and farther up the estuary. This separation would be a record of the two-layer flow, which is superimposed on tidal flows and is therefore called the *nontidal flow* of the estuary.

Cause of the two-layer flow We can determine the rate of nontidal flow without using bottles by calculating the difference between the ebb and flood velocities at each depth. The differences are shown on the right side of Figure 4-20. When Pritchard calculated the average velocities of nontidal flow for the entire estuary, he found that the upper layer is moving seaward somewhat faster than the lower is moving landward. This difference could be accounted for exactly by including the inflow of water from the James River. Evidently, addition of river water at the upper end of the estuary causes the estuary surface to slope slightly toward the sea, and this hydrostatic head (vertical difference) presses the upper layer seaward.

But what drives the lower layer? The basic clue already mentioned is that the salinity of the upper layer gradually increases toward the sea. This change is caused by the two layers mixing slowly across their boundary, due mainly to the turbulence of tidal currents. As a result, the upper layer transports a large volume of sea water back to the sea and this must be replenished continuously to keep the estuary's volume in balance.

The required sea water flows in along the bottom because it is saltier and therefore denser than estuarine water. As shown in Figure 4-21, it slowly mixes with the upper layer, finally disappearing at what is called the *salinity edge*. Both layers are thus driven, basically, by the river, but the total circulation requires the mixing produced by tidal currents. Estuaries thus derive their dynamics as well as their substances from both rivers and the sea. The nontidal flow is one of their basic characteristics.

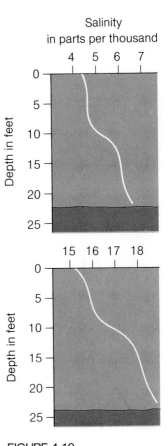

FIGURE 4-19
Variation of salinity with depth for a station in the upper part of the James estuary (top) and one in the lower part. From D. W. Pritchard (Reference 6).

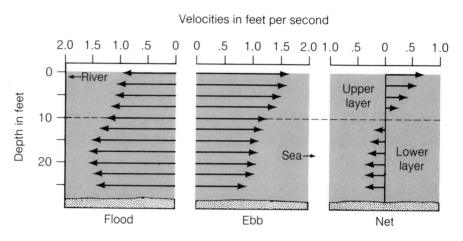

FIGURE 4-20

Average velocities of tidal currents in the James estuary. Differences between ebb and flood velocities are shown on the right—expressing the directions and rates of nontidal flow. Data from same source as Figure 4-19.

Accumulation of Sediment in the James Estuary

Pritchard's discovery greatly affected ideas on the transport of sediment to the sea. Consider the course taken by river sediment entering the estuary of Figure 4-21. At first it is borne seaward in the upper layer, but when it settles into the lower layer, it is carried back landward and is likely to be deposited far up the estuary. Because sewage and industrial wastes ride mainly with the suspended sediment, this accumulation of river mud in estuaries—often in harbors and alongside heavily populated shores—is a problem. To understand this process, Robert H. Meade integrated data available on large United States estuaries and found that at least two, San Francisco Bay and Savannah estuary, had indeed been trapping sediment much as predicted by Pritchard's model (Reference 7). Meade found evidence, however, of other processes that accumulated sediment in estuaries, and he cautioned that further quantitative studies should be made. These studies have now been made for the James estuary by Maynard M. Nichols and others at the Viriginia Institute of Marine Science (Reference 8).

Nichols repeatedly measured the sediment suspended at various locations and various depths in the estuary in the hope of determining where it was moving and why. He found that much fine sediment was eroded off the bottom when tidal currents were moving at a maximum rate, most of it being stirred into the lower half of the channel and no higher. About 80 percent of this sediment settled back to the estuary bottom when the current slackened, so that transport of sediment was distinctly stepwise. His most important finding was that about twice as much sediment was stirred up by flood currents as by ebb currents, due to the greater velocities of flood currents along the bottom (Figure 4-20). More sediment is thus transported up-estuary, stepwise, than down-estuary. Net transport by the two-layer system is toward the land rather than the sea.

By averaging the amount of suspended sediment over one to two complete tidal cycles, Nichols obtained the distribution shown in Figure 4-22. Note that sediment is by far most concentrated near the salinity edge. Note, too,

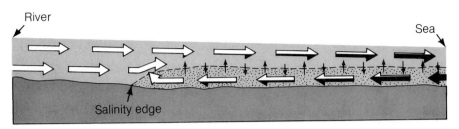

River

Sea

Salinity edge

FIGURE 4-21

Vertical section along the main
channel of an estuary. The
small arrows indicate the mixing
across the two-layer boundary.
The black parts of the arrows
indicate the proportion of sea
water, the white parts river water.

that it is concentrated near the bottom. This result is exactly as predicted by Pritchard's model and by Meade's synthesis of data from other estuaries.

Nichols's studies went on to show where most of the sediment has been accumulating on the bottom. The estuary, like many, has a main channel bounded by shelf-like shallows—probably the original river channel and its bounding floodplain (Figure 4-23). Most of the sediment is accumulating on the shallow shelves near the channel edge and high on the channel sides. Nichols suggested that these sites may be favored because they are near the level between the lower, more saline layer and the upper layer and will feel weaker tidal currents and less turbulence than the main channel. The mechanics would be analogous to deposition from a river moving on its floodplain: estuary mud stirred up by the swift currents in the deep channel would tend to be deposited when moved out onto the shallows.

But when detailed surveys of the bottom were compared to surveys made 70 years earlier, it was found that most of the sediment had been deposited 3 to 15 miles down-estuary from the suspended sediment concentration of Figure 4-22. There is no certain explanation for this offset position. Possibly, the 70-year accumulation took place during periods of river floods. The river carries most of its sediment during floods and the increased input of fresh water should press the salinity edge toward the sea—though how far is not known. It seems that the entrapping circulation would work more strongly

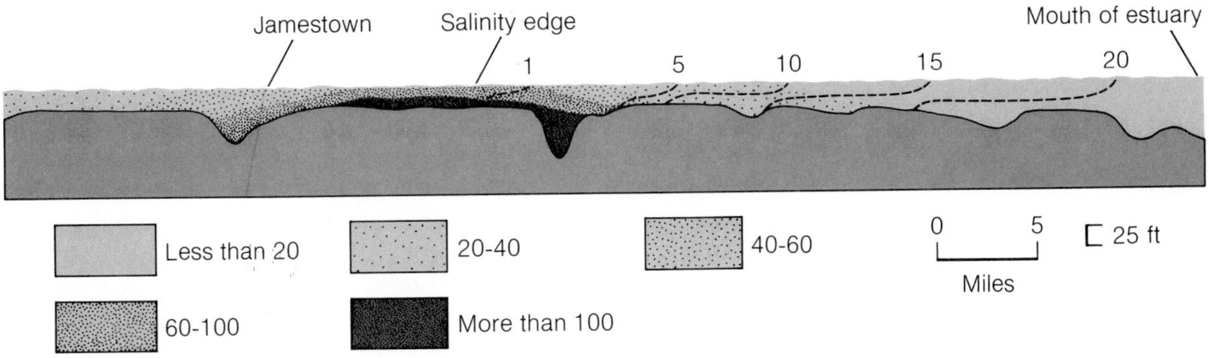

Jamestown   Salinity edge   Mouth of estuary

1   5   10   15   20

| Less than 20 | 20-40 | 40-60 |

0   5   ⊏ 25 ft

Miles

| 60-100 | More than 100 |

FIGURE 4-22
Vertical section along the main channel of the James estuary with dotted patterns showing the distribution of suspended sediment, in parts of sediment per million parts of water. The distribution of dissolved salt is indicated by the dashed lines numbered above the section in parts per thousand. The measurements were taken during March 1st to May 15th, a period of large river flows and large sediment loads. The vertical dimensions have been increased tremendously, as shown by the scale bars, and thus all slopes are unnaturally steep. From Maynard M. Nichols (Reference 8).

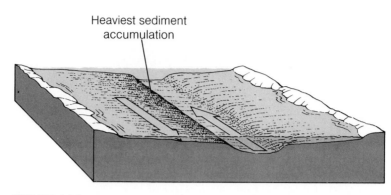

Heaviest sediment
accumulation

FIGURE 4-23

Diagrammatic view up the James estuary showing how the area of greatest sediment accumulation is related to the main channel, the adjoining shallows, and the nontidal flow along the bottom. From Maynard M. Nichols (Reference 8).

during floods, because, as we have seen, the circulation is driven by the river's flow.

Another process that adds sediment to the areas of maximum accumulation is the coagulating action of shellfish. Oysters, for example, ingest muddy water for detrital food and excrete the undigested mud as small fecal pellets. They also accumulate mud on their gills and discharge it as aggregates held together by mucous. The pellets and aggregates discharged by oysters have a resistance to erosion because of their weight. For this reason, oysters effectively deposit suspended sediment. Shellfish also contribute their shells to the bottom, locally forming layers several feet thick. No accurate accounting of the total rates of these kinds of accumulations has been made, but from the measured rates of pelletization of mud by oysters, it appears that shellfish deposit something like 10,000 tons (dry weight) of sediment per year in the main part of the estuary.

But roughly a million tons of sediment are brought to the estuary each year (on the average) by the James River, and perhaps another hundred thousand tons by the Appomattox River and lesser streams. The 70-year accumulation already mentioned probably cannot be measured precisely enough to determine the average annual increment, but it appears to be very much larger than the contributions from shellfish and wave actions. The two-layer circulation system thus appears to be the main cause of sediment accumulation, although the amount of sediment trapped compared to the amount going out to sea is yet to be determined.

One reason the sediment budget is difficult to balance is that sediment has been added to the estuary from the sea! Nichols found that sediments in the lowest 12 miles of the estuary are dominated by sand that contains mineral grains characteristic of Atlantic beaches and not of the James River. Moreover, tiny shells of an ocean-dwelling species of the protozoan *Elphidium* (a foraminifer) were found in sediment many miles up the estuary. These provide striking evidence of transport up the bottom of the estuary and also of a long-standing linkage with the sea.

Summary

That linkage will be explored in the next chapter. Before turning to it, we should summarize the origin and main attributes of the estuaries themselves:

1. They were formed by a rise in sea level, one of a series of changes caused by episodic glaciation in the recent past; they are now delicately adjusted to sea level.
2. They are the places where fresh water from land mixes with salt water from the sea.
3. Living things affect the broad design of estuaries as well as almost all processes within them; where the water is warm and clear, animals and plants may dominate estuarine processes, especially the formation of sediment.
4. Sediment transported by rivers also affects the broad nature of estuaries as well as many details in them; some river sediment is deposited directly in the estuaries, some is carried first to the sea.
5. Tides are a major connection between estuaries and the sea, influencing all living forms as well as all other physical processes.
6. The water entering through rivers drives a broad circulation that commonly moves saline water up the bottoms of estuaries, trapping sediment that would otherwise have been carried to the sea.

Because each attribute is dependent on the others, perhaps you can see the value of visiting these amazing systems in many places. By synthesizing a variety of studies, we can begin to get a feel for the whole—for the essence of estuaries in general.

REFERENCES CITED

1. John and Mildred Teal. *Life and death of the salt marsh.* Boston: Little, Brown and Co., 278 pp., 1969.
2. Eugene P. Odum and Armando A. de la Cruz. Particulate organic detritus in a Georgia salt marsh-estuarine ecosystem. pp. 383–388 in *Estuaries*, G. H. Lauff (editor), American Association for the Advancement of Science Publication 83, 1967.
3. Eric J. Heald. The production of organic detritus in a south Florida estuary. Univ. of Miami, Sea Grant Technical Bulletin No. 6, 110 pp., 1971. William E. Odum. Pathways of energy flow in a south Florida estuary. Univ. of Miami, Sea Grant Technical Bulletin No. 7, 162 pp., 1971.
4. Robert N. Ginsburg. Environmental relationships of grain size and constituent particles in some south Florida carbonate sediments. *Bulletin of the American Association of Petroleum Geologists*, vol. 40, pp. 2384–2427, 1956.
5. John H. Hoyt. Barrier island formation. *Geological Society of America Bulletin*, vol. 78, pp. 1125–36, 1967.
6. D. W. Pritchard. Salinity distribution and circulation in the Chesapeake Bay estuarine system. Sears Foundation, *Journal of Marine Research*, vol. 11, pp. 106–123, 1952.

7. Robert H. Meade. Transport and deposition of sediments in estuaries. pp. 91–120, in B. W. Nelson (editor), *Environmental framework of coastal plain estuaries.* Geological Society of America Memoir 133, 619 pp., 1972.
8. Maynard M. Nichols. Sediments of the James River estuary, Virginia. pp. 169–212, in B. W. Nelson (editor), *Environmental framework of coastal plain estuaries.* Geological Society of America Memoir 133, 619 pp., 1972.

ADDITIONAL IDEAS AND SOURCES

1. Besides the various articles cited in the text, there is a wealth of recent information on estuaries; see, for example, Memoir 133 of the Geological Society of America (References 7 and 8). Several papers in it treat aspects of the James estuary and Chesapeake Bay. The glossary in the back of this book may help in reading them.
2. Topographic maps and nautical charts are available for most estuaries; however, vertical aerial photographs are often more valuable because they show waves and currents as well as a host of other details. Photo index sheets are especially valuable because they cover large areas by groups of photoprints. These sources and materials are explained in Appendix A.
3. It might be interesting to examine Figure 4-5 again. By carefully comparing the patterns of turbid (bay) and clear (ocean) water, can you see that the tide has been at flood for some time before the photo was taken?
4. An interesting study is that of obtaining aerial photographs taken many years apart and noting the actual changes with time—especially those resulting from human works. A study using this method as well as sediment-coring and various measurements of sediment accumulation is described by William H. Kanes in his paper "Facies and development of the Colorado River delta in Texas." This is one of several interesting papers in the volume *Deltaic sedimentation, modern and ancient* (James P. Morgan, editor; Society of Economic Paleontologists and Mineralogists, Special Publication 15, 1970). The rapid growth of well-defined deltas into the estuaries of the Gulf Coast is an important topic in that region.
5. If you live near a marsh or shallow estuary, collect jars of water at different tidal stages to see if the amounts of suspended sediment do indeed change during each tidal cycle. A procedure for separating and weighing the sediment is given in Additional Ideas and Sources for Chapter 3. If you do not have access to a balance and equipment for filtering, you can simply let the sediment settle in each jar and compare the amounts visually.

A day of distinct drift (slow water movement) alongshore toward the foreground, caused by the oblique arrival of the waves. The sand, too, is being transported in that direction, for each wave swirls it into suspension and carries it a short distance alongshore.

5. The Sea near Shore

The Sea near Shore

Movement of ocean sediment into estuaries and onto beaches is a serious topic in the area shown in Figure 5-1. Note the gray ribbon of the Hudson River, which is outlined in ink because its waters are so turbid that they look almost like land. A mass of turbid (lighter gray) water can be seen extending from the river into Lower Bay, and a broader but more diffuse mass can be seen in the ocean southeast of the bay. Of particular interest are the distinct streaks of turbid water well offshore, which are probably the results of wastes dumped from barges. Similar streaks can be seen in Earth Resources Technology Satellite (ERTS) images taken at other times and are only a hint of the massive transfers of waste materials from the metropolis to the sea. An assessment by the Army Corps of Engineers has shown that 65 million cubic yards of sewage sludge, city and industrial wastes, and harbor dredgings (largely derived from sewage) were dumped from barges off the mouth of Lower Bay during the period 1965–70 (Reference 1). In addition, nearly 2 billion gallons of sewage were discharged daily into this part of the sea in 1970, but only 57 percent of it received even primary treatment. In primary treatment, the solids are decomposed and more or less innocuous, but only about one-third are thoroughly oxidized.

The total amount of sediment is greater than that carried by all Atlantic seaboard rivers combined and is increasing each year. When the Coastal Engineering and Research Center (of the Army Corps of Engineers) surveyed

$10^6 m^3 \approx$ one cal football stadium full

which

FIGURE 5-1

The coastal regions of New Jersey and New York, a view transmitted from Earth Resources Technology Satellite 1 (ERTS 1) on April 23, 1973.

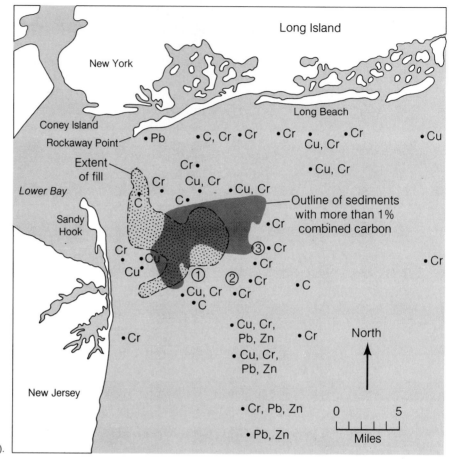

FIGURE 5-2

Map showing designated dumping sites (1 = dredge spoils 2 = dry wastes, and 3 = sewage sludge) and location of principal despoits (dotted) and of sediments containing high contents of combined carbon (shaded). Dots with letter symbols are places where an analyzed sample showed unusually high amounts of carbon (C), copper (Cu), chromium (Cr), lead (Pb), or zinc (Zn). Data from George Pararas-Carayannis (Reference 1).

the principal area of dumped materials, they found that it covers roughly 30 square miles of seafloor and is as much as 50 feet thick (rising at its peak more than halfway toward the surface from the original seabottom). An alarming discovery was that the mound does not lie under the sites officially designated for dumping (Figure 5-2). One must conclude either (1) that barges had regularly been emptied closer to shore than the designated sites, or (2) that the dumped materials had shifted position considerably, mainly toward the mouth of Lower Bay but also in a lobe pointing toward the New Jersey shore.

Tracing organic wastes   The results of studies by the State University of New York at Stony Brook and by the Sandy Hook Laboratory (contracted by the Coastal Engineering and Research Center) suggested that organic wastes, too, had been dispersed widely from the dumping sites. Bottom sediments with more than 1 percent of combined carbon form the patch shaded in Figure 5-2; several other sample sites with unusual amounts of carbon are indicated separately in the figure. Because normal sediments on this part of the seafloor contain only traces of combined carbon, the samples indicated in the figure suggest contamination by carbon-rich sewage sludge or harbor dredgings. Additionally, metals (such as copper, lead, zinc, and chromium) that are concentrated in sewage sludge but exceedingly scarce in the normal bottom sediments of the area were found to be very abundant in the shaded area and moderately abundant at the scattered sample sites shown in Figure 5-2. The metals have particular significance in that they are toxic to many forms of life when ingested regularly or in large amounts; however, only local contamination of fish and shellfish has been detected so far. Bacteria, including varieties typical of human feces, also are more abundant and widespread than expected normally. A concern not yet tested is the possibility that some may carry diseases to sea life and thence to humans.

As suggestive as these data are, however, they do not prove firmly that the dumped wastes have spread widely and are an immediate danger. We cannot be sure, for example, that the various outlying sample sites were not affected by other sewage outlets or by the sewage and bilge water released from the thousands of vessels putting in and out of New York Harbor. More studies are being made to assess these and other possibilities, and their results will doubtless provide some surprises. What we can do, however, is examine other studies of the sea near shore and thereby judge whether the wastes are *likely* to move onshore. We can investigate a number of specific questions. Do sediments on the bottom move in the same directions as the turbid water seen on the surface? If so, why do they move? What is happening on the seafloor seaward from the shore and estuaries—beneath the opaque gray of the ERTS image? The best way to find answers is to start at the very shoreline and move oceanward step by step.

Action of Waves near Shore

The ERTS image does not show the waves and current lines that would help us interpret movements of sediment near shore, but we can easily find views of the Atlantic coast that do. The photographs in Figure 5-3A, for example, were taken looking straight down on sand islands similar to though smaller than those in Figure 5-1. The location is 20 miles north of the mouth of Chesapeake Bay, and the islands are labeled in Figure 4-1. The ocean and its waves are our first interest, and waves least affected by the coast can be seen approaching from offshore in the upper right part of Figures 5-3A and B. They are typical winter waves, arriving from the northeast and measuring about 400 feet from crest to crest. As they approach the *shoals* (shallows), marked by white tracks of breaking waves, they become more closely spaced because they are slowed down due to dragging against the sea bottom. They also become more distinct because they are more peaked. They change direction locally and form interfering (intersecting) patterns over the shallower areas. Note how the waves converge on the flanks of the small island labeled X in Figure 5-3B, meeting obliquely behind it in a criss-cross arrangement.

Effects of tidal channels   The relation of wave spacing to water depth is emphasized by the fact that the most widely spaced waves are those moving along the channel approaching Great Machipongo Inlet (Figure 5-3). Such channels are by far the deepest parts of the sea near shore and are eroded and

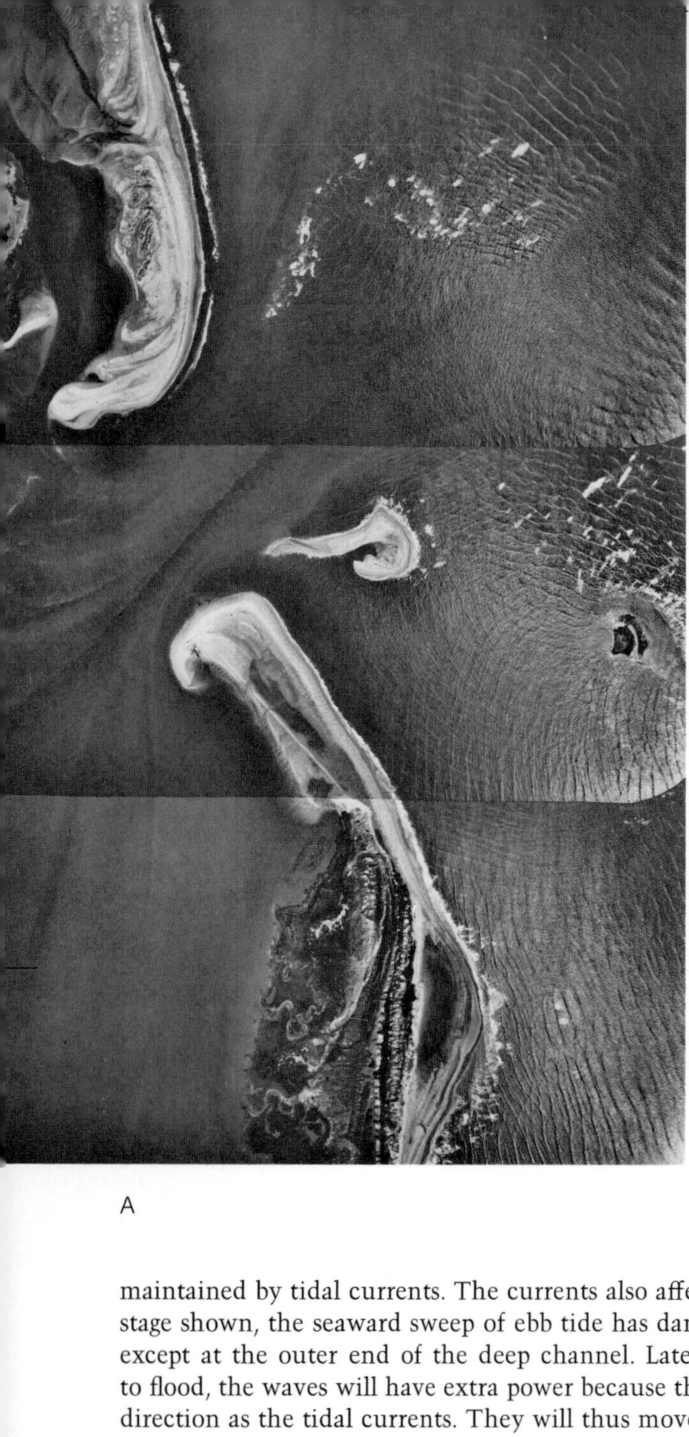

A

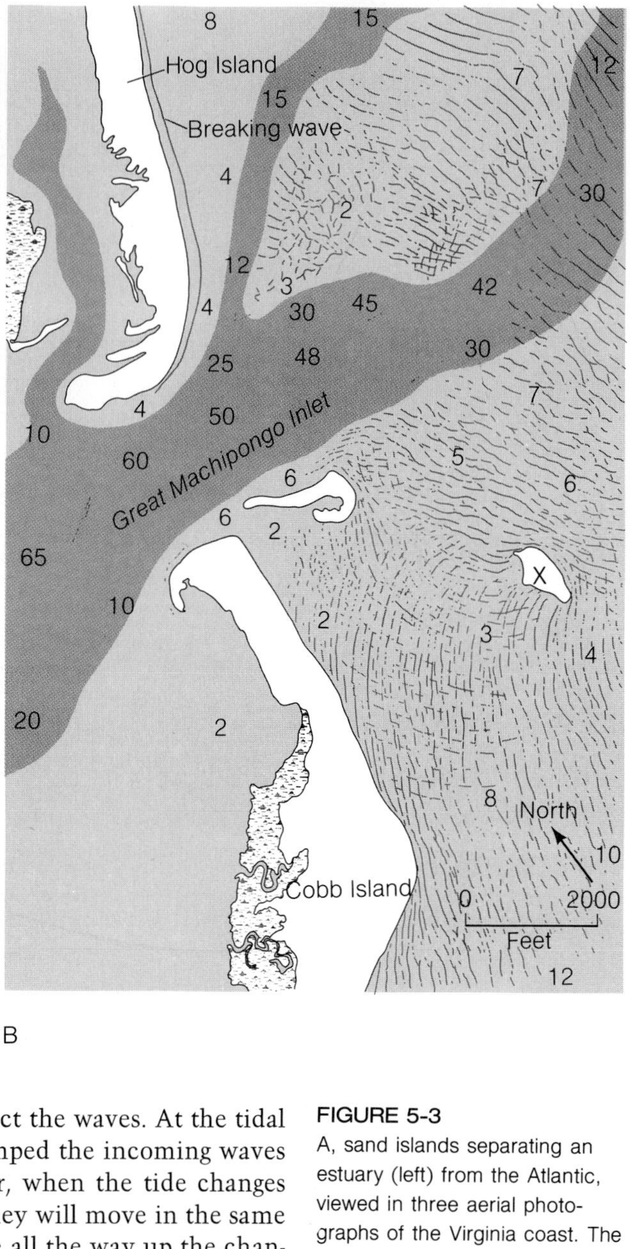

B

maintained by tidal currents. The currents also affect the waves. At the tidal stage shown, the seaward sweep of ebb tide has damped the incoming waves except at the outer end of the deep channel. Later, when the tide changes to flood, the waves will have extra power because they will move in the same direction as the tidal currents. They will thus move all the way up the channel and into the estuary. This action emplaces sand in the estuary, as shown by the distinct sand hooks on the two large islands and on the small island just outside the inlet (Figure 5-3A).

Transport of sand along shore   Sand on the bottom is set in motion as each wave passes, because the water beneath the wave moves in elliptical paths at depths as much as seven times the height of the wave (Figure 5-4A). Small waves thus rock the sand offshore into symmetrical ripples, and large waves swirl the sand forward with each surge. The turbulence of waves that steepen and break near shore puts large amounts of sand in suspension. Each wave thus sweeps sand toward the beach and back again. If the approach of

FIGURE 5-3

A, sand islands separating an estuary (left) from the Atlantic, viewed in three aerial photographs of the Virginia coast. The white patches are foam from breaking waves, and the dark streaks in the estuary are ebb currents. U.S. Geological Survey photographs taken January 12, 1967. B, map showing the wave crests and the distribution of deep water (dark gray). Depths in feet are from a National Ocean Survey chart.

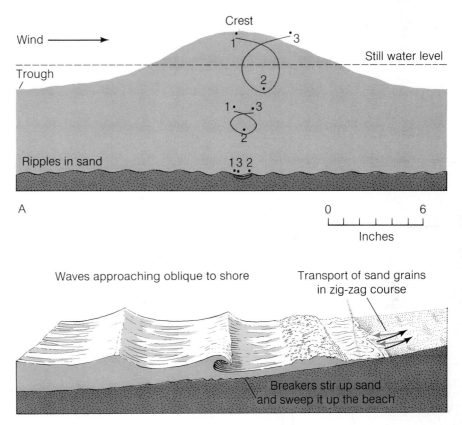

FIGURE 5-4

A, vertical section through a wave photographed in a laboratory flume. Water particles at positions 1 will move in the dotted courses—to positions 2 as the trough passes and to positions 3 as the next crest arrives. From John C. Harms, "Hydraulic significance of some sand ripples," *Geological Society of America Bulletin*, vol. 80 pp. 363–396, 1969. B, diagram of waves steepening and breaking near shore, and sweeping up the beach obliquely so that many sand grains would describe the course shown during the passage of two waves.

the waves is oblique to the beach, as in the view on the chapter opening page, the back-and-forth movements will carry sand grains in a zig-zag course along the beach (Figure 5-4B). This sort of transport can be predicted for Cobb Island (Figure 5-3), because the waves north and south of the most seaward projection are arriving onshore obliquely. Sand must thus be moving gradually northward on the long straight stretch of beach, finally being carried around the north end of the island and into the estuary. The rate of transport has not been measured here, but measurements made on other Atlantic beaches by the Army Corps of Engineers suggest that it is somewhere between 2 and 6 million cubic feet of sand per year.

The shapes of the deposits that compose Hog Island show that southward transport of sand has long been typical along this beach. Compare Figures 5-3A and 5-5A. The series of rudely parallel bands are accretions of sand that have built the island southward, forming a spit. Note how similar additions have all but closed a small tidal channel through the island. Note, too, that sand has been carried beyond the spit accretions into the shallow part of the estuary behind the island. These relations express a general movement of ocean sand to estuaries: first along the shore by waves, and then into the estuary by the action of waves during flood tide.

Cobb Island provides a somewhat different kind of evidence of long-term sand transport. By examining Figure 5-3A carefully, we can see the island is made of a series of sand-dune ridges, the older of them largely covered by

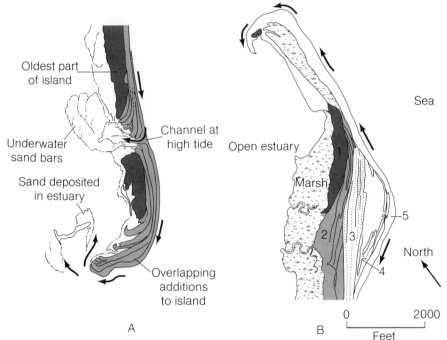

FIGURE 5-5
A, southern end of Hog Island, showing how old, sand-dune ridges have been enlarged by accretions that express transport of sand southward and into the estuary as shown by the arrows. B, succession of sand deposits in the northern end of Cobb Island, numbered in order of their deposition. Note that the island grew progressively seaward but was trimmed back along its northern edge, the sand being transported as shown by the arrows.

Oldest part of island

Underwater sand bars

Sand deposited in estuary

Channel at high tide

Open estuary

Sea

Marsh

5

North

Overlapping additions to island

0   2000

Feet

A

B

plants (the dark areas). Each dune ridge was partly eroded by the sea before the next formed. It is therefore possible for one to recognize their relative ages, as numbered in Figure 5-5B. Note that they tend to curve toward the tidal inlet at their northern ends and are cut off obliquely by the band of the beach. This relation indicates a long-term trimming of the shore in that area and a transport of sand toward the inlet, as shown by the arrows.

Now look back at Figure 5-1 and note that all the sand barriers along the southern shore of Long Island have spit ends pointing toward New York Harbor. The western end of Fire Island has shifted westward approximately 6 miles since 1825. Note, too, that Sandy Hook is a spit extending into Lower Bay. Thus, although this coastal area is much larger than that of Figure 5-3, it is similar with respect to its interactions with waves. One can thus predict that waves will similarly move sediment along the shores of New Jersey and Long Island toward New York Harbor.

Transport of Sediment Offshore and Onshore by Waves

A further question relative to the transport of dumped wastes to the New Jersey beaches is what happens to bottom sediment that lies well offshore but at depths shallow enough to be affected by waves. For an answer, we need to look in detail at the seafloor from the beach to at least several miles offshore. To date, such detailed examinations have been made in only a few places. The one described in this section was made south of Chesapeake Bay, on a stretch of sea bottom closely similar to that off New Jersey. This coast

104

extends southward from Cape Henry, shown in Figure 4-1. Indeed, Cape Henry has been eroded much like the northern part of Cobb Island. Sand is moving northward along the shore and into Chesapeake Bay—one of the sources of the ocean sand lodged in the James estuary.

The study of this coast was made by Donald J. P. Swift and several co-workers (Reference 2). Primarily, they measured water depths along lines extending from the beach for 2.5 to 14 miles out to sea and collected samples of bottom sediments at regular intervals along these lines. They also examined the sediments of the beach and estuary, as well as older deposits underlying the area. The method they used for measuring water depths is called *continuous sonic profiling*. A ship is navigated along a prescribed line, and an instrument aboard emits repeated percussion-like sound waves that travel to the bottom and are reflected back to a receiver on the ship (Figure 5.6). Because the speed of sound in water is known, including variations due to temperature and salinity, the instrument can be adjusted to measure depths precisely. In the shallows, depth data and samples were collected with the aid of scuba equipment. Samples offshore were collected from shipboard by a clam-like device called a *grab sampler*, which is lowered on a wire rope and snaps shut automatically when it touches bottom, grabbing a scoop of sediment. Each sample was later run through a laboratory device that separates sediment into fractions according to grain sizes. Each fraction was then weighed, and the average grain size of each sample was calculated from these data.

Relations between topography and sand sizes Figure 5-7 shows results from four surveyed lines that are representative of the eleven lines overall. Note how the bottom is steepest within an area 1 to 2 miles offshore, sloping approximately 40 feet per mile or roughly 0.5°. This part of the seafloor is called the *shoreface*. Note that the bottom farther offshore varies from slightly irregular (line 10) to distinctly irregular (line 3). Its average slope is very slight—only 5 feet per mile or about 0.06°. Figure 5-7 also shows that the bottom sediments become finer as one progresses seaward down the shoreface. Farther offshore they vary complexly, the more so where the bottom is distinctly irregular.

FIGURE 5-6

Measuring water depths by timing sound waves that are reflected from the bottom back to the survey ship.

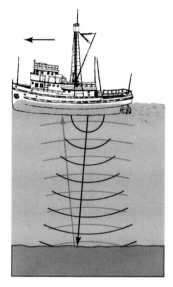

FIGURE 5-7

Four profiles of the seafloor south of Cape Henry, showing average grain sizes (dots) of sand collected directly beneath. Data from Swift, et al. (Reference 2).

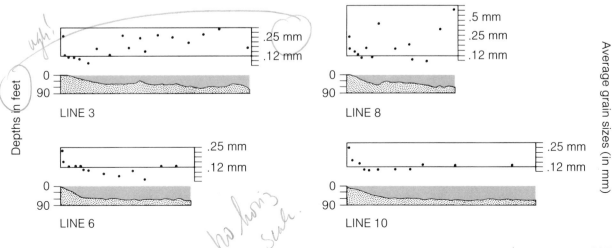

LINE 3

LINE 8

LINE 6

LINE 10

Depths in feet

Average grain sizes (in mm)

The data indicate a genetic relation between the various parts of the sea-floor and the sediments that comprise them. This relation is further emphasized by the degree of sorting of the sediments (the degree to which the grains in a sample approach uniformity of size). We have already noted that the samples were separated into fractions of a narrow size range, each of which was weighed. These data were composed into diagrams, like those in Figure 5-8, called histograms. Each vertical bar in a histogram represents a size fraction. The height of each bar shows the amount of sediment in that fraction, expressed in percent of the total sample. Degree of sorting is thus shown by the height of the tallest one or two bars and by the spread of the other bars—the higher and narrower the array, the better the sorting. The beach sand in Figure 5-8 is thus less sorted than the sand in the surf zone, and both are less sorted than the sands on the shoreface below the surf zone. The latter are sorted to a remarkable degree, probably as well sorted as natural sediments get. Two relations on the figure are true for all the sampled lines: (1) degree of sorting increases from the surf zone down the shoreface and then becomes variable, and (2) sediment gradually gets finer down the shoreface and varies greatly in size seaward of the shoreface.

By drilling and excavating through the sands on the beach and shoreface, Swift and coworkers found them to be a relatively thin blanket over older, eroded deposits (Figure 5-9, top). These underlying deposits become exposed during storms. It is common for the shoreline and shoreface to be cut back by powerful storm waves that erode the sand blanket and part of the older

FIGURE 5-8

Histograms showing the amounts of the various size fractions in samples collected along sounded profile 3. The lines indicate sample locations and the numbers above the histograms show the depth at each location. Data from Reference 2.

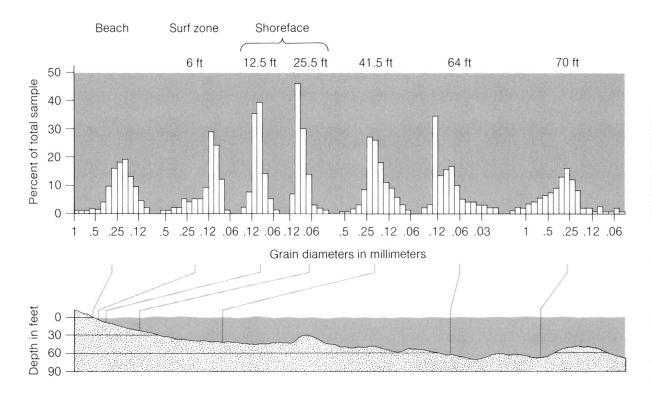

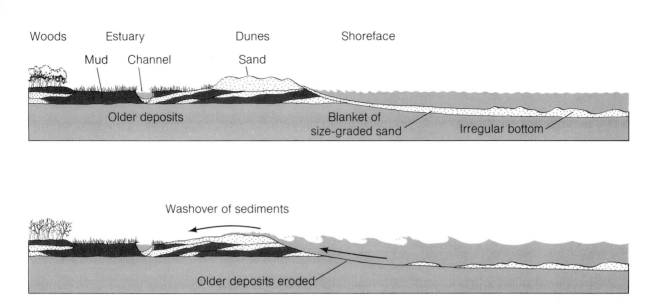

Woods   Estuary   Dunes   Shoreface
      Mud   Channel   Sand
                      Older deposits   Blanket of   Irregular bottom
                                               size-graded sand

Washover of sediments

Older deposits eroded

FIGURE 5-9
Vertical sections across estuary, dune ridge, and near shore part of the sea, showing how sand deposits formed during fair weather (top) are eroded and displaced during exceptionally strong storms (bottom).

deposits (Figure 5-9, bottom). The storm waves sweep much of the sediment over the dunes into the estuary and transport the remainder offshore. As a storm subsides, the newly eroded materials that were carried seaward are deposited as a new blanket over the seafloor, and fairweather waves slowly transport and sort these materials. Because wave motion becomes weaker with depth (on the average), the sands shifted down the shoreface become finer with depth, and because those at greater depths travel farther, they become better sorted by size.

Rates of erosion by waves Evidence that these processes are contemporary and are going on at a considerable rate is shown by measured and estimated rates of landward erosion. By surveying the coast for 12 miles south of Cape Henry during 1932–46, the Army Corps of Engineers measured 29.5 feet of coastal retreat, indicating an exceptionally rapid rate of 21 inches per year. A radiocarbon age determined on an old forest stump at the shore 15 miles farther south (reported in reference 2) suggests that if the edge of the forest was as far inland then as it is now (see Figure 5-9) the rate of inland erosion has been at least 17 inches per year since about 1200 A.D., the date determined from the stump. For the coast as a whole, old maps suggest something like 9 inches of retreat per year since the mid-1880s. These data are comparable in magnitude and range to those determined elsewhere in eroded parts of the Atlantic coast, including the New Jersey shore.

In conclusion, waves of exceptional storms put bottom sediment in motion for at least 1 to 2 miles offshore along the coast south of Chesapeake Bay and wash large amounts of it onshore and into estuaries. These powerful waves also erode inland into sediments that include the sandy washover deposits of former storms. After each storm, lesser waves sort the eroded sediment down

the shoreface and probably along the shore. If similar processes affect the New Jersey shore, it appears that polluted materials now lying within a few miles of shore (as in Figure 5-2) will be swept onshore during major storms.

A Sand Plain Beneath the Sea

Thousands of soundings like those made by Swift and coworkers show that the gently sloping, irregular sea bottom beyond the shoreface continues seaward for tens of miles, finally breaking to a steeper slope (one of about 5°) that is the side of the ocean basin proper. This steeper incline is called the *continental slope*. The nearly flat surface between it and the shoreface is called the *continental shelf* (Figure 5-10A). The thousands of samples of sediment taken from the shelf between Long Island and southern Florida include a variety of materials besides sand: mud, pebbles, gravel composed largely of shells, solid rock outcrops, even relics of coral and algal reefs, the latter near the outer edge of the shelf off Florida and Georgia. By and large, however, if one could stand at its center, the continental shelf of the Atlantic would extend to the horizon on all sides as a vast sand plain.

The idea of standing on the continental shelf is not completely absurd. During several periods in the last million years, the shelf has, in fact, been dry land. As already noted for the estuaries, these withdrawals of the sea were due to the formation of major glaciers. The last great glaciation lowered the sea about 350 feet, exposing the entire shelf and part of the continental slope for approximately 8,000 years (from approximately 20,000 to 12,000 years ago). This is proven by dated land plants collected from the seafloor and by the fact that the atmosphere oxidized the uppermost marine sediments, staining them a reddish or brownish color. Rivers also eroded valleys across the shelf and into the upper parts of the continental slope (Figure 5-10B). These various effects have indeed been "seen"—not just by closely spaced samples and soundings but also by examining the seafloor from submersible research vessels (small submarines).

Sand swells and hollows on the shelf   These data have shown that the plain is characterized by elongated swells that are typically one half to 2 miles wide and tens of miles long. These swells are so subdued that they rise only 10 to 50 feet above the surrounding hollows and flats. Swells form the irregularities beyond the shoreface that were described in the last section. When only a few had been discovered, they were thought to be beach-dune ridges submerged beneath the sea when it rose after glaciation. But when Elazar Uchupi compiled depth data for the entire shelf, the alignments of the swells suggested otherwise (Figure 5-11). Note that nearly all the swells north of Cape Hatteras are oblique to the present beach, even those that lie close to it, and that swells south of the cape show strongly oblique alignments, even radial groupings.

The linear swells that are almost perpendicular to the coast, for example those along the Georgia coast, may possibly have been sand shoals near river mouths or tidal outlets. Figure 5-12 left shows a large shoal at low tide. As the sea rose slowly and the shoreline shifted landward, shoals like this would

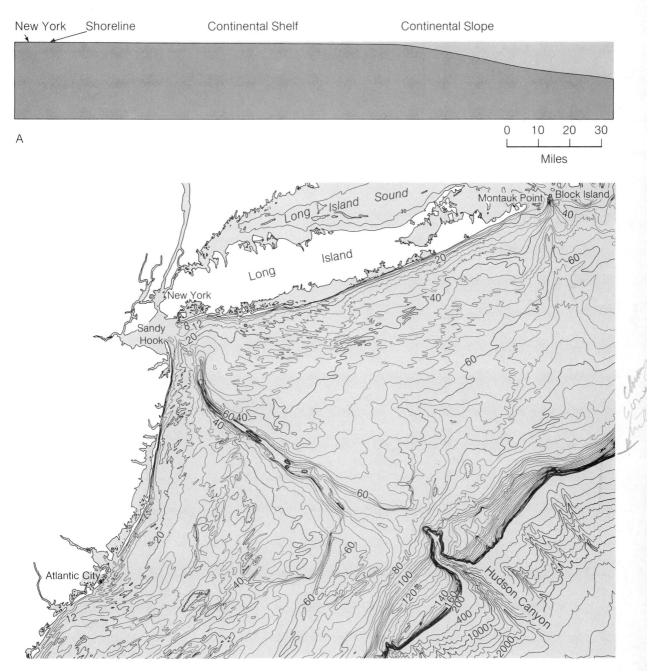

A

B

FIGURE 5-10

A, vertical profile from the coastal lowlands near New York City to the ocean basin. B, the continental shelf and upper part of the continental slope south of Long Island, showing the distinct valley extending seaward from the Hudson estuary and the upper part of Hudson submarine canyon. The contours are numbered in meters below sea level, and the interval between contours increases to 200 meters on the continental slope (which would appear black otherwise). Contours are explained in Appendix A. From Elazar Uchupi, "Atlantic continental shelf and slope of the United States—deep structure," U.S. Geological Survey Professional Paper 529-I, 44 pp., 1970.

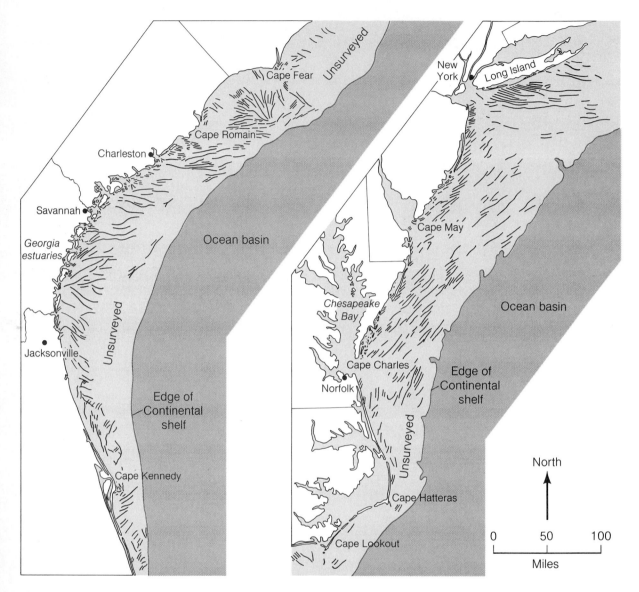

FIGURE 5-11

Continental shelf between Florida (lower left) and Long Island (upper right) showing the orientations of the larger sand ridges. From Elazar Uchupi, "Atlantic continental shelf and slope of the United States—physiography," U.S. Geological Survey Professional Paper 529-C, 30 pp., 1968.

also have been extended slowly landward and could have become miles long. Although the idea seems plausible, the same coast shows smaller sand ridges that are also perpendicular to the shore, yet are almost certainly forming and reforming by the actions of waves and coastal currents today (Figure 5-12, right). Other swells near shore in Figure 5-11 may thus also be dynamic (forming and moving today) rather than relics.

Movements of sand on the shelf   If the major swells of the shelf are indeed dynamic, and if their dynamics were known, maps like that of Figure 5-11 could be used to interpret movements of sand over large areas. We are not nearly at that state of knowledge, but at least some of the large swells are in motion, as determined by several studies comparing soundings made be-

110

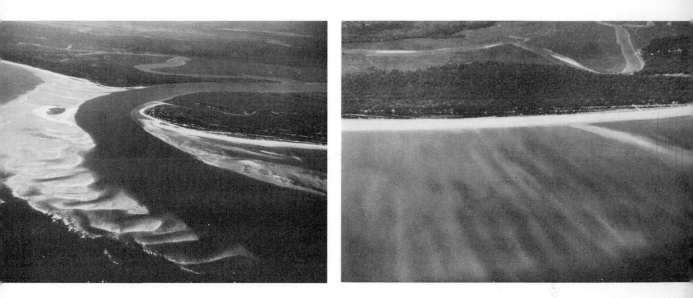

tween 1850 and 1922 with soundings made recently. At four widely separated localities—Bethany Beach, Delaware; Ocean City, Maryland; False Cape, Virginia; and Cape Kennedy, Florida—these measured movements have averaged 10 to 20 feet per year (Reference 3). A particularly informative discovery was that the swells off Bethany Beach migrated 250 feet between 1961 and 1963, a period that included a great storm in March 1962 (Reference 4). The storm eroded the beach and dunes severely and at the same time shifted sand across the seafloor at depths as great as 80 feet. Studies by the Army Corps of Engineers have provided evidence that major swells off the Florida coast shifted systematically over long periods: the swells are systematically steeper on their southeastern sides, and their internal layers have low slopes in the same direction. Some of the swells thus have forms and internal layers somewhat like very broad, low dunes.

Other evidence of sand presently moving across the shelf comes from studies of the mineral species that make up the sand. Grains of calcium phosphate on the beaches and inner shelf of North Carolina evidently are derived from older deposits lying 25 to 30 miles offshore and at depths of around 100 feet (Figure 5-13). Along the shore and inner shelf of the central Florida coast, Michael E. Field and David B. Duane found distinctive, round calcium carbonate grains (called *ooliths*) that must have been derived from older rocks exposed 15 miles and more offshore. Moreover, the grains are so soft and easily abraded in the surf that their presence implies that they are arriving onshore from time to time today (Reference 5).

That the sand on the swells near shore is also in motion is suggested by the considerable degree of polish and abrasion of the grains and by the fact that they are well sorted by size. The sand on the hollows and flats between the swells is coarser and more poorly sorted. Thus, the fine- and medium-sized sand that makes up the swells appears to have been swept up (winnowed) from the surrounding seafloor, leaving the coarser grains behind.

FIGURE 5-12

Left, elongated sand shoal at the mouth of a tidal channel between two sand islands of the Georgia coast, at an exceptionally low tide. Right, sand ridges (pale gray) visible underwater during an unusually low tide, northeast of Sapelo Island, Georgia. Note that the smaller ridges are almost perpendicular to the shoreline but the large one is oblique to it.

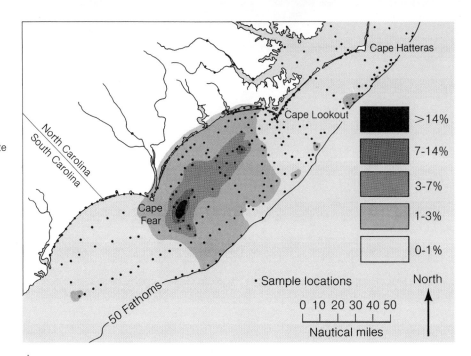

FIGURE 5-13

Distribution of calcium phosphate grains eroded from older sediments that lie near the black area off Cape Fear. The patterns on the landward side of the coastline show abundances on the beaches. Data from J. L. Luternauer and Orrin H. Pilkey, "Phosphorite grains—their application to the interpretation of North Carolina shelf sediments," *Marine Geology*, vol. 5, pp. 315–320, 1967.

On the outer parts of the shelf, in water deeper than 80 feet, all the sand has brown stains and coatings formed by exposure and alteration during periods of glaciation. The shell fragments have pits made by bottom-dwelling organisms. These features indicate little or no abrasion and, therefore, that the grains have not been transported recently. Perhaps the swells on the outer part of the shelf formed long ago but in the same ways as those on the inner shelf; they were then immobilized as the water deepened and the shoreline migrated landward.

We can conclude that major storms probably put sand in motion over much of the inner half of the shelf—to depths of about 80 feet. This action has great significance with regard to the wastes dumped off New York Harbor: these materials are likely to be moved during storms. The directions of movement are not fully understood; in some cases, movement is toward shore and in others away from shore. With many of the swells being aligned about parallel to the approach of winter waves, it is possible that sand travels mainly along the swells rather than across them. This is an intriguing topic for further studies.

Landward Movement of Fine Sediment

So far we have treated only the movement of sand, whereas the polluted wastes being dumped off New York Harbor (and in many other parts of the world) are largely mud. What is known about the movements of fine sediments on the shelf? We have already noted that sediments on the shelf are mainly sand; indeed, most include only traces of finer grains. This finding was formerly thought to mean that waves and currents winnowed (sorted)

a new idea

out the fines and worked them across the shelf to the deeps of the ocean basin. Some materials doubtless take that route, but evidently most of them do not. One line of evidence is that extremely little land-derived sediment is ordinarily suspended in shelf waters. Samples collected widely over the Atlantic shelf show that amounts of suspended sediment decrease rapidly away from land—from about 1 part per million of sea water (0.0001 percent) at sites within 3 to 10 miles offshore to less than 0.1 part per million at sites 30 miles out (Reference 6). Moreover, the suspended sediment 30 miles from land is almost entirely organic detritus rather than land-derived mineral grains. Just where the fine sediment is going, and why, were shown by an important study initiated to resolve a problem of the coastal fisheries rather than the sediment question (Reference 7).

The young of certain commercial fishes, notably the menhaden, are common in Chesapeake Bay, yet their eggs have rarely been found there. The valuable fish are abundant in some years and scarce in others, and in 1959 the Virginia Institute of Marine Science undertook to determine why. By netting bottom waters at various places off the mouth of the bay, it was discovered that menhaden larvae are less and less mature the farther offshore they are taken. Large numbers of eggs were finally netted 65 miles east-northeast of the mouth. Evidently, the fish spawn far out on the shelf, and their young move slowly landward. Remarkably, they somehow find and enter the narrow mouth of the bay to begin their principal stage of growth in the estuary.

Tracing currents at the surface and on the bottom These findings were especially intriguing because *drifts* (sluggish currents) measured at the ocean surface were known to be typically away from the mouth of the bay rather than toward it. However, these movements were not well known, and it was decided to make a thorough study of movements of water on the seafloor as well as at the surface, covering an area extending 80 miles out to sea and 120 miles north and south of the mouth of the bay. Surface currents were tracked by floating bottles. Bottom currents were tracked by seabed drifters, plastic structures that ride buoyantly along the seafloor (Figure 5-14). Sets of both were air-dropped once in each of 17 consecutive months at 110 points spaced equally over the study area. Each bottle and drifter carried a mail-return card to report the position and date of discovery, and special searches were made on beaches not frequented by the public. The fact that whims of discoverers did not spoil the study is indicated by the large number of bottles (292 out of 660) reported just after periods of *onshore winds* (winds blowing toward land) and the small number (11 out of 660) reported just after periods of *offshore winds* (winds blowing away from land). The timing between air drops and discoveries showed that most devices moved quite directly from drop point to discovery point; therefore, this direction could be used to chart the approximate directions of currents at that time.

Approximately 1,600 of the 9,500 seabed drifters were discovered and reported, 124 of them in Chesapeake Bay. One moved 58 miles up the bay, and many beached near the entrance to the James estuary. Many others doubtless lie in the bottom muds of the bay or elsewhere in the estuary complex, as evidenced not so much by the recoveries in the estuary as by the directions of drift plotted from all discoveries (Figure 5-15). The convergence of bottom

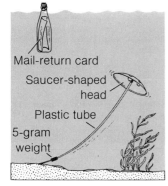

FIGURE 5-14
Surface drift bottle and seabed drifter (Woodhead type), the latter drifting to the right.

Mail-return card
Saucer-shaped head
Plastic tube
5-gram weight

drift on the mouth of the bay is striking, as is the southward bottom drift on the outer shelf.

Velocities of the bottom currents could be approximated from some of the 62 seabed drifters netted by fishermen at sea by simply comparing the time and place of drop with time and place of discovery. One seabed drifter moved 40 miles in 15 days, or at an average rate of 2.7 miles per day; several

FIGURE 5-15
Directions of bottom drift derived from all recoveries of seabed drifters in the 1963–64 study. From Reference 7.

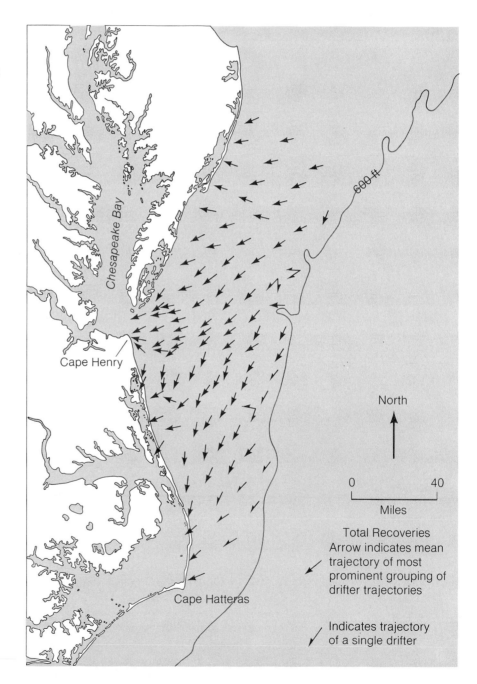

others moved at rates of between 1 and 2 miles per day. Because all drifters must have zig-zagged to some degree, those that appear to have moved fastest should give velocities most nearly like those of the actual bottom currents. This is around 2 inches per second—too sluggish to erode even the finest sediment from the bottom but capable of transporting sediment stirred into suspension by waves, tidal currents, or animals. If we consider the size of the area affected, it is evident that the bottom currents must often move large quantities of fine sediment toward the bay.

Causes of the currents   During the period of the study, the shoreward bottom current was strongest and most consistent when winds blew consistently offshore. The bottom current thus appears to have been a return flow to replace surface water pressed seaward by the wind. As shown in Figure 5-16, the winds of January 1964 were dominantly offshore, so much so that no surface bottles were recovered during the month. Onshore discoveries of seabed drifters during that period were far above average and showed the pattern of currents depicted by the arrows. Note, then, the directions on the map for August 1964 (Figure 5-16), which is typical of summer months. The dominantly northerly winds moved surface bottles northward and in a broad gyre (circle), whereas seabed drifters were moved southward and generally not onshore. The opposite directions of surface and bottom currents during both periods thus strongly support the idea of a two-layer circulation driven by the wind—water at the surface moves with the wind and that on the bottom returns to replace it.

Observe, however, that seabed drifters dropped close to the mouth of the bay moved toward it even during August. This movement suggests that outflow from the bay was also generating a return of water along the bottom. Salinities measured during August show that water from the bay, carrying 29 to 30 parts of salt per thousand of water, streamed seaward over a water layer that contained somewhat more than 32 parts of salt per thousand of water. The surface water slowly became more saline as it moved out over the shelf and mixed with the underlying water, but it could be identified by its lower than average salinity for 30 miles out to sea! The denser, more saline bottom water was presumably flowing toward Chesapeake Bay to replace the saline water mixed upward into the outgoing layer. The circulation is thus exactly like that in the James estuary and connects the estuarine circulation described in Chapter 4 to the entire near shore part of the sea.

The estuary-driven circulation was less important than wind-driven circulation during the period of the study, because river flows were unusually low. Such periods may be those when few young menhaden arrive in the bay. In any case, the study proved that the larvae do not "find" the mouth of the bay but are conveyed to it in bottom currents produced by offshore winds and seaward flow of estuarine water. No wonder Maynard Nichols found shells of open-sea protozoans far up the James estuary.

What, then, of similar studies made off New York Harbor? As the contours on Figure 5-17 indicate, more than half the seabed drifters recovered in Lower Bay or other parts of New York Harbor were those released over the mound of dumped wastes, or nearby. Seabed drifters released northeast of

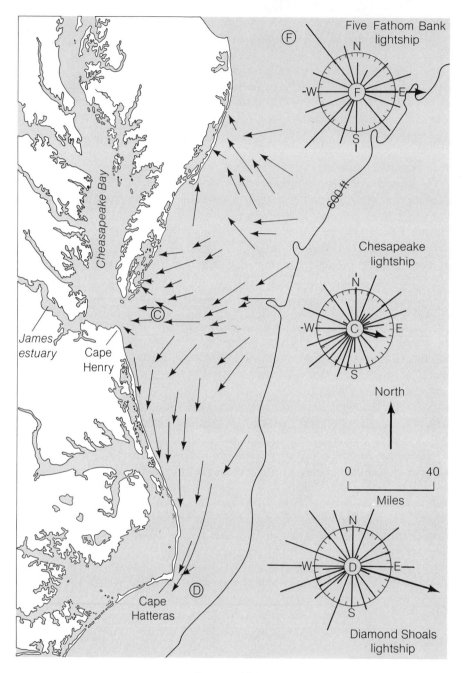

January 1964

FIGURE 5-16

Currents derived from seabed drifters (black arrows) and surface drift bottles (gray arrows) recovered during the periods shown. The circled F, C, and D show the positions of the lightships on which the wind data (right) were collected. The arrows of the wind diagrams are the resultants of the measured winds (shown by lines) and their lengths are proportional, more or less, to wind energies. From Reference 7.

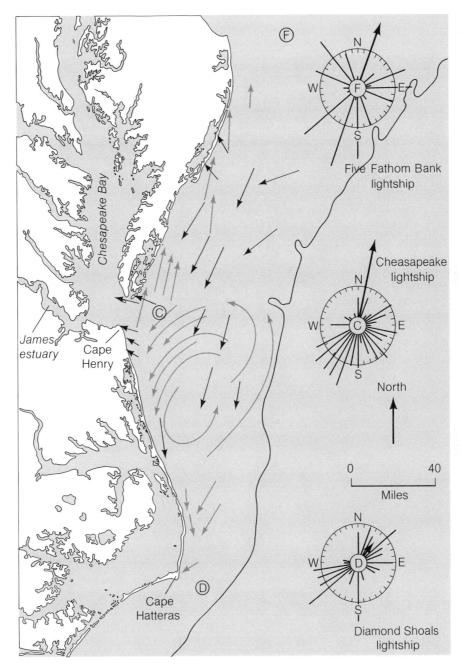

Chesapeake Bay

James
estuary

Cape
Henry

Cape
Hatteras

(F)

Five Fathom Bank
lightship

Cheasapeake
lightship

North

0   40

Miles

Diamond Shoals
lightship

(C)

(D)

August 1964

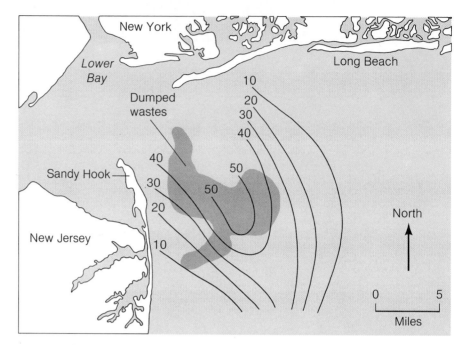

FIGURE 5-17
Percentages of seabed drifters recovered in Hudson River estuary, according to the sites where they were released (for example, more than 50 percent of all the drifters found in the bay were released inside the line numbered "50"). Data from the Sandy Hook Marine Laboratory, reported in Reference 1.

that area moved mainly to the shore of Long Island, and those released southwest of the dump area moved mainly to the New Jersey shore. It has been argued that seabed drifters are not mud particles and may move in different directions, but this seems unlikely. The general lack of fine sediments on the shelf and the abundant remains of shelf animals in the muds of estuaries indicate that the drifters are going in much the same directions as the fine sediments.

Summary

Our examination of the near shore parts of the Atlantic Ocean has led to a variety of studies and some important results:

1. Waves arriving obliquely to the coast transport sand laterally along the shore, constructing or eroding beaches, spits, and sand islands.
2. Sand is thus moved into deep tidal channels at the mouths of estuaries, where waves may transport it into the estuaries, especially at flood tide.
3. Waves sort sand down the shoreface during fair weather or moderate storms, causing the sand to become finer grained and better sorted offshore.
4. Waves of strong storms erode the beach and the entire shoreface, carrying some sand into deeper water and sweeping some onshore and into estuaries.
5. Most low-lying parts of the Atlantic Coast are thus being eroded inland at rates averaging 9 inches and more per year.
6. The broad sand plain of the Atlantic continental shelf is characterized by low linear swells that are activated by waves during great storms, at least to depths of approximately 80 feet.

7. Persistent winds generate slow surface drifts of water and thereby slow return flows in the opposite direction along the bottom; the persistently offshore winds of the Atlantic region thus tend to move fine bottom sediment toward land.

8. Fine sediment is also transported onshore and into estuaries, because low-salinity water flowing from the estuaries generates a return flow of more saline water along the seafloor—an extension of the two-layer flow typical of many estuaries.

9. The bottom drifts have great practical significance because the pollutants dumped on the shelf typically consist of mud and lightweight organic particles that can be suspended by waves and then carried in the drift currents.

REFERENCES CITED

1. George Pararas-Carayannis. *Ocean dumping in the New York bight: an assessment of environment studies.* Coastal Engineering Research Center, U.S. Army Corps of Engineers, Technical Memorandum No. 39, 159 pp., May 1973.

2. Donald J. P. Swift, Robert B. Sanford, Charles E. Dill, Jr., and Nicholas F. Avignone. Textural differentiation on the shoreface during erosional retreat of an unconsolidated coast, Cape Henry to Cape Hatteras, western North Atlantic shelf. *Sedimentology*, vol. 16, pp. 221–50, 1971.

3. D. B. Duane, M. E. Field, E. R. Meisburger, D. J. P. Swift, and J. S. Williams. Linear shoals on the Atlantic inner continental shelf, Florida to Long Island. In *Shelf sediment transport: process and pattern*, D. J. P. Swift, D. B. Duane, and O. H. Pilkey, editors. Stroudsburg, Penn.: Dowden, Hutchinson and Ross, 656 pp., 1972.

4. D. W. Moody. Coastal morphology and processes in relation to the development of submarine sand ridges off Bethany Beach, Delaware. Unpublished thesis, Johns Hopkins Univ., 167 pp., 1964.

5. M. E. Field and D. B. Duane. Geomorphology and sediments of the inner continental shelf off Cape Kennedy, Florida. U.S. Army Coastal Engineering and Research Center, Technical Memorandum 39, 1973.

6. Frank T. Manheim, Robert H. Meade, and Gerard C. Bond. Suspended matter in surface waters of the Atlantic continental margin from Cape Cod to the Florida Keys. *Science*, vol. 167, pp. 371–76, 1970.

7. W. Harrison, J. J. Norcross, N. A. Pore, and E. M. Stanley. Circulation of shelf waters off the Chesapeake bight. Environmental Sciences Services Administration, Professional Paper 3, 82 pp., 1967.

ADDITIONAL IDEAS AND SOURCES

1. The Atlantic continental shelf and slope are subjects of extensive research published by the U.S. Geological Survey in a series of professional papers (numbers 529-A through 529-M) issued between 1966 and 1973. Several such papers are referred to in this chapter, and the subjects of the others can be obtained from the Geological Survey.

2. A very readable source on near-shore processes is the book *Waves and beaches* by Willard Bascom (Anchor Books, Doubleday & Co., New York, 267 pp., 1964). Waves are also described clearly and illustrated beautifully in several articles in the September–October 1975 issue of *Oceans* (vol. 8, no. 5). One article describes tsunamis, giant sea waves generated by movements on faults, by landslides, and by volcanic explosions.

3. To measure the rate at which water is being displaced along the shore by the oblique arrivals of waves described in the chapter, throw a grapefruit or orange into the surf. It will flow low enough so that the wind will not affect it, yet it is easy to see.

4. The transport of sand in and out of bay entrances usually forms underwater dunes and bars (shoals) that become stabilized when ebb currents and flood currents become concentrated along certain parts of the channel. In fact, the shapes of the dunes and bars can then be used to interpret the positions of the current courses. An important example is described by John C. Ludwick in "Tidal currents and zig-zag sand shoals in a wide estuary entrance." *Geological Society of America Bulletin*, vol. 85, pp. 717–26, 1974.

Fish attracted to a can of bait (measuring 1 foot across) at a depth of 3,200 fathoms (19,000 feet) at 34° 03′ N, 163° 59′ E in the northwest Pacific. The dark lumps are manganese nodules—a possible source of several valuable metals. This is one of thousands of similar photographs of fishes at the bottom of the deep sea made by the Marine Life Research Group at Scripps Institution of Oceanography. Courtesy of John D. Isaacs.

6. The Deep Atlantic

The Deep Atlantic

Circulation of the Atmosphere
    Heating by the sun
How the Wind Moves the Sea
    The circulation in summary
Mixing by the Gulf Stream
    The Gulf Stream as a river
    Mixing of specific water bodies
Deep Currents and Sediments of the Western Margin
    Evidence of a southward bottom current
    Source and nutrient content of the bottom water
Transport of Sediment to the Deep Basin
    Pelagic sediment of the western basin
    Transport related to submarine canyons
    Turbidity currents in the deep sea
    Turbidity currents from debris flows
Summary

We have traced the flow of water and solid materials from within the continent to the edge of the deep sea—the largest region on earth. The western Atlantic basin alone is vaster than all the systems we have examined so far and is but one of the many basins forming the continuous world ocean. These basins are the least known parts of the world but also the parts least depleted of resources. We now sorely need the ocean's store of minerals, fuels, and food. Although our realization of that supply was only a hope twenty years ago, it is now a possibility based on notable advances in technology and science. The National Research Council has reported that 28 of the 31 minerals expected to be scarce by the year 2,000 can be obtained on the ocean's floor (Reference 1). They occur there in large quantities and almost certainly will be mined when methods are somewhat more advanced. Fuels can now be sought and recovered nearly to the edge of the continental shelf, and continuing improvements in drilling methods should someday make it possible to reach huge sedimentary accumulations along the sides of the deep basins. As a food resource, too, the main ocean has recently yielded promising surprises. The photograph on the chapter opening page, for example, shows large fish attracted almost immediately to a can of bait lowered to a depth of 19,000 feet—a region once assumed to be populated thinly, if at all, by large animals (Reference 2).

These resources of the deep sea are doubtless linked in many ways with its currents, its bottom configurations, and its sediment distribution. To win any sizable amount of the resources, we will have to constantly enlarge our view of oceanic processes. To get some idea of these processes and their interrelations, we will look in this chapter at the western basin of the North Atlantic Ocean. This is the best known ocean basin in the world and connects directly with the water systems described in Chapters 4 and 5. We will want to find out how and why water and sediment are cycled through it and something of their connections to sea life and recent earth history. We will want especially to see the ocean as an integrated system, one with specific connections to other systems.

Circulation of the Atmosphere

Our gains from the sea must take into account the effects of wind and weather, for the air and the ocean are closely and powerfully interdependent. We are so unused to this world that it may seem foreign to us. Figure 6-1, for example, is scarcely recognizable as the earth shown on maps, although after a careful search one can find the northern half of South America in the lower central half of the photograph and, perhaps, see that the cloud-free (dark) area in the upper left includes most of western and central North America. By and large, however, the surface is dominated by the dark sea and by the motions of the atmosphere and its clouds.

The close connections between the atmosphere and ocean are shown by the atmosphere's movements over the central North Atlantic, which is the dark oblong shape extending west-to-east across the central part of the view.

FIGURE 6-1

View of the earth transmitted on August 27, 1974 from NASA's Applications Technology Satellite 3, 22,500 miles (approximately an earth diameter) away. The solid arrows show probable wind directions, and the large open arrow indicates the movement of a major storm (the cloud spiral). Photograph by NOAA, National Environmental Satellite Service.

The clouds on the west (left) side of it are moving northeastward along its margin and then are sweeping eastward in the great white belt across its northern side. The long white fingers curling in from the right edge are cloud trains moving westward along the southern side of the ocean; they will eventually sweep northward to complete the circulation at the western edge. The nearly cloud-free area in the center is the Sargasso Sea, a region typified by calms and slow currents that (according to old-time mariners) set derelict ships circling endlessly. Around this ocean core, swifter currents wheel in a pattern similar to the wind-driven clouds—northward along its western edge, eastward in a broad band across its northern part, and westward in the southern part of the ocean.

Heating by the sun This circulation is due primarily to the sun, which heats the equatorial regions more than the polar regions. The main reason for the unequal heating is the geometric relation illustrated in Figure 6-2: the sun's radiation is more concentrated near the equator. Another reason is that snow, ice, and clouds reflect much of the radiation into space (note the whiteness of the north-polar area in the satellite image). The overall result is that each square mile within 20 degrees of the poles receives, on the average, somewhat less than one-third the energy received by each square mile within 20 degrees of the equator.

Little of the sun's radiation heats the atmosphere directly; instead, heat is absorbed by the land and sea and is then transmitted to the lower atmo-

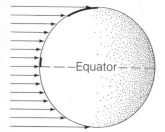

FIGURE 6-2

Solar energy (arrows) is more concentrated on equatorial regions than on polar regions. The angular relations are for an equinox, when they are average for the year.

124

sphere. The sea is especially important in this linkage because the warm equatorial region is 80 percent water. The sea heats the air partly by contact with it and partly by evaporating (gaining heat from the sun) and then condensing as low-lying clouds (transferring the heat to the atmosphere).

The heated air of the equatorial region thus expands and rises, and cooler (denser) air flows in along the surface to replace it. This starts a broad circulation run by alternate heating and cooling, as depicted diagrammatically in Figure 6-3A. For an entire hemisphere, the air generally moves in three interconnected cells, as shown in Figure 6-3B, and each cell generates one of the principal winds at the earth's surface (Figure 6-3C).

Figure 6-3C illustrates a universal characteristic of moving bodies of air, namely the curving paths of the winds. Air that starts moving toward the equator shifts toward the west, and air that starts moving toward the poles shifts toward the east. These deflections are an effect of the earth's rotation, commonly called the *Coriolis effect*. The deflections are not due to any real forces but rather to our referring motions near the earth's surface to points and lines on the rotating sphere. As seen from outer space, the winds would appear to move in straight lines, from north to south or from south to north, as implied by Figure 6-3B. As earth residents, however, it is easier for us to deal with them as curving courses—to chart them with respect to our immediate frame of reference. The results are simplified by being systematic: moving bodies in the northern hemisphere curve to the right of their initial directions of movement, and those in the southern hemisphere curve to the left. These relations are shown by the curving lines in Figure 6-3C, and a complete right-handed cell is illustrated by the overall circulation in Figure 6-1.

FIGURE 6-3

A, vertical section showing how heated air rises near the equator and generates a circulation in both hemispheres. B, vertical section of typical circulations in the northern hemisphere. C, typical wind directions on the earth's surface, the results of movements like those in B combined with the earth's rotation.

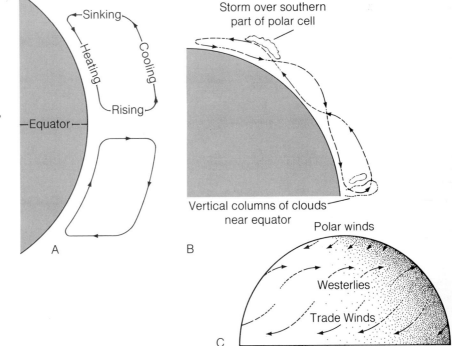

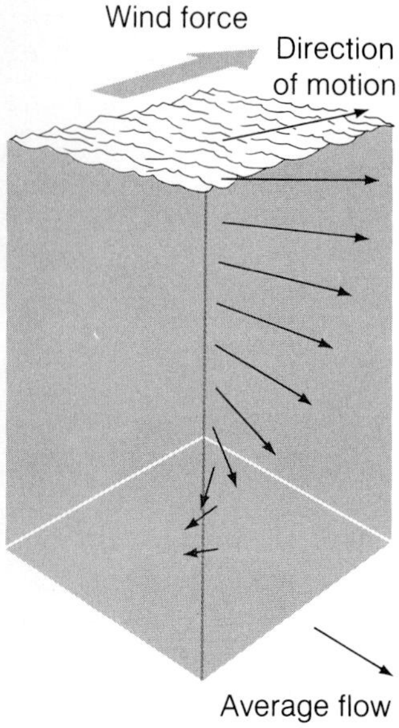

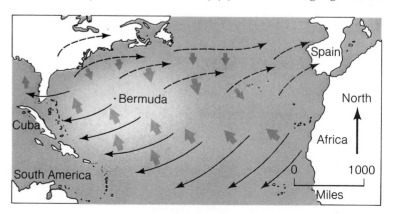

Wind force
Direction of motion

Average flow

FIGURE 6-4

Diagram of the upper few
hundred feet of the ocean under
the Trade Winds, showing
how the direction of flow and
the rates of flow (which are
proportional to the lengths
of the arrows) vary with depth.
Called an *Eckman spiral*, this
results in an *average* flow
due right of the wind's direction.
From R. W. Stewart, "The
atmosphere and the ocean,"
Scientific American, vol. 221,
pp. 76–86, Sept. 1969.

How the Wind Moves the Sea

The Coriolis effect also influences the movements of the sea. When the Trade Winds in the northern hemisphere drag against the sea's surface, the upper few inches of water flow somewhat to the right of the wind's direction. This upper layer drags against the water just under it, which flows somewhat to the right of the direction of the surface layer. The same effect is handed down through several hundred feet of water, finally dying out with depth because of the frictional resistance of water (Figure 6-4). Averaged through the entire body of moving water, flow is due right of the wind direction. The Trade Winds therefore press the upper part of the ocean toward the northwest, and the westerly winds press it toward the south, though not as strongly (Figure 6-5). This creates a vast mound of water about 4 feet high in the west-central part of the Atlantic.

The weight of the mound forces water to flow outward, but this flow, too, is turned to the right by the earth's rotation, so that the final result is the clockwise (right-handed) circulation shown in Figure 6-6. The flow lines in the figure are essentially parallel to level lines (contours) around the water mound.

All these right-hand turnings may seem so unnatural that you may wonder if there is direct evidence that the ocean's circulation is driven by the water mound rather than by the direct action of the winds. For one thing, the ocean currents do not increase and decrease with short-term variations in the winds. Even month-long periods of exceptionally light Trade Winds may not reduce flows at all (Reference 3). A second kind of evidence comes from the vertical dimensions of the moving water. The circulation extends down for thousands of feet in some parts of the ocean—depths far too great to be affected by the drag of wind on the surface.

The circulation in summary   Before examining some of the effects of the circulation, let's briefly review its linkage to the sun: (1) solar energy heats the sea near the equator more than that near the poles, and the sea, in turn, heats the air; (2) this heating generates broad circulations of air masses which start southward and northward along the earth's surface but are turned westward and eastward by the earth's rotation; (3) as the air drags against the

FIGURE 6-5

Directions of the Trade Winds (black thin arrows) and the westerlies (dashed thin arrows) over the Atlantic, and the consequent press of water toward the west-central part of the ocean (thick, outlined arrows). The broad summit of the resulting mound west of Bermuda, is about 4 feet higher than sea levels along the southeastern shore of the United States.

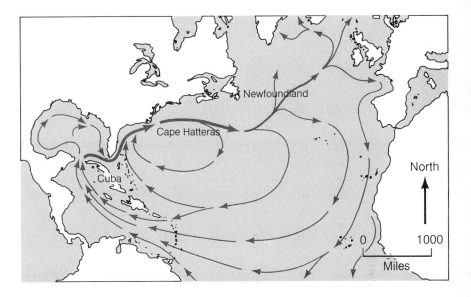

FIGURE 6-6
Generalized circulation of
water near the surface (chiefly
above a depth of 500 feet)
in the North Atlantic. Sverdrup,
Johnson and Fleming, *The
Oceans* © 1942 renewed 1970.
Reprinted by permission of
Prentice-Hall Inc. Englewood
Cliffs, New Jersey.

sea, the upper waters flow to the right (in the northern hemisphere), forming a low mound in the ocean; (4) the weight of the mound presses the water outward; and (5) this movement is turned to the right to form the broad circulation of the North Atlantic.

Mixing by the Gulf Stream

The greatly concentrated current around the west and northwest sides of the Atlantic is the Gulf Stream, shown as the thickest line in Figure 6-6. Its velocities at the surface average 5 feet per second across stream widths of 10 to 20 miles and are locally (or temporarily) as swift as 7 to 10 feet per second (5 to 7 miles per hour). As Figure 6-7 shows, the velocities decrease rapidly down to a depth of about 4,000 feet and then decrease gradually to the seafloor, here 10,000 feet beneath the surface. The great depth of the stream indicates how concentrated the ocean's circulation is here, for currents elsewhere affect only the upper 500 to 1,000 feet of water. Note, too, the "lean" of the stream toward the edge of the ocean basin, shown in Figure 6-7A by the position of the swifter currents. This is due to the water mound of Figure 6-5 being pressed against the western margin of the ocean by the earth's rotation.

The western margin of the stream itself is often marked by a line of foam and Sargasso weed (an alga characteristic of the Sargasso Sea) that gather where the outer part of the Gulf Stream turns outward and downward against the water on the shelf and slope. The stream is also visible because its exceedingly clear, blue water contrasts with the greenish (slightly turbid) water of the shelf and slope. The clarity of the Gulf Stream has important attachments. The water is essentially barren because *phytoplankton* (minute marine plants) consume all its nutrients during the long crossing of the equatorial Atlantic, and the plants are eaten or sink out of sight by the time the water reaches the Straits of Florida.

127

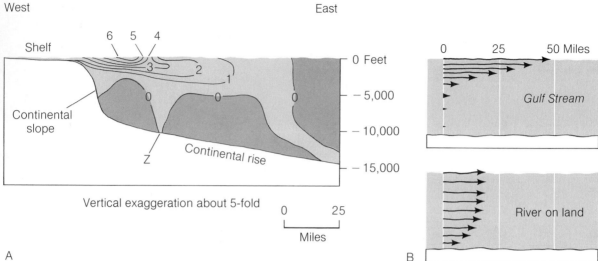

West   East

FIGURE 6-7

A, vertical section across the Gulf Stream east of Cape Hatteras, showing velocities in feet per second. Water in the dark gray regions is moving opposite to the stream (toward the reader). Based on data from Philip L. Richardson and John A. Knauss, "Gulf Stream and Western Boundary Undercurrent observations at Cape Hatteras," *Deep-Sea Research*, vol. 18, pp. 1089–1109, 1971. B, section parallel to the stream, at the same scale as A, passing obliquely from the maximum current at the surface down to point Z on the bottom. Lengths of arrows show distances traveled in 12 hours, and the diagram beneath shows how different the velocity distribution is from that of a river on land.

The Gulf Stream as a river The Gulf Stream has banks comparable to those of an ordinary river where it passes between the Florida Keys and Cuba, in the Straits of Florida, but from there to Cape Hatteras it has a bank only on the left side, and beyond that point it is a river with banks of water. Where it also has a bed of water, as in Figure 6-7B, velocities are distributed differently from those of a river on land, for the "bed" moves slowly with it, and the velocities increase continuously toward the surface. Another characteristic is a remarkable increase in discharge (amount of water passing through a given section) as the stream flows northeastward—from about 1.2 billion cubic feet per second at the east end of the Straits of Florida, to 2.5 billion cubic feet per second off Cape Hatteras, to somewhere between 3 and 6 billion cubic feet per second south of Newfoundland. Surface currents joining the Gulf Stream (as the lines shown in Figure 6-6) do not appear nearly great enough to explain these increases, which suggests that water is being pressed into the stream at depth—perhaps, again, a result of the water mound.

One riverlike characteristic discovered recently is that the stream is sinuous and that its bends increase in size downstream. Donald V. Hansen has described a study in which the bends were tracked from Cape Hatteras for nearly 1,000 miles eastward and were found to travel progressively downstream at rates of 2 to 5 miles per day (Figure 6-8). They also grew progressively in width, becoming complex loops called *meanders* that carried Gulf Stream water for as much as 150 miles on both sides of the stream's average position. Other studies have shown that the loops are occasionally closed and

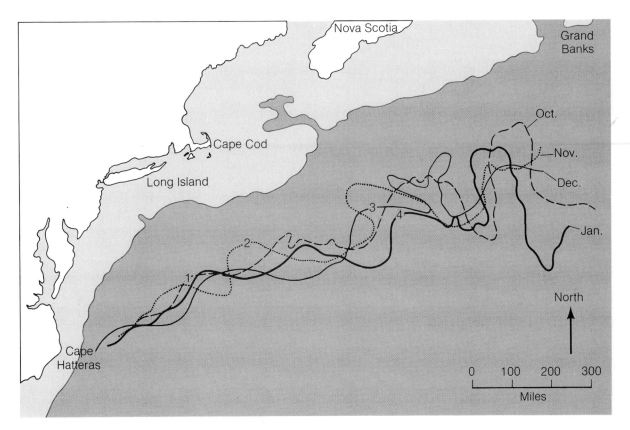

FIGURE 6-8

Charted courses of the center line of the Gulf Stream, mapped during the periods 5–13 October 1965, 9–16 November, 9–17 December, and 7–14 January 1966. Thus, one bend moved through positions 1, 2, 3, and 4 during the four months. From Donald V. Hansen, "Gulf Stream meanders between Cape Hatteras and the Grand Banks," *Deep-Sea Research,* vol. 17, pp. 495–511, 1970.

cut off (Figure 6-9). The isolated rings then continue to rotate counterclockwise, sidling southwestward into the Sargasso Sea for as much as 1,000 miles before their rotational energy is dissipated by friction—1 to 5 years after the rings formed!

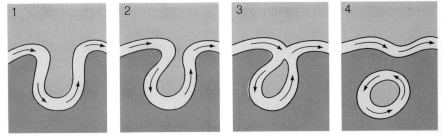

FIGURE 6-9

Steps in the cutting off of a meander loop and the formation of a rotating ring of water, the latter enclosing water originally on the north side of the stream. After Charles E. Parker, "Gulf Stream rings in the Sargasso Sea," *Deep-Sea Research,* vol. 18, pp. 981–93, 1971.

Mixing of specific water bodies Note that the ring in Figure 6-9 mixes Gulf Stream water as well as water from the northwest side of the stream into the Sargasso Sea. This mixing is part of a general movement of waters from the Atlantic shelf to the central ocean, and it can be traced back to its source by means of the compositions and temperatures that serve to label the different waters. Water from the Gulf Stream can be recognized by high salinity (somewhat more than 36 parts of salt per thousand parts of water), high temperatures (varying seasonally between 20 to 28°C), and low dissolved oxygen content. These characteristics are due to the stream's source in equatorial regions, where high temperatures drive out much of the dissolved oxygen and evaporate the water so that salt is concentrated at the surface. Waters on the shelf north of Cape Hatteras, on the other hand, have salinities mainly in the 31–33 parts per thousand range, temperatures that vary seasonally from 4 to 25°C, and an abundance of dissolved oxygen.

Winds from the northeast often force Virginia shelf water southward around Cape Hatteras and down the Carolina shelf for as much as 100 miles. These southward flows, coupled with the northward flow of the Gulf Stream, produce *gyres* (circular currents) over the shelf and slope (Figure 6-10). The cool, relatively fresh Virginia shelf water is thus mixed with the southern shelf water and the two mix partly in the Gulf Stream.

The waters of the shelf may also be carried to the deep basin after temporary incursions of the Gulf Stream onto the southeastern part of the shelf, as detected by repeated measurements of salinities and currents. Measurements of temperature, too, have indicated a shift of cool shelf water into the western edge of the Gulf Stream, such that it either mixes with the stream or moves northward alongside it.

In broad view, these various exchanges are intermediate links in the flow of water from the continent to the deep basin, illustrated in Figure 6-11. The rivers north of Cape Hatteras (for example, the James River) mix with sea water in estuaries, and water from the estuaries mixes on the shelf, as described in Chapters 4 and 5. The shelf water generally shifts southward and

FIGURE 6-10

Flow of water around Cape Hatteras from the Virginia shelf (the short, heavy arrows) on 21–25 June 1966. The consequent circulation on the Carolina shelf is shown by the thin dashed arrows. From Unnsteinn Stefánsson, Larry P. Atkinson, and Dean F. Bumpus, "Hydrographic properties and circulation of the North Carolina shelf and slope water," *Deep-Sea Research*, vol. 18, pp. 383–420, 1971.

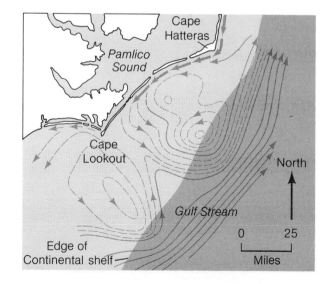

FIGURE 6-11

Generalized flow of waters from the continent to the shelf, and thence to the Gulf Stream, or next to it, and out into the open ocean. The rings formed from the cut off of meander loops, as in Figure 6-9, are drifting southwestward into the Sargasso Sea.

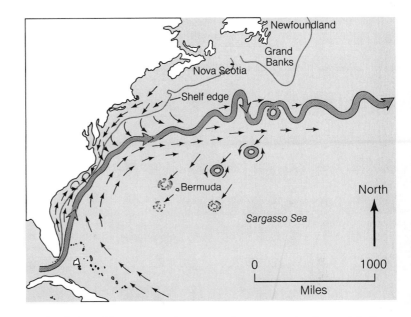

out to sea north of Cape Hatteras, or it crosses the cape and mixes with Carolina shelf water as just described. The waters are then entrained in the stream or along its western margin, coursing out into the Atlantic. They are finally mixed by meanderings of the Gulf Stream or are carried in cutoff rings into the Sargasso Sea. Here evaporation will emplace water in the atmosphere, which will circulate it back to land and thus start another cycle.

Deep Currents and Sediments of the Western Margin

If the Gulf Stream mixes river waters into the deep ocean, does it carry land-derived sediment there too? We noted in Chapters 4 and 5 that fine-grained sediments are trapped largely in estuaries and that samples of sea water from the outer shelf carry extremely little sediment derived from land. Still, the water samples from the shelf were collected at certain times and thus do not disprove occasional movements of turbid shelf water into the open ocean. Figure 6-12 evidently shows such an occasion. Taken from a satellite at an altitude of 134 miles, the picture shows masses of turbid (lighter toned) water that had flowed from Pamlico Sound and from the Virginia shelf to the north (the latter much as in the case illustrated in Figure 6-10). As the map shows, the outer part of the turbid water had moved out over the continental slope and was being swept northeastward in the Gulf Stream. The sediment will thus settle eventually to the bottom of the deep Atlantic.

That the Gulf Stream has long transported sediment to the deeps has been proven by studies of foraminifers carried in it. William F. Ruddiman, for example, has counted the shells (*tests*) of warm-water species in 152 samples of bottom sediments and found that they are concentrated in a band extending into the northern part of the Atlantic under the Gulf Stream (Reference 4). Morever, each sample included materials deposited as much as 2,500 years ago, so that the Gulf Stream has been in about its present position for at least that long.

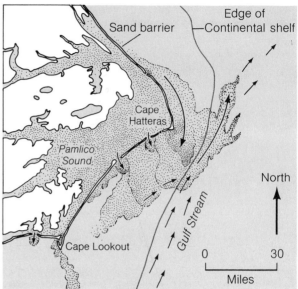

FIGURE 6-12

Atlantic Ocean and estuaries near Cape Hatteras, photographed 12 March 1969 from the Apollo 9 satellite manned by James R. McDivitt, David R. Scott, and Russell L. Schweikart. The pale tones in Pamlico Sound are due to mud stirred up by waves, and the turbid water can be seen leaving the estuary through several tidal channels (short, solid arrows). A surge of turbid water (the longest arrow) has also rounded Cape Hatteras and is pressing the shelf waters out into the Gulf Stream and the open ocean. Photograph courtesy of NASA, Johnson Space Center, Houston.

Evidence of a southward bottom current Ruddiman found, however, that the tests are also concentrated in a band that extends southward from Cape Hatteras. This concentration is due to a deeper current that crosses beneath the Gulf Stream near Cape Hatteras and flows southward, evidently transporting foraminifer tests that fall into it from the Gulf Stream waters. The deep current was first discovered by tracking floats weighted to ride buoyantly at depth, each equipped to broadcast radio signals to receivers on ships. The velocities thus measured were 0.3 to 0.6 foot per second, and measurements made elsewhere in the same current have been similar, locally approaching 1 foot per second. The current forms a diffuse stream, perhaps 100 miles wide, that hugs the bottom but extends upward through some 5,000 feet of water. A cross section of the current at a specific time is shown by the two darker areas under the Gulf Stream in Figure 6-7A.

An especially thorough study of the current and its effects was made by Gilbert T. Rowe and Robert J. Menzies, who took 3,000 photographs of the seafloor at 150 sites off North Carolina (Reference 5). The area extended from the edge of the shelf to the deep *abyssal plain*—a nearly smooth surface sloping very gradually to the lowest part of the basin. Each photograph includes a closeup view of a compass attached to the frame holding the camera, so that the geographic orientations of current-formed features can be determined (Figure 6-13A and B). Still other things showing current directions are trails of turbid water kicked up by the lowering frame, bending of attached, plant-like animals, and attached animals that grow facing the current. Photographs of areas unaffected by the bottom current show delicate burrows and trails (Figure 6-13C). Such features are scoured and smoothed where the current has been active recently.

The sediments themselves were sampled by means of a rigid tube mounted on the lowering frame. The tube plunged into the bottom sediment when the frame was lowered and retained a core (a cylindrical sample of sediment) when the frame was hoisted back to the ship. The cores could generally be pressed out intact and thus studied in various ways, such that the sediment at each site could be identified. This information was combined with the compass directions to give a three-dimensional view of bottom currents and kinds of sediment across the area studied (Figure 6-14). Note that the current extends generally from just under the Gulf Stream to the upper part of the continental rise but that it is divided into separate streams. The fact that the varieties of sediment are distributed in bands parallel to these streams indicates that the sediments were deposited or reworked by them. Rowe and

A

FIGURE 6-13

Indications of current conditions at the bottom of the deep Atlantic. A, ripples in thin layer of sediment over rock (partly exposed) indicating a current moving toward the east (lower right). Depth is 12,300 feet (3,631 meters). B, sediment banked in streamers behind mounds, indicating current moving toward the northeast (lower right). Depth is 17,900 feet (5,260 meters). C, animal tracks and burrows in silt and clay not recently affected by a bottom current. View is 4 feet across and at a depth of 7,200 feet (2,129 meters), 300 miles east-southeast of New York Harbor—in the general path of the south-flowing bottom current. A and B are from Gilbert T. Rowe and Robert J. Menzies (Reference 5.) Photographs courtesy of John G. Newton, Duke University Marine Laboratory. C is from Richard A. Pratt. Atlantic continental shelf and slope of the United States—physiography and sediments of the deep-sea basin. U.S. Geological Survey Professional Paper 529–B, 44 pp., 1968.

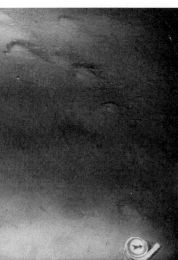

B

C

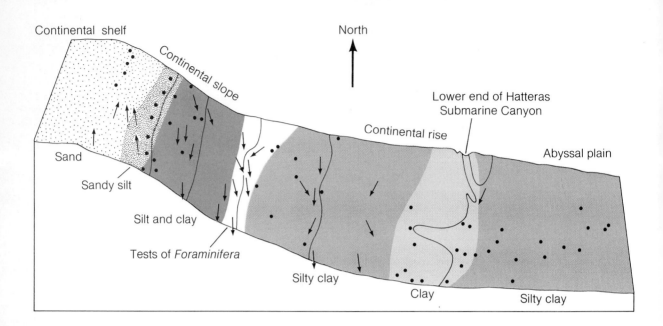

Continental shelf

North

Continental slope

Lower end of Hatteras
Submarine Canyon

Continental rise

Abyssal plain

Sand

Sandy silt

Silt and clay

Tests of *Foraminifera*

Silty clay

Clay

Silty clay

FIGURE 6-14

Part of the seafloor studied by
Rowe and Menzies, with
arrows showing current directions on the bottom and
patterns indicating the extent
of various bottom sediments,
as labeled. The heavy dots
mark photographed sites that
showed no current features.
Note how closely the current
arrows parallel contour lines.
The vertical scale is exaggerated
about 20 times. Data from Rowe
and Menzies (Reference 5).

Menzies also made systematic counts of the kinds of animals visible in the
photographs and found that animal communities form bandlike arrays parallel
to the current. The current evidently provides an ideal environment for
attached and burrowing animals, carrying to them the organic particles that
have fallen from the overlying water and have thus been concentrated along
the bottom.

Source and nutrient-content of the bottom water Data collected by various workers show that the south-flowing current is cold (2.5
to 6°C) and contains 5 to 6 parts per thousand of dissolved oxygen, about
twice as much as surface waters of the southwestern Atlantic. It is also less
saline than the surface waters, containing 34.9 parts of salt per thousand parts
of water compared to about 36 parts per thousand near the surface. Surface
waters in seas near the Arctic develop these characteristics, and the current
has indeed been traced back for 2,500 miles to the Norwegian Sea (Figure
6-15). Waters there are chilled until they become so dense that they sink to
the bottom and flow southeastward (downslope) into the Atlantic basin. The
earth's rotation then sends the water toward the right (west) side of the basin,
causing it to follow an approximately level course along the east side of the
Reykjanes Ridge, then through a gap between that ridge and the Mid-Atlantic
Ridge (Figure 6-15). From there the water flows to the continental slope and
rise of the western Atlantic, always moving along level (horizontal) lines because the earth's rotation presses it against the side of the basin.

The upward press of the deep water is of tremendous importance to sea
life, for it moves dissolved nutrients toward the surface, where they can nur-
ture oceanic plants that are the basis of the ocean's food chain. Recall that
the waters of the Gulf Stream are barren of nutrients and plankton where
they enter the southwestern edge of the ocean. As the Stream sweeps
northward, deeper waters pressed into its lower parts gradually add nutrients

Explain better

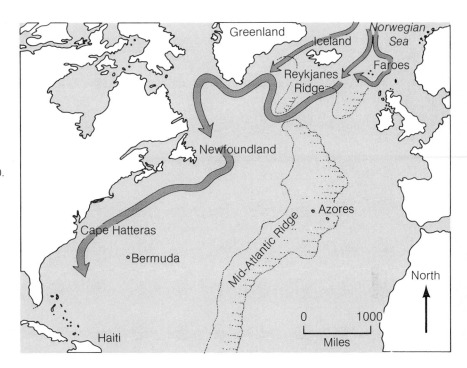

FIGURE 6-15

Course of Arctic waters that sink to the bottom of the Norwegian Sea and form a deep current around the northern and western perimeter of the Atlantic basins. Chiefly from L. V. Worthington, "The Norwegian Sea as a mediterranean basin," *Deep-Sea Research*, vol. 17, pp. 77–84, 1970.

to it. The principal gain of nutrients, however, takes place when the surface waters sink at the northern end of the system and flow slowly back along the bottom, dissolving substances from organic particles on the bottom or falling to the bottom. The renutrified water then mixes upward and the mixtures are carried onto the shelf by intrusions. Carried thus to the lighted layer of the ocean, the water nurtures plants and thence animals. The fisheries on the outer parts of the shelf and on the continental slope depend on this source.

Transport of Sediment to the Deep Basin

The deep Atlantic is divided into two halves by the Mid-Atlantic Ridge (Figure 6-15). The broad subsea mountain range acts as a barrier to bottom currents, such that sediments carried eastward from the North American continent are not intermixed with those from Europe and Africa. The ridge further simplifies interpretations of the sediments near the center of the ocean by providing a hard rock base on which they have accumulated. The hills are volcanoes, most of which are now inactive but formerly erupted lava called *basalt* (Supplementary Chapter 2). The skin of each lava flow cooled and congealed because of contact with sea water, but molten lava burst repeatedly through this shell in spurts and tongues that solidified into forms called *pillows* (Figure 6-16). The forms are exceedingly useful to oceanographers who must examine the sea bottom from 10,000 feet above it, for the pillows can be recognized easily in photographs (Figure 6-16). Note the striking contrast between the dark pillows of basalt and the sediment that has fallen on them and drifted like snow among them.

135

FIGURE 6-16

Left, steps in the emission of lava pillows as a flow moves slowly from left to right.
Right, view downward to the seafloor showing basalt pillows partly buried by younger sediment. Near the center of the Mid-Atlantic Ridge, at 36° 38.4′ N latitude and 33° 36.2′ W longitude. Photograph by Woods Hole Oceanographic Institution, courtesy of Elazar Uchupi.

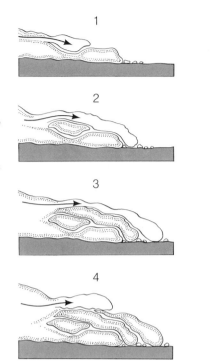

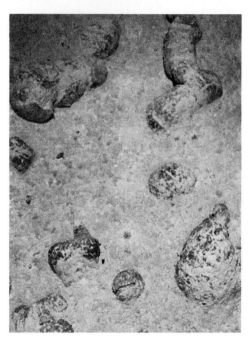

Pelagic sediment of the western basin The sediment in Figure 6-16 is said to be *pelagic*—literally "made in the ocean." The principal constituent in this view is calcium carbonate, occurring chiefly as tests of foraminifers (Figure 6-17A) and smaller (dust-size) coccoliths, the latter being hard parts of microscopic plants of the plankton (Figure 6-17B). The plankton also includes organisms that secrete shells of opaline silica (SiO_2): sponges, single-celled animals called radiolarians, and minute plants called diatoms (Figure 6-17C). These siliceous particles make up perhaps 2 to 15 percent of the sediment in the North Atlantic. All the microscopic plants and many of the small animals that yield the sediment live in the upper few hundred feet of the ocean, and although their total mass is very large (because of the size of the ocean), their mass in each cubic foot of water is small. Their solid remains therefore accumulate slowly on the seafloor, typically at 0.4 inch to 1.2 inches (1 to 3 cm) per 1,000 years. The rates of accumulation vary because some parts of the sea are more productive than others, due to amounts of nutrients, temperatures, and stirrings by currents near the surface. If you examine Figure 6-18 carefully, you can find areas where calcium carbonate has been concentrated under the path of the Gulf Stream. Another cause of the variations shown in Figure 6-18 is the solution of calcium carbonate particles in the deep sea. The cold waters that dominate the ocean below depths of 12,000 to 15,000 feet contain carbon dioxide (CO_2) derived from decaying organic materials. The water is thus somewhat acidic and slowly dissolves calcium carbonate. As Figure 6-18 shows, this solution results in a

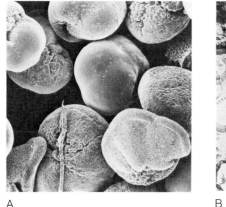
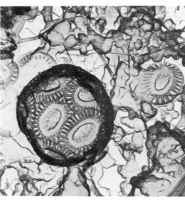
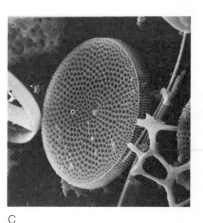

A   B   C

FIGURE 6-17

A, foraminifer shells (tests) forming ''Globigerina ooze,'' the typical pelagic sediment of tropical waters. The shells are 0.5 to 1.0 millimeter across. Courtesy of James C. Ingle, Jr.; scanning electron microscope image by Gerta Keller. B, coccosphere of the living Coccolithophorid *Emiliania huxleyi,* from the Gulf Stream. The sphere will eventually break up into separate plates (coccoliths) like those near it. The other materials are protoplasm and detritus of the plankton. Taken with a scanning electron microscope, the objects are enlarged about 6,000 times. Courtesy of Doctors Hisatake Okada and Andrew McIntyre of Lamont-Doherty Geological Observatory of Columbia University. C, *Coscinodiscus noduliter,* a common Atlantic diatom measuring approximately 0.1 millimeter across. Scanning electron microscope image courtesy of Lloyd H. Burckle.

general decrease of calcium carbonate with depth, such that the sediments of the abyssal plains typically contain less than 1 percent.

Where calcium carbonate has been dissolved, the sediment is mainly clay, with varying amounts of the siliceous skeletons of diatoms and radiolarians. The clays on the low rises within the ocean basin (such as the one around Bermuda in Figure 6-18) consist of grains that have fallen into the sea from the atmosphere: wind-blown dust from the continents, fine particles erupted from volcanoes, and fragments of meteorites. The clay on the abyssal plains and on the seafloor between the plains and the continent, however, has probably been transported within the sea from land sources. Several kinds of evidence support this: (1) the sea bottom slopes quite consistently from the North American continent to the far edge of the abyssal plains, gradually flattening in that direction; (2) isolated undersea volcanoes rise abruptly from the plain, their lower parts covered by the surrounding sediment; and (3) the pelagic sediment that has fallen on top of the volcanoes contains far less clay than do the sediments on the abyssal plains around the volcanoes. These relations suggest that clays from the continent have been transported down the gently sloping seafloor by currents hugging close to it.

Transport related to submarine canyons   The problem is how sediment can be carried down slopes of only a foot or two per mile for distances of 800 to 1,500 miles! The difficulty is compounded by the fact that the only persistent bottom currents known in the deep ocean are those that move paral-

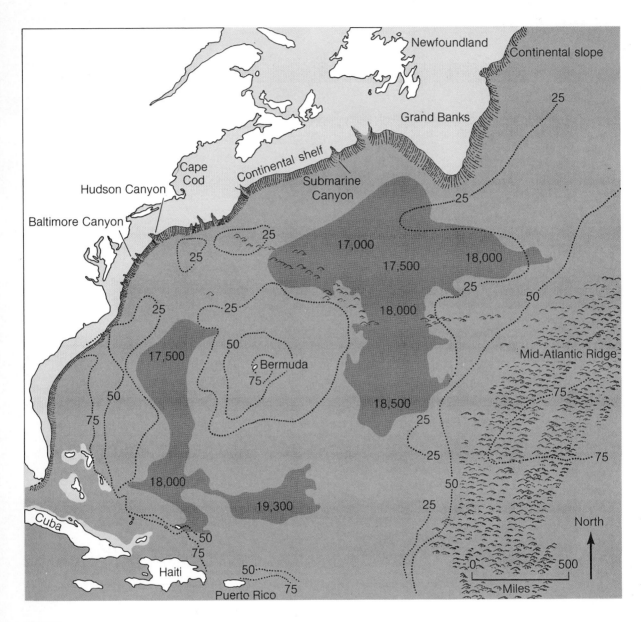

FIGURE 6-18

Map of the western Atlantic basin (the North American basin) showing the broadest features of the topography and the amounts of calcium carbonate in the bottom sediments (given in percentages by the numbers on the dotted lines). The dark gray areas are the abyssal plains with approximate depths indicated in feet. Note the many indentations (submarine canyons) in the edge of the continental shelf. After Richard M. Pratt, ''Atlantic continental shelf and slope of the United States—physiography and sediments of the deep-sea basin,'' U.S. Geological Survey Professional Paper 529–B, 44 pp., 1968.

lel to the western boundary, described in the last section. The first solutions to these problems were a result of interest in features that might seem unrelated —the submarine canyons that indent the continental slope (Figures 5-10 and 6-18). When the canyons were fairly thoroughly charted in the 1930s, a major puzzle arose because many were found to connect to major rivers on land. How might rivers control the positions of canyons cut deep under the sea?

An unusually perceptive scientist, Reginald A. Daly, suggested a mechanism in considerable detail (Reference 6). He reasoned that during the glacial periods, when the sea had fallen to a level near the top of the continental slope, rivers carried abundant sediment from melting glaciers to the edge of the slope, depositing it there in great undersea mounds. The deposits were unstable because of the steepness of the slope and occasionally slid down it and mixed with sea water. This turbid mixture was more dense (heavier) than clear water and therefore flowed downslope along the seafloor. Daly proposed that such flows gradually eroded submarine valleys into the side of the basin and carried sediment out onto the abyssal plain.

As Daly noted, this kind of underwater flow was known from observations of muddy waters of the Rhone River that plunged to the bottom of Lake Geneva and moved for many miles as a bottom current, driven by their greater density and by the gentle slope of the lake bottom. He could also have cited many cases of similar underwater flows along the bottom of Elephant Butte Reservoir on the Rio Grande, which had been described by that time. These sorts of flows came to be known as *turbidity currents*, meaning that their greater density was caused by suspended sediment.

Turbidity currents in the deep sea Some persons argued against the idea, noting that the cases in lakes and reservoirs were miniscule compared to those suggested for the Atlantic basin—that the velocities and turbulence of diffuse turbidity currents would not be great enough to carry sediment over the distances required. But the idea seemed logical in other ways, and two oceanographers, Bruce C. Heezen and Maurice Ewing, reasoned that large turbidity currents in the deep ocean might be recognized because they would break submarine telephone cables as they swept over them. They searched company records and found many cases of breaks that might well have been caused in this way. A particularly convincing case was caused by an earthquake in 1929 on the upper continental slope south of Newfoundland (Figure 6-18). The quake's immediate effect was a landslide or slump measuring some 1,300 feet in thickness and 3,600 square miles in areal extent (Figure 6-19). The slide instantaneously broke the upper five cables shown in Figure 6-19 and then generated a turbid flow of some sort that swept down the continental rise and onto the abyssal plain, breaking five more cables in a timed succession. Even by conservative reckoning of its course, the current plummeted at least 65 feet per second (44 miles per hour) down the steepest slope and was still moving at 20 feet per second 400 miles from the base of the slide, from which point it continued for an unknown distance, presumably at lesser and lesser speeds. Cores taken afterward at points 1 and 2 in the figure show a surface deposit of silt, 2 to 4 feet thick, that is somewhat coarser at the base than at the top. The coring tube could not penetrate the deposit at points 3, 4, and 5, but

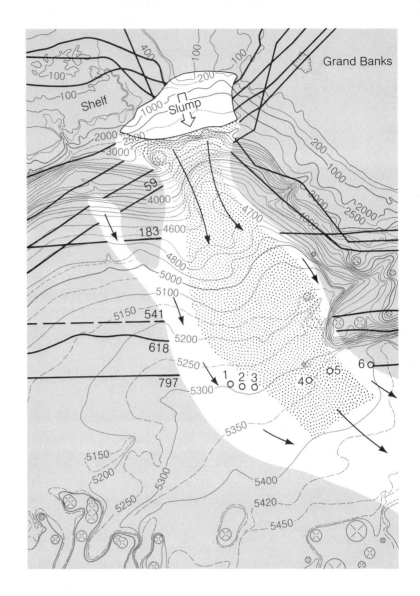

FIGURE 6-19

Map of the seafloor south of Newfoundland, showing the slump, the turbid-
flow deposits, and the telephone cables
(heavy lines) broken by the flow. The
numbers above the cable lines are the
numbers of minutes between the slumping
and the breaking of the cables as the
flow swept past. Samples collected at
the circled stations indicate that the
central part of the deposit is sand
(dotted) and the outer part mainly silt.
Based on figures in *The face of the deep*
by Bruce C. Heezen and Charles D.
Hollister (New York: Oxford University
Press, 659 pp., 1971). Topography from
Richard A. Pratt, ''Atlantic continental
shelf and slope of the United States—
physiography and sediments of the
deep-sea basin,'' U.S. Geological Survey
Professional Paper 529–B, 44 pp., 1968.

the materials recovered indicate that it consists of firm sand. Because the
sediment elsewhere in this part of the basin is typically clay, the deposit of
sand and silt proves that a large mass of sediment was transported here over
the abyssal plain.

Was this a once only event? Perhaps other turbid flows move less swiftly,
but study of many deep-sea cores has shown numerous layers of sand and silt
interlayered with typical deep-sea clay, some of them several inches to several
feet thick. Many deposits contain remains of land plants and shallow-water
organisms that indicate a source near the margins of the ocean. Most layers
have sharply defined boundaries against the underlying clay and grade upward
from coarser grains near their base to finer ones near the top of the layer,
finally to clay. This gradation is suggestive of sediment that traveled as a
suspension from which the largest grains fell first, the medium ones next, and

the finest last (recall from Chapter 1 how the settling velocities of grains vary with their sizes).

Turbidity currents from debris flows   Although these observations support the idea of large-scale turbidity currents, cautious oceanographers have noted difficulties with the entire concept. For one, attempts to generate turbidity currents by dislodging sediment on the sloping seafloor have not been successful. Another difficulty is that some flows have carried particles as large as pebbles for hundreds of miles across a gently sloping floor, a phenomenon that would not be possible for the diffuse turbid flows observed in reservoirs. Some deposits are also totally chaotic—unsorted and ungraded by size—and this too is not possible for diffusely turbid flows.

Some promise of resolving these difficulties has come from observations and experiments by Monty A. Hampton, who noted that landslides of thoroughly wetted materials, on land, disaggregate into mixtures of mud and coarser materials called *debris flows* (Reference 7). These flows are much like wet concrete, moving readily downslope but having enough "body" (viscosity) to keep large fragments from sinking in them. The greater the thickness of the flows, the faster they move. Judged from the velocities of thin flows measured on land, a debris flow 100 feet thick would travel down a slope of 5° at a velocity somewhere between 20 and 50 feet per second. Using a trough that could be immersed in a large tank, Hampton experimented with small debris flows in water. Clay-rich slurries poured into the upper end of the trough flowed down it as underwater debris flows. The significant discovery, however, was that the sweep of water around the bullet-shaped snouts of the slurries generated a diffuse turbid cloud that became a turbidity current, flowing down the trough after the debris flow (Figure 6-20).

The experiments, coupled with other observations, suggest this overall mechanism for the Atlantic's deposits: (1) most submarine turbidity currents start as debris flows (initially landslides) on the steep margins of the basin; (2) large debris flows move swiftly down submarine canyons because the canyon walls contain them as thick bodies; (3) those flows large enough to move out onto the rise carry coarse materials far into the basin, locally forming chaotic layers; (4) each debris flow generates a turbidity current that carries fine materials beyond the ends and sides of the debris flow; and (5) numerous but small debris flows have no more effect than to send diffuse turbidity currents out on the lower rise and abyssal plain, and these turbidity currents might build up the predominantly fine sediments in those parts of the basin.

These speculations round out our look into the deep sea, for we have traced sediment and water movements as far downslope as we can go. It is interesting, in hindsight, that the shifts of materials from the upper ends of river systems to the deepest ocean have generally been driven by gravity, although local water movements capture large amounts of sediment en route—on floodplains, in estuaries, and on the upper part of the continental rise. We may also note broad similarities between water systems on land and in the deep sea. Both have their year-to-year, measurable processes, such as the gradual erosion of soil on land and the slow rain of organic and atmospheric particles to the seafloor. These sediments result from the average yearly flows within river systems and the predictable sweep and drift of air and oceanic currents. But both

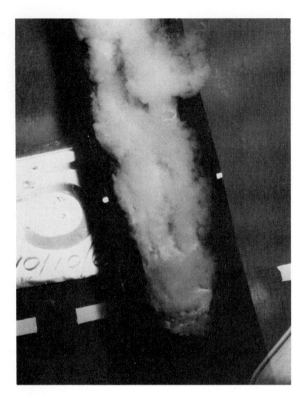
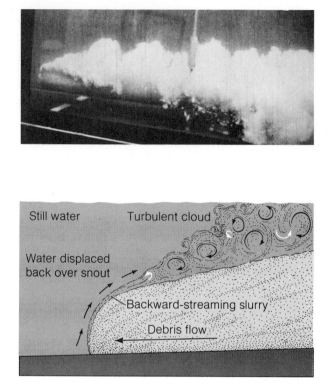

Still water   Turbulent cloud

Water displaced
back over snout

Backward-streaming slurry

Debris flow

FIGURE 6-20

Top view (left) and side view
(upper right) of one of
Hampton's clay-slurry flows in
an underwater flume, showing
the turbid cloud it generated.
The drawing shows a vertical
section through the snout of
the flow, with a thin layer of
slurry that is swept up and
back—mixing with water to
form the cloud of the turbidity
current. Photographs and
data courtesy of Monty A.
Hampton.

kinds of systems also have cataclysms of sliding and chaotic mixing that move
giant batches of suspended materials for hundreds of miles. Because these
events are infrequent on land and because we cannot easily observe them in
the deep sea, we do not know whether or not they are dominant. As we have
seen, the very nature of the seabottom flows remains puzzling. Like the great
floods of rivers on land, however, we must understand them if we are to use
the sea's resources as fully as seems necessary.

Summary

Because we have covered a large area and a variety of features and movements,
it may be useful to summarize the more important relations in the western
Atlantic Ocean:

1. Waters in the upper part of the ocean are put in motion by the atmosphere,
   which circulates due to solar heating, to gravity (driving vertical movements of light and heavy air masses), and to the earth's rotation.
2. The prevailing winds force the upper water into a low mound in the west-
   central part of the ocean, from which water is pressed outward and is thus
   caused to circulate to the right (clockwise) by the earth's rotation.
3. This circulation is concentrated along the western side of the ocean in
   the Gulf Stream, which flows so swiftly and meanders so widely as to

entrain water from the shelf and mix it into the central part of the ocean.

4. The Gulf Stream also carries continental sediment over the deep basin, and much of the sediment that sinks from the stream is carried southward in a bottom current that hugs the continental slope and the upper part of the continental rise.

5. The bottom current is cold and is enriched with oxygen because its source is in the surface waters of the Arctic (the Norwegian Sea). It also gains plant nutrients as it flows slowly around the western margin of the basin, being pressed along that course by the earth's rotation.

6. The mixing of these waters up into the Gulf Stream and onto the continental shelf is vital to the sea life in those regions.

7. The sediment in the deeper parts of the basin consists of plant and animal shells and skeletons that have fallen chiefly from the plankton and of mineral grains from the continent that have been carried great distances by the wind and by bottom currents.

8. The bottom currents that move sediments to the deep basin are not entirely understood, but many appear to start when sediment high on the continental slope slumps and forms debris flows that pour down submarine canyons and generate turbidity currents on the continental rise and abyssal plains.

9. In the sea as on the land, gravity thus drives cataclysmic movements as well as gradual continuous ones; the cataclysmic events seem unusual but may be dominant in the long run.

REFERENCES CITED

1. Panel on Operational Safety in Marine Mining. *Mining in the outer continental shelf and in the deep ocean.* National Academy of Sciences, Washington, D.C., 119 pp., 1975.

2. John D. Isaacs and Richard A. Schwartzlose. Active animals of the deep-sea floor. *Scientific American,* vol. 233, pp. 85–91, October 1975.

3. Henry Stommel. *The Gulf Stream, a physical and dynamical description,* 2nd ed., pp. 141–42. Berkeley and Los Angeles: University of California Press, 248 pp., 1965.

4. William F. Ruddiman. Historical stability of the Gulf Stream meander belt: foraminiferal evidence. *Deep-Sea Research,* vol. 15, pp. 137–48, 1968.

5. Gilbert T. Rowe and Robert J. Menzies. Deep bottom currents off the coast of North Carolina. *Deep-Sea Research,* vol. 15, pp. 711–19, 1968.

6. Reginald A. Daly. Origin of submarine "canyons." *American Journal of Science,* vol. 231, pp. 401–20, 1936.

7. Monty A. Hampton. The role of subaqueous debris flow in generating turbidity currents. *Journal of Sedimentary Petrology,* vol. 42, pp. 775–93, 1972.

ADDITIONAL IDEAS AND SOURCES

1. Excellent descriptions of deep-sea processes, illustrated by many photographs of the seafloor, are given by Bruce C. Heezen and Charles D. Hollister in *The face of the deep* (New York: Oxford University Press, 659 pp., 1971). The deep southerly current along the western edge is especially well described, and another interesting subject that is discussed is the origin of the

great fields of manganese-rich nodules on the seafloor—one of the unusual mineral resources of the ocean basin.

2. The occurrences and interrelations of living forms in the sea are presented in an unusually complete and interesting article by John D. Isaacs, "The nature of oceanic life," in the September 1969 *Scientific American* issue on the oceans.

3. Another article in the *Scientific American* just mentioned, "The atmosphere and the ocean," by R. W. Stewart, gives a clear, well-illustrated description of the Coriolis effect and how the winds generate the oceanic circulation.

4. Drilling from the specially designed ship *Glomar Challenger* has shown that the upper continental rise of the western Atlantic is a broad mound of muddy sediment banked against the side of the basin. Data obtained by drilling in it off the central Atlantic states are reported in *Initial reports of the deep-sea drilling project*, Volume XI, by C. D. Hollister, J. I. Ewing, and others (U.S. Government Printing Office, Washington, D. C., 1077 pp., 1972). A drill hole going to a depth of 3,439 feet into the rise showed that the lower 2,264 feet consist of clayey muds deposited between approximately 25 million and 3 million years ago and that the upper 1,175 feet are coarser (silty and sandy) muds deposited in the past 3 million years, the period marked by the repeated glaciations mentioned in Chapter 4. The comparatively rapid rate of deposition during the glacial period appears to correlate with rapid erosion of the northern part of the continent and shelf at the same time. The sediment forming the rise thus appears to have been transported southward from the glaciated region, indicating that the southerly bottom current has been active for at least 3 million years.

Evidence of the last great ice sheet—hills of sediment in northern Saskatchewan that were streamlined by ice flowing over them from upper right to lower left (northeast to southwest), as recently as 9,000 years ago. The view is 7 miles across. Original photograph supplied by the Surveys and Mapping Branch, Department of Energy, Mines, and Resources, Ottawa.

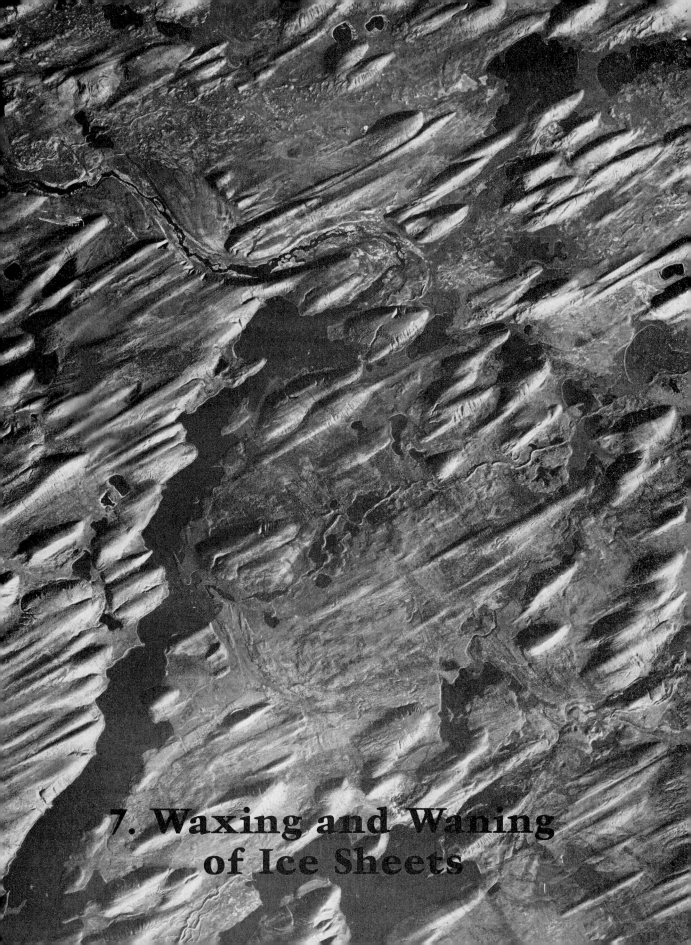

7. Waxing and Waning
of Ice Sheets

Waxing and Waning of Ice Sheets

We have already noted how glaciers have affected oceans and estuaries, and we will see their broad effects on other earth systems in several chapters that follow. In this chapter we will explore major glaciers in their own right—what they are, how they work, and why they might have waxed and waned repeatedly in the recent geological past. Their nature as active systems is depicted superbly in Figure 7-1, which shows the most northerly tip of Canada—Ellesmere Island—in July 1950. The message is undeniable: snow and ice have accumulated on the mountain ranges like frosting on giant cakes and are spilling outward and downward through valleys in a series of ice streams. Three of the streams have flowed so voluminously that they have formed great puddles (*lobes*) of ice on the floor of a valley beneath.

Where the snow and ice in the picture are highly (almost strangely) reflective, as in the lower left part of the view, snow has accumulated faster than it has evaporated and melted (*ablated*). The ice streams, on the other hand, are somewhat grayish and are scored by numerous cracks called *crevasses*. The change in tone is due to the winter's snow having ablated totally from the ice streams and lobes, exposing the underlying solid ice. Note that the change takes place at about the level at which the surrounding slopes are generally free of snow in summer. The system is thus divided into an accumulating upper part and a dwindling lower part, much as the Rio Grande starts in well-watered mountains but becomes a dwindling trunk river. One can even note specific analogies to the distribution of moisture in river systems.

FIGURE 7-1
The Viking ice cap and surrounding glaciated terrain of Ellesmere Island, Northwest Territories. Part of a second ice cap, in the foreground, lies directly under the aircraft. Original photograph supplied by the Surveys and Mapping Branch, Department of Energy, Mines, and Resources, Ottawa.

For example, ice streams flowing from the north-facing slopes of the accumulation area (the three with ice lobes) bring far more ice to the valley floor than streams flowing down the opposite slope, which faces the summer sun.

The moving ice is also like a river in that it erodes its bed and walls. The erosion is indicated by the dark bands around the lower ends of the ice streams, which consist of rock debris carried from the mountains and released as the ice wastes away. Sediment deposited from ice is called *till*; a deposit that forms low ridges around the ends of glaciers (an especially nice one can be seen around the lobe on the right) is called an *end moraine*.

The air on Ellesmere Island is so dry and cold that much of the snow and ice simply evaporates, but enough melts in summer to form streams that can be seen flowing along the ice-free parts of the valley. The streams carry sediment from the glaciers and deposit part of it on the valley floor to build up a flat surface called an *outwash plain*. If you examine the valley floor to the left of the large ice lobe near the center of the photograph, you will find another feature typical of glacial valleys—a lake dammed behind the ice lobe and frozen white except at its edges. If you look closely at the far side of the lake you can see a pale gray band—a high-water mark showing that the lake was formerly deeper and therefore that the glacier was larger sometime in the recent past.

Flow in Ice; Formation of Ice from Snow

The most intriguing parts of this glacier system are the ice streams. They constitute a paradox basic to all glaciers: ice is solid yet it can flow! Indeed, if you were to stand on one of the streams it would not seem at all like soft frosting on a cake. The angular crevasses and pinnacles and the occasional groaning under your feet would more likely impress you with the brittleness of ice. Down beneath, nonetheless, the ice would be flowing as though it were a very viscous (nearly stiff) liquid.

Ice flow depends on the bonding among its constituent molecules. These molecules form layers in which the atoms are repeated in hexagonal (six-sided) groups, an arrangement that gives rise to the hexagonal symmetry of snow and ice crystals (Figure 7-2A and B). The layers in the crystals are stacked on top of one another with the six-sided groups superimposed (Figure 7-2C). The bonding between layers is weaker than that within each layer. Thus, when a crystal is pressed hard, as suggested by the arrows in Figure 7-2D, each layer remains intact but slips over the layer beneath. If the slip takes place in certain directions, the displaced layers can end up with their six-sided groups again superimposed (Figure 7-2D). Repeated displacements of this sort permit ice to flow without changing the solid, crystalline (orderly) nature of its constituent grains. This is called *solid-state flow*. It is the sort of flow that permits metals to be drawn out into wire or be hammered into various shapes. As we will see later, it also permits some rocks to flow without melting.

Many glaciers also move by sliding on their beds. The melting temperature of ice is lowered by increasing the pressure. Thus, the deepest parts of a glacier may be at the melting point even though they are colder than 0°C. Moreover, temperatures deep in the ice may be raised slowly by solid-state flow (an effect

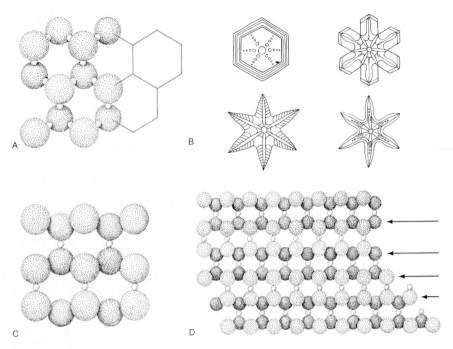

FIGURE 7-2

A, geometric arrangement within a layer of H_2O molecules in ice. The large spheres represent atoms of oxygen and the small ones hydrogen. B, snow crystals, oriented parallel to the atomic arrangement in A. C, edge view of three layers like that of A. Note that there are fewer hydrogen atoms between the layers than within them, with the result that the layers are not held together as strongly as are the atoms within the layers. D, edge view of many layers like those of C, showing how applied forces (arrows) may cause the layers to glide laterally over one another, but that they may end up in the same geometric relation as before.

much like frictional heating), and heat may also be added from the earth beneath. The combined effects often result in local melting, especially where pressures are concentrated at the upstream faces of obstacles ("bumps") in the bed. The ice in contact with such obstacles melts and then freezes again on the downstream side of the obstacle because the pressure there is reduced. The overall effect is to permit the overlying glacier to move along unimpeded. Only a few drill holes and ice tunnels have been used to measure this type of flow. However, most glaciers (all but the coldest) probably owe something close to half their flow to melting and freezing at the base and thus to sliding on their beds.

Conversion of snow to ice   Figure 7-1 suggests another basic question: how does the loose snow turn into solid ice in the first place? Two processes are involved. The dominant one in very cold environments is the compaction and recrystallization of dry snow under the increasing load of subsequent snowfalls. The load compresses the buried snow, drives air out, and deforms the intricate snow crystals so that they tend to recrystallize to simple equidimensional grains. Deeper burial causes the aggregates to become more and more tightly interlocked, finally forming solid ice that typically contains small air bubbles as relics of its snow source. The slowness of this process was demonstrated in a cold part of the glacier that covers most of Greenland by a drill hole that did not reach ice until a depth of 250 feet—a part of the deposit that is at least 100 years old.

The other process by which snow is changed into ice is common in moderately cold environments, as in the lower parts of the snow fields shown in Figure 7-1. The snow melts at the surface during summer and refreezes after it has percolated down into the packed snow. These changes totally rearrange the snow crystals into solid ice, generally within a few years after deposition and often within a few feet of the surface. Both processes go on simultaneously in many glaciers, dry compaction being dominant in the higher, colder parts of the accumulation area and melting and freezing in the lower parts.

A Major Ice Sheet

Imagine the ice cap in Figure 7-1 enlarged ten thousand times in surface area and thickened so that its upper surface would lie 1 to 2 miles above the highest peaks. Such is the Greenland Ice Sheet, the last relic of similar ice sheets that once covered much of the land area of the northern hemisphere (Figure 7-3). Note that the Greenland Ice Sheet lies approximately in the center of the array of large former sheets, a location that suggests conditions there should be

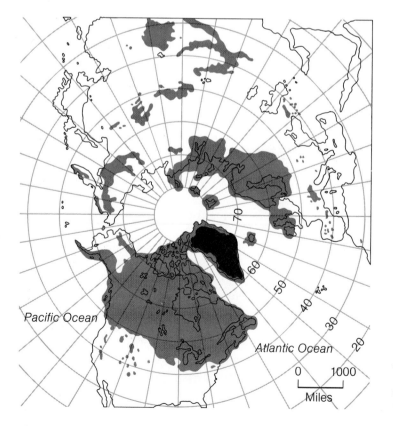

FIGURE 7-3

Pole-centered map of the northern hemisphere, showing the position of the Greenland Ice Sheet (black) relative to major glaciers of the last glacial period, shown at their maximum extent (gray). After Figure IV-6 in *Snow and ice on the earth's surface* by Malcolm Mellor (Cold Regions Research and Engineering Laboratory publication II-C1, 1964).

significant in interpretations of all of them. Indeed, studies of these conditions have proven crucial in understanding former ice sheets as well as glaciers in general, and many people have been involved in such studies. Many have been made by glaciologists and engineers associated with the Army's Cold Regions Research and Engineering Laboratory, and much of what follows is based on their findings.

Mapping of the Greenland Ice Sheet has shown that its surface rises gradually in the form of an elongate, somewhat irregular shield (Figure 7-4). The lower of the two cross sections is at natural scale and shows how thin the giant ice mass is compared to its breadth. The upper cross section is exaggerated in its vertical dimension to show that the surface steepens toward the margins to form a shape typical of large masses of viscous materials lying on more-or-less flat surfaces (the ice lobes in Figure 7-1 probably have similar shapes). The Greenland Ice Sheet is asymmetrical because it is held back by a large mountain range along its eastern margin and by lesser mountains in some other sectors. As suggested by the cross section in Figure 7-4, the mountains cause the sheet to be extra thick for its extent, which is probably one reason why it has survived long after the other great ice sheets disappeared. (They extended over flatter surfaces.) Some of the Greenland ice flows to the sea in *outlet glaciers*, which are thick ice streams that traverse deep valleys passing through the mountains, much as the ice caps in Figure 7-1 spill out through mountain valleys (Figure 7-5). Much of the Greenland ice ablates before reaching the sea, however, and some parts of the sheet's margin are ice cliffs or smooth slopes on land (Figure 7-5).

Effects of precipitation and temperature Outward flow and ablation are counterbalanced by precipitation, which varies from one part of the ice sheet to another (Figure 7-6A). Although all the precipitation is in the form of snow except for minor rainfalls near the coast, the map shows it as inches of water to permit comparisons with other parts of the world. On this basis, the wettest parts have a rate of precipitation about equivalent to that of the eastern seaboard of the United States and the driest to that of the more arid American deserts. The high rate of precipitation on the southern half of Greenland is probably due to moist air masses that move northward from parts of the Atlantic warmed by the Gulf Stream and are cooled as they rise over the ice sheet. Other important sources of snow are winds that sweep eastward across northern Canada, gather moisture from the Labrador Sea, and intersect Greenland at intermediate elevations along its western slopes, especially in the vicinity of Thule. These transfers of moisture from the north Atlantic system must be very important, for the ice sheet extends far south of other cold regions that have little if any ice (Figure 7-3). Because the former great ice sheets also centered around the north end of the Atlantic, the air currents and water exchanges linked to that ocean must have been essential to those glaciers also.

In addition to a dependence on precipitation rates, the survival of snow and ice is linked to temperature, which decreases with increasing altitude and increasing latitude (nearness to the pole). As Figure 7-6B shows, temperature is the principal control on patterns of ablation of the annual snowfall in Greenland. In the coldest parts of the ice sheet, snow melts only rarely, and in the warmest parts, along the coasts, it melts completely by the end of the

FIGURE 7-4

The Greenland Ice Sheet and ice caps of neighboring areas. The contours show elevations above sea level in meters, and the cross sections are aligned roughly east-west through Station Centrale. Cross sections based on data of H. Malzer and B. Brockamp in "An analysis of the relation between the surface and bedrock profiles of ice caps" by W. F. Budd and D. B. Carter, *Journal of Glaciology,* vol. 10, pp. 197–209.

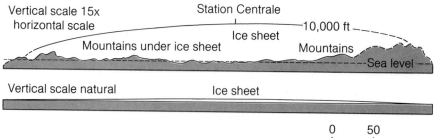

FIGURE 7-5

Left, outlet glaciers of the Greenland Ice Sheet flowing into fjords eroded when the glaciers were more extensive. Photograph courtesy of Army Cold Regions Research and Engineering Laboratory. Right, margin of the Greenland Ice Sheet near Thule. The mounds of rock fragments are an old end moraine. Detritus has been distributed by melt water streams across the outwash plain in the foreground. Photograph by Mel Griffiths.

FIGURE 7-6

A, amounts of precipitation shown in inches of water per year. From Steven J. Mock, "Calculated patterns of accumulation on the Greenland Ice Sheet," *Journal of Glaciology,* vol. 6, pp. 795–803. "Snow accumulation studies on the Thule Peninsula, Greenland," *Journal of Glaciology,* vol. 7, pp. 59–76, 1968. B, average annual temperatures in degrees Celsius. *White*—snow rarely melted in summer; *light gray*—snow melts during summer and melt-water refreezes in snow; and *darker gray*—all snow and some ice ablated during summer. From S. J. Mock and W. F. Weeks, "The distribution of 10 meter snow temperatures on the Greenland Ice Sheet," *Journal of Glaciology,* vol. 6, pp. 23–41, 1966. Data of C. S. Benson reported in *The Greenland Ice Sheet* by Henri Bader (Cold Regions Research and Engineering Laboratory publication I-B2, 1961).

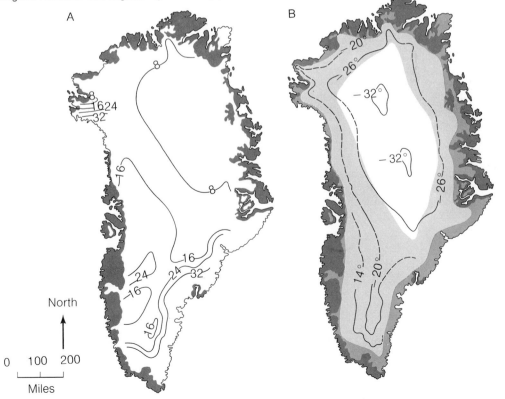

153

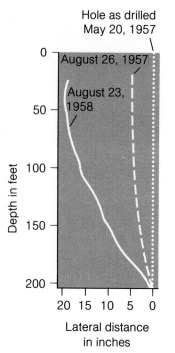

Hole as drilled
May 20, 1957

August 26, 1957

August 23, 1958

Depth in feet

Lateral distance
in inches

FIGURE 7-7

Vertical section through the
ice sheet near Thule, showing
deformation of a drill hole due
to flow of the glacier. From
a figure in *Snow and ice
on the earth's surface* by
Malcolm Mellor (Cold Regions
Research and Engineering
Laboratory publication II-C1,
1964) based on a report by
S. D. Wilson, 1959, for the
Army Corps of Engineers.

summer. In the broad areas between (light gray in the figure), the degree of summer melting increases gradually downslope and toward the southern end of the sheet.

Rates of glacier flow   The glacier has been about the same size for several thousand years, so losses by ablation must be balanced by snow accumulation and outward flow. Rates of outward flow, in turn, vary with the thickness of the ice and with its temperature. Where the outer part of the glacier is only 200 feet thick and frozen to its bed, near Thule, a bore hole has shown that the rate of lateral flow increases gradually downward (Figure 7-7). The overall effect is to move objects on the surface of the glacier laterally at rates of about 0.3 inch per day. At Site 2 (labeled on Figure 7-4), the ice is 6,800 feet thick, and its upper surface is moving westward 1 to 2 inches per day. The upper surface west of Station Centrale, where the ice is about 10,000 feet thick, is evidently moving at roughly 16 inches per day (Reference 1). Perhaps this sector of the ice sheet feeds into Jacobshaven Glacier, an outlet glacier that flows at the remarkable rate of 66 feet per day. Rates of flow of nine other outlet glaciers range from 10 to 88 per day—among the fastest flowing ice streams anywhere.

In summary, then, the ice sheet is spreading slowly outward as snow accumulates on its higher parts, and flow near the margins is unusually swift where ice is concentrated in thick, valley-filling streams.

Origin of the end moraine   The progression of slow burial, lateral flowage, and ablation causes the oldest and deepest parts of the glacier to become exposed eventually at the margin. These basal parts of the ice sheet carry rock debris eroded from the bed, and the arrival of the debris-laden ice at the edge of the sheet has been studied near Thule by Roger LeB. Hooke. Figure 7-5, right, shows part of this area and Figure 7-8 identifies the features he studied. The ramplike body of snow and ice along the margin (left) of the ice sheet accumulated as windblown snow. The distortions in its layering were caused by the press of the advancing ice sheet. Hooke measured the rate of this advance by setting poles in the ice sheet along courses perpendicular to its margin and surveying their positions in 1966 and again 2 years later. He found that all the poles moved toward the margin but at rates that decreased regularly from 15 feet per year (0.4 inch per day) at the pole farthest from the margin to 6 feet per year at the pole nearest the margin. As shown below in the diagram, the ice flowed laterally toward the margin but also rode obliquely upward, an action that helps explain how the older ice is exposed toward the outer part of the ice sheet.

The detritus in the ice exposed near the margin is often concentrated in bands or layers separated by clean ice. Most of the dirty ice contains only 1 part of rock material in 1,000 to 10,000 parts of ice, and the particles are so fine that they cannot be seen with the naked eye. This ice appears amber. Other layers contain much more detritus, and a few are essentially sand and gravel cemented by ice. As the detritus-rich ice moves upward and ablates at the margin, it deposits a moraine. By measuring the amounts of detritus arriving in the ice and also measuring the amounts in the modern moraine and in the drifted snow in front of the ice sheet, Hooke calculated that the

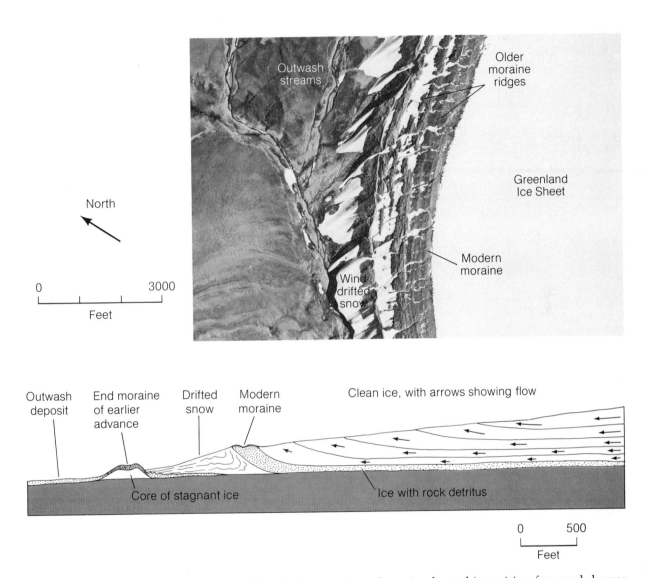

North

0   3000

Feet

Outwash streams

Older moraine ridges

Greenland Ice Sheet

Modern moraine

Wind drifted snow

Outwash deposit   End moraine of earlier advance   Drifted snow   Modern moraine   Clean ice, with arrows showing flow

Core of stagnant ice   Ice with rock detritus

0   500

Feet

FIGURE 7-8

Top, air view looking down vertically on the border of the Greenland Ice Sheet near Thule (located on Figure 7-4). Photograph by U.S. Navy. Bottom, vertical section at right angles to the edge of the ice sheet, showing similar features. Generalized from data of Roger LeB. Hooke, ''Morphology of the icesheet margin near Thule, Greenland,'' *Journal of Glaciology*, vol. 9, pp. 303–24, 1970.

margin of the glacier must have been in about this position for much longer than 50 years. The thicker moraine isolated beyond the ice sheet in Figures 7-5 and 7-8 thus has explicit meaning: it must have formed when the ice sheet was more extensive and must record a much longer period during which the margin of the sheet was at that place.

An end moraine thus records a glacier's history. Each ridge records the position of its margin at that time and the duration of flow to that margin is recorded by the size of each deposit. As we shall see, the ice sheets of the past commonly piled up whole series of ridges and mounds like the outer ones shown in Figures 7-5 and 7-8. These show that the margin of the ice sheets advanced or wasted back short distances and remained in each position for long periods of time. Judged from the studies of the Greenland Ice Sheet, these changes were due to the enlargement or shrinkage of the ice sheets, which were caused by changing rates of accumulation and flow within the broad ice systems.

155

Border Zones of the North American Ice Sheet

The extent and dynamics of former ice sheets may thus be judged by their effects on the land that lay beneath them. These glaciers, however, covered immense regions, so that mapping and other studies have only gradually unfolded their histories. A major discovery has been that the sheets did not spread in a simple way to the outer limits shown in Figure 7-3, but rather that their outer parts often consisted of distinct lobes that advanced episodically. As we shall see in this section, these relations were worked out by tracing end moraines such as that just described for the Greenland Ice Sheet and by mapping other kinds of features, some formed by erosion and some by deposition beneath the ice.

Erosional features of the ice sheets Glacial features formed by erosion are common in eastern Canada, where the former ice sheets were thickest. In the United States they are best seen in the hilly parts of the northeastern states, where ice flowed across the gently sloping margin of the continent and onto the continental shelf, scraping the soil from ridges to expose bare rock in many places. The bare rock was commonly scraped and planed by sand and stones carried at the base of the glacier, and it was also broken and plucked by ice flowing away from the downstream side of prominent outcrops (Figure 7-9, right). The resulting asymmetric forms, which are typical of many larger hills in the region, indicate the direction in which the ice moved. At closer range, the eroded forms may also show polished and scratched surfaces that record the details of glacial abrasion (Figure 7-9, left).

Elongate glacial lakes and intervening ridges impart a linear "grain" to the entire terrain, as shown in the topographic map of Color Plate 15. The Maine

FIGURE 7-9

Right, outcrop of granite on Mt. Desert Island, Maine, with a vertical section showing how the streamlined upper surface and broken downstream surface were caused by ice flowing from left to right. Left, rock surface that was grooved, polished and scratched by a glacier flowing over it from left to right. Mt. Desert Island, Maine.

Glacier ice

Abrasion by fragments of rock in the ice

Granite

Wedging and plucking by ice

coast, also, is characterized by deeply indented bays and elongate rocky head-lands that express the scour and flow of the ice. These features contrast with the subdued coast of the southern States and the elongate mound that forms Long Island—the end moraine of the same ice sheet (Figure 5-1).

Deposition of ground moraine   In addition to end moraines, the ice sheets deposited what is called *ground moraine*—a layer of till laid down partly as the ice moved over its bed and partly when the ice melted and re-leased the materials carried above the bed. The ground moraine is thus com-parable to the deposit of sand and gravel that lies under most parts of active rivers, just as eroded forms such as those in Figure 7-9 are comparable to places where rivers flow over cleanly eroded rock. Ground moraine is gen-erally less obvious than features eroded in rock because it develops a soil and plant cover, but it often has the value of recording more than one ice advance. Each successive ice sheet deposits its own layer of till, and all may be preserved in a layered succession, the oldest at the bottom and the youngest at the top.

The deposited layers are of further value because some can be dated by the radiocarbon method, using plant materials deposited with them. Studies of tills in the northern United States have thus unraveled a remarkable history of advances and retreats of glaciers, including records of interglacial periods warmer than the present. Probably the best way to understand this history is to take one sector across the glaciated margin and examine its features in some detail. I have selected central Minnesota, because its deposits and land-forms are largely undisturbed and have been studied in various ways, chiefly by Herbert E. Wright, Jr., and his coworkers and students (Reference 2).

Depositional features in Minnesota   One of the first sets of features Wright studied turned out to be the oldest in that part of Minnesota—a series of long, low hills that have a distinctly smoothed-over, streamlined appearance (Color Plate 1A). Seen from the air, their shapes would appear similar to those on the Chapter opening page; however, the hills in Minnesota are closer to-gether. The hills are also similar in shape to some eroded ridges in Maine, but deep cuts show that they consist entirely of unsorted mixtures of stones and finer sediment (Color Plate 1B). Many stones, such as that in Figure 7-10, were scraped and scratched by being dragged against the glacier's bed or other stones, showing that this deposit is indeed till and not a debris flow or some other kind of unsorted deposit. Proof that the hills were deposited by ice rather than eroded by streams from older till is given by small but deep ponds that lie at hill crests. These ponds must have formed where masses of ice were buried in the till and later melted to leave water-filled depressions.

When plotted on a map, the hill forms constitute an array that opens fan-like toward the west and south, with the ridges becoming smaller toward the perimeter (Figure 7-11). Wright measured the orientations of hundreds of platy and elongate stones in the till and found that the longer axes of most stones are aligned roughly parallel to the length of the hill in which they occur. He found, too, that the stones point downward at moderate angles toward the north and east (Figure 7-11). These he took to be the directions from which

FIGURE 7-10
Fragment of limestone from one of the elongate hills in central Minnesota, with one face scraped almost to a plane and scored by several sets of scratches, the fine, white ones being the youngest. The fragment is 4 inches across.

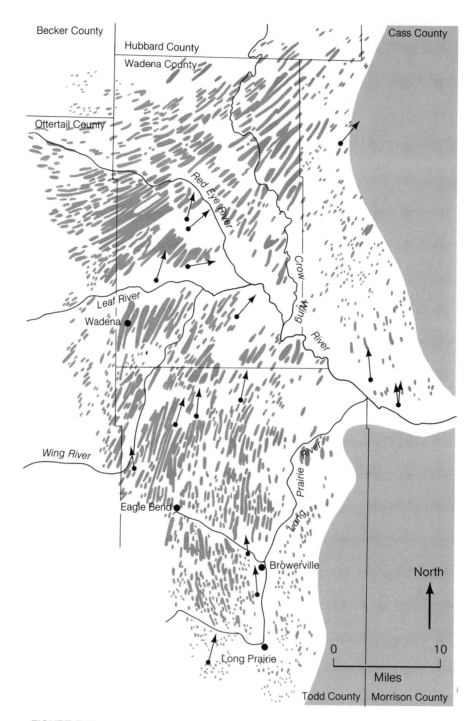

FIGURE 7-11
Ice-streamlined hills (gray) in and near Wadena County, Minnesota. The shafts of the arrows show the alignments of elongate stones in the hills and the arrowheads indicate the direction in which the stones point downward—at angles of around 25°. The shaded area is a younger till deposited by ice that flowed into the area from the right. From H. E. Wright, Jr. (Reference 2).

the ice had flowed, reasoning that the inclined stones were oriented by obliquely upward flow in the lower part of the glacier, like the flow noted by Hooke in the Greenland Ice Sheet (Figure 7-8). Hills of this sort are called *drumlins*, and they have been used in many other places to judge the directions in which ice sheets moved. The fan array of the elongated drumlins described here shows that the ice sheet was spreading southwestward and southward from a source in the northeast.

The specific tills   Wright came to recognize the till of this area by its pale yellow-brown color, by its large sand content, and by abundant limestone fragments like that in Figure 7-10. Even where melt water in ice tunnels sorted the till into sand and gravel, the yellow-brown color can be recognized (Color Plate 1C). Distinctive glacial deposits are often given a local geographic name, and Wright named this one the Wadena till, after Wadena County.

Other distinctive tills lap over the Wadena till on all sides. To the east it is overlain by a brown till that contains many large stones but none of the limestone found in the Wadena till. To the southeast, a distinctive reddish till overlies the Wadena till, and the red till is overlain by a still younger yellow-brown till (Color Plate 2A). The latter is the same color as the Wadena till but contains many fragments of mudstone (shale) not present in the Wadena till.

The characteristics of the two yellow-brown tills can be seen more completely by sawing and grinding samples of each until they are very thin slices called *thin sections*, about 0.001 inch thick. These sections are mounted between thin glass plates so that they can be examined and photographed under a microscope (Figure 7-12). Besides its fragments of mudstone, note that the younger till has a more abundant matrix of clay. The clay was derived from mudstone that was ground up thoroughly by the glacier that eroded it. If you now look back at the views of the same tills in Color Plates 1 and 2, you can get a feel for the unsorted nature of till in general—the presence of abundant pebbles as well as much sand, silt, and clay-size material. This mixture results from the material's being dropped from the ice without sorting, either by sinking to the base of the glacier to form the ground moraine or by dropping out as ice melts.

Mapping advances of ice lobes   Each end moraine consists of till like that in the associated ground moraine, but it can be identified by an irregular, hilly surface that extends across the country in a broad band (Color Plate 2B). The hills were once till ridges like those at the margin of the Greenland Ice Sheet (Figures 7-5 and 7-8), only most are larger. Their size indicates that the former ice sheet advanced to certain positions in Minnesota and remained there for long periods. These positions can be recognized on maps, because hundreds of lakes and ponds fill depressions left in the end moraines by the melting of masses of ice at the margin of the glacier (Figure 7-13). The relative ages of the end moraines can be determined by mapping the extent of the different tills. As the figure shows, the exposure of Wadena till is only part of a more extensive ground moraine that was left exposed by the wasting back of the ice that deposited it and then partly overrun by an ice sheet that advanced from the east and deposited the end moraine that passes through

FIGURE 7-12
Top, a thin section, shown at actual size. Center, part of a thin section of Wadena till, 1/10 inch across, showing the abundant sand grains of quartz (the clear, white grains). Bottom, part of a thin section of the youngest till, also 1/10 inch across, showing fragments of shale (speckly gray) and the abundant matrix of clay (dark). Thin sections and photographs by Ruperto Laniz.

159

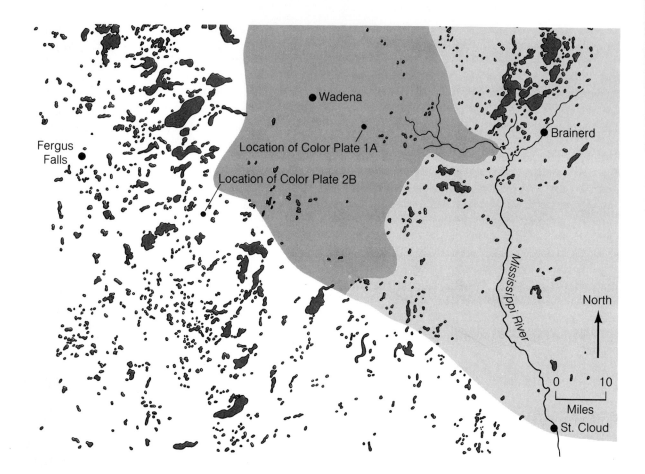

Fergus
Falls

● Wadena

Location of Color Plate 1A

Location of Color Plate 2B

● Brainerd

Mississippi River

North

0   10

Miles

● St. Cloud

FIGURE 7-13

Glacial lakes and ponds in
central Minnesota. Their
distribution defines the two broad
end moraines that lap over the
Wadena till (the darker shading).
The shaded moraine on the right
is the older because it is overlapped by the end moraine of
youngest till (unshaded). Based
on the Twin Cities aeronautical
chart (1:500,000) of the National
Ocean Survey, 1975.

Brainerd (Figure 7-13). That ice then wasted back and another ice sheet advanced from the west to deposit a moraine that laps across both of the older
moraines.

Wright worked out the course of each ice advance by tracing rocks found
in the till to their eroded source in the bedrocks of the region (Figure 7-14A).
The curving course of the ice lobe that deposited the Wadena till, shown in
Figure 7-14B, is based on the presence of limestone fragments brought from the
Winnipeg area, along with the absence of the shale that lies south of Winnipeg.
Other kinds of rocks in the till show that parts of the ice flowed into central
Minnesota from the northeast. The orientations of the drumlins help to com-
plete the picture.

After the Wadena ice lobe had wasted back to the position shown in
Figure 7-14C, the brown and red tills were deposited by ice that moved from
the northeast, as shown by stones in the tills. The red till, for example, can be
traced back to red sandstones and other distinctive rocks in the Lake Superior
area (Figure 7-14A). Finally, the course of the youngest ice advance was
charted by abundant shale fragments eroded from the lowlands of the Dakotas
(Figure 7-14D). This remarkable lobe of ice, called the Des Moines lobe, has
been mapped in unusual detail because its end moraines were nowhere
covered by later advances of ice.

Studies of tills, ground moraines, drumlins, and end moraines have thus established a relation that has been found in other carefully mapped glacial regions: the history of glaciation was distinctly episodic. Lobes of ice arising from a parental ice sheet in Canada advanced hundreds of miles southward, maintained certain margins for long periods of time, and then wasted back over long periods of time. Judged by the small recent variations in the Greenland Ice Sheet, these large changes must record truly major variations in accumulation and ablation and thus suggest major changes in climate. The Minnesota record thus points toward the basic meaning of the glacial record.

FIGURE 7-14

A, map of Minnesota and surroundings, showing the distribution of some of the rocks that can be used to determine the direction of flow, assuming the glacier carried them more or less directly from their source to the place where they are now found in till. B, general form and flow directions of Wadena ice lobe, the oldest of the lobes. Note how ice flowing from the vicinity of Winnipeg was turned southward by ice flowing from the northeast. C, flow pattern of the second of the lobes, showing also the front of the wasted Wadena ice at that time. D, the youngest ice lobe at its maximum extent, and the small ice lobe in the Lake Superior basin. After H. E. Wright, Jr., Charles L. Matsch and Edward J. Cushing (Reference 2).

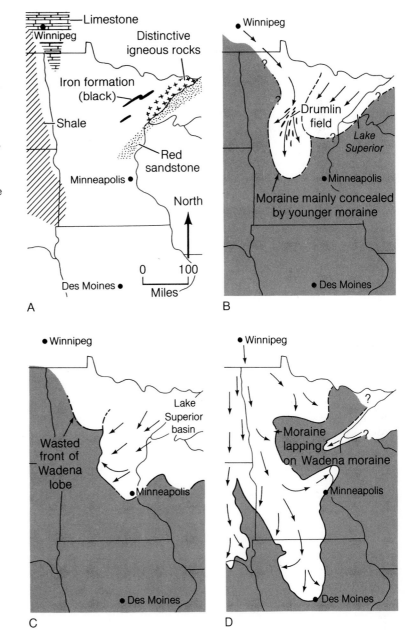

161

Glacial Cycles, Past and Present

The glacial history of Minnesota has been dated by the radiocarbon method, using plant remains buried in the various deposits. This history is shown diagrammatically in Figure 7-15A by a line indicating southward advances and northward retreats of the ice lobes. The age of the Wadena lobe is known only to be greater than 40,000 years, but the other ice advances have been dated quite exactly. The maximum advance of the Superior lobe is shown in Figure 7-14C and that of the Des Moines lobe in Figure 7-14D.

Figure 7-15B shows the same sort of information for the eastern Great Lakes region, which also exposes several tills. The arrows indicating flow directions are based on drumlins, ridges eroded in rock, end moraines, and till sequences, much as in Minnesota. More plant materials have been dated here than in Minnesota, and the ages and approximate geographic positions of these samples are shown on the diagram. The advance-and-retreat curves are based on these points. The figure indicates that an ice sheet covered the St. Lawrence lowland from about 65,000 to less than 12,000 years ago and spread as much as 800 miles to the southwest of that area. More important, Figure 7-15 shows that the dated advances and retreats in Minnesota took place at the same time as those in the eastern Great Lakes region. Because the lobes in Minnesota flowed from a part of the parental sheet 1,000 miles west of the St. Lawrence lowland, the two records indicate exceedingly widespread, basic changes in the great ice sheet that covered most of Canada and part of the United States.

Temperature histories from oxygen isotopes That sheet, of course, is now completely gone, but we can gain some additional idea of its history by examining a remarkable record of temperature variations preserved in the Greenland Ice Sheet. The method of study is based on the fact that a small percentage of the oxygen atoms in water are heavier than average, with an atomic weight of 18 rather than the usual 16 (the two isotopes abbreviated ^{18}O and ^{16}O). Because water molecules made of ^{18}O are heavier than those made of ^{16}O, they condense more readily from the atmosphere. Snows formed at somewhat higher than average temperatures thus contain somewhat greater proportions of the heavier molecules. The proportions of ^{18}O and ^{16}O have been measured at closely spaced intervals in a core of ice obtained by drilling through the ice sheet at Camp Century, and the age of each part of the core has been calculated from rates of accumulation of snow and of spreading (flow) of the ice. The oxygen data can thus be cast into a history of colder and warmer periods, going back some 126,000 years (Figure 7-16A). Note that the simplified version of Figure 7-15B on the right shows that the cooler periods recorded in the Greenland ice have counterparts in the glacial history of the continent. Each cooler period evidently began before the southward advances of ice on the continent. This relation seems reasonable, for it should have taken some time after the beginnings of cooling for the ice sheet to be thickened and enlarged laterally by hundreds of miles.

Another method of obtaining temperature histories, in this case for the upper part of the ocean, is to compare the numbers of warm-water and cool-

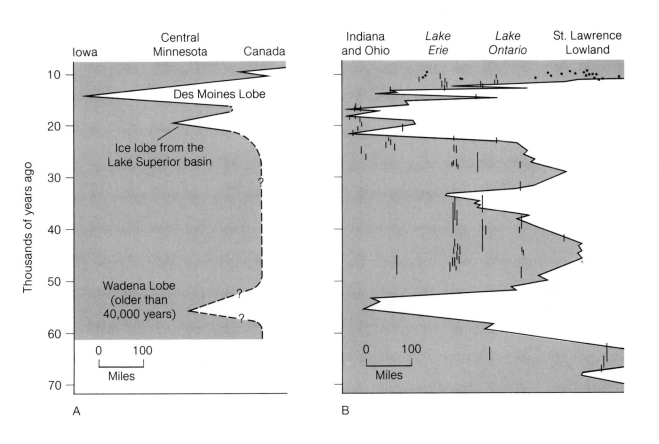

FIGURE 7-15

A, dated advances (to the left) and retreats of the fronts of ice lobes across Minnesota, with locations shown at the top of the diagram. B, similar diagram for the region shown in the map at the right, which shows flow directions of the ice. Dated samples are indicated in the advance-retreat diagram by dots and lines, the latter showing the uncertainty in the age. Note, again, the place names at the top of the diagram. Data for B from A. Dreimainis and R. P. Goldthwait, "Wisconsin glaciation in the Huron, Erie, and Ontario Lobes," pp. 71–106 in *The Wisconsinan Stage*. Geological Society of America, Memoir 136, 334 pp., 1973).

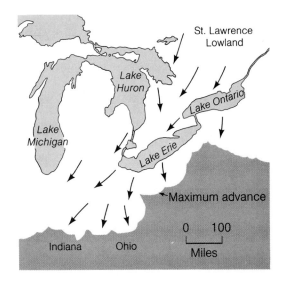

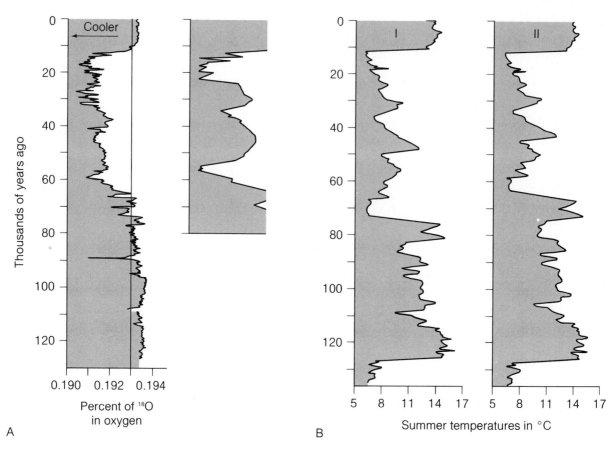

FIGURE 7–16

A, variations in content of ^{18}O in ice recovered from a vertical drill hole in the Greenland Ice Sheet, indicating fluctations in atmospheric temperatures over the past 126,000 years. To the right is a generalized diagram of the ice advances and retreats in the eastern Great Lakes region shown in Figure 7-15B. Greenland curve from S. J. Johnson, H. B. Clausen, and C. C. Langway, Jr., "Climatic record revealed by the Camp Century ice core," pp. 37–56 in *The late Cenozoic glacial ages*, K. K. Turekian, editor. New Haven: Yale University Press, 606 pp., 1971. B, fluctuations in the upper layer of the Atlantic Ocean, determined from the proportions of cold-water and warm-water foraminifers in a core of bottom sediment. Curve I is the authors' preferred dating of the history and Curve II a less likely dating. From Constance Sancetta, John Imbrie, and N. G. Kipp, "Climatic record of the past 130,000 years in the North Atlantic deep-sea core V32-82: correlation with the terrestrial record," *Quaternary Research*, vol. 3, pp. 110–116. 1973.

water animals whose remains have accumulated on the seafloor. Planktonic (surface-living) foraminifers have proven especially useful in this regard, because the ranges of temperature needed by closely related living forms are known. The various species are counted in samples taken from deep-sea cores, and the age of each part of the cored sediment is calculated from rates of sediment accumulation, matched where possible with ages determined in other ways. Figure 7-16B shows the most likely history of temperature changes based on a thoroughly studied core and also a less likely history suggested by the authors. The exact correspondence of these episodes of cooling to those recorded in the Greenland Ice Sheet are thus somewhat problematical, but the general fit between the data suggests a close connection among cooling of the atmosphere, cooling of the seas, and major advances of the ice sheets.

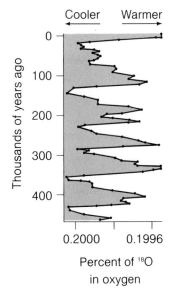

Cooler   Warmer

Percent of ¹⁸O
in oxygen

FIGURE 7-17

Temperature fluctuations indicated by content of ¹⁸O in tests of foraminifers in a 36-foot core of pelagic sediment from the Caribbean Sea. The original data were determined on two species at each sample point, but the results are so similar that I have averaged them into one set of points. Data from W. S. Broecker and J. van Donk. Insolation changes, ice volume, and the ¹⁸O record in deep sea cores. *Review of Geophysics and Space Physics*, vol. 8, pp. 169–98, 1970.

How long has this episodic history been going on? Unusually long sediment cores from the Caribbean Sea indicate long cycles of change, each equivalent to the total record discussed so far, going back some 450,000 years! Figure 7-17, for example, shows proportions of ¹⁸O in the oxygen making up calcium carbonate remains of foraminifers. These data give a measure of the oxygen composition of the water in which the animals grew. The proportion of ¹⁸O in the water changes with temperature because the heavier (¹⁸O-bearing) molecules tend to be concentrated during evaporation, an effect that increases as temperatures are lowered. Considering the number of data points in Figure 7-20, we see that the variations during the past 120,000 years were much like those derived from the other studies. The figure also shows that the variations are only the latest of a series of major cycles, each presumably causing a major period of glaciation. Older records in deep-sea sediments are more difficult to date but indicate several additional periods of cooling during the 2 million years preceding the record of Figure 7-17. Still older sedimentary rocks indicate several more periods of cooling during the preceding 40 million years; however, these periods were of much longer duration and probably not so intense as the later periods.

A possible cause of glaciation   Many causes have been suggested for this cyclical history, but so far no theory has explained all the facts. One promising idea is based on variations in the orientation of the earth's axis of rotation and in the distance between the sun and the earth—both of which cause cyclical variations in the amounts of solar energy received at a given latitude. These variations can be computed far back in time, and some variations for northern latitudes correlate closely with parts of the glacial history. One can speculate from there. If such variations were to cool the northern part of the northern hemisphere, the resulting increase in snow cover, sea ice, and clouds would reflect increasing amounts of solar energy back into space and thus intensify the initial cooling at northern latitudes. The temperature contrast between this region and the warm seas to the south would increase the flow of moisture northward—by the sea and air currents described in Chapter 6—and this would lead to the formation of ice sheets around the northern Atlantic (Figure 7-3). But the sea would slowly cool due to evaporation and to additions of melt water from the glaciers, so the process would tend to be self-arresting in the long run. Whether the strength and timing of such variations would explain the major periods of glaciation and also the lesser cycles within each major period remains to be determined.

A new ice age?   Whatever its cause, the cyclical record of glaciations suggests strongly that the present interglacial period is due to end soon and will be followed by the increased cooling of another glacial period. Exactly when this will happen cannot be predicted until the causes of glaciation are known more fully, but some scientists have noted changes that may represent the first wobble of cooling toward the next glaciation (Reference 3). Besides an estimated drop in average temperature of 1 to 2°C, the 25-year period since 1942 has seen an increased cover of snow on arctic and subarctic lands and an increased amount of ice on adjacent seas. A study of the armadillo, an animal whose circulatory system requires a moderate climate, has shown them to

165

have migrated to southern Kansas from former ranges in Nebraska. Climates farther to the south appear also to be affected, evidently because of southward shifts of the polar and intermediate circulation systems shown in Figure 6-2B. Studies of pollen preserved in the sediments of Florida lakes, for example, indicate a distinctly drier climate during the last period of glaciation, and recent rainfalls suggest a return to those conditions (Reference 4).

The cyclical record of Figure 7-16 suggests that a new ice sheet is not likely to form in eastern Canada and spread southward for tens of thousands of years. However, the world's food supply would be affected drastically by even a moderate cooling that lasted for tens of years, because the world's main grain sources are in its relatively cool, northerly parts. The hybrid corns raised in the northern states, for example, require a long growing period and even now mature so late that they must often be dried artificially. An added need for fuel would thus be imposed where there is already a serious shortage. The natural food resources of the estuaries would be depleted seriously, even if sea level were lowered a few feet by the beginnings of a new glaciation.

Studies of glaciers have thus linked the past to a climatic cycle that will affect humanity profoundly. It is vital to discover the basic causes of the waxing and waning of ice sheets.

Summary

Discovery of the causes of glaciation will depend on understanding all aspects of glaciers, modern as well as former ones. We may briefly review the main points in our present state of knowledge:

1. Glaciers are bodies of slowly flowing ice that range in size from small valley-filling streams to large ice caps and giant ice sheets.
2. Glaciers flow in the solid state because the molecules making up crystalline ice are dislocated systematically by moderate forces. Glaciers also slide bodily over their beds by melting and freezing at the base.
3. Glacier ice forms slowly when snow is compacted under the weight of additional snowfalls; it forms much faster when water melted from the snow at the surface becomes frozen in the packed snow beneath.
4. Snow and ice accumulate most rapidly where precipitation rates are high and temperatures are low, in the most elevated parts of the glacier. Higher temperatures and lower rates of precipitation make ablation dominant on the lower parts of glaciers, so that the glacier thins near its margin and the oldest (deepest) parts of the ice become exposed there.
5. Rock debris is carried in the lower parts of ice sheets, and it is deposited partly as a ground moraine under the sheet and partly as an end moraine where the ice ablates completely at the margin of the glacier.
6. Ice-eroded ridges and valleys and scraped, striated rock surfaces indicate that former glaciers were most erosive under their faster flowing (generally thicker) parts.
7. On the other hand, vast areas within hundreds of miles of the margins

of the ice sheets are characterized by depositional features such as ground moraine, drumlins, and ice tunnel deposits.

8. The advances and recessions of the lobate margins of the former ice sheets can be mapped by end moraines, distinctive tills, drumlin orientations, and rock fragments transported from known sources.

9. The resulting episodic history has been dated by the radiocarbon method and correlated with temperature variations indicated by oxygen isotopes in modern ice sheets and in the remains of oceanic animals.

10. The age relations suggest that the atmosphere and sea were cooled prior to the start of each major glacier advance, a change that may have been due to reduced solar radiation at those times.

11. Patterns of oceanic circulation were evidently also important, for the former ice sheets were grouped around the northern part of the Atlantic Ocean, and the Greenland Ice Sheet lies at the center of this array.

12. The duration of the dated cycles of glaciation indicate that the present interglacial period is due to end, and some indications of climatic changes suggest the next glacial period has begun.

REFERENCES CITED

1. From a study reported by George Wallerstein. Movement observations on the Greenland ice sheet. *Journal of Glaciology*, vol. 3, pp. 207–10, 1958.

2. The data that follow are mainly from Wright's paper, "Role of the Wadena lobe in the Wisconsin glaciation of Minnesota." *Geological Society of American Bulletin*, vol. 73, pp. 73–100, 1962. Additional data and references can be found in "Superior and Des Moines Lobes," by H. E. Wright, Jr., Charles L. Matsch, and Edward J. Cushing (pp. 153–85 in *The Wisconsinan Stage*, edited by Robert F. Black and others, Geological Society of America, Memoir 136, 334 pp., 1973).

3. The few data mentioned here are from a series of papers in *Quaternary Research*, vol. 2, no. 3 pp. 261–445, 1972, and subsequent numbers of this journal.

4. Joseph M. Moran. Return of the ice age and drought in peninsular Florida? *Geology*, vol. 3, no. 12, pp. 695–96, 1975.

ADDITIONAL IDEAS AND SOURCES

1. An excellent way to get a feel for the main ice sheet of the last glaciation is to examine the superb Glacial Map of Canada, which shows, at a scale of 1:5,000,000, erosional and depositional features indicating directions of flow, end moraines formed at various stages, linear deposits in ice tunnels, positions of huge glacial lakes, distribution of modern glaciers, and more. Of special value is a listing of the many published sources from which the data were abstracted. The map is number 1253A, published in 1968, available from the Director, Geological Survey of Canada, Ottawa.

2. As the ice sheet wasted back and the climate became warmer, a succession of plant communities followed the retreating ice front, one replacing the other at any one locality. A record of these changes is described in "Vegetation history of the southern Lake Agassiz basin during the past 12,000

years" by Creighton T. Shay, on pages 231–52 in a book presenting many other interesting accounts of glacial effects on south-central Canada: *Life, land and water*, edited by William J. Mayer-Oakes (University of Manitoba Press, Winnipeg, 414 pp., 1967).

3. Dramatic effects of the ice sheet on a bordering river system are given in the well-illustrated booklet *The channeled scablands of eastern Washington—the geologic story of the Spokane flood*. Prepared by the U.S. Geological Survey, it is stock number 2401-02436, for sale by the Superintendent of Documents, Washington, D.C., 20402, for 65¢.

4. Because of their relative importance, I have presented information on the great ice sheets rather than on lesser glaciers, which occurred abundantly in high mountains throughout the western United States and may still be seen at many places in Alaska, western Canada, and the northwestern United States. Truly beautiful places where glaciers can be visited easily include the Canadian Rocky Mountains, Glacier National Park, and Mount Rainier National Park (see Figure 12-1 for a view of ice on the latter mountain). Even easier for many persons to reach are the countless places where the effects of recent valley-filling glaciers can be interpreted from well-preserved moraines and eroded surfaces. Perhaps the most dramatic and famous is Yosemite National Park, the glacial features of which are described in "Geologic history of the Yosemite Valley," by Francois E. Matthes (U.S. Geological Survey Professional Paper 160, 137 pp., 1930). This is a classical account of how a river-valley system was modified by repeated glaciation.

5. A connective of glaciers to the ocean is described in an account of the sea-floor sediments and topographic forms off southeastern Canada and the northeastern United States, given in "Atlantic continental shelf and slope of the United States—gravels of the northeastern part" by John Schlee and Richard M. Pratt (U.S. Geological Survey Professional Paper 529-H, 39 pp., 1970).

Dark prairie sod exposed in a gully near Canton, Kansas. Grass roots make the sod more resistant to erosion than the paler deposit beneath, which is a layer of silt deposited by the wind during the last period of glaciation.

8. Origin of the Soil

Origin of the Soil

Working of Glacial Sediment by the Wind
    Wind dunes of the Great Plains
  Deposition of loess
    Soil erosion by the wind
A Clay-rich Soil on Granite
    Determining the parent of the main soil layers
    Creep of the upper layers
The Clay Minerals
    The clay mineral kaolinite
    Properties of the expandable clays
    Formation of clay minerals
Chemical Changes That Produce Soil
    Altering rock to soil
    Laboratory studies of soil minerals
Plants—The Other Parents of the Soil
    Organic alterations along cracks
    Actions of soil microbes
    Organic processes and soil layers
Factors in the Origin of Soils
    Soil depletion due to human uses
Summary

The chapter opening page shows a small part of North America's most important single resource—the prairie sod of the Great Plains. This soil extends from Texas northward to central Alberta and from the foot of the Rocky Mountains eastward to the Mississippi Valley and the woodlands of Manitoba. This vast region is the continent's heartland, covering an area of approximately 1,100,000 square miles and including a fourth of the total area of the United States. Now largely under crops, it contains most of the farms of North America and produces most of the wheat and related grains, corn, soybeans, livestock, and dairy products that feed the continent's human population as well as many people elsewhere.

This crucial resource was formed over a period of thousands of years by grasses! Indeed, this kind of soil is called a *grassland soil*, one naturally supporting grasses because it was constructed by them. The relationship is remarkable and simple. To adapt to climatic conditions that are too harsh for most plants, grasses developed a life cycle based on underground storage of nutrients. Because their above-ground parts are likely to be eaten, trampled, burned, or blown away by the wind, grasses grow an extensive system of roots, one that is renewed annually. The roots from former years decay slowly in place, building a rich, porous bank of nutrients at exactly the depth the new roots can reach easily each spring. As a result, growth in the late spring and early summer can be exceedingly vigorous, for the new plant feeds on a mixture of substances preselected by its forebears. It is no wonder that grass crops (grains, corn) have been so successful in the plains region.

Grassland soil has two notable characteristics besides its richness. First, it is held together by the closely spaced grass roots and is thus resistant to erosion. In the photograph on the chapter opening page, the sod forms an overhanging layer because it is tougher than the sediment beneath. Grassland soil is also characterized by an upper layer colored black, dark gray, or brown by the organic materials formed from the roots. Note in the photograph that the soil is dark near the surface and gradually becomes lighter downward. The gradation is due to most grass roots occurring near the surface but some tapping more deeply.

The photograph was taken near the center of the Great Plains, where the thickness of the soil is about average for the region. To the east, where the annual rainfall increases, the grasses are taller and ranker; their roots and the black soil go deeper—to depths as much as 5 feet or so. To the west, where the annual rainfall decreases, grasses and their root systems are sparser and the dark layer thins progressively to 6 inches or so. The layer also becomes lighter and browner to the west because it is dry much of the time, and oxygen from the atmosphere decomposes the black organic materials to carbon dioxide (CO_2) and water. Another geographic variation in the soil is a gradual change to orange or reddish tints in the southern plains, due basically to an increase in temperature toward the south. Higher temperatures increase the rate at which the black organic materials are oxidized by the air, and they also promote the formation of iron oxides, which are orange or red.

The soil of the plains is thus constructed everywhere by grasses but varies from place to place because of differences in climate. In this chapter we will examine some of these relations further and explore other factors that influence the origin of soil. We will find that soils constitute a thin but remarkable layer, one in which the atmosphere, the solid earth, and living things combine in a tightly interwoven system. We will see, too, that this skin on the earth tends to be delicate—that our uses of it have typically diminished it, even though we have long realized that it is our most basic resource.

Working of Glacial Sediment by the Wind

A factor in the origin of all soils is the nature of their parental materials—the materials from which they were formed. The grassland soil just described comes mainly from sediments produced during the last period of glaciation. In the parts of the plains once covered by the ice sheet, the sediments are mainly till that was reworked locally by melt water as the ice sheet wasted back to the north. A large area of rich soil in Canada was formed from silt and clay deposited in a huge lake that lay alongside the ice sheet (Figure 8-1). Silty

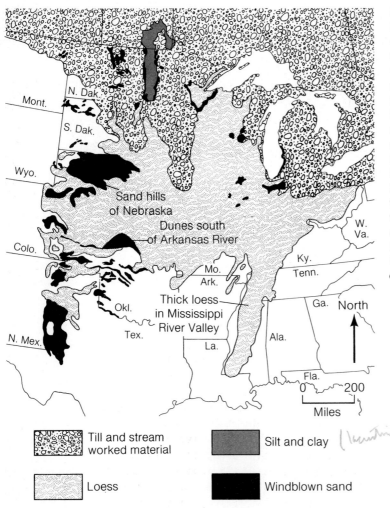

FIGURE 8–1

Parent materials of soils in the central part of the continent. The circle pattern indicates till and stream-worked materials associated with it; the gray area is silt and clay deposited in a lake near the margin of the ice sheet; the dotted area is loess (windblown silt); the black areas windblown sand, and the unpatterned areas older sediments and rocks. Generalized from map of Pleistocene Eolian Deposits of the United States, Alaska, and parts of Canada, National Research Council Committee for the study of eolian deposits, published by the Geological Society of America, 1952, and from *Soils in Canada,* edited by Robert F. Legget (Royal Society of Canada Special Publications, no. 3, 240 pp., 1961, revised edition, 1965).

Till and stream worked material

Silt and clay

Loess

Windblown sand

PLATE 1

Top, view in central Minnesota showing the simple, streamlined ground surface over one of the elongate hills (called drumlins) and the valley adjacent to it. Middle, freshly exposed surface of the unsorted till (the Wadena till) making up the elongate hills in central Minnesota. The pebbles and coarse sand have been washed clean by a brief rain. The largest pebble is 1.5 inches long. Bottom, vertical wall in a gravel-and-sand pit dug into one of the elongate hills (drumlins) in central Minnesota. These materials lie along the crest of a ridge and therefore must have been sorted and deposited by a stream flowing in an ice tunnel under the ice sheet. The largest pebble is 2.5 inches long.

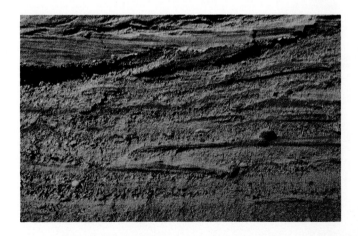

PLATE 2

Left, vertical face of a quarry in St. Paul showing the red till that is younger than the Wadena till, and the still younger yellow-brown till on top of it. Several of the stones in the red till are distinctive rocks transported from the northern edge of Lake Superior. Below, view southward along the hilly end moraine of the youngest of the three tills in central Minnesota. The gray stones are granite from northern Minnesota. The top of the moraine is here 300 feet above the plain underlain by Wadena till, in the left distance. The locality, Inspiration Peak, is labeled in Figure 7-13.

A

B

C

PLATE 3

A, roadcut on State Highway 79, northeastern Georgia, showing the upper 15 feet of a thick soil (the entire roadcut). At the crossing of Broad River, Elton County. B, closer view of a deeper exposure at the same place, showing soil and altered granite, with sheets of white clay along former cracks. C, a still deeper part of the same roadcut as in Figures A and B showing granite altered along cracks, some carrying sheets of white clay.

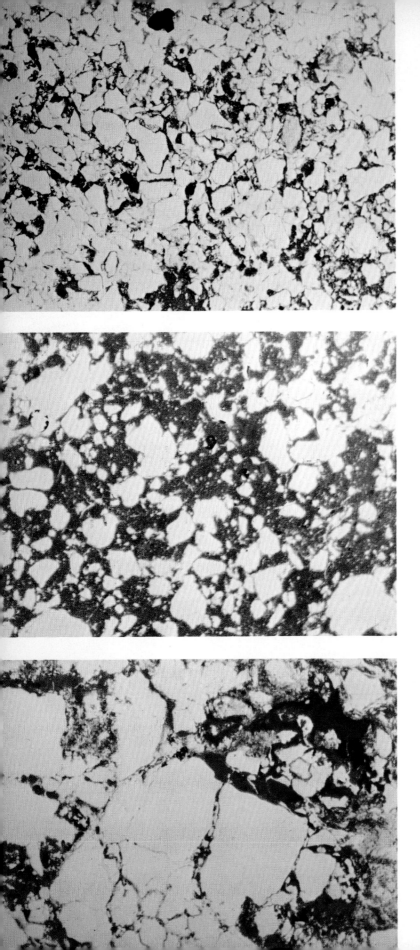

PLATE 4

Thin sections of the Georgia soil, each view being $\frac{1}{2}$ inch across. The colorless grains in all three are quartz.

A is the topsoil, at a depth of 6 inches. The black and gray materials are organic substances and clays pigmented by them.

B is from the middle of the deep red layer in color plate 3A, often called the subsoil. The reddish substances are clays and iron oxide, and the black ones are residues of roots.

C is from a depth of 6 feet in the saprolite (the lower part of color plate 3A). The pale and medium reddish substances are clay, and the deep red ones are iron oxide.

deposits are also the most important sediments on the plains of the United States, but these were deposited by the wind to form a distinctive sediment called *loess*. Loess consists of the fine materials that fall and accumulate from clouds of dust. It forms a layer that ranges from a few inches to perhaps 200 feet in thickness and typically shows no signs of internal layering. Loess is firm enough to form steep banks, as on the chapter opening page, but can be pulverized easily in the hands, giving a gritty feeling because of its silt and very fine sand content. Seen under a microscope, many of the grains are irregular and many are minute flakes of mica or feldspar. They appear to have been dropped helter-skelter, for there is much space between them. The openings, in fact, are just the right size to admit fine grass rootlets, and the loose nature of the sediment permits the roots to enlarge easily. The looseness of the soil makes it easy to till (plow), and the larger openings permit it to drain readily. At the same time, the smaller spaces between the grains retain moisture because water sticks to mineral grains and tends to be held where grains lie closely together. These various properties make soils on a loess unusually nutritive and valuable.

Wind dunes of the Great Plains   To see why the loess occurs where it does, we must determine the wind directions during the glacial period. We can do this by studying the sand deposits of the plains region, the larger areas of which are shown in Figure 8-1. These deposits are sand dunes that are largely stabilized by grass and bushes but become active if the protective cover is removed. The wind then erodes the sand with amazing rapidity and creates a shallow depression called a *blowout*, which may have a wind-fluted surface and a residue of pebbles too large for the wind to transport (Figure 8-2). The eroded sand accumulates again nearby to form dunes that start as low mounds but evolve to forms much like the river dunes of the Rio Grande and finally to more peaked crescent-shaped dunes (Figure 8-3). More commonly, larger dunes develop on the downwind side of blowouts, with crescent shapes pointing in the other direction with respect to the wind (Figure 8-4). The wind directions indicated by the modern dunes vary from one place to another but are typically from south to north.

The dunes composing the Sand Hills of Nebraska cover an area of 20,000 square miles, the largest dune field in the western hemisphere (Figure 8-1). The principal dunes there are much larger than those just described and have been inactive since about 10,000 years ago. Each is a ridge nearly a mile across and elongated for many miles in an east-west direction. The steeper sides of the ridges face south-southeast, indicating winds blowing mainly in that direction (Figure A-4, Appendix A). South-blowing winds are also indicated by the fact that most of the other dune fields in the region lie just south of river valleys, where sand blown from the floodplains would tend to accumulate. The wind direction during the glacial period thus appears to have differed from that now, presumably because of the presence of the ice sheet to the north.

Deposition of loess   This wind direction also fits well with the occurrences of loess around the Sand Hills. The loess on the south and southeast sides of the hills is unusually thick, and Figure 8-1 shows that there is none on

FIGURE 8-2
Fluted surface cut by a wind blowing from left to right, and low dunes deposited locally at the same time. Three inches of sand were eroded here by a 30 mile per hour wind in approximately 2 hours.

which way now?

173

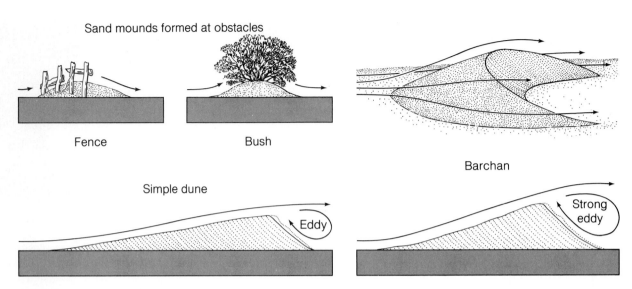

Sand mounds formed at obstacles

Fence

Bush

Barchan

Simple dune

Eddy

Strong
eddy

FIGURE 8–3
Progression of growing dune forms, the simple dune being similar to river dunes and the barchan more peaked because of the increasing effects of the eddy on its downwind side. Seen in three dimensions (upper right), the eddy tends to hold back the highest part of the barchan, while sand streams around the two ends, constructing a crescentic form.

FIGURE 8–4
Active blowout in an old dune field partially stabilized by grass, as in the distant part of the view. This and nearby blowouts are on the downwind sides of large, older dunes and thus appear to be due to turbulence caused by eddying behind the sand hills. One mile south of the Arkansas River, western Kansas.

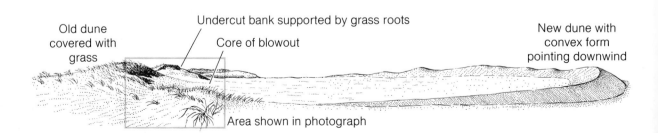

Old dune
covered with
grass

Undercut bank supported by grass roots

Core of blowout

New dune with
convex form
pointing downwind

Area shown in photograph

174

the north side. Evidently sand and silt eroded from the region to the north were transported southward; the sand accumulated nearer the source, while the dusty silt was carried farther. This winnowing action can be seen at any blowout during a period of strong winds. The sand rolls and skips along the surface or near it but accumulates on new dunes nearby, whereas the silt and finer grains are suspended in a cloud of dust that trails off beyond the dunes.

The sand in the Sand Hills, however, must have been rolled and skipped a long way, for it is well sorted by size and unusually abraded and rounded for grains this small (Figure 8-5). Some of the sand and silt was probably blown from the outwash plain in front of the ice sheet, and some was probably eroded from older deposits north of the Sand Hills (Figure 8-6). Additional sources along the major river valleys are suggested by the fact that the loess is typically thicker next to the river valleys than farther away. Winds blowing southward apparently whipped up dust from the floodplains of the rivers, and, as they carried it over the surrounding plains, they dropped much of it near the river valleys. The thicker parts of the loess are thus analogous to the low mound of sediment built up on river floodplains near the river's banks (Figure 2-14).

FIGURE 8–5

Sand from the Sand Hills of Nebraska, showing how nearly it is sorted to one size (approximately 0.2 millimeter) and how abraded the surfaces have become due to impacts during transport. Scanning electron microscope photograph by Gerta Keller.

FIGURE 8–6

Map of the principal areas of windblown sand and silt, with arrows indicating wind directions during the glacial period. Especially thick accumulations of loess occur southeast of the Sand Hills and along the east side of the Mississippi River valley. River floodplains oriented about parallel with the wind were eroded especially strongly.

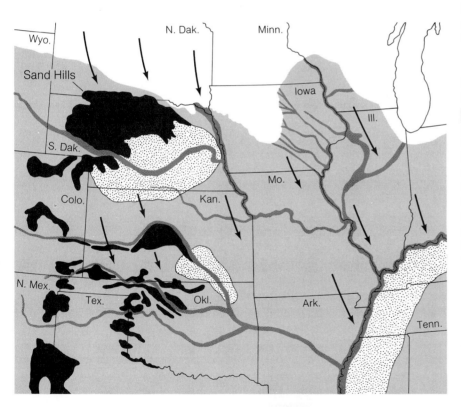

| | Ice sheet at greatest extent | | Dune fields |
| | Principal wind directions | | Exceptionally thick loess |

0   100
Miles

175

The accumulation of the loess is linked to the grasses, which almost certainly acted to check the wind along the ground and thus caused the silt to be deposited permanently. Otherwise it would have been blown episodically south and off the continent, as is the case in regions without vegetation, such as the Sahara Desert. That the plains were not a desert is proven further by shells of land snails found in the loess, remains of species that require a cool climate and abundant plants for food.

Soil erosion by the wind Recent history has confirmed that grass-grown regions can be the scenes of towering dust clouds. The 1930s, the decade of the "dust bowl," is probably the best known case, although wind erosion during that period was brought on largely by human uses. This erosion was set up by the development of heavy farm machinery in the 1920s, which led to speculative dry-wheat farming and to the breaking of 15 million acres of natural sod during the period 1924–1929 (Reference 1). A drought in the 1930s thus caught a tremendous area exposed by plowing, much of it farmed by persons who had little understanding of it. The crop failures that followed led to year-round erosion by the wind, such that dust storms must have seemed endless between 1933 and 1940. In Liberal, Kansas, 128 days of dust storms were counted in 1938 alone, many days turning dark as night in the midafternoon (Reference 2). Where hooves of cattle cut the sod over old sand dunes, wind eroded the sand and spread it over adjacent grasslands. When the winds stopped, the depleted sod was eroded rapidly and the land was scored by countless gullies.

In summary, grasses had three major roles in the formation of the world's greatest soils: they accumulated windblown dust into a layer of loess; they slowly constructed a rich soil from the loess and other deposits; and they protected that soil for thousands of years. The entire plainsland fauna and a large segment of the human population were based on that soil system. Study of the loess and the grass-covered dunes could have foretold what would happen when the protective sod was farmed during a period of drought.

A Clay-rich Soil on Granite

The soil we have just considered is so young and so dominated by organic materials formed from grass roots that it is difficult to appreciate some of the other things happening in it—things basic to the origin of all soils. We need to examine an older, more strongly developed soil, and the beautiful one shown in Color Plate 3A is ideal for our purpose. It is an important soil, one typical of the southern part of the Piedmont of the eastern seaboard states and of other parts of the world where climates are warm and wet. Precipitation in this part of Georgia is approximately 46 inches per year, twice the annual precipitation in central Kansas. However, there are more important differences between this site and the one on the chapter opening page: (1) the area is a woodland rather than a grassland, and (2) the soil is distinctly layered. We will find in due course that these characteristics are interrelated, but first we must determine the parental material of the Georgia soil.

We can best start at the surface and go down. The uppermost layer, called the *topsoil*, is largely hidden in the shadow at the top of the photograph,

but it is a brown, sandy layer 8 inches thick. The evenly reddish layer beneath looks like sandy clay deposited by a river on its floodplain; however, it extends across country with no correspondence to river valleys or any other part of the landscape.

The parentage of these two upper layers is not obvious, but the lowermost part of the soil shows direct evidence of its derivation. Deeper cuts than that of Figure A show that the lower materials grade down into granite that is partly altered to soil along cracks (Color Plate 3C). Between Figures A and C, the soil contains relics of granite—evidently those parts lying farthest from the cracks (Color Plate 3B). Note how the hard relics grade out into crumbly gray granite, which grades to reddish soil such as that forming the lower half of the exposure in Figure A. Evidently this part of the soil, at least, has been formed by progressive alterations of granite.

We can learn more by looking closely at the lower part of the soil and its crisscrossing white bands. Figure 8-9, a drawing made at the road cut, shows that the lower reddish soil and parts of the white bands are mixtures of clay and quartz grains. The quartz grains in the soil have sizes, shapes, and abundances identical to those of quartz grains in the granite. The white bands are sheetlike bodies of clay, and many have thin layers of quartz-free clay along their centers (Figure 8-7). Where the white sheets can be traced down to granite, they connect with cracks in the granite that also contain clay sheets (Color Plate 3C). The quartz-free clay thus appears to be filling what were once cracks, and the array of white bands in the lower half of Figure A must therefore show the positions of cracks in what was originally granite. This relation is important because it proves that the lower part of the soil remained intact while it was being formed from granite and that it hasn't moved since. Such clay-rich soil, showing ghost-like but intact features of the parent rock, is called *saprolite.*

Determining the parent of the main soil layers The topsoil and the red layer beneath it, however, have no white bands (Figure A). To determine whether or not they were formed from the granite, I collected chunks of each and also chunks of the saprolite and granite. In the laboratory, I crushed about 30 grams of each to silt-size powder and stirred the powders separately into methylene iodide, a liquid of high density—3.3 grams per cubic centimeter. Most of the grains (quartz and clay) were so light that they floated on this liquid, but small numbers of grains were heavy enough to sink in it. Some of them were a mineral called zircon (zirconium silicate), described in Supplementary Chapter 1. Zircon occurs sparsely in granites and is so chemically inert that it is unchanged by soil-forming processes. Many studies of zircons have shown that their sizes and shapes are similar among all samples from a given rock but that they vary from one rock to another. The zircons separated from the upper two soil layers are similar to those from the saprolite and granite, suggesting strongly that all the soil was derived from the underlying granite.

Very thin slices (thin sections) were then made from the sample chunks in order to examine each layer with a microscope. The sections are so thin (.001 inch) that one can see through most of the mineral grains or photograph them with transmitted light (Color Plate 4). Compare the photographs care-

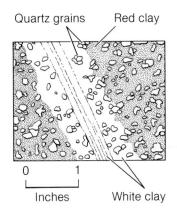

Quartz grains   Red clay

0   1

Inches   White clay

FIGURE 8-7
Closeup of one of the white clay seams of Color Plate 3A. Note that the central, laminated part is free of quartz grains.

SAPROLITE

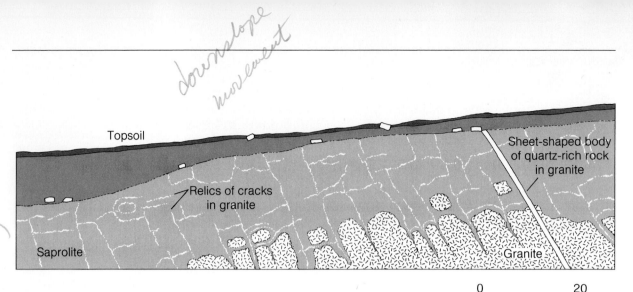

Topsoil

Relics of cracks in granite

Saprolite

Sheet-shaped body of quartz-rich rock in granite

Granite

0   20

Feet

FIGURE 8–8
Relations shown in a steep road cut near where Highway 601 crosses Lynches River, Lancaster County, South Carolina. The blocks of rock strewn through the upper soil layers came from the elongate (sheet-shaped) body shown on the right.

fully. Note that the quartz grains sketched in Figure 8-7 show up clearly and in detail in the thin section of the saprolite. They are like those in the granite. If the quartz in the upper layers (Figures A and B) came from the granite, however, they have been broken and moved about considerably.

Creep of the upper layers We already noted one indication of movement in the upper layers: the white clay sheets have been obliterated in them. Even better evidence for movement can be found in other places. Figure 8-8 shows a Piedmont soil that is like the Georgia soil except for a sheet-shaped body of fine-grained granite that extends up through the saprolite. The sheet-shaped body, called a *dike*, is only a little younger than the main granite but has resisted soil formation because it is almost impervious to groundwater. Note that it is undeformed in the saprolite but is broken and strewn downhill in the upper soil layers. Note also that the upper soil thickens downhill. These relations indicate that the soil above the saprolite has moved downhill somewhat like the Rio Grande soils measured by Leopold and coworkers, only here a thicker soil was involved. This movement probably occurs widely in the upper soil layers and explains why the quartz grains in Color Plate 4A and B have been broken and stirred.

Our findings so far may be summarized as follows: (1) the Georgia soil has probably all formed from the underlying granite; (2) the alterations of the parent rock started along cracks; (3) the thick lower part of the soil, the saprolite, was altered completely but remained in place and retained its original volume; and (4) the two upper layers have moved gradually downhill, much as though they had flowed.

The Clay Minerals

Color Plate 4 demonstrates a major aspect of the soil-forming process: Quartz has remained unaltered, but the more abundant minerals of the granite have been changed to clay. What exactly is clay? Broadly, it is what pots are made of—the earthy stuff that is slippery when wet, plastic (pliable) when moist, and hard when dry. When thoroughly stirred into water, clay breaks up into grains so small that they remain suspended for hours. We noted the grain-size definition of clay in Chapter 1; it is the sediment less than .002 millimeter in diameter (Figure 1-8).

Of greatest significance, however, are the constituents of the clay. These are largely clay minerals—tiny flakes and rods made up chiefly of oxygen, hydrogen, silicon, and aluminum, in proportions depending on the specific mineral. Each clay mineral forms under certain conditions. By studying the specific clay minerals, we can thus interpret the origin of individual soils.

The clay mineral kaolinite The simplest clay mineral, called *kaolinite*, is the dominant mineral of most soils in the southern Piedmont. Studies using X-rays have shown that kaolinite consists of layers of tightly joined atoms. As Figure 8-9A shows, silicon atoms lie in a plane extending through half of each clay layer and aluminum atoms lie similarly in the other half, all of them joined to oxygen atoms. A hydrogen atom (a tiny bump in the figure) is attached to each oxygen atom on one side of each layer. Viewed from the top, as in Figure 8-9B, silicon repeats in hexagonal groupings within each layer, which is why kaolinite crystals grow in hexagonal shapes.

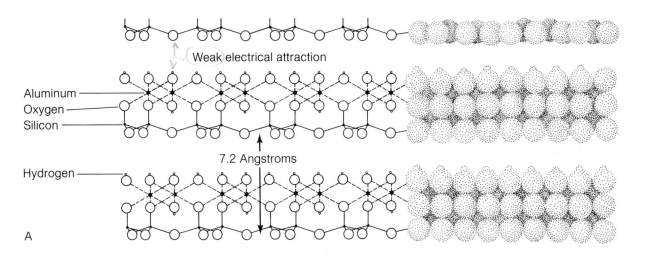

Aluminum
Oxygen
Silicon

Weak electrical attraction

7.2 Angstroms

Hydrogen

A

FIGURE 8-9

Arrangement of atoms and ions in kaolinite. A, edge-view of a layer, with lines representing bonds and dotted forms the space traversed by electrons. An angstrom is one ten millionth of a millimeter. B, diagram of a view down onto a layer showing the arrangement that leads to crystal-shapes like those in the electron microscope image (enlarged 12,000 times). Photograph by the Smithsonian Institution, courtesy of Kenneth M. Towe.

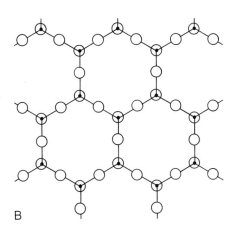

B

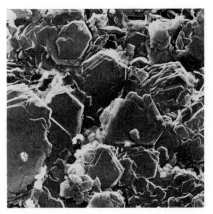

Each layer is electrically neutral and is held to adjoining layers by the local electrical attractions between facing oxygen and hydrogen atoms (Figure 8-9A). The attractions are so weak that the layers can slide against one another quite easily, making kaolinite feel somewhat slippery even when dry. This weak cohesion also makes kaolinite soils loose, which is one reason why the Piedmont's forest soils were so easily eroded when they were cleared and farmed (Chapter 2).

Properties of the expandable clays Considerably different from kaolinite are clay minerals called *expandable clays*. The makeup of a typical one is shown in Figure 8-10. Compared to kaolinite, the layers have two planar arrays of silicon rather than one and contain other elements—chiefly magnesium and iron—in place of some silicon and aluminum atoms. Because of these substitutions, the layers have a negative electrical charge that attracts positively charged ions into the spaces between them (Figure 8-10). The ions bring attached water molecules (recall the electrical polarity of water) and thus force the layers to lie farther apart than those of kaolinite.

The distance between the layers in expandable clays is so large, in fact, that additional water molecules squeeze in when the clay is wetted, causing it to expand (Figure 8-11, right). This property is important in several ways.

FIGURE 8-10

Arrangement of atoms and ions in an expandable clay, showing on the right how sodium ion and water molecules spread the layers. The lines and dotted forms have the same significance as those in Figure 8-9A.

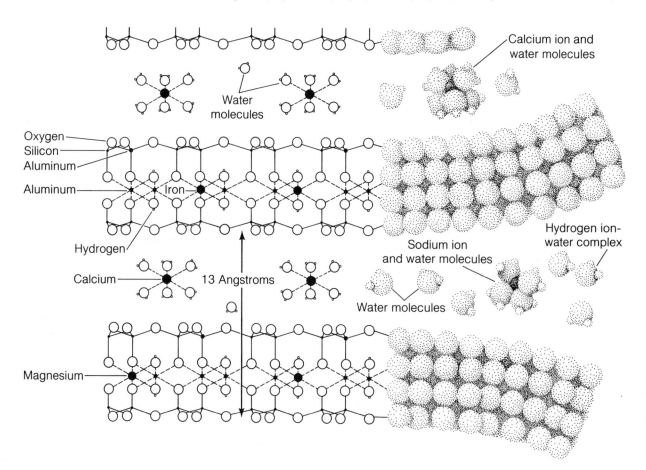

180

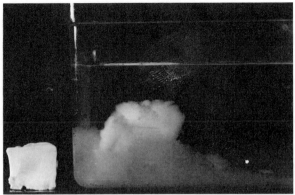

FIGURE 8–11

Effects of wetting and drying expandable clay. Right, two pieces of clay that were of equal size before one was immersed in water for 12 hours. Clay courtesy of Kendrick Fink. Photograph by Ruperto Laniz. Left, cracked clay-rich soil in northern Utah, dried but showing parts of the swollen forms due to the last wetting. The soil creeps down the slope toward the viewer and slides down the steep stream bank in the foreground.

It gives soils a tendency to swell tight during rains, thus increasing the proportion of water that runs off the surface. The fluffing and dispersion of expandable clays in water also increases the capacity of running water to transport coarser grained sediment, as noted for the Rio Grande (Figure 3-8). The alternate swelling and shrinking caused by expandable clays also results in downhill creep of soils (Figure 8-11, left).

Another important property of expandable clays is their ability to exchange their ions with others in the soil water. They commonly contain ions useful to plant growth that are released when plants reduce the concentration of the ions in the soil water. In contrast, kaolinite has few ions adhering to its simple layers, and once they are taken up by plants they must be replenished from other sources, either from fertilizers or from naturally decaying plant materials.

Formation of clay minerals How do the clay minerals form? The Georgia soil provides an interesting example in that the mineral called *mica* in the granite appears to have been altered first to expandable clay and then to kaolinite. I say "appears," because the changes have not been seen taking place—they are interpretations based on grains seen in thin sections.

The mica in the unaltered granite is black and shiny and in thin sections appears dark (Figure 8-12A). In somewhat altered granite it is paler and reddish, evidently because iron is removed from the grains and iron oxide is deposited in minute cracks next to them (Figure 8-12B). The layers in mica are much like those in expandable clays but they are held together by abundant potassium ions (Figure 8-12E). During soil formation, hydrogen ions and calcium ions replaced the potassium ions and thus converted parts of the

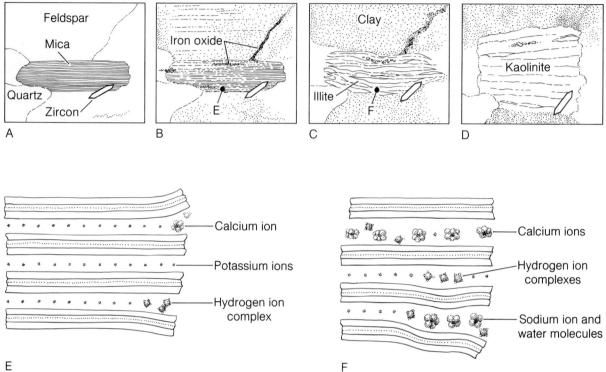

A   B   C   D

E   F

FIGURE 8-12

Stages in the alteration of mica to kaolinite (A to D). E shows the layer structure of mica at an early stage of alteration to layers of expandable clay, and F shows an intermediate stage, a mixed layering resulting in a mineral called *illite*.

mica to expandable clay (Figure 8-12C). As noted in the figure, the mixture is generally considered a separate clay mineral called *illite*—sometimes called *"soil mica"* because it may retain many properties of mica. Illite is important to plants because it can supply one of their vital nutrients, potassium, as well as other nutrient ions. It is abundant in many areas of the grassland soils already described, accounting for much of their fertility.

As the drawings suggest, the mica grains are enlarged as they are changed to illite and finally to kaolinite, but the layers in the clays retain the orientations of the layers in the mica. This relation must mean that new layers grow parallel to the original ones and, therefore, that atoms of aluminum and silicon are added from other parts of the rock. Enlargement tends to force the grains apart. Rocks thus tend to disintegrate mechanically when they are converted chemically to clay minerals.

Chemical Changes that Produce Soil

Kaolinite's simple composition implies the removal of many kinds of ions from the parent rock. Kaolinite thus forms readily in wet, warm climates, in places where rainwater moves easily through soil and rocks. We have just noted these effects on mica, which is a common mineral in many rocks. Similar changes affect other minerals, notably the most abundant minerals of all—the feldspars, described in Supplementary Chapter 1. In addition to oxygen, silicon, and aluminum, feldspars contain sodium, potassium, and

TABLE 8-1
Characteristics and Mineral Compositions (in Percent by Volume) of the Layers in the Soil of Plates 3 and 4 and its Granite Parent.

Layer	Color	Consistency when moist	Depth of base	Quartz	Clay	Iron oxide	Feld-spar	Mica
Topsoil	Brown	Crumbly, sandy	8 inches	73	20	2	0	0*
Subsoil	Deep red	Plastic (pliable)	5 feet	25	71	4	0	0
Saprolite	Red mottled with white	Crumbly to plastic	15 to 30 feet	24	70	3	2	1
Granite	Gray	Hard rock	Miles	30	0	1	63	6

*Plant material makes up about 5 percent of the topsoil at the depth sampled (6 inches); the proportion increases upward to nearly 100 percent in the uppermost inch.

calcium, all three of which tend to be removed by soil-forming alterations. The abundant oxygen, silicon, and aluminum remain to form kaolinite and other clays.

To determine these changes more exactly in the Georgia granite, I measured the minerals in it and in each layer of the soil by counting their amounts in thin sections. The results are shown in Table 8-1. The data show that all the feldspar and mica were altered to clay by soil-forming processes. Because feldspar contains calcium, sodium, and potassium and kaolinite contains none, these elements were evidently removed. Feldspar also contains more silicon than does kaolinite, so that silicon, too, was depleted. Evidently even the quartz was partly dissolved, for the granite contains 30 percent quartz and the saprolite 24 percent. Note that the soil-forming processes also concentrated iron oxide in the saprolite and the deep-red layer under the topsoil (often called the *subsoil*).

Altering rock to soil   Some of the dissolved substances were picked up by roots and used in plant growth, but most must have been removed from the soil by the downward flow of groundwater. It is not difficult to connect this leaching process with some other things we have observed. Recall from the James River studies that groundwater is visibly mobile in cracks in granite and similar rocks (Figure 2-5). Recall, too, that groundwater and surface water contain dissolved substances much like those in the rocks nearby. These facts fit well with the occurrences of clay along cracks in the Georgia granite. We can thereby propose an explanation: soil is formed when groundwater seeps down through rocks, gradually dissolving some substances and adding others, such that most minerals are changed to clay and iron oxide. Chemically inert minerals such as quartz and zircon are left more or less unchanged.

This scheme seems reasonable and is doubtless partly correct, but it dodges some important complications. Table 8-1 shows that clay and iron oxide are produced when granite is altered to saprolite, but it also shows that they decrease markedly in the topsoil. Quartz grains from the granite are thereby concentrated in the topsoil, as is apparent in Color Plate 4. The double role of rainwater seems contradictory—it forms kaolinite and iron oxide at the base of the saprolite but destroys them in the topsoil.

Laboratory studies of soil minerals A second related complication is based on experiments. If rainwater leaches chemical substances from mica and feldspar, changing them to kaolinite, we should be able to do the same thing in the laboratory. The problem is that we haven't, at least not in a way comparable to the scheme just mentioned. When water has been percolated through granite and similar materials in the laboratory, ions have been dissolved and removed, but kaolinite has not been formed.

A single set of experiments, however, stands apart. J. Linares and F. Huertas recently made kaolinite for the first time in the laboratory, under conditions similar to those in soils (Reference 3). They treated a solution of aluminum ions and silica with fulvic acid, an organic substance typical of decayed plant materials in many topsoils. Molecules of fulvic acid evidently form complex ions with aluminum, which may move in solution and deposit aluminum elsewhere, forming kaolinite. The experiments thus suggest a plausible explanation for the double role of clay in the Georgia soil: organic acids dissolve clays in the topsoil and carry aluminum to the lower part of the soil, where it is added to newly forming clays.

Some experiments made by John D. Hem indicate a similar mechanism for the transport of iron from the upper to the lower soil (Reference 4). He tested the mobility of iron in various solutions of tannic (digallic) acid, which is formed when the common plant substance called tannin is dissolved in water. Hem found that weak solutions of the acid, similar to those normally expected in groundwater, have little effect on iron oxide, but that fairly concentrated solutions (0.05 percent) dissolve and transport iron, apparently by forming complex ions with it. As he pointed out, only in topsoil is tannin likely to be concentrated enough to do this. Iron thus dissolved in the topsoil would be deposited again in the lower soil, either because the tannic acid would be diluted or because the organic parts of the complex ions would become oxidized.

These mechanisms for transporting iron and aluminum are not completely understood, but they are doubtless very important. Organic substances, and therefore life forms, are thus crucial to forming the beautifully layered Georgia soil and probably most other soils.

Plants—The Other Parents of the Soil

Some of the most direct evidence that soil is formed by organic processes comes from studies of the small scaly plants called *lichens*. These plants are symbiotic (intersupportive) mixtures of fungi and algae and are typically the first life forms to colonize freshly broken surfaces on rocks. Eugene T. Oborn, who was especially interested in the ability of lichens to dissolve iron from rocks, found that the first lichens to appear, the tightly clinging patches called crustose lichens, contain between 1.5 and nearly 4 percent iron, which may be compared to the 0.002 percent he determined to be present in such plants as spinach (Reference 5). Evidently the lichens dissolve iron and other elements directly from the rock by the action of organic acids, partly along roots that penetrate as much as several millimeters into minute cracks.

Lichens also attack rocks mechanically, disintegrating them by shrinking and swelling caused by drying and wetting of the lichen's gelatinous substances. Oborn dried similar plant gelatins on smooth glass plates and found their adhesion and shrinking so powerful that it rent flakes of glass from the surfaces! Crustose lichens thus prepare ground for larger, *foliose* (leaf-like) lichens and for mosses, the earlier forms decaying into nutrients and a thin granular soil that supports larger and larger plants. The plants also entrap dust to add to the thin soil layer.

Figure 8-13 illustrates a case in point: a surface eroded down to unaltered granite by the ice sheet that advanced across Maine during the last glaciation. The surface was exposed to the atmosphere about 8,000 years ago, at which time it must have been scraped as smoothly as the one shown in Figure 7-9, left. The latter surface has been changed little since then because the rock consists of tightly interlocking grains of quartz and other relatively inert minerals. The granite of Figure 8-13, on the other hand, has been affected by chemical reactions like those described in the last section and by the mechanical effects of frost-wedging (the opening of water-filled cracks by expansion of freezing water). The surface has also been used by lichens and mosses, which have eroded it and formed a thin soil in pits and cracks.

Organic alterations along cracks   The lines of larger plants are growing along what were originally cracks in the granite, which are hidden by the plants but can be seen clearly where the granite has been excavated (Figure 8-14). The cracks—called *joints* by miners and geologists—form when rock bodies are deformed slightly. The joints parallel to the ground surface, for example, formed because the rock expanded upward when erosion removed the rock load that once lay over it. Joints are important to the origin of soil

FIGURE 8-13

Horizontal surface on Mount Desert Island, Maine, eroded by glaciers on granite and exposed to the weather and plant actions approximately 8,000 years ago. The small, dark patches on the granite are crustose lichens, and the gray patches are pits corroded by former colonies (as in the lower left corner). The light and dark patches in the cleft just above the knife are foliose lichens. Note how plants and dark organic substances have been concentrated along the cracks.

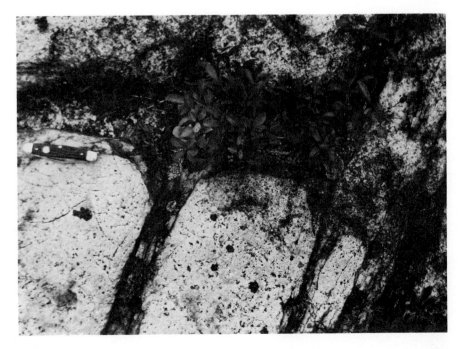

FIGURE 8-14

Major cracks (joints) in granite exposed in a shallow excavation on Mount Desert Island, Maine. The locality, only 3 miles from that of Figure 8-13, has a thicker soil and a forest cover because a thin layer of till covered the granite when the glacier melted and also because it is more sheltered from the wind.

because they provide channels for the larger roots of trees. Recall that joints are also the principal channels for groundwater, and that soil-forming alterations are most intense along similar cracks in the granite in Georgia. The joints in Figure 8-14 show comparatively little alteration at depth, because the surface has been exposed for a relatively short time. Figure 8-13 shows, nonetheless, that plants and the first-formed soil are concentrated along them, suggesting a source of organic acids that will slowly change feldspars to clays along the cracks beneath.

Actions of soil microbes Bare rock is scarce in regions with deeply developed soils like those of Georgia, but the importance of organisms in such areas is suggested by their numbers alone. A gram (a good-sized pinch) of rich topsoil is likely to contain tens or hundreds of millions of bacteria, millions of fungi, and thousands of algae. This buried microflora weighs about 2 to 5 tons per acre—roughly the same weight as the living plant protoplasm (leaves and live bark) above ground.

The importance of these microbes in dissolving mineral substances was demonstrated in further studies made by Eugene T. Oborn and John D. Hem, who were interested primarily in the solution and transport of iron in soils (Reference 6). When they mixed pure water, sand, plant detritus, and manure, and kept the mixtures sterile (microbe-free) for two weeks, the iron dissolved in the water generally amounted to less than one-hundredth of one part per million. When microbes were permitted to live in the mixtures, however, the dissolved iron increased about one hundred times. Moreover, when the activity of the microbes was increased by raising the temperatures of the laboratory soils, the dissolved iron also increased. Similar results were obtained with natural soils.

The iron-dissolving actions were probably twofold. First, the microbes broke down iron-bearing organic substances and thereby released soluble

TABLE 8-2

Concentrations, in Parts Per Million, of Important Nutrients in Loblolly Pine and in a Typical Forest Soil of the Southern Piedmont. From U.S. Forest Service Southeastern Experiment Station Papers 17 and 28.

Material Analyzed	Potassium	Calcium	Magnesium	Phosphorus
Pine needles	4,000	3,000	1,000	1,000
Pine stembark	3,000	2,000	600	200
Pine stemwood	1,000	600	300	200
Upper part of topsoil	200	1,300	220	14
Lower part of topsoil	80	300	120	3
Upper part of subsoil	70	150	120	1
Lower part of subsoil	80	200	150	1

iron complexes. Second, the microbes' life processes produced carbon dioxide that dissolved in soil water to form carbonic acid, like in carbonated drinks. The acid added hydrogen ions to mineral grains and released metal ions from them, as in Figure 8-12E. Iron is thus dissolved in the Georgia topsoil (where microbes are by far most abundant) and transported to the subsoil and the saprolite.

Organic processes and soil layers Soil microbes in woodland areas depend basically on the trees, which use solar energy and soil nutrients to synthesize new plant materials. When these materials fall as litter they are slowly consumed by the microflora. The litter is also eaten by soil animals, some of which also feed on the trees, on the microflora, or on each other. A gram of rich soil contains thousands of protozoans, and each square foot of litter covers hundreds of roundworms and minute arthropods. Larger animals such as pill bugs, millipedes, earthworms, insects, reptiles, and mammals are less numerous but so energetic as to stir, transport, and break up soil particles almost continuously. Soil is also moved by plant roots, which wedge grains and rocks apart along cracks and even lift them to the surface when trees are blown over by the wind.

All these actions produce soil mechanically and contribute to its downhill creep. The soil in the southern Piedmont remains too moist and too warm to creep downhill because of wetting and drying or freezing and thawing. Additional evidence that living things cause the creep is circumstantial but strongly suggestive: the rates of creep are greatest near the surface, where roots and organisms are by far most abundant.

The density of roots and soil organisms also correlates with the soil layers, and Figure 8-15 suggests how the layers relate to the cycling of dissolved nutrients. The tree gains newly dissolved ions by its deep roots, but the great concentration of roots in and just under the topsoil shows that it feeds mainly on the breakdown products of its own litter. This conclusion is supported strongly by chemical analyses of nutrient elements in the tree and in the soil layers (Table 8-2). The data show that trees and their litter are storage places for most of the potassium and phosphorus that become available in the soil.

In summation, plant roots are zoned with the soil layers in woodlands and create the soil in grasslands. In both cases, the upper soil layer coincides with

FIGURE 8–15
Cycling of substances through a
forest soil like that of Georgia.

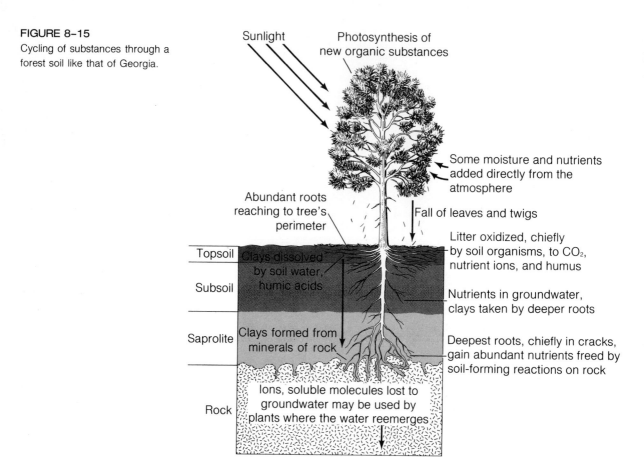

Sunlight

Photosynthesis of
new organic substances

Some moisture and nutrients
added directly from the
atmosphere

Abundant roots
reaching to tree's
perimeter

Fall of leaves and twigs

Litter oxidized, chiefly
by soil organisms, to CO_2,
nutrient ions, and humus

Topsoil

Clays dissolved
by soil water,
humic acids

Subsoil

Nutrients in groundwater,
clays taken by deeper roots

Saprolite

Clays formed from
minerals of rock

Deepest roots, chiefly in cracks,
gain abundant nutrients freed by
soil-forming reactions on rock

Ions, soluble molecules lost to
groundwater may be used by
plants where the water reemerges

Rock

the greatest concentration of organic detritus, of living organisms, and of available nutrients. Mechanical stirring there is by far the greatest. The experiments of Hem and Oborn and of Linares and Huertas link organic substances with the chemical movements of iron and aluminum, and thereby the development of clays and iron oxide. From the standpoint of mineral content, clays and iron oxide define the layered soil, and it is an intriguing fact that these substances then support the living things that made them!

Factors in the Origin of Soils

Together the soils of grasslands and woodlands constitute the earth's main agricultural resource and cover all but a small part of its temperate regions. We have looked at certain aspects of each soil, and in this last section we might try to compare them briefly. What factors have caused their differences? Which soil-forming processes are similar in the two kinds of soil?

The main differences appear to be due to plants. The fertility and much of the physical character of grassland soils result directly from the annual production of dense root systems by grasses. In contrast, the annual litter of a forest lies on the surface, and the nutrients carried downward cause a pronounced zonation of roots. Organic acids formed from the litter dissolve aluminum and iron in the topsoil and carry them down to construct clays and iron oxides at depth, resulting in a distinctly layered soil. The clays are commonly rich in kaolinite, whereas grassland soils contain mostly illite and expandable clay.

Because time is a major factor in soil development, age must be considered carefully in making comparisons between soils. The beautiful Georgia soil is at least one million years old, which is perhaps the main reason why it is so much more developed than the other soils described, in particular why it has a thick saprolite. All the other soils began to form at about the close of the last glaciation and are thus of comparable ages—between 8,000 and perhaps 15,000 years old. On that basis, the woodland soils appear to have developed more rapidly than the grassland soils, for iron and aluminum have been markedly dissolved and transported downward in the woodland soils in Maine. This more rapid development is presumably due to the organic acids formed from woodland litter.

Both soils are produced under a wide range of temperatures and both show geographic variations due to temperature. Recall that the grassland soils become leaner and more orange or reddish to the south, due to increased oxidation and the formation of iron oxides at higher temperatures. The same general change can be seen in woodland soils. Topsoils and subsoils in Maine and other northern woodlands are gray, whereas those to the south become progressively browner and, in the southern States, orange and red. In tropical areas, for example, Puerto Rico, iron and aluminum oxides are so concentrated in the subsoil that it can be mined locally as a source of those metals.

Of the other climatic factors, rainfall is often cited as a principal cause of differences between soils, and we have noted its effect in producing thin as opposed to thick grassland soils. How much of this difference is due to the plants developed and how much to water passing downward through the soil is difficult to determine. Although most grassland regions are drier than most woodland regions, large areas of each have nearly the same amounts of precipitation.

Differences among parental materials cause major differences among young soils, but these effects become much less important with increasing age. Porous sediments are doubtless the major cause of thick young soils. The deep, porous loess of the plains permitted grasses to develop a thick soil rapidly. The soil in Figure 8-14 is thicker than that in Figure 8-13 because of a layer of till that was not present at the latter place. In contrast, the old woodland soils of the eastern states not covered by the ice sheet are so thick and so similar that it is difficult to tell when one has passed from one parental rock to another.

Soil depletion due to human uses The factor affecting soils most drastically today is human use. The loose topsoil of woodlands has been eroded widely, and because the trees and their litter have been cleared and burned, the soils that remain require repeated fertilizing. We have noted the susceptibility of the grassland soils to wind erosion when their sod was broken. Even where the grassland has not been eroded, its dark sod has faded in place. Depletions in carefully tended, uneroded fields are depicted graphically in

FIGURE 8-16

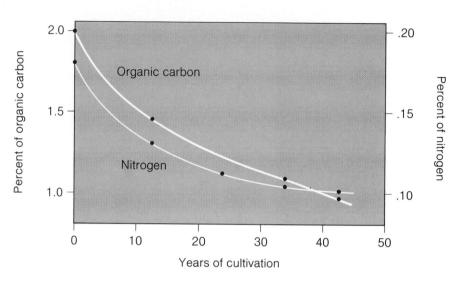

Figure 8-16. Note that about half of the original organic materials was lost in the first 42 years of agriculture, presumably because of removals by crops and exposure of the organic materials by plowing.

Depleted soils can be kept productive but only by expenditures of energy, materials (fertilizers and chemicals), and, often, imported water. Because these things are becoming scarcer and more costly each year, it is crucial to understand the nature of each local soil and to discover the crops it will yield abundantly without undue depletion. We may note some results in 74 plots in northwestern Kansas used to test different crops and tilling practices (Reference 7). A plot in 4-year rotation to milo, corn, winter wheat, and rye (the latter plowed in as green manure) showed a 31 percent decrease in soil nitrogen and a 52.9 percent decrease in organic carbon during the period from 1916 to 1946. The same soil plowed once in the spring and planted to oats lost only 5 percent of its nitrogen and was enriched by 2 percent of organic carbon during the same period. Here is a specific relation expressing what we found broadly by exploring the origin of soils: soils and plants are intricately interrelated. To maintain a productive soil resource, we must continue to explore these relations in untilled soils as well as in soils under agriculture.

Summary

Soils form by the combined effects of plants, animals, moisture (principally rainfall), heat, specific properties of the parental materials, and age. Some important relations among these effects can be summarized briefly:

1. The agriculturally vital grassland soils consist of a tough sod and a rich, dark layer formed by the annual growth and decay of grass roots.
2. This young soil formed rapidly because of the vigor of grasses in somewhat harsh, windblown environments and because of the porous nature of recent glacial deposits, especially of the wind-deposited silt called loess.

3. Grassland soils may be destroyed rapidly when their sod is broken, for the underlying loess and windblown sand are highly erodable.

4. Rich soils may also be developed from hard rocks, partly by the chemical action of water descending along cracks (joints) and partly be the actions of lichens, microbes, plant roots, soil animals, and freezing and thawing of soil water near the surface.

5. Organic acids derived from plant materials are particularly effective in converting rocks to soil; the acids dissolve aluminum and iron from rock-forming minerals and carry them downward in the groundwater, forming clay and iron oxide in the subsoil and the saprolite.

6. Clay minerals are the most important substances formed in soils; their chief varieties are: (1) kaolinite, a simple aluminum mineral formed in well-drained soils occurring in moist, warm regions; (2) expandable clay, an aluminum mineral with a variety of additional ions, formed in dry climates, young deposits, or poorly drained soils; and (3) illite, a potassium-bearing aluminum mineral formed under intermediate conditions from potassium-rich minerals.

7. Among the substances dissolved from rocks and soils and carried away in the groundwater are potassium, calcium, magnesium, and phosphate; plants must gain a major share of these nutrients from the decay of their own litter.

8. Soils slowly become segregated into a topsoil, where roots and other organic materials are concentrated and where clays and iron oxide are depleted, and a subsoil, where clays and iron oxide become concentrated.

9. Human uses have generally depleted soils, partly due to the erosion of the loose topsoil when its plant cover is removed and partly to oxidation of organic substances exposed to the sun and air by tilling.

REFERENCES CITED

1. Hugh H. Bennett. *Soil conservation*. New York: McGraw Hill Book Co., 993 pp., 1939.
2. H. T. U. Smith. Geological studies in southwestern Kansas. University of Kansas Bulletin 34, 212 pp., 1940.
3. J. Linares and F. Huertas. Kaolinite: synthesis at room temperature. *Science*, vol. 171, pp. 896–97, 5 March 1971.
4. John D. Hem. Complexes of ferrous iron with tannic acid. U.S. Geological Survey, Water-Supply Paper 1459-D, pp. 75–94, 1960.
5. Eugene T. Oborn. A survey of pertinent biochemical literature. U.S. Geological Survey, Water-Supply Paper 1459-F, pp. 111–90, 1960.
6. Eugene T. Oborn and John D. Hem. Microbiologic factors in the solution and transport of iron. U.S. Geological Survey, Water-Supply Paper 1459-H, pp. 213–35, 1961.
7. J. A. Hobbs and P. L. Brown. Nitrogen and organic carbon changes in cultivated Kansas soils. Kansas State College of Agriculture and Applied Science, Manhattan, Agricultural Experiment Station Technological Bulletin 89, 48 pp., 1957.

ADDITIONAL IDEAS AND SOURCES

1. In addition to the *regional* variations described, soils vary *locally* for the same reasons (differences in vegetation, moisture, temperature, parent materials, and age). For example, the soils on the two sides of the canyon in Figure 3-11 are distinctly different. Try examining soils near where you live to see if you can see reasons for the local variations. When large fields are freshly plowed, you may see obvious differences in color or tone that you can examine further.

2. To determine whether soils have been mapped in detail nearby, write the Soil Survey, Soil Conservation Service, U.S. Department of Agriculture, Washington, D.C. These surveys include aerial photographs or maps with the areas of different soil types shown on them. County planning commissions and agricultural colleges may also have soil maps you can examine.

3. Some terms used in soil surveys (and by soil scientists generally) are different from those I've used. They call the layers *horizons*—the topsoil being the A horizon, the clay-enriched (or reddened) subsoil the B horizon, and altered materials below that level the C horizon. I have included the more common of these terms in the glossary in this book, but if you want to read the soils literature more extensively you may obtain a copy of the *Glossary of Soil Science Terms* (at a cost of $1.00) from the Soil Science Society of America, 677 South Segoe Road, Madison, Wis. 53711.

4. In sampling soils, try to collect them moist and keep them that way—by using thin plastic bags wrapped tightly around the sample (which should be cut out intact, to preserve its features). You can later crumble part of each sample and make an approximate size analysis by the settle-and-decant method described in Chapter 1. This will give you cleanly washed and sorted samples of the coarser material (sand, gravel), which are useful indicators of the parent material.

5. Color Plate 3 shows an interesting relationship for persons using roundness of boulders and cobbles to interpret amounts of abrasion during transport in streams. The granite relics have been rounded beautifully with no transport whatsoever! This effect is common enough to be worth some thought. The white clay seams show clearly that the granite was originally segmented into angular masses by the cracks. Why don't the relics simply remain angular as alteration progresses outward from the cracks?

Death Valley viewed from the south. Photograph by John S. Shelton.

9. Landscapes and Soils of a Desert

Landscapes and Soils of a Desert

To further understand the effects of climate on the earth's solid surface, we move farther west and south to a true desert, part of which is shown in Figure 9-1. This is a fascinating, highly visible place. The mountains have an endless variety of craggy forms, all ending abruptly at the edge of smoothly sloping plains. Space seems limitless and plants are so sparse that the entire scene is colored by soils and rocks—partly gray but chiefly tints of orange, red, and lavender. In Figure 9-1, all the rocks from the foreground to the distance are granite. In that sense the area is comparable to the parts of Georgia and coastal Maine described in Chapter 8, and the large and obvious differences are caused primarily by rainfall and evaporation. The area in the photograph receives only a few inches of rain annually, and the sun shines on it nearly 4,000 hours per year—90 percent of the total of daylight hours!

The scene is not a rarity. Similar desert country, receiving up to 10 inches of rain a year, extends from the Rio Grande drainage of west Texas and New Mexico across all of southern Arizona to California, and thence northward across most of Nevada to eastern Oregon. In all, this desert region covers about 200,000 square miles. Its dryness is created by high mountains to the west, especially the Sierra Nevada, that cause moist air moving eastward from the Pacific Ocean to rise to great heights in passing over them. The air is thereby cooled and precipitates most of its moisture on the mountains, creating a dry area—a *rain shadow*—to the east.

The region is one of interposed plains and mountain ranges—about 200 ranges in all. The largest are about 80 miles long and 20 miles wide, but most are about a third that size. Between the ranges are gently sloping plains, such as that in the photograph, that interconnect as a sort of matrix to the region. Through-going highways zig-zag moderately to stay on the plains but also rise and fall appreciably, because the plains have various slopes and low divides between one segment and the next. Some parts of the plains are so completely enclosed by mountains and divides that they form separate basins or troughs. These basins are typically marked by a *playa* (an ephemeral lake) that catches their inward-directed drainage and in some cases forms a salt-encrusted flat.

FIGURE 9-1

A typical desert mountain range and adjoining basin, part of the Orocopia Mountains, southeastern California.

Although the region is sparsely inhabited, it is not a wasteland. Strong winds occasionally winnow dust and sand from the plains, but sand-dune scenes such as those of the movies are exceedingly scarce. Indians grew crops in even the most southerly areas, and the high temperatures and abundant sunlight make modern agronomists dream of importing water for year-round controlled agriculture. The solar radiation also has tremendous potential as a source of energy. M. King Hubbert, who has studied energy sources and energy capabilities more than anyone else I know of, has pointed out that 10 percent of the area of Arizona could have generated all the electricity used in the United States in 1972 (Reference 1). This source of energy is undiminishing and its use would probably be far less destructive to water and clean air than the sources now being used. Deserts may thus see great changes in the next few decades. They are, however, our last fund of relatively unspoiled wilderness, so they should first be studied and understood with care.

Faults and Sediment Fans

Any plans for using the desert region must take into account its instability. The basins have been displaced bodily downward relative to the ranges, typically along steeply inclined faults that have broken the region into elongated blocks (Figure 9-2). In addition, some of the basins have sagged downward and some of the ranges have arched upward, perhaps as a result of small displacements along many closely spaced faults. Without question faulting has taken place, for several marked displacements have occurred in the past 100 years, each causing an earthquake and producing a low cliff near the foot of a range. The historic displacements show that the basins and ranges were not generated in one movement but rather by repeated offsets of typically less than 100 feet or so over thousands of years. Some basins and ranges have formed more recently than others and some more rapidly than others. Their relative ages are indicated by the degree to which the ranges are eroded and the basins filled. The parts of Nevada, California, and Oregon east of the Sierra Nevada and Cascade Range have been especially active recently. Within that subregion, Death Valley is particularly suitable for studies of recent faults and of rapid erosion and deposition, for it has sunk so rapidly and so recently that it is partly below sea level. As the photograph on the chapter opening page shows, the sparseness of the vegetation makes the land forms marvelously apparent.

The most easily studied features in Death Valley are sediment deposits at the foot of the mountains. In their simplest form, the deposits are one-half of a low cone with its apex at the mouth of the canyon (Figure 9-3, left). Note how a single stream has created two complementary land forms: it has

FIGURE 9-2

A, idealized basin-and-range blocks, shown as if faulted so rapidly that they were not eroded. B, typical basins and ranges resulting from erosion and episodic faulting. Arrows show relative motion along each fault.

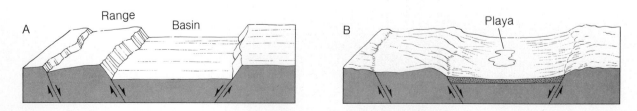

FIGURE 9-3
Left, sediment cone on the east side of Death Valley, with salty mudflat of the playa in the foreground. The fault along which the basin subsided lies at the base of the steep mountain slope. Right, view of the same cone from a point about halfway up the mountain. The canyon mouth is hidden by the rocky ledges in the center foreground. The playa, with its white salt deposits and dark gray mudflats, extends to the base of larger fans on the far (west) side of the valley.

cut a deep canyon in the mountains, and it has spread the detritus out on the floor of the trough to form an evenly graded cone. To get a better feel for this spreading, one may climb the steep mountain side and look down onto the cone (Figure 9-3, right). Each fine, wiggly line is the bank of a dry stream channel. The radial pattern expresses the spreading of sediment from the mountain, as does the cone's curving perimeter—its edge against the nearly horizontal playa that occupies the center of the basin.

Because of their radial lines and semicircular shape, such deposits are called fans, specifically *alluvial* (stream-deposited) *fans.* They are widespread and obvious in arid and semiarid regions and are also common in moister regions. Indeed, the largest fans in the world are formed in ocean basins, at the lower ends of submarine canyons! A fan originates for one basic reason: a stream confined by canyon walls can spread and flow in various directions when it empties onto the adjoining plain. As the stream does so, it typically widens and therefore becomes shallower, thus reducing its velocity and turbulence and thereby its transporting power. The stream therefore deposits bars of sand and gravel that tend to deflect it, sending it off along a new radius of the cone-shaped surface. Over thousands of years, the stream shifts widely over the fan, as suggested by Figure 9-3, right.

Visiting the fan's surface, one finds deposits of sand and gravel criss-crossed by dry stream channels that are at most a few feet deep. Here and there, however, are distinct gravel ridges that curve sinuously down the fan and have crests 3 to 10 feet above the rest of the surface (Figure 9-4). These ridges are remnants of debris flows—mixtures of coarse gravel, sand, and sparse mud that streamed down the fan like tongues of wet, gravelly concrete. They contain far less mud and plant materials than the James River debris flows but probably also originated from landslides during rains in the mountains. Debris flows move as welts or lobes in the shape of a streaming tear (Figure 9-4). The lateral edges of the moving lobe tend to become immobilized before the central part, which flows on to leave lateral ridges like that in the photograph. The ridges occur only on the upper half of this particular fan, probably because debris flows are more viscous than ordinary streams. The upper part of the fan thus slopes about 10°, whereas the lower half slopes 5° and less.

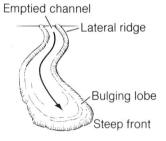

FIGURE 9-4
Top, ridge of sandy gravel about one-third the way down the fan of Figure 9-3, deposited by a debris flow. Bottom, moving debris flow, looking upstream, with lateral ridges like the one in the photograph.

Emptied channel
Lateral ridge
Bulging lobe
Steep front

FIGURE 9-5

Cobbles of impure marble (mainly calcite but with iron-magnesium silicate minerals) from the fan of Figure 9-3. The one on the left is from the modern channel, and the one on the right from the oldest part of the fan. Each is 5 inches in diameter.

FIGURE 9-6

Desert pavement on the lower part of a large alluvial fan 50 miles southwest of Phoenix. Wind and sheet flows have removed the fine sediment near the surface, leaving a lag one pebble thick.

Distinguishing older fans Young fans grow actively across their entire surface, but older ones become segmented into active and inactive parts. In Figure 9-3, right, the lightest stripes are active, and the dark area on the left has long been inactive. The darkening results from the weathering (surface alterations) of the stones on the fan surface; the degree of weathering can be judged by selecting one kind of rock and comparing weathered pieces with freshly abraded pieces in the active channels (Figure 9-5). The dark patches on the darker cobble were originally a nearly white mineral (a silicate of iron and magnesium) that became stained when the iron in it was oxidized to ferric oxides. The lighter, pocked areas are calcite ($CaCO_3$) that has been partly dissolved, especially along fine cracks that were enlarged in the process. The amount of solution and the darkening both give a measure of the age of this part of the fan.

The degree of surface erosion also indicates the relative ages of parts of the fan. In Figure 9-3, right, note that several light-colored stream courses cross the older (left) part of the fan and that recent streams have eroded a high bank curving down the center of the fan. In some large fans, a deep channel is gradually eroded across the entire fan, and the stream is then constrained in that channel. The apex of fan deposition is thus shifted to the foot of the fan, where a new fan begins to form. This relation may be caused by deepening of the canyon above the original fan apex or by tilting (steepening) of the original fan due to faulting or warping along the mountain front. Whatever the cause, the older parts of the fan are then subject to erosion and may become deeply incised by gullies.

Where old fan surfaces are not eroded away, rain, wind, and soil creep slowly smooth out the stream channels and eventually erase all signs of them. Fine sediments are blown away by the wind or washed away by occasional sheetflows. These actions tend to concentrate pebbles and cobbles at the fan surface to form a *desert pavement*. No part of the fan in Figure 9-3 is that old, but Figure 9-6 shows a typical desert pavement on the lower part of a large fan in Arizona. Note that the traffic of small animals (there are no cattle in the area) has cleared off the larger stones, showing that the pavement is essentially one stone thick.

FIGURE 9-7

Relics of desert mountains and part of an extensive plain on the west side of the Ajo Mountains, Arizona.

Eroded Plains Called Pediments

When faulting slows down or ceases, all aspects of the landscape begin to change. The mountains are eroded lower and lower, which reduces the amount of precipitation on them as well as the potential energy (vertical fall) of runoff. The fans thus receive less and less sediment, and a stage is reached when streams flowing from the mountains carry so little sediment that they consistently erode the fans. Rainfall on the fans contributes to this dissection. Meantime, because the basins have stopped sinking, they fill with sediment, and runoff that would have been held in a playa flows to the next adjoining part of the plains and thence to another part. Eventually the drainage becomes integrated into a river system such as that of the Colorado or Rio Grande.

The most stable parts of the desert region—southern Arizona and contiguous areas—are in this stage of landscape evolution. Plains there cover two to three times as much area as do mountain ranges, whereas basins and ranges are of about equal extent in the more recently active (faulted) part of the region already described. The desert plains in southern Arizona have been enlarged by erosion of the mountain ranges, and one can see many stages in this enlarging process. Figure 9-1 shows a case in which the nearly straight, faulted mountain front has been dissected greatly. The main canyons of the range probably once looked like the one in Figure 9-3, but they are now broad floodplains that divide the range into relict spurs (subsidiary ridges) pointing toward the main adjoining plain. Figure 9-7 shows a later stage, in which each spur has been reduced to a series of island-like peaks.

A search of the plains near the mountains generally discloses exposed solid rock here and there, showing that these parts of the plains formed by erosion rather than deposition. Such plains are thus basically different from alluvial (deposited) plains such as Death Valley and are called *pediments* (Figure 9-8). The surface eroded on rock is covered by sand and gravel a few inches to 10 feet thick, so that pediments superficially resemble alluvial plains. Junctions between the two become apparent when wells are drilled for water

FIGURE 9-8

Vertical sections through a basin and its bordering mountain ranges, the left one representing a stage when faulting was frequent and the right one a stage when faulting was so infrequent that a pediment had been eroded across parts of the ranges.

Alluvial plain   Playa

Pediment   Alluvial plain   Added sediment

Fault inactive

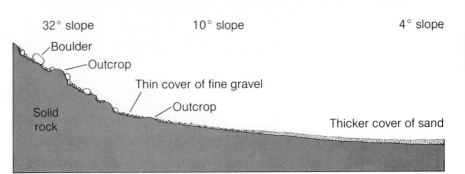

32° slope   10° slope   4° slope

Boulder
Outcrop
Thin cover of fine gravel
Outcrop
Solid
rock   Thicker cover of sand

FIGURE 9–9

Vertical section through the base of a mountain and part of a pediment like those shown in Figure 9-7. Note that the inclinations of the slopes correspond to the sizes of the particles that cover them.

or where parts of the plain have been uplifted and dissected enough to expose the rocks (or older deposits) underneath.

But how is a pediment incised across hard rock? Why do the mountain relics retain steep sides as they are eroded back? These questions apply broadly to mountains and adjoining plains, for they could be asked of temperate as well as of desert regions. The open nature of the desert, though, makes it an ideal place to look for clues. Let's explore the landscape of Figure 9-7 in some detail.

Correlation of slopes and sizes of particles   The steep slope in the foreground is exactly like the steeper slopes on the two small peaks in the distance. Some of the large rock bodies are part of the solid mass of the mountain, but most are separated boulders of the same rock, lying among smaller fragments. These bouldery slopes are inclined at 25° to 32° to the horizontal. The slopes forming the lower right half of the large peak in the distance are more gentle, being inclined at 15° to 20°, and the rock fragments there are all smaller than fist size. The pediment, too, shows a variation in slopes and in grain sizes. The upper part, as in the right foreground, slopes at 10° away from the base of the steep, bouldery slope; the larger fragments there are approximately an inch in diameter (Figure 9-9). The gently sloping parts of the pediment, which are inclined at 2° to 4°, are covered mainly by fine gravel and sand. Studies in other parts of the desert region (made chiefly by others) support these findings so repeatedly that one can make this generalization: the angles of slopes correspond to the grain size of fragments on them.

A related generalization derives from the area of Figure 9-7, as well as from many others: very few boulders move from the steep slopes to the pediment. How, then, do the mountains get eroded back to form the pediment? We need to search more closely for something that gives a feel for the actual progress of erosion, and of great value is a dark substance celled *desert varnish* that slowly coats exposed rock surfaces (Figure 9-10). It consists mainly of iron oxide and manganese oxide deposited in minute increments from rain, dust, and the rocks themselves. Bacteria and microscopic plants may play an important role in its formation.

An important relation is that the varnish covers outcrops and boulders to different degrees. In the case shown in Figure 9-11, right, the darkest, most heavily varnished part of the surface must have been exposed the longest; the medium-gray parts must have been exposed more recently; and the pale-gray

FIGURE 9–10

Pieces of granite from a steep (32°) slope in the Maricopa Mountains, Arizona: top, fresh rock hammered from the core of a boulder; center, natural cobble coated by desert varnish; bottom, fragment that has crumbled so rapidly as to be unvarnished.

FIGURE 9-11
Left, pediment surface sloping 9° toward the viewer. The locality is 200 feet from the base of the slope in the foreground of Figure 9-7. The gravel forms a thin layer over rock like that forming the irregular outcrops. Right, outcrop of granite at the foot of the slope from which the pieces shown in Figure 9-10 were collected. All the rock in sight is the same pale granite, coated to various degrees by desert varnish.

parts, which are like the fragment on the top in Figure 9-10, must have been exposed so recently as to be unvarnished. The detailed shapes of the surface help complete our interpretation: flakes and curved slabs have broken piecemeal from the outcrop. They can be seen at the base of the outcrop and in the foreground and are in various stages of decay into fine gravel and sand, many being similar to the cobble on the bottom in Figure 9-10.

Development of the pediment   Sand and fine gravel such as that in parts of Figure 9-11, right, are also found in rivulet channels that spill out onto the pediment, which is covered largely by the same materials. It thus appears that the mountain slopes are eroded back as fast as the outcrops and boulders on them disintegrate into grains small enough to be washed down onto and across the pediment. As the grains are washed away from the steep slopes, more rock is exposed and more boulders are formed. As a result, the slopes maintain their character as they are eroded back.

The pediment is also very slowly being eroded downward. Figure 9-11, left, shows a nearly planar part of it with rock outcrops that stand only an inch above the gravel surface. Most of the gravel on the right is moving across the pediment from a steep mountain slope behind the view, but note that the outcrop, too, is disintegrating into similar gravel. Combining this evidence with what we learned of the steeper slopes, the progress of pediment cutting might proceed over many thousands of years as shown in Figure 9-12.

This prolonged development might suggest that once pediments have started to form, the mountains and plains are forever stable, but this is not true. Our findings indicate that pediments formed on similar materials should have similar slopes. To see if they actually do, Jacqueline Mammerickx studied forty-nine pediments in California and Arizona, all cut on granite or similar rock (Reference 2). She measured the average slope on the upper 3,500 feet of each pediment and found the slope angles surprisingly different, ranging from 1.5° to 4.5°. These differences suggest strongly that some pediments had been tilted, warped, or affected by unknown climatic differences.

She found other indications that they had been tilted or warped. Some of the pediments have young alluvial fans lapping onto their upper slopes, a relation showing that the pediments have either been displaced downward by faults near their upper edges or tilted downward by local warping. Several

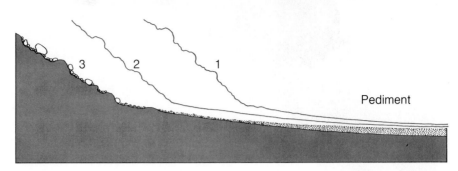

pediments are being dissected by stream systems that are forming new pediments, typically with steeper slopes than those of the older pediments. This relation suggests that the older pediments were tilted, which rejuvenated the erosive action of the streams. Her results are unifying in that they show the region to be one with continuing interplay between deforming movements and the processes of erosion and deposition. Broadly, this is probably true of all landscapes. Each mountain range and valley must be examined with respect to its individual history.

Dating Land Surfaces by Their Soils

Knowing something of desert landforms, we can turn to a remarkable study that links them to their soils. The study came to be called the Desert Soil-Geomorphology Project and was made by the Soil Conservation Service. Geomorphology is the science of landforms, and this study of them was started by Robert V. Ruhe and Leland H. Gile in 1957 due to the general importance of soils formed on desert plains and related landforms (References 3 and 4). The area studied is along the Rio Grande near Las Cruces. Somewhat as in the Middle Valley described in Chapter 3, the river has cut a wide valley in plains that slope gently toward it from adjoining mountains (Figure 9-13). Note the playa, Isaacks' Lake, that marks a closed basin 300 feet above the river floodplain. Note, too, that the plains near the mountains are partly pediments and partly alluvial fans, the latter coalescing downslope into the nearly smooth surface of an alluvial plain. Finally, note the fault along which the west side of the valley has been dropped (probably 50,000 or more years ago). All these things make the area comparable to the other parts of the desert region, but the area has special significance because it includes a major, through-going river.

Ruhe's initial study treated both the river valley and the surrounding landforms, and the same broad approach was used in later, intense studies by Leland H. Gile, John W. Hawley, and R. B. Grossman (Reference 4). The separate deposits forming the fans and alluvial surfaces were delineated by walking over the surface and examining gullies, road cuts, and other exposures. The extent of each deposit was thus drawn accurately on maps and aerial photographs. The deposits were found to have the form of thin tongues that extend downslope from the canyons and steep fans where they originated. That the younger tongues lap over the older ones was confirmed by digging trenches across many parts of the alluvial deposits. The trenches also showed that the older deposits have much thicker soils than the younger ones, and the

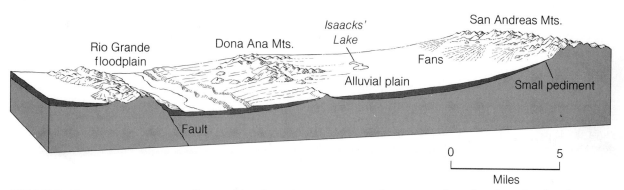

FIGURE 9–13
Landforms near Las Cruces, New Mexico, viewed from the southwest. Thicknesses of sediments (shown in black) are known only approximately. The slopes are somewhat exaggerated.

soils on older deposits are commonly preserved under deposits that lap over them (Figure 9-14).

In all, five episodes of sediment deposition and soil development could be recognized in this way. Organic materials are scarce in the deposits, but occasional fragments of carbonized wood can be used to date deposits by the radiocarbon method. It was thereby found that the youngest deposits started to accumulate about 6,500 years ago and in some places continued to accumulate until 1,000 years ago or later. The oldest deposits could be dated only roughly, but their accumulation appears to have ended somewhere between a million and a half million years ago.

Rates and periods of soil formation   The soils on the younger, well-dated deposits give a definite measure of the rate of soil formation, for they must have started to form at the time the deposit was completed and continued forming up to the present. Since the trenched deposit in Color Plate 5,A stopped accumulating about 4,000 years ago, its soil represents about 4,000 years of soil formation. The unaltered parts of the deposit beneath the soil can be seen in the lower one-fourth of the photograph. They are a poorly sorted mixture of angular pebbles, sand, and finer grains, transported approximately 8 miles from a body of monzonite, a rock closely similar to granite. The various mineral grains—quartz, feldspar, and mica—are no more altered in the sediments than they are in the source rock, so the parent material of the soil consists of the same minerals as the parent of the Georgia and Maine soils described in Chapter 8.

The young desert soil shows vastly less alteration than does the Georgia soil, but it nonetheless has begun to differentiate into layers. The upper two inches protected by the grass clump are a fairly loose, sandy topsoil. It is largely eroded away (for example, where the knife lies)—probably due to the causes

FIGURE 9–14
Diagrammatic view of three deposits on a small part of the alluvial plain, each with a soil (black) formed during long periods when sediment was not being deposited. The soils are numbered in order of their ages. Note that each was eroded locally before the next layer was deposited.

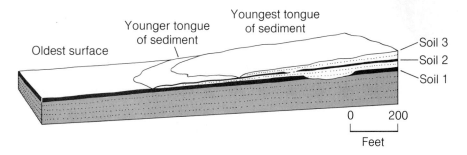

described in Chapter 3. Beneath it is a 9-inch layer comparable to the deep-red subsoil in Georgia, and it, too, is somewhat reddened by finely divided iron oxide and made slightly coherent by the addition of small amounts of clay. Beneath this is an 18-inch layer unlike anything in the Georgia soil—a somewhat white-tinged layer that effervesces (fizzes) when dilute hydrochloric acid is dropped on it. The escaping gas is carbon dioxide (CO_2), confirming that the white mineral is calcium carbonate ($CaCO_3$) (Supplementary Chapter 1). The layer forms in dry climates because rainwater carrying small amounts of dissolved calcium carbonate soaks down only a short distance before evaporating and therefore depositing the calcium carbonate in soil pores.

Color Plate 5B shows an older soil developed from exactly the same kind of parent material as that of Color Plate 5A. This soil started to form somewhere between 10,000 and 15,000 years ago. The pit is only a few hundred feet from that of A, so the differences between the two soils must be due to age alone. The older soil has a somewhat looser topsoil (again, largely eroded away) and a somewhat thicker subsoil. The latter is distinctly reddened by iron oxide and held together quite firmly by clays deposited in soil pores. The most obvious difference, though, is the greater abundance of calcium carbonate in the lowest layer, much of it forming hard, white nodules that protrude from the side of the trench.

Still older soils have thicker calcium carbonate deposits. The nodular bodies in the oldest soils often coalesce to form a remarkable layer that is locally as solid and smooth on top as a concrete highway and commonly many feet thick (Color Plate 6A). The red subsoil in this ancient soil is also strongly developed and often contains enough expandable clays to be toughly coherent and to generate distinct shrinkage cracks when it dries (Color Plate 6B). Indeed, this layer is locally so filled with clay that water would rarely sink through it were it not for the cracks, occasional root holes, and burrows of animals, notably those of ants.

The increasing thicknesses of the layers in the older soils may seem to indicate that the desert soils have developed at some steady rate with respect to time, but interestingly enough this is not the case. The evidence is shown schematically in Figure 9-14: the older soils are almost as thick where they were covered at an earlier stage as where they have been exposed continuously to the present time. Thus, the older soils must have formed mainly during earlier periods, which implies that rains soaked the soils more deeply and frequently during those periods. These moister periods are indicated in other ways. The youngest soils contain pollen and animal remains much like those of today, indicating that ground moistures were about the same. The next older soil, however, was affected by more moist conditions, and the oldest soil contains plant and animal remains that indicate a distinctly moister climate. Playa lakes were larger at that time, and glaciers formed in high mountain valleys about 90 miles northeast of the area studied. Some of these periods almost certainly correlate with the periods of glaciation described in Chapter 7, but dating is not complete enough to match the periods in detail.

Determining a climatic history   Gile and Hawley linked the events to the Rio Grande by comparing the soils of the fans and alluvial plain with

PLATE 5

Left, soil developed on one of the youngest alluvial deposits trenched by Gile and Hawley. The base of the layer in which $CaCO_3$ was added is a short distance above the large patch of shadow. Unaltered parent sediment extends from there to the base of the photograph. The knife is 4 inches long. Right, the next oldest of the soils, with the deposit from which it developed shown in the lower right. The reddish material extending down vertically on the left side of the photograph is subsoil that filled an opening left by the tap root of a yucca plant. The red, clay-enriched layer on the right is 14 inches thick, and the whitened calcium carbonate layer is 21 inches thick.

PLATE 6

Above, remnant of the oldest of the desert soils in the Las Cruces area, stripped down to the calcium carbonate layer, the upper part of which forms the resistant ledge along the skyline. Just beneath is a 4-foot-thick layer of calcium carbonate that is divided into vertical columns. The remainder of the bluff exposes horizontally layered alluvial sediments—the parental material of the soil. Right, horizontal surface cut across the clay-enriched subsoil of the oldest soil, showing a system of vertical cracks caused by shrinkage of expandable clays. Similar cracks may have controlled the vertical columns in the calcium carbonate layer shown above.

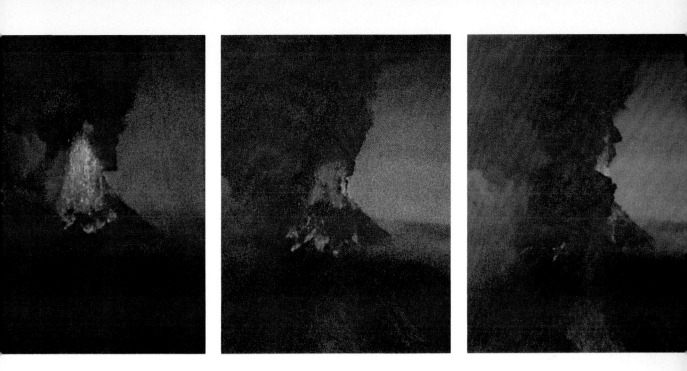

PLATE 7

Sequence of photographs showing the eruption of hot fragmental flows from Mayon Volcano early in the morning of May 2, 1968. The three upper views show glowing fragmental lava that was erupted upward and then fell onto the summit of the volcano. The hot avalanches that then raced down the mountain are shown in the lower view, at the time they reached the base of the steep slope. U.S. Geological Survey photographs by James G. Moore.

PLATE 8

Above left, flow-wrinkled top of a basalt flow at Mount St. Helens, erupted in approximately 70 A.D. Above right, similarly wrinkled lobes of a moving basalt flow of 1973 on the Island of Hawaii. Note that it is so liquid that it has not pushed the small tree over even though it has surrounded and ignited it. The latter photograph is by Wendell Duffield, U.S. Geological Survey. Right, steep-sided domes formed by the extrusion of viscous (silica-rich) lava in the Coso Range, California. The degree of erosion indicates that the farthest is the oldest and that the middle and nearest are of approximately the same age (possibly the nearest is the younger).

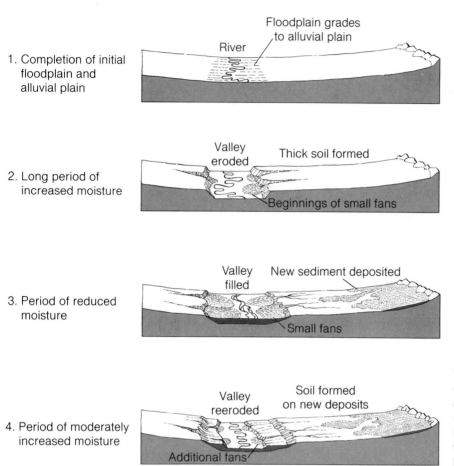

1. Completion of initial floodplain and alluvial plain

Floodplain grades to alluvial plain
River

2. Long period of increased moisture

Valley eroded   Thick soil formed
Beginnings of small fans

3. Period of reduced moisture

Valley filled   New sediment deposited
Small fans

4. Period of moderately increased moisture

Valley reeroded   Soil formed on new deposits
Additional fans

FIGURE 9–15
The first four steps in the development of the Rio Grande valley and its flanking alluvial plains near Las Cruces. Faults possibly broke the surface at various times, but their traces are hidden by the younger deposits.

soils developed on a series of surfaces formed next to the river by tributary arroyo systems. These surfaces are the gently sloping tops of step-like forms called *terraces.* Each was formed in two stages: (1) one during which the valley was partly filled and (2) one during which the sediment fill was partly eroded away (Figure 9-15). Deposition took place during periods of drier-than-average climate. Recall from the studies of the Middle Valley (Chapter 3) that dry periods are characterized by weak plant growth and by summer rains that erode hillslopes and deposit sediment in the river valley. Possibly we are now seeing the beginning of a period of valley filling such as that of stage 3 in Figure 9-15.

In any case, soil began to form on each terrace when it was stabilized. Gile and Hawley found that the terrace soils show the same sequential history as the soils of the fans and alluvial plain (Figure 9-16). We can conclude that the cutting and filling by the Rio Grande were caused by the same climatic changes that affected the fans and alluvial plain just described. To affect a major river, however, the climatic fluctuations must have been regional. Gile and Hawley's detailed study of local soils and landforms thus helps us interpret the climatic history of the entire desert region.

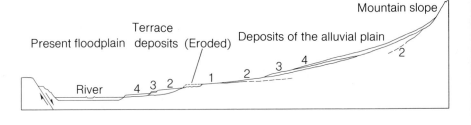

FIGURE 9-16

Section showing the deposits of the Las Cruces area, numbered order of their formation. On the left are the terraces, and on the right are the deposits of the alluvial plain. The width of the alluvial plain has been reduced and the slopes of the deposits exaggerated.

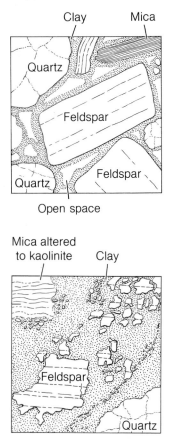

FIGURE 9-17

Microscope views comparing the altered grains in the Georgia saprolite (bottom) with the unaltered grains of the desert soil (top). Note that clay was deposited around the grains in the desert soil but formed from feldspar and mica in the saprolite. Both views are 1 mm (.04 inch) across.

A Broad Cycling of Soil Materials

One discovery of the desert soil project is quite astounding: in parent materials that contained little or no $CaCO_3$, most or all of the accumulated $CaCO_3$ came from the atmosphere! Clay from the atmosphere has also been added to some soils, particularly those on gravelly parent materials. The main evidence for these additions comes from examining the soils microscopically and comparing them with soils in which clay has definitely formed from the minerals in the parent material (Figure 9-17). Note the small, irregular relics of feldspar in the Georgia soil compared to the distinct sand grains in the desert soil. Note, too, that clay in the Georgia soil occupies space once occupied by parts of the feldspar grain, whereas the clay in the desert soil forms shells deposited around the sand grains, with some unfilled space between the grains. The clay must have been added from the outside, because it is as thick around grains that cannot change into clay (such as quartz) as it is around grains that can (such as feldspar).

Thin sections of the underlying layer show that $CaCO_3$, too, was added to small openings in the soil. It could not have been deposited by the underlying groundwater, for the water table (the top of the groundwater reservoir) lies several hundred feet underground (Reference 5). Neither could the clay and $CaCO_3$ have been leached and transported downward from mineral grains in the topsoil, for the grains are scarcely altered, even in the oldest soils in the area.

The additions of clay and $CaCO_3$ took place throughout the region during the entire period of soil formation, which suggests a continuous and widespread atmospheric process. To check one possibility, Gile and others measured the dust falling at several places by putting out trays covered with marbles to serve as proxies for a porous soil surface. Placed 3 feet above the ground (above the level of most blowing sand), the trays caught about 25 grams of dust per square meter of tray surface during the dry season alone, and the dust included about 8 grams of clay. This deposit is probably much greater than that laid down during former periods of greater moisture and before extensive grazing and agriculture. Nonetheless, only a small fraction of the clay caught in the dust traps would account for the clays added to the various soils. It was also found that organic substances were about ten times more abundant in the dust catches than in nearby topsoils. Perhaps these substances formed organic acids that helped move clays down into the soils, as described in Chapter 8.

Solids added to soil from the rain   The calcium carbonate additions from dust were only about one-tenth the amount needed to account for the $CaCO_3$ deposited in the soils, but rain may have supplied the difference. Calcium carbonate and other salts are carried in the atmosphere as minute particles, which are gained from evaporated spray of the ocean or lakes or from dust eroded on land. Chemical analyses of rain show that it contains enough dissolved $CaCO_3$ to place 1.5 to 2 grams on each square meter of ground each year. This amount plus that in the dust would account for the most slowly accumulated carbonate layers, but not for the older, more rapidly accumulated ones. Possibly the greater rainfalls of former periods brought down more calcium carbonate. It is also likely that local erosion of older soils provided calcium carbonate to soils forming at that time.

To check on additions from the atmosphere, we can look at the overall additions and transport of dissolved substances in the Rio Grande system. Besides calcium and carbonate ions, rain and snow contain dissolved salts of sodium, magnesium, potassium, chloride, and sulfate ions in small but measurable quantities (about one part per million). We have already noted that some of these substances arise from spray evaporated over oceans and salty playas or are eroded from bare ground. An ever increasing part of the modern load comes from the smokes and vapors of human works and from cultivated fields. Chemical analyses of rain and snow in the Southwest suggest that somewhere between 1 and 1.5 million tons of dissolved substances fall on the Rio Grande system each year. Chemical analyses of the river near its mouth show that it carries about 900,000 tons of dissolved substances to the Gulf of Mexico each year. These data thus support the idea that the desert soils are concentrating calcium carbonate from rain water.

The numbers seem to say, in fact, that the Rio Grande landscape is expanding rather than diminishing, and some parts certainly must be. In other parts, however, winds doubtlessly blow away more Rio Grande dust than they deliver. Nor are the dissolved substances simply moved through the river system without being altered. Chemical analyses show that the proportions of dissolved ion species in rain and snow differ from those in river water. As described in Chapter 8, some ions are used in the alteration of minerals to clay, others are set free by the alterations. Some ions are taken up preferentially by plants, for example, potassium, calcium, magnesium, nitrate, and phosphate. The exchanging machine is a complex one that involves all the attachments of the soil system.

Effects of concentrating sodium ion   One exchange, that of sodium ions in and out of expandable clays, has been highly detrimental to users of arid lands and has caused major agricultural problems in ancient as well as modern times. Sodium ions in concentrations greater than about 100 parts per million parts of water are detrimental to most crops. Rain and snow in northern New Mexico contain only .25 parts per million of sodium ions, but the Rio Grande at the head of the Middle Valley (at the Otowi gaging station) contains 25 parts per million, increasing to 68 parts per million just above Elephant Butte Reservoir and to 90 or more parts per million at El Paso.

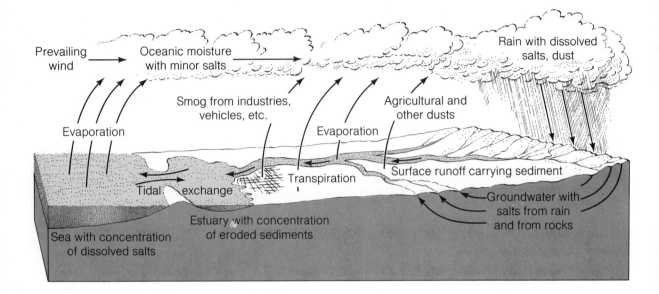

Prevailing wind

Oceanic moisture with minor salts

Rain with dissolved salts, dust

Smog from industries, vehicles, etc.

Agricultural and other dusts

Evaporation

Evaporation

Transpiration

Tidal exchange

Surface runoff carrying sediment

Groundwater with salts from rain and from rocks

Sea with concentration of dissolved salts

Estuary with concentration of eroded sediments

FIGURE 9–18

Broad cycling of water, atmospheric solids, and dissolved substances—a typical earth system today.

The increases are due partly to evaporation of water and partly to additions of deep, saline groundwater. An increasing proportion, however, has been coming from sodium-bearing clays in irrigated fields.

As we have already noted, the small amounts of rainwater soaking through the natural soil concentrate clay and calcium carbonate in layers. The rainwater also carries the very soluble sodium ion to greater depths—either to the groundwater or into expandable clay below the calcium carbonate layer. But when repeated irrigation saturates the soil and the sediment and rocks beneath it, the water becomes an interconnected system that evaporates steadily at the surface. The evaporation concentrates sodium ions near the surface, to the point that they start displacing other ions from the expandable clay in the soil. This process may continue as long as an interconnecting water body feeds sodium ions upward. Sodium may eventually become so abundant at the surface that it forms thin crusts of white salts, especially sodium bicarbonate, by which time little will grow in the soil.

The land may be reclaimed by spreading calcium carbonate on the surface and flooding it repeatedly with sodium-poor water. Calcium ions dissolved from the calcium carbonate exchange for sodium in the clays. The sodium-enriched water must then be drained off; however, its return to the river may cause downstream users to suffer the same effects. This problem is also affecting large agricultural areas in the Colorado River system as well as other rivers in the southwestern United States. Irrigation uses may thus spoil large parts of arid regions—the desert may be forced to bloom but perhaps only temporarily.

Cycling of solids through the atmosphere Studies of desert soils have thus disclosed some important exchanges of solid substances between the atmosphere and the earth. Such exchanges are a final addition to our scheme of soil formation. All the exchanges may be expressed as a broad cycling of materials, as shown in Figure 9-18. This sort of cycling was once thought to involve only the movement of water from sea to atmosphere to land and back to sea—the so-called water cycle. Soils were made (and eroded) in the process, and dissolved substances were carried to the sea. We can no longer view earth systems so simply. An important realization is that our own contributions to the cycling of atmospheric solids are very large. Many of their effects are yet to be discovered.

Summary

Some relations among landforms, soils, and the atmosphere are particularly apparent in deserts but are probably important in many other parts of the earth. We may summarize the relations in deserts briefly:

1. The capture of moisture by high mountains lying in the path of prevailing winds has created a large desert region in the western United States.
2. Much of the region is also unstable, being transected by many steep faults along which broad valleys have sunk relative to adjoining mountain ranges.
3. Sediment eroded from the steep, faulted edges of the mountains is deposited where streams leave mountain canyons and spread out on the floors of the valleys, forming alluvial fans.
4. Where fault movements do not take place over long periods of time, streams and sheetwash erode the mountains back to form gently sloping, rock-floored surfaces called pediments.
5. The slopes of the mountains and pediments correlate generally with the sizes of the larger fragments on the surfaces, the coarser fragments lying on the steeper slopes.
6. Dissection or superposition of fans and pediments shows that faulting or warping has recurred from time to time during the long history of the region and that the movements have varied from one part of the region to the next.
7. The soils in the region typically consist of a thin sandy topsoil (much of which has been eroded in historic times), a subsoil enriched with clay and iron oxides, and a distinct layer beneath cemented by calcium carbonate.
8. The soils generally show an increasing development with age, and unusually thick older soils were formed during periods of increased moisture that probably correlate with periods of glaciation farther north.
9. The soils on the desert surfaces can also be matched with soils on terraces along the Rio Grande, thus relating them to a climatic history of the region.
10. The desert soils have received major additions of clays from atmospheric dust and additions of calcium carbonate from rain, implying a broad cycling of solids along with the water that is carried from the ocean to land and back again.

11. A specific chemical interchange that is destructive to desert agriculture is the concentration of sodium in the soil due to intensive irrigation.

REFERENCES CITED

1. M. K. Hubbert. Man's conquest of energy: its ecological and human consequences. Unpublished address to the Institute of Energy Studies, Stanford University, October 8, 1973.
2. Jacqueline Mammerickx. Quantitative observations on pediments in the Mojave and Sonoran deserts (southwestern United States). *American Journal of Science*, vol. 262, pp. 417-35, 1964.
3. R. V. Ruhe. Geomorphic surfaces and surficial deposits in southern New Mexico. State Bureau of Mines and Mineral Resources, Memoir 18, 66 pp., 1967.
4. L. H. Gile, J. W. Hawley, and R. B. Grossman. Soils and geomorphology in a basin and range area of southern New Mexico. Guidebook, Soil Conservation Service, in preparation.
5. W. E. King and J. W. Hawley. Geology and ground-water resources of the Las Cruces area, New Mexico. *Las Cruces country*, New Mexico Geological Society Guidebook, 26th field conference, pp. 195–204, 1975.

ADDITIONAL IDEAS AND SOURCES

1. See Figure 2-13 to be convinced that alluvial fans form in regions with abundant rainfall. The farmhouse near the center of the view was built at the apex of a small alluvial fan! Figure 2-18 also shows part of a fan—at the foot of the mountain valley facing the observer.
2. If you want to test ideas on why alluvial fans form, try some experiments with miniature streams such as those described by Roger LeB. Hooke in "Processes on arid-region alluvial fans" (*Journal of Geology*, vol. 75, pp. 438–60, 1967).
3. Pediment formation has been studied in detail in miniature forms developed in so-called *badlands* (intricately dissected, completely exposed clay-rich terrains). Two papers by Stanley A. Schumm are especially pertinent and interesting: "The role of creep and rainwash on the retreat of badland slopes" (*American Journal of Science*, vol. 254, pp. 693–706, 1956) and "Erosion on miniature pediments in Badlands National Monument, South Dakota" (*Geological Society of America Bulletin*, vol. 73, pp. 719–24, 1962).
4. The outcrop in Figure 9-11 shows a way in which large rock masses are rounded by weathering. The separation of flakes and plates is called *exfoliation* and probably results from alteration and swelling of mineral grains exposed to water near the surface. The corners and angular edges exfoliate first because they are soaked more thoroughly (from more than one side).

The lower Mississippi River and its active delta, an image relayed from Earth Resources Technology Satellite 1. U.S. Geological Survey photograph.

10. From Soft Mud to Hard Rock

From Soft Mud to Hard Rock

Much of the soils eroded from the continent are carried to estuaries and the sea, and their arrival is monitored by photographs such as the one on the chapter opening page. This area is part of the Mississippi Delta, the huge deposit of mud formed where the river empties into the Gulf of Mexico (Figure 10-1). Note the river's active distributaries, radiating outward like the toes on a bird. On the day the satellite image was taken, they were emplacing huge quantities of turbid water (white toned) into the gulf. Some of the muddy plumes were moving 30 miles and more to the south and northeast. As additional satellite views are taken, we will learn whether this pattern is typical or unusual.

Studies of the sea bottom show that much of the sediment has been deposited near the ends of the distributaries, which thus grow outward into the gulf. Additional mud is distributed over the floor of the gulf by waves and currents, and some of it doubtless comes back onshore and into the marshes on the delta surface, as we noted for other estuaries. Sediment is also added to swamps and lakes on the emergent part of the delta when the river overflows its normal banks during floods. The delta is thus receiving sediment both above and below sea level, and the entire complex is growing into the

FIGURE 10-1

Map of the Mississippi Delta and surroundings, showing the area covered by the photograph on the chapter opening page and the features and sample site referred to in the text.

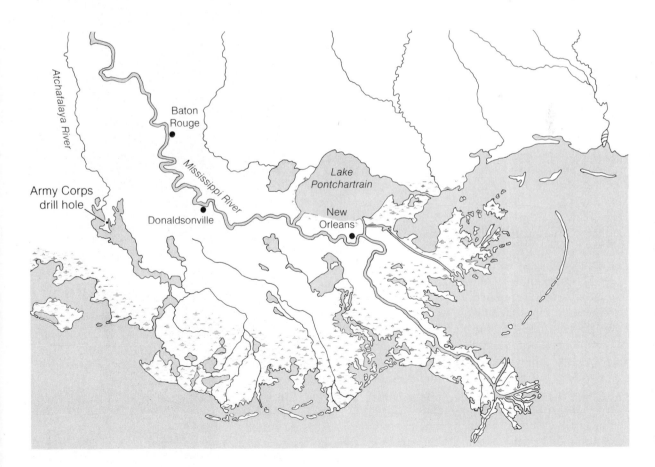

gulf. Satellite images will trace this growth precisely, for example, by tracking the highly visible white lines, which are offshore sand bars sensitive to changes in the entire delta.

An intriguing fact is that the sediment deposits are also affected by changes within the delta—changes that are hidden except for local surface effects. Of particular interest are the many circular or elliptical lakes and bays, which are probably caused by subsidence of the surface. Sediments under these areas have either shrunken or flowed away, and one would like to know if these actions are typical and exactly what causes them.

We must also be concerned with whether the changes are shallow or deep-seated, for under the delta is a huge accumulation of sediment—between 30,000 and 50,000 feet thick! The deepest parts were deposited at least 150 million years ago and have slowly subsided as layers of sediment were laid on top of them. The deeper layers have long since changed to solid rock, and a major purpose of this chapter is to find out how. We will start by looking at a shallow hole in the top of the delta and then go downward in stages.

Samples from the Delta

How do you collect a sample of sediment without disturbing its various features? You can tap a pipe down into the upper few feet of loose mud and pull it up with the mud inside. If you then press the mud out with a plunger, you will have a cylindrical sample cut at right angles to the nearly horizontal layering. Deeper holes must be made with a motor-driven rotating drill that cuts a circular hole essentially like that cut by the pipe. As this type of drill cuts downward, sediment is enclosed in a barrel, which can then be brought to the surface to remove the sediment core. Depending on the drilling method, cores can be collected from loose sediment or hard rock and from depths as great as 5 miles.

The cores described in this section were drilled by the Army Corps of Engineers in the Atchafalaya Basin, an older part of the delta west of the Mississippi River (Figure 10-1). The basin includes hundreds of square miles of lakes and swamps that receive fine mud when the river spills over its levees. Radiocarbon dating showed that the sediment at the bottom of the drill holes, 120 feet beneath the surface, was deposited approximately 12,000 years ago. The cored mud that accumulated slowly since was studied by Clara Ho and James M. Coleman, who were interested in what had happened to it as it was gradually buried (Reference 1).

The cores were sealed by the Army Engineers as soon as they were hoisted from the drill holes in order to retain the water that occurred naturally in the sediments. The amounts of water were determined by weighing pieces of the core, drying them in an oven, and then weighing them again. The differences in weights show that the sediments near the surface are about 40 percent water and those at a depth of 120 feet about 25 percent, with the amounts between showing an irregular gradation (Figure 10-2). The irregularities are due to small-scale differences in grain size, content of organic materials, and other things. The general change, though, is clear: mud tends to lose water as it is buried more and more deeply.

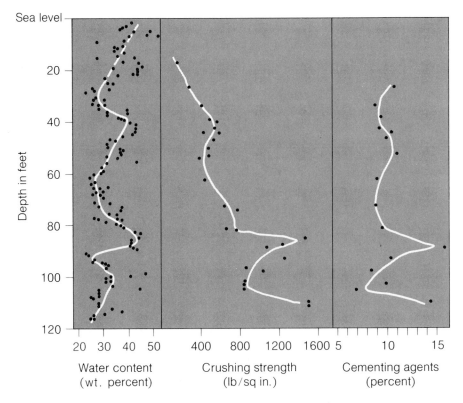

FIGURE 10-2

Water content, strength, and amounts of cements at various depths in the Atchafalaya muds in the Army Corps of Engineers boring. The circles show the measured data, and the lines are my rough averages of the variations. Data from Army Corps of Engineers and Reference 1.

FIGURE 10-3

Loose sand grains of $CaCO_3$ (top) and a surface cut through the same kinds of grains cemented by $CaCO_3$ added in pores. The pore space on the top is 42 percent, that on the bottom 25 percent. From Robert A. Robertson ("Diagenesis and porosity development in Recent and Pleistocene oolites from southern Florida and the Bahamas," *Journal of Sedimentary Petrology*, vol. 37, pp. 355–64, 1967).

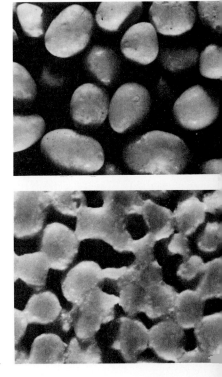

One reason for the loss is that water is pressed out by the weight of the overlying mud, much as water can be pressed from a sponge. The process is called *compaction* because it increases the amount of grain-to-grain contact, such that the muds become firmer and stronger. The Army Engineers systematically tested the strength of the cored muds by placing samples in a press and measuring the load (force) needed to crush each one. This load, called the crushing strength of the sample, is generally measured in pounds per square inch of sample area. As Figure 10-2 shows, it increases generally with depth, although there are exceptions.

Cementation of sediments   One exception points up another way in which mud becomes rock. Note that several of the stronger samples occur at depths of about 85 to 90 feet, yet they contain more water than average for the deeper parts of the core. Ho and Coleman found that these muds were strengthened by mineral grains deposited from solution between the grains of sediment. The deposited minerals are called *cement*. The graph on the right in Figure 10-2 shows that they are distinctly more abundant in the strong muds at depths around 90 feet and in the deepest sample tested than in the mud elsewhere in the core.

Cement in muds is difficult to photograph, but the relations are probably basically like those in other cemented sediments. Figure 10-3 shows an especially clear case of a sand consisting of calcium carbonate grains cemented by additional calcium carbonate. Note that because the resulting rock is still

A

B

C

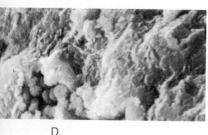

D

FIGURE 10-4

Fine-grained rocks consisting of silica: diatomite (A) and chert (C), each above a scanning electron microscope image of similar rocks enlarged 1,000 times. Note the shells of diatoms in B and the pervasively cemented nature of D. The images are of core samples drilled from sedimentary layers in the San Joaquin Valley, California, the diatomite from a depth of 3,300 feet and the chert from 5,800 feet. Electron microscope images courtesy of Larry Beyer, U.S. Geological Survey.

very porous, it can contain a good deal of water even though the grains are bridged solidly by the cement. Figure 10-4 shows a sediment and a solid rock that are as fine grained as the Atchafalaya muds, but they have been derived entirely from the shells of diatoms. The somewhat loose, powdery diatomite may later be cemented by additions of silica (SiO_2) that is probably dissolved from parts of the diatom shells. When the silica grains are thoroughly bound together, the rock is called *chert*, and it is so hard and tough that it can be worked into durable tools such as the one in the figure.

Ho and Coleman used X-rays to determine that the cements are unevenly distributed in the Atchafalaya muds but that they generally increase with depth. Small lumps of mud are quite distinctly cemented just under the modern ground surface; the lumps get larger and firmer with depth and in the deepest parts of the deposit form hard nodules that can be broken only with a hammer. Two abundant cements are calcium carbonate and iron oxide. Analyses show that the cemented layers and lumps contain the most organic materials, chiefly residues of plants. This correlation suggests that bacteria and organic compounds associated with plant remains were important in forming the cements—a situation reminiscent of some soil-making processes described in Chapter 8. Indeed, we can note the additions of calcium carbonate in the desert soils (Chapter 9) as an example of cementation associated with soil formation.

Ho and Coleman's study is important and intriguing because it shows that cementation may change the commonest of sediments into rocks surprisingly near the surface. Their data also illustrate a general result of the burial of most sediments: compaction and cementation go hand in hand to consolidate sediments, and these effects generally increase with depth.

Movements of Fluids in Deeply Buried Sediments

The Mississippi Delta and its surroundings have been explored with many deeper drill holes, some reaching as much as 30,000 feet beneath the surface. About 3,000 holes were drilled in southern Louisiana and southern Texas during 1972 alone! The primary purpose of the drilling was to find and produce oil and gas, but it also provided an immense sampling of rock-and-fluid systems. These data were used to guide further drilling. The petroleum industry *has* to understand what happens in buried sediments. Drilling, however, is costly (as much as 2.5 million dollars for one deep well), so the results are valued highly and thus are not typically available. Among the exceptions is a well drilled by the Chevron Oil Company to a depth of 16,450 feet in the delta. Cores from it were studied in detail by Charles E. Weaver and Kevin C. Beck, whose research provides some real answers to the question of how mud becomes rock (Reference 2).

The shallowest cores studied came from a depth of 4,233 feet and show that the clay is made up of the same minerals found in the recent muds of the Mississippi River. There is little cementation such as that found by Ho and Coleman, perhaps because the muds were deposited in the sea rather than in organic-rich swamps. The mud has nonetheless been consolidated by compaction and is now a rock called *mudstone* or *shale*. This rock is not

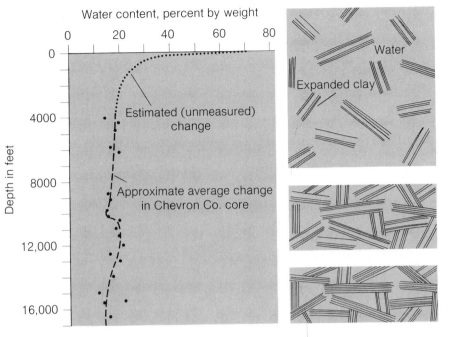

FIGURE 10-5

Left, changes in water content with depth, determined in the Chevron well study (Reference 2). Right, diagrammatic representations of clay flakes in newly deposited mud (top), mudstone at 4,000 feet (middle), and mudstone at 16,000 feet.

hard; it can be scraped easily with a knife and broken in the hands when moist. Its firmness is due to the grains lying together more tightly than those in the shallow muds of the Atchafalaya Basin. The resulting changes in water content are shown on the left side of Figure 10-5.

Compaction of clays Weaver and Beck found additional reasons for the mud's becoming firmer with depth, as illustrated on the right side of Figure 10-5. The top diagram depicts mud as first deposited in the sea. It is about 60 percent water (by weight), which lies mainly between the grains but partly in fluffed-out expandable clays. This oozy mud rapidly loses most of its water when buried, as in the Atchafalaya cores. At depths around 4,000 feet, the mud has been compacted to the stage shown in the middle diagram, but note that its clays remain unchanged.

Below that level, however, Weaver and Beck found that water in the clay grains is gradually squeezed out as the load increases. The water moves to the spaces between the grains and thence slowly out of the mudstone. At depths around 10,500 feet, water is evidently expelled from the clays faster than it can flow away, for the amount of water between the grains of mudstone increases abruptly (Figure 10-5, left). The constrained water in these rocks is under very high pressure—high enough to push a water column in a well for thousands of feet above the ground surface. The high pressure occasionally causes "blowouts" of wells drilled into porous sands in the pressurized mudstone. Below 12,000 feet, however, the mudstones continue to compact and lose water normally, and the expanded clays are further compressed, as shown in the bottom diagram. The deepest cores have only about 18 percent of water, by weight, and only about 12 percent of the clays remain expanded with internal water. Thus, the mud has become mudstone because of changes in the clays themselves in addition to compaction of the grains.

Origin and movement of oil and gas   An important aspect of these changes is their connection to occurrences of oil and gas. These substances are formed from remains of plants and animals, especially marine plankton, that accumulate in fine muds and decompose into a variety of organic compounds soon after being deposited. The compounds tend to stick to the clay flakes and thus are not carried away in the great quantity of water expelled during the early stages of compaction. But temperatures in the delta sediments increase at a rate of about 6 to 7 degrees Celsius (formerly called Centigrade) per thousand feet of depth, so that the organic substances are slowly warmed as they are buried. Heat breaks them down into lighter molecules that become the constituents of petroleum. This conversion is well underway at a depth of 3,000 feet and is probably most effective between depths of 5,000 and 8,000 feet, possibly in part because of the changes in the clays at those depths. Below 10,000 feet, temperatures are generally so high (over 80°C) that many organic compounds begin to change to natural gas and asphaltic residues. However, some compounds continue to change to petroleum at greater depths.

Like the water in the rock pores, petroleum and gas are pressed into motion by compaction and slowly stream upward as tiny drops mixed with water.

FIGURE 10–6

Map and enlarged vertical section (A-B) of the buried sandstone of the Milbur Field, Texas. The drill holes from which the data were obtained are shown by small circles, the black ones being those that produced oil. The lines pass through points where the sandstone has the thicknesses indicated. The numbered views show stages in the reservoir's history. From Stewart Chuber (Reference 3).

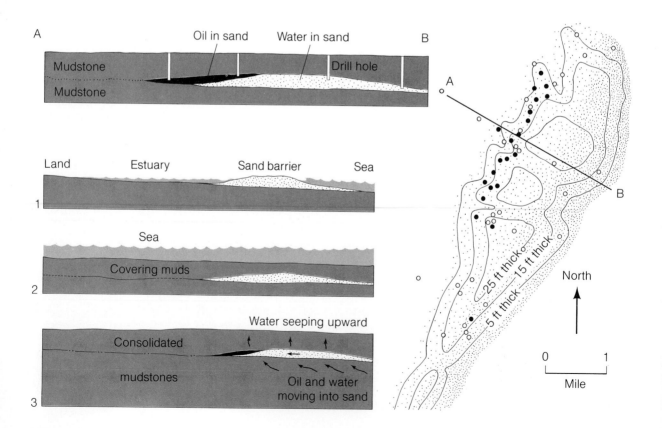

The fluids tend to be channeled along porous layers of sand or limestone, such as that of Figure 10-3, and may be trapped behind relatively impermeable layers of rock. The oil droplets then segregate upward beneath the trapping layer, partly because they are lighter than water and partly because water can move slowly through some trapping layers even though petroleum cannot, since it is more viscous. Eventually, the petroleum may form a continuous, though filamented, body in the porous rock, to become what is called a *petroleum reservoir.*

Steward Chuber has described a reservoir that shows some interesting linkages with the sedimentary accumulation in which it developed (Reference 3). The reservoir is located near the Gulf Coast, about 100 miles northwest of Houston. Cores and rock chips brought up by drilling show that the oil lies in what was originally an offshore sand bar, similar to the sand barriers off the Georgia coast or the bars off the Mississippi Delta (the white lines in the photograph on the chapter opening page). The map in Figure 10-6 shows the elongate form of the sand body, and the vertical cross section (upper left) shows how it thins, wedge like, toward what were then sea and land. The fine mudstones on the landward side contain layers of plant remains now altered to coal, suggesting that this was once a marshy estuary or bay. As illustrated by the numbered diagrams, the sand bar eventually subsided beneath the sea and was covered by mud. When oil and water from deeper layers flowed up the inclined sand layer (diagram 3), the mudstone cap slowly let water (and possibly gas) through but held back the petroleum. A reservoir was thus formed in the landward, upper part of the sand body.

Fluid reservoirs in porous sands   The sand just described forms an ideal reservoir because 34 percent of its volume consists of large, interconnecting spaces between grains. The *porosity* (percent of pore space) is about intermediate between that of the cemented and uncemented sands in Figure 10-3. The large pores permit oil to be pumped out easily. Indeed, the search for oil is largely a matter of finding porous reservoir rocks, and we have thereby learned a good deal about why rocks are porous. The high porosity of the rock in the Milbur field results from its grains being well sorted by size (Figure 10-7A). Such sands are typical of beaches and bars, and we have already noted their occurrence in windblown dunes (Figure 8-5). Sands that have not been winnowed of fine particles, for example those deposited by debris flows or turbidity currents, have much less pore space (Figure 10-7B). As already noted (Figures 10-3 and 10-4), another factor influencing porosity is the degree to which the spaces between grains have been filled by cementing minerals. This effect is difficult to predict without considerable drilling, for many sediments are cemented unevenly. In Figure 10-8, for example, the more resistant (jutting) parts of the exposure are the more cemented parts. The large ovoid bodies, called *concretions,* indicate that cements tended to accumulate around nuclei as they did in the nodules studied by Ho and Coleman.

Still another factor affecting the pore space in sandstones is the amount of compactive loading. Weak grains become squeezed into the spaces between strong ones, and strong grains may be dissolved and interlock at points of contact (Figure 10-9). Sands are thus changed to sandstones in three ways:

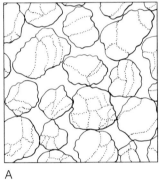

A

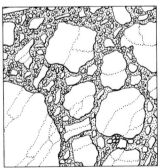

B

FIGURE 10-7
Well sorted sand (A) has more pore space than poorly sorted sand (B). Note, too, that the openings in A are much larger, so that fluids will flow through them much more rapidly.

FIGURE 10-8

Variably cemented sandstone, with hollows eroded in the least cemented parts and very strongly cemented bodies (concretions) protruding like boulders. Three hundred yards away, the same sandstone is cemented so weakly that it forms a meadow. Monterey County, California.

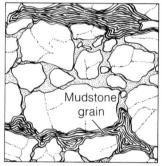

Mudstone grain

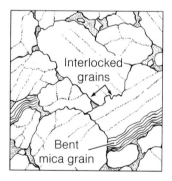

Interlocked grains

Bent mica grain

FIGURE 10-9

Strongly compacted sandstones, the one on the top with grains of mudstone squeezed between quartz grains, and the one on the bottom with quartz and feldspar grains that have been pressed together and partly dissolved along their margin so that they interlock (as at the arrow).

by cementation, by squeezing of soft grains, and by solution of parts of hard grains. The changes result in water and petroleum being forced out slowly —typically flowing to higher, less consolidated layers. This general evolution is essentially the same as that for the muds. It means that the search for oil tends to be limited by depth, both because of drilling costs and because of the decreasing abundance of porous reservoirs in deep-seated rocks.

Compaction Due to Pumping of Fluids from Wells

The evidence for consolidation considered so far relates to things that have already happened. Are there any on-going changes, any measured processes, that prove that deep-lying sediments do indeed compact and lose water and other fluids? In fact, many cases of on-going compaction have been measured, but they were not planned as useful experiments and we have only recently come to realize their significance. The rather alarming discovery has been that the land surface is subsiding (sinking) over reservoirs being pumped for water or oil. The sinking in some cases has been rapid and exceedingly damaging (Reference 4). An area of 50 square miles in Mexico City has subsided as much as 28 feet since 1938. The industrial heart of Tokyo has sunk as much as 14 feet since 1920, placing much of the city below sea level and therefore in considerable danger from flooding during earthquakes or typhoons. The southern part of the San Francisco Bay area has sunk below sea level since 1920, the maximum subsidence being 13 feet and the total area affected covering 250 square miles.

Each area of measured, recent subsidence coincides exactly with an area in which fluids have been pumped in unusually large quantities just before and during the periods of subsidence. Evidently, the removal of part of the fluids in underground reservoirs has caused the grains to be pressed more closely together, just as in the natural process of compaction. The compaction reduces the thickness of the sedimentary layers, and the land surface sinks. To get a more complete view of this widespread effect, we can examine a thoroughly studied case, one due to excessive pumping of groundwater.

Compaction of a groundwater system   The world's most extensive case of subsidence due to groundwater withdrawals is in California, where 5,200 square miles of the San Joaquin Valley sank as much as 29 feet between 1930 and 1972 (Figure 10-10). Extensive studies by Joseph F. Poland and other hydrologists of the U.S. Geological Survey have shown that the cause was excessive pumping of groundwater from deep sand layers when huge areas of semiarid land on the west side of the valley were brought under irrigation. The proof lies partly in the coincidence of the areas of pumping and of subsidence but even more convincingly in the fact that the amounts of water pumped have correlated closely with the amounts of subsidence. The studies went further to show exactly what was happening and why. The results are important because they not only help us understand how rocks compact but also illustrate the workings of a groundwater system typical of deep-seated systems in many other parts of the world.

FIGURE 10-10

Map of central California, showing the areas that subsided 1 foot or more because of groundwater withdrawals (light shading) and those that subsided 8 feet or more (darker shading). From J. F. Poland, B. E. Lofgren, R. L. Ireland, and R. G. Pugh (Reference 5).

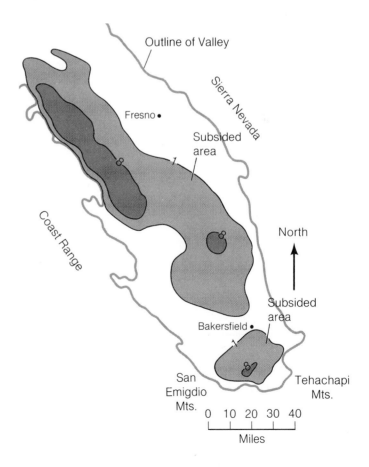

221

As Figure 10-11A shows, the San Joaquin Valley groundwater system consists of layers of porous sand that rise toward the mountains along the two sides of the valley. Because the west side of the valley is comparatively dry, most of the water has been fed into the sands from the Sierra Nevada catchment area, on the east side. Note especially that the layers of sand called *aquifers* are separated and covered by mudstone layers. The latter contain water but are so slightly permeable due to the minute size of their water-filled openings that they effectively seal off the water in the sand layers. Wells drilled into the sand layers thus "feel" the water pressure resulting from the elevation of the collecting parts of the system, as indicated in Figure 10-11B. Water in wells rises to a lesser height because of the "stickiness" of the aquifer—that is, the small openings in the sand layer make it a somewhat inefficient pipe. Nonetheless the water rises far up the wells toward the surface, making this a so-called *artesian* water system.

After years of intensive pumping, however, the water pressure was so reduced that water rose hundreds of feet less (Figure 10-11B). The reduction in water pressure also lessened the buoyant effect of the water on the sand grains in the aquifer, and they became pressed together somewhat more

FIGURE 10–11

A, vertical cross section of the San Joaquin Valley, showing the sedimentary deposits and some of the deep wells used to irrigate the semiarid west side of the valley. Vertical scale exaggerated about 3 times. B, diagram with vertical scale greatly exaggerated to show the components of the artesian water system.

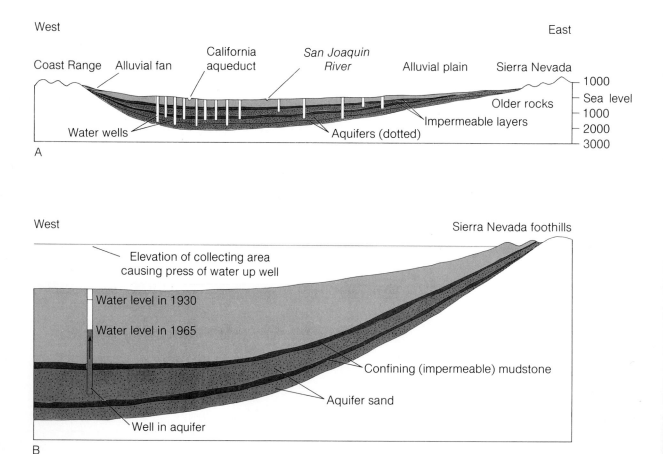

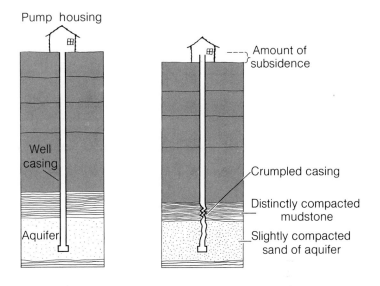

Pump housing

Amount of subsidence

Well casing

Crumpled casing

Distinctly compacted mudstone

Aquifer

Slightly compacted sand of aquifer

FIGURE 10-12

Casing and pump housing of a well before and after compaction of the mudstone layers. Amount of subsidence exaggerated (it was typically 8 to 20 feet for wells 1,000 to 3,000 feet deep).

closely. A far greater effect of the reduced pressure, however, was that water in the mudstone layers flowed out slowly but continuously to replace part of the water pumped from the sand layers. In effect, the overlying load of sediments compacted the mudstone layers far more rapidly than it would have naturally.

As Figure 10-12 illustrates, the compaction (thinning) of the mudstones led to subsidence at the surface. The thinning also resulted in the crushing of thousands of well casings (a pipe sealing the drill hole against water losses). It also thrust many casings upward through their pump housing (Figure 10-12). There were additional problems typical of regions undergoing subsidence. Irrigation systems were set off grade because of unequal sinking of the surface. Communities were affected by the tilting of drainage canals and of sewage disposal lines. The most alarming effect, however, was that huge canals being constructed to import surface water into the area were also set off grade. The canals were nonetheless completed at considerable expense, and the fact that they carried water to the San Joaquin Valley resulted in a drastic reduction in the pumping of groundwater. This led to a final proof of the cause of subsidence, for subsidence stopped as soon as pumping was reduced.

In conclusion, these cases of induced subsidence provide firm evidence that compaction of sediments and withdrawal of water are closely related phenomena. The pumping of fluids hastened the one-way process of consolidation. The open spaces that closed will not supply water again. We can also understand why the surface of the Mississippi Delta is scored by numerous lakes and bays. The muds beneath are compacting as they are loaded by more mud. Subsidence thus provides room for the accumulation of more sediments. Perhaps we can now see that all components of such systems —deposition of sediment, loading, compaction, upward flow of fluids, and subsidence of the surface—are interconnected. Every use of a region underlain by sediments must take these connections into account.

Changes Due to Heating at Great Depth

Another major reason why rocks become less porous and harder with depth is that heating eventually causes them to grow new mineral grains that interlock tightly. We have already noted that temperatures under the delta increase at 6 to 7°C per thousand feet (21° per kilometer). Part of this increase is due to the escape of heat from the earth's interior and part to the radioactive decay of elements in the rocks themselves. Measurements made elsewhere in the earth show a considerable range of values, averaging perhaps 9° per thousand feet (30° per kilometer).

When Weaver and Beck determined the proportions of the various clay minerals in the Chevron well, they found that the proportions change at about 15,500 feet, where the temperature is about 110°C. The mudstones above that level contain about 84 percent expandable clay, 10 percent illite, and 3 percent kaolinite, whereas those at 15,500 feet contain 52 percent expandable clay, 35 percent illite, and 10 percent kaolinite. Evidently 110°C is just hot enough to cause the original clay minerals to interact to form new ones. This process is similar to heating a mixture of chemicals to the point at which they begin to react and recombine.

I can give no pertinent data from deeper holes in the delta, but Weaver and Beck also studied Oklahoma mudstones that were originally 24,000 feet deep. At an average increase of 7°C per thousand feet, the base of this thick sequence would have been heated to about 170°C. These deep-seated rocks show some additional changes in their clay minerals but nothing major or abrupt.

Because there are no deeper drill holes, we must use other means to explore changes at greater depths. An area in which layered sedimentary rocks have been arched upward and eroded greatly (Figure 10-13) provides an ideal location for study. The original burial depth of a rock at a given site is at least as great as the thickness of all the layers that can be projected above it, as shown by the gray lines in the figure. To study the sequence, one walks across the eroded edges of the layers, measures their thicknesses to determine the original depth of burial, and collects samples to examine for mineral changes.

A truly thick sedimentary sequence that has been examined more or less in this way occurs in western Montana and Idaho. It is called the Belt Series and includes many rock layers that were originally mudstones. The clays and related minerals were studied by Dwight T. Maxwell and John Hower, who

FIGURE 10–13

A thick sequence of layered rocks, arched and eroded so that the complete sequence is exposed over a distance of many miles.

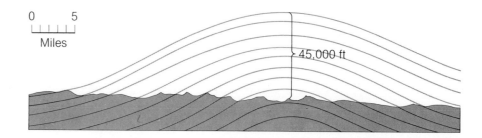

0   5

Miles

45,000 ft

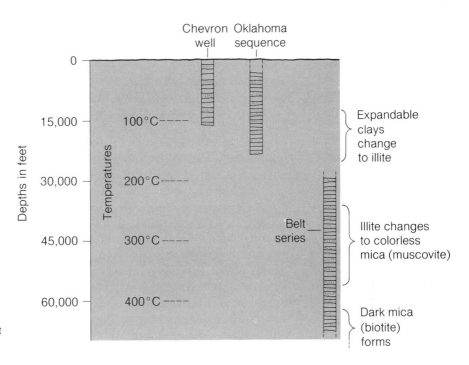

found that those in the upper layers of the sequence are almost entirely illite rather than expandable clay and kaolinite, indicating a progressive change beyond those found by Weaver and Beck (Reference 6). A 38,000-foot sequence of Belt Series rocks measured in northwestern Idaho showed another stage of mineral evolution: illite clay was found to diminish downward and be replaced by mica. Colorless mica (muscovite) became the dominant mineral about halfway down in this thick section, and another new mineral, biotite (dark mica), appeared near the base.

Figure 10-14 suggests how these changes may be related to the depths and temperatures already noted for the rocks from the Mississippi Delta and Oklahoma. This information for the gap between the deepest Oklahoma mudstones and the shallowest Belt Series rocks can be guessed only roughly; however, the diagram gives some idea of how mineral changes progress with depth. It definitely shows that a thick sequence of rocks must have lain above the Belt Series when the micas formed.

Metamorphic rocks   Once new mica forms in rocks, geologists speak of them as *metamorphic* rather than sedimentary. This nomenclature may seem rather arbitrary, but the practical fact is that the rocks do change in appearance and firmness when mica forms. They become harder, or tougher, and generally darker. The minute mica flakes commonly grow parallel to one another, with the result that the rock splits in that direction and is somewhat lustrous on broken surfaces. These varieties of metamorphic rock are called *slate* or *phyllite*. If the mica flakes grow in random orientations, the rock breaks in tough, angular fragments and is called *argillite*. In both rocks, the amount

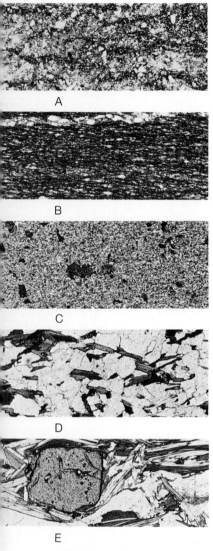

FIGURE 10–15

Thin sections of mudstone (A) and metamorphosed mudstones from the Belt Series, Idaho. Each view is 2 mm across. A is from a depth of 15,044 feet in the Chevron Company well; B is a slate from the Belt Series in Idaho (courtesy of Bruce R. Clark); C is an argillite with dark grains of biotite; D is a schist consisting of biotite (dark) and quartz; and E is a mica-quartz schist with a large garnet grain. C, D, and E courtesy of Allan B. Griggs.

of water-filled space is reduced greatly, typically to less than 1 percent. Metamorphism thus causes water to flow out of the rocks and upward to join the groundwater nearer the surface.

The nature of metamorphic changes is more obvious in rocks that have been more strongly metamorphosed. Such rocks can be found where the Belt Series was more deeply buried and strongly heated, southeast of the rocks just described. These variants have been studied by Anna Hietanen, Allan Griggs, and others of the U.S. Geological Survey, and Figure 10-15 shows how typical samples look in thin sections. A Mississippi Delta mudstone and a Belt Series argillite are included for comparison.

The most apparent change is in grain size. Clays in the delta mudstone (A) are too small to be seen, being only .005 millimeter across. Illite and first-formed mica in the argillite (B) are about .01 millimeter across. The new biotite grains in C are about 0.1 millimeter in diameter, and those in D about 0.2 millimeter. The garnet grain in E is 0.5 millimeter across, and many such grains in the same rock are two to three times that large.

Many metamorphic grains can be seen without using thin sections and a microscope. The new biotite grains of Figure 10-15C appear as tiny, shiny specks in a fine, black matrix. The biotite in Figures 10-15D and E forms easily visible shiny flakes, and the garnet in E appears as clear, pink grains among the mica flakes and abundant glass-like grains of quartz. Flaky metamorphic rocks in which the mineral grains are large enough to be seen without a microscope are called *schists.* The most metamorphosed rocks of the region are too coarse grained to be shown usefully in photographs of thin sections. These rocks, called *gneiss* (pronounced nice), are mainly granular but retain flaky groups of mica grains along which the rock splits fairly readily (Figure 10-16). The gneiss shows a still more advanced mineral change: much of the mica (especially muscovite) has changed (metamorphosed) to feldspar.

That these various rocks do indeed represent the progressive metamorphism of mudstone can be proven only by walking across tilted and eroded layers such as those in Figure 10-13. One thereby crosses layer after layer of light rocks (originally sandstones and limestones) and layer after layer of dark rocks (former mudstones). By examining the dark rocks especially, one notes that the mineral grains gradually get larger and that biotite, garnet, and feldspar appear in a more or less systematic progression. The important thing to imagine now is that these layers are again approximately horizontal, such as the thick sequence beneath the Mississippi Delta. By traversing the tilted layers, you have been descending, in effect, deeper and deeper into the earth! You started from a position comparable to the base of Figure 10-14 and have been moving into ever hotter layers of rock. The mineral changes you saw are the record of this increasing temperature. Perhaps you can thus see that these rocks, now so handily exposed, afford a view of what is happening far below the surface today.

Experimental Metamorphism and Melting

We cannot accurately add the changes just described to the bottom of Figure 10-14 because it is difficult to determine the depths at which the rocks were buried. Metamorphic rocks are typically deformed, so measurements of thickness may be misleading, and temperatures may not continue to increase downward systematically. We can, however, heat minerals in the laboratory and thus determine the temperatures at which the metamorphic minerals form.

In the method generally used, a small sample of powdered or melted minerals or chemicals is placed in a nonreactive (platinum or gold) capsule. Water is added to help the growth of new minerals grains, and the capsule is enclosed in a strong steel jacket. Temperatures and pressures in the capsule are then increased and regulated closely for the weeks needed for complete reaction within the sample. Finally, the capsule is cooled rapidly, after which the sample is removed and examined with a microscope and with X-rays to determine what new minerals have formed.

Many such experiments have shown that clay in mudstone can be expected to change to muscovite at around 380 to 400°C if pressures are equivalent to those at depths of 20,000 feet (6 kilometers) or more. This result is similar to the conditions suggested in Figure 10-14. Some experimenters have succeeded in forming biotite at around 525°C. The change of mica to feldspar has been found to take place at about 620°C when pressures are equivalent to those at depths of at least 20,000 feet (6 kilometers).

Granite from mudstone At somewhat higher temperatures, 650-700°C, mudstones heated in the laboratory begin to melt. One set of melting experiments is especially interesting, because the starting material was a clay consisting largely of illite and therefore comparable to the mudstones already described (Reference 7). Salt water was added to the capsules to serve as proxy for the sea water buried with marine muds. The clays began to melt at 670°C and were 44 percent molten at 675°C. When the capsules were cooled, the molten part of the sample (which set as a glass) was analyzed chemically and found to have the composition of granite. The remaining (unmelted) material contained most of the iron and magnesium of the original clay.

These experiments tie nicely to naturally metamorphosed rocks, for schist and gneiss are commonly laced with small bodies of granite where they are very strongly metamorphosed (Figure 10-17). Note that some granite appears to have been emplaced along former cracks, as though it had been injected there as a molten liquid. Other granite bodies, however, are irregular and highly diffuse, suggesting that parts of the metamorphosed mudstone were converted to granite without first being melted into a liquid (Figure 10-17). At the high temperatures required to produce these mixed rocks, atoms and ions evidently move about so readily that coarse patches of granite can grow in the solid state, with only local melting of mudstone. If the temperatures indicated by the experiments are correct (around 670°C) and if temperatures in this region increased downward at a rate of 9° per thousand feet (30° per kilometer), the rocks were buried at depths of about 14 miles (22 kilometers).

Although these numbers should be taken as approximations, they permit three broad and important conclusions that bring together other findings in this chapter. First, metamorphic rocks now exposed at the surface must have been uplifted many miles since they were metamorphosed. Second, their

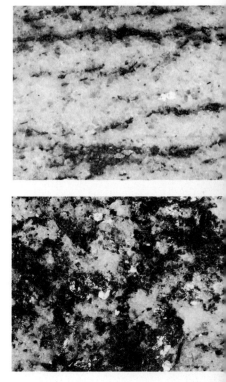

FIGURE 10-16

Sample of gneiss, 1 inch across, sawed perpendicularly to its platy groups of mica grains (above) and broken parallel to them (below).

227

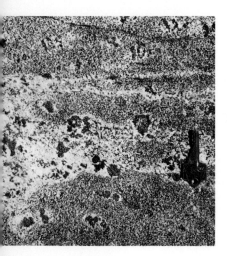

FIGURE 10-17

Left, metamorphosed mudstone (the finer grained rock) with patches of granite that appear to have formed in place rather than by intrusion of molten material. An initial stage is suggested by the small, vague patches of granite just below the upper edge of the photograph. Right, granite formed from metamorphosed mudstone, showing several sharply bounded sheets, as those just below and to the right of the pencil. The intimate sort of mixture in the two photographs is often treated as a separate kind of rock called *migmatite* (literally, "mixture-rock"). Santa Lucia Range, located on the map on the inside cover.

former "cover" must have been eroded away and deposited elsewhere as a thick sequence of sedimentary layers, presumably initiating metamorphism at that new site. Third, the progressive changes of mudstone to metamorphic rock are the opposite of the changes that make rock into soil; mica and feldspar in soil are changed to clay and dissolved ions, and clays and ions are changed back to mica and feldspar in the earth's depths. We have even discovered a process that might have generated the Georgia granite whose clay-rich soil we examined so closely! This suggests some important interconnections among the processes we have examined so far (Figure 10-18). Note that the linkages imply a broad lineage over long periods of time: a given parcel of earth material may have been cycled more than once—from exposed rock to soil to eroded sediment to metamorphic (perhaps molten) rock, and back again.

Summary

By starting at the surface of the Mississippi Delta and noting changes in muds as they are buried more and more deeply, we have discovered systematic gradations from soft sediments to hard rocks, even to partly melted rocks. Our principal findings may be summarized as follows:

1. The water surrounding grains in newly deposited sediments is gradually pressed out by the load of accumulated materials; this process, called compaction, is quite predictable and regular.
2. Sediments are also consolidated (made firmer) by additions of minerals that cement the grains together, but this process is often irregular.
3. Compaction and cementation cause water, oil, and gas to flow slowly toward the surface, chiefly through sandstones that have abundant spaces (pores) between grains.
4. The porous rocks become reservoirs of fluids when less permeable layers of rock trap the fluids beneath them.
5. When reservoirs are drilled and fluids are pumped out faster than they can be replaced by flow at depth, the porous rocks compact and the surface subsides.

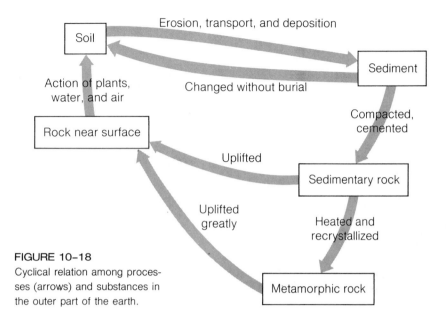

FIGURE 10–18
Cyclical relation among processes (arrows) and substances in the outer part of the earth.

6. Because temperatures increase downward at rates typically between 6 and 12°C per 1,000 feet, the heat in deeply buried sedimentary rocks may change them to metamorphic rocks, the first step being the conversion of clays to micas.

7. Graduated progressions of metamorphic changes may be observed where layered rocks have been arched upward over large areas and erosion has cut deeply into them.

8. The metamorphic changes may also be studied in laboratory vessels that confine samples at controlled temperatures and pressures.

9. Laboratory studies combined with studies of rock exposures suggest that many metamorphic rocks have been heated to temperatures of more than 500°C and were probably buried to depths of 10 miles or more.

10. Extensive exposures of metamorphic rocks provide evidence of uplift and erosion, which implies the deposition of more sediments and the beginnings of additional progressions from mud to rock.

REFERENCES CITED

1. Clara Ho and James M. Coleman. Consolidation and cementation of recent sediments in the Atchafalaya Basin. *Geological Society of America Bulletin*, vol. 80, pp. 183–92, 1969.

2. Charles E. Weaver and Kevin C. Beck. Clay water diagenesis during burial: how mud becomes gneiss. Geological Society of America, Special Paper 134, 78 pp., 1971.

3. Stewart Chuber. Milbur (Wilcox) Field, Milam and Burleson Counties, Texas. Pp. 399–405 in *Stratigraphic oil and gas fields*, R. E. King, editor. American Association Petroleum Geologists, Memoir 16 (Society Economic Geologists Special Publication 10), 687 pp., 1972.

4. J. F. Poland. Subsidence and its control. Pp. 50–71 in *Underground waste*

management and environmental implications. T. D. Cook, editor. American Association of Petroleum Geologists, Memoir 18, 412 pp., 1972.

5. J. F. Poland, B. E. Lofgren, R. L. Ireland, and R. G. Pugh. Land subsidence in the San Joaquin Valley, California, as of 1972. U.S. Geological Survey, Professional Paper 437-H, 78 pp., 1975.

6. Dwight T. Maxwell and John Hower. High-grade diagenesis and low-grade metamorphism of illite in the Precambrian Belt Series. *American Mineralogist,* vol. 52 pp. 843–57, 1967.

7. H. G. F. Winkler, Experimentelle Gesteinsmetamorphose-I. Hydrothermale Metamorphose karbonatfreier Tone. *Geochimica et Cosmochimica Acta,* vol. 13, pp. 42–69, 1957.

ADDITIONAL IDEAS AND SOURCES

1. Superb descriptions of river-related processes and subsidence on the Mississippi Delta are given by Richard J. Russell in *River plains and sea coasts* (Berkeley: University of California Press, 173 pp., 1967). Russell worked for years on these subjects, and his interests and concerns come through clearly.

2. You can observe the initial compaction of clay-rich sediments by the following procedure: (1) Disaggregate a handful of moist potters clay in a quart jar of water, and let it settle until the water is clear enough so that you can see the water-sediment interface. (2) Tilt the jar and note that the sediment is a slurry—its upper surface remains horizontal. (3) Now follow the course of compaction by making a mark on the jar at the initial interface and then again after a period of several hours. (4) Tilt the jar carefully to see if the sediment still flows like a liquid, and continue marking the position of the interface and letting the sediment compact until the sediment remains firmly in place when tilted. You can now discover some effects described in the next chapter by pouring off the water, tilting the jar about 45° from horizontal and tapping it repeatedly on a table top (earthquake!). This will typically liquefy the sediment—it will flow as a slurry until its upper surface is again horizontal. If you then let it stand for a few minutes it will set back into a firm body; however, you can reliquefy it by tapping it again on the table! Evidently the compaction of clay-rich sediments is a particularly vital subject in places where earthquakes may be expected.

Surface trace of the San Andreas fault (the central furrow in the view) where it crosses an arid plain in central California. Segments of the same fault near San Francisco and Los Angeles pose serious threats to those communities because they will almost certainly generate major earthquakes in the near future. U.S. Geological Survey photograph by Robert E. Wallace and Parke D. Snavely, Jr.

11. Earthquakes: Causes and Controls

Earthquakes: Causes and Controls

The water in sediments and rocks is also related closely to earthquakes and to the fault displacements that cause them. The most obvious relation is that earthquakes may jar water-saturated sediments into extensive motion. We saw one such case in the great sediment flow and turbidity current started by the 1929 Grand Banks earthquake (Chapter 6). On land, too, sediment movements are the most destructive effects of many earthquakes and a major cause of sediment transport. Figure 11-1 shows an important case caused by the 1964 Alaska earthquake. Other parts of Alaska were affected by thousands of additional slides and flows, and water-saturated sediments also lurched, cracked, and spouted sandy water into the air. Similar effects could be produced any day in seismically active (earthquake-prone) places such as California, and one can only imagine the resulting damage. No place, moreover, is truly safe from major earthquakes. From 1811 to 1812, for example, some of the strongest earthquakes in history caused sliding, sinking, and jumbling of sediments over thousands of square miles of the Mississippi Valley in parts of Missouri, Arkansas, Kentucky, and Tennessee—a part of the world we do not ordinarily think of as seismically active.

Perhaps the most important relation connecting water, sediments, and earth movements, however, is the recent discovery that water under high pressure triggers the fault movements that cause earthquakes. This discovery, in fact, points to possible ways of controlling the faults that produce major earthquakes! To understand this important possibility clearly, we will examine (1) how and why water-saturated sediments move when shaken, (2) the immediate causes of earthquakes, especially of the major Alaska earthquake, and (3) the effects of water on the faulting of deep-seated rocks.

FIGURE 11-1

A small part of the immense Turnagain Heights landslide, Anchorage, a result of sediment flow caused by the 1964 Alaska earthquake. Photograph by the U.S. Army.

Slides and Flows Caused by Earthquakes

Of the many slides in Anchorage caused by the 1964 earthquake, one of intermediate size, called the Fourth Avenue slide, showed typical relations between form and movement. As Figure 11–2A shows, the slide was oval in shape and measured about 1,000 by 1,800 feet. It moved downslope (to the right in the figure), yet drilling disclosed that the main zone of offset at the base is almost horizontal (Figure 11-2B). The material at the leading edge (toe) was even thrust upward as well as outward, forming a series of giant slab ends and wrinkles called *pressure ridges.* These features suggest that parts of the mass moved like very viscous fluids. At the same time, other parts of the slide broke in a way typical of brittle materials, cracking and collapsing along shallow faults such as those labeled in Figure 11-2B. Especially typical is the steep-walled depression along the upper (left) part of the slide, which is called a *graben* (the German word for trench). The steeply inclined fault that forms the back (left) wall of the graben formed when the slide pulled away from the rest of the hill; the fault on the right side of the graben formed when the upper part of the slide collapsed into the opening gap. The graben and the many other cracks thus indicate the lateral movements of the materials forming the slide. Hundreds of similar cracks and small faults formed

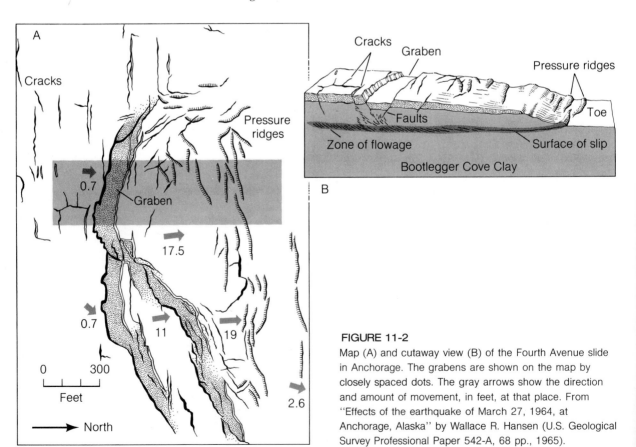

FIGURE 11-2

Map (A) and cutaway view (B) of the Fourth Avenue slide in Anchorage. The grabens are shown on the map by closely spaced dots. The gray arrows show the direction and amount of movement, in feet, at that place. From "Effects of the earthquake of March 27, 1964, at Anchorage, Alaska" by Wallace R. Hansen (U.S. Geological Survey Professional Paper 542-A, 68 pp., 1965).

elsewhere in the city, and slides that moved farther, such as the one in Figure 11-1, broke into many tilted masses and grabens.

Flow of clay due to quaking   Drill cores showed that most of the viscous movement was in a layer of clay, named the Bootlegger Cove Clay, that extends under most of Anchorage (Figure 11-2B). The clay in a 20 to 50 feet thick zone near sea level apparently liquefied—that is, it flowed as though it were a liquid. To determine why, many clay samples from the drill cores were studied in the laboratory. They were put under a pressure equivalent to the load they bore underground and were shaken at a frequency similar to that of the earthquake. Samples from well below sea level and from the upper part of the clay layer remained firm, but those from the zone near sea level collapsed after a few minutes of shaking, spreading out like viscous liquids. Left to stand without shaking, the samples quickly returned to their original, firm state, which may explain why the slides generally stopped moving when the earthquake stopped.

This flow-and-set type of behavior, called *thixotropy*, has been studied considerably in other clays and soils. In this case, it almost certainly relates to how the clay grains were held together. As was shown in Figures 8-9 and 8-10, oxygen atoms form the tops and bottoms of clay flakes, whereas metal atoms are buried within the flakes. The tops and bottoms thus have a slightly negative electrical charge, and the edges of the flakes have a slightly positive electrical charge. Because the Bootlegger Cove Clay was originally deposited in the sea, positive ions were attracted to the negatively charged surfaces of the flakes, while negative ions were attracted to the positively charged edges (Figure 11-3B). The ions thus acted as a cement that held the loosely compacted clay flakes together. The clay, however, was uplifted, with the result that fresh water flowed through the part near sea level and removed the ions. This action probably caused the clay flakes to become rearranged as in Figure 11-3B—their positively charged edges being attracted to the negatively charged surfaces of neighboring flakes. This bonding, however, is so weak that the house-of-cards arrangement tends to collapse when heavily loaded. Because the lower part of the salt-free clay was most loaded (bore the greatest weight), it liquefied when shaken by the earthquake (Figure 11-3C). When the shaking stopped, the flakes probably again assumed an arrangement like that in Figure 11-3B.

Liquefaction of sand   Sand layers may also become liquefied during earthquakes and flow as sand-water mixtures. Liquefaction is probably due to the fact that the grains near the bottom of loosely compacted sand layers are jiggled and pressed more closely together during a quake. The water thus released rises through the overlying part of the layer, producing a loose, watery mixture that may flow laterally or become injected into adjoining muds. Figure 11-4 shows features formed in horizontal layers that were shaken briefly during the San Fernando Valley earthquake of 1971 (Reference 1). Of great concern in cities built around estuaries, on deltas, or over former swamps or glacial lakes is the possibility that sand layers or weak clay layers would liquefy similarly and cause lateral sliding of large tracts of land.

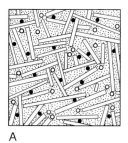

A

B

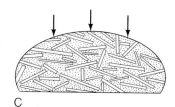

C

FIGURE 11-3
A, clay flakes with interspersed ions (positively charged ones, black; negatively charged ones, open); B, flakes joined in edge-to-face arrangement; C, the result of jiggling under load, with arrows representing the load.

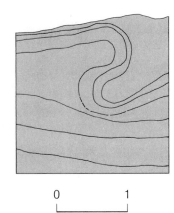

0   1

Inches

FIGURE 11-4

Vertical section through sediments behind Van Norman Dam, Los Angeles, that were horizontal layers before the San Fernando Valley earthquake of February 1971. Photograph by John D. Sims (Reference 1).

The damages thus caused are often double: (1) to any buildings on the slides and (2) to water mains, electrical cables, and sewage lines that may supply large sectors of the city. Slides caused by the San Francisco earthquake of 1906 severed water mains that could otherwise have been used to extinguish fires started by the quake. The slides were thus the greatest single cause of damage by that earthquake.

Cause and Nature of Earthquakes

Earthquakes are caused by sudden displacements of rock masses along faults. Certain parts of the rock masses move almost instantaneously, impacting and wrenching the surrounding rock and thus generating wave motion in it (Figure 11-5). The effect is somewhat like that of using a hammer (the moving masses of rock) to strike a bell (the adjoining parts of the earth) to set it vibrating and ringing (the earthquake).

The potential of the "hammer" to move is generated by slow distortion of the rock masses on the two sides of the fault. This distortion is elastic and is comparable to the slow compression of a steel spring or the bending of a bow or a leaf spring (Figures 11-6A, B, and C). Slippage does not occur at the start of this deformation because the two sides of the fault are somewhat irregular and are pressed together tightly, or possibly because they have been cemented together by mineral grains. Regardless of what causes the fault to stick initially, the increasing elastic distortion builds up forces that eventually overcome the resisting forces. At that instant the rock masses snap past one another, simultaneously producing a fault offset and an earthquake (Figures 11-6D and E). This process is often called the *stick-slip mechanism* of generating earthquakes—a fault sticks for some time but eventually slips.

The earthquake itself is a series of waves that are propagated outward through the rocks, transmitting the energy of the initial impact to other parts of the earth. The waves are analogous to those generated by a great storm, which sweep across the ocean for thousands of miles and crash with great energy on distant coasts. Ocean waves and earthquake waves are also analogous in that water rises and falls many times as a series of waves passes but

FIGURE 11-5

Generating an earthquake by displacement along part of a fault. The unshaded bodies of rock move suddenly in the directions indicated by the two heavy arrows, impacting and distorting the surrounding rock and thus causing it to vibrate. The displaced bodies may be anywhere from a fraction of an inch to hundreds of miles long, and the strength of the earthquake will vary accordingly.

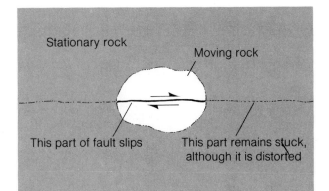

Stationary rock

Moving rock

This part of fault slips

This part remains stuck, although it is distorted

236

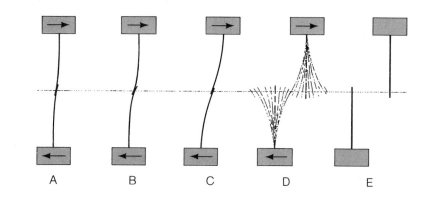

FIGURE 11-6

Two overlapping leaves of spring steel gradually bent by the displacements indicated (A, B, C) until the elastic bending is greater than the resistance holding them together. Just after snapping, they vibrate freely for a few moments (D) before coming to rest in their new, offset positions (E). The arrows indicate the forces that cause the bending and therefore the offset.

A   B   C   D   E

later returns to its original position; rocks likewise vibrate as they transmit an earthquake's energy but they return finally to their original state. Rocks transmit energy especially effectively because they are elastic—a fact you can verify by throwing a hard, round pebble against a hard rock surface. The pebble will bounce like a rubber ball. Earthquake waves thus travel through rocks as highly energetic, rapid vibrations. In rocks found in the outer parts of the earth they travel at velocities of 2 to 8 kilometers (1 to 5 miles) per second!

Kinds of earthquake waves   The two principal kinds of earthquake waves result from two specific responses to the "hammer." One kind of wave is generated most forcefully in front of and behind the moving masses (Figure 11-7A). The rock in front is suddenly compacted or compressed and that behind is rarificated (stretched). Because rocks are elastic, the initial motions are followed by an oscillation in the opposite direction, such that each compaction is followed by a rarification and then another compaction, again and again. Meantime the first motion is propagated outward from the source, followed by a series of oscillating pulses (Figure 11-7B). The waves are like sound waves; earth scientists call them *compressional waves* or *P-waves* (an abbreviation for *primary waves*, meaning that these waves travel fastest and therefore are received first at earthquake recording stations).

FIGURE 11-7

A, two moving rock masses, depicted as blocks, compact the rock ahead and stretch the rock behind, as suggested by reference lines that were spaced equally before the offsets. B, this propagates a series of compactive and rarifactive impulses (seismic waves) away from the source, depicted here as a steel spring.

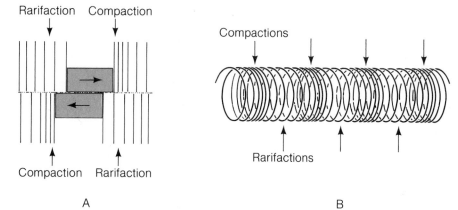

Rarifaction   Compaction

Compaction   Rarifaction

A

Compactions

Rarifactions

B

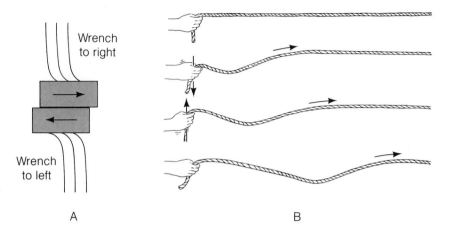

FIGURE 11-8

A, moving rock masses also wrench the rocks beside them, bending one segment one way, the other the opposite way. As shown in B, these motions are propagated outward just as a wave can be propagated along a tautly held rope by an up-and-down flick of the hand.

Wrench to right

Wrench to left

A

B

The generation of the second kind of seismic (earthquake) wave is illustrated in Figure 11-8A. The fault motion wrenches the adjacent rocks in the direction of the arrows and creates motion at right angles to the direction of propagation. The elastic nature of the rocks again causes them to oscillate back and forth as the series of waves travels outward. Comparable waves can be formed by holding a rope taut and giving one end a sudden up-and-down or sideways shake (Figure 11-8B). Such waves are called *shear waves* or *S-waves* (an abbreviation for *secondary waves,* meaning that they arrive after the *P-waves*).

Seismographs and seismograms Earthquake waves are recorded by *seismographs,* instruments that include a heavy mass that is hung and pivoted so that it remains steady when the earth and the rest of the instrument vibrate (Figure 11-9A). The motions are amplified, relayed, and scribed on a drum that revolves at a precisely fixed rate. The arrival times of the seismic waves can thus be read from the resulting *seismogram* (Figure 11-9B). The volocities of seismic waves have been determined from the elastic properties of rocks and also from seismograms of explosions that are timed and located exactly. The velocities give a basis for constructing time-distance curves such as those in Figure 11-10A. If we measure the differences in arrival times of P-waves and S-waves at a given station (as the number of seconds between P and S in Figure 11-9B), we can determine the distance to the point of origin of an earthquake as indicated in Figure 11-10A. When these measurements are made at three or more stations, arcs with radii equal to the distances can be constructed on a map (Figure 11-10B). Their mutual intersection marks the *epicenter* of the earthquake—the point on the earth's surface directly over the point of origin.

Earthquake magnitudes Seismograms may also be used to determine the *magnitude* (strength) of an earthquake because the amplitudes (heights) of waves on seismograms are a measure of the strength of the impact at the point of origin. This relation is analogous to the effect of stones thrown

FIGURE 11-9

A, seismograph oriented to measure horizontal ground motions. B, seismogram removed from the drum and spread out parallel to the direction of rotation (the time axis of the diagram). Note that the wiggles made as each set of waves arrives fall off in amplitude after a few vibrations, because rocks are not perfectly elastic and oscillations in them are damped rapidly.

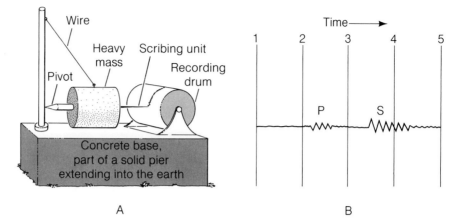

FIGURE 11-10

Locating the epicenter of an earthquake. A, time-distance curves constructed from measured velocities of P- and S-waves give a means of determining distance by using the difference in arrival times of the waves. B, distances determined at three or more stations can then be used to construct arcs that intersect at the epicenter.

in a pool—the larger the stone, the higher the waves that spread from the point of impact. Like the waves in a pool, however, seismic waves diminish in amplitude away from the point of origin. To compare the strengths of earthquakes we must therefore measure the waves at some standard distance from the epicenter, and this measurement is taken at 100 kilometers (60 miles). When the seismograph is at some other distance from the epicenter, the amplitude at 100 kilometers is calculated from the measured amplitude and the rate at which amplitudes are known to diminish with distance.

The magnitudes thus determined are reported in numerical units on the *Richter scale*, named after the seismologist Charles F. Richter. An increase of 1 unit on the scale is equivalent to a tenfold increase in the amplitude of waves at a distance of 100 kilometers from the point of origin. The increase in the amount of energy released is even greater; an increase of 1 unit on the scale represents a 31-fold increase in the energy released (Table 11-1). The

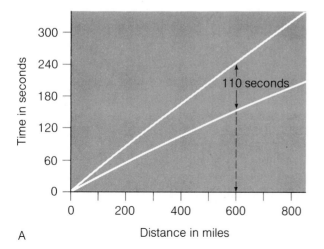

A

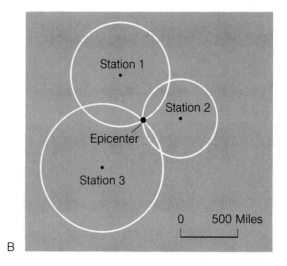

B

TABLE 11–1

Comparative Wave Amplitudes, Amounts of Energy Released, and Typical Intensities for Earthquakes of Different Magnitudes.

Magnitude (Richter scale)	Comparative wave amplitudes, 60 miles from epicenter	Comparative amounts of energy released	Intensities near the epicenter (abridged greatly from the Mercalli Intensity Scale in general use)
1	1	1	I: Felt by few if any persons; detected by seismographs
2	10	31	
3	100	960	II–III: Barely felt by persons indoors, especially on upper floors of buildings
4	1,000	30,000	
5	10,000	920,000	IV–V: Felt by nearly everyone; dishes rattled; loose objects overturned
6	100,000	29,000,000	VI–VII: Damage slight to well built structures but appreciable to poorly built ones
7*	1,000,000	890,000,000	VIII–IX: Damage great to poorly built structures and considerable in most other structures
8	10,000,000	28,000,000,000	X–XI: Damage great in most structures XII: Few structures remain standing; ground broadly fissured and disrupted

*Intensities of earthquakes of magnitude greater than 7 vary greatly with kind of underlying material.

1964 Alaska earthquake, which had a magnitude of approximately 8.4 on the Richter scale, released about twice the energy released by the 1906 San Francisco earthquake, which had a magnitude of approximately 8.25.

Earthquake intensities The intensities of earthquakes, described briefly in Table 11-1, are a different sort of measure. Intensity refers to the amount of motion and damage at a given location. Intensities increase with the magnitude of a quake and with nearness to its epicenter, but they also depend on the kind of material beneath the surface. We have seen a specific example of this in the case of the slides in Anchorage. Generally, extensive layers of water-saturated sediment are put into exaggerated wave motion by the energetic P-waves and S-waves, surging rhythmically back and forth like a pan of gelatin that has been given a sudden jolt. This kind of motion contributes greatly to local intensities, being notably destructive to tall buildings shaped in such proportions that they wobble with the same frequency as the body of sediment beneath them. The same building constructed on solid rock might well be undamaged by a major earthquake. Forecasting intensities at a given site is thus a major factor in planning communities in seismically active regions.

First-motion solutions Seismograms also show the direction of the first motion of the ground at the instrument, and this information can be used to interpret the faulting that caused the earthquake. The first motion of the ground is shown by the direction of the first wiggle on the seismogram. Since most seismograph stations have instruments oriented to record vertical as well as horizontal motions, the wiggles can be resolved into actual ground motions of up *vs.* down and right *vs.* left for both P-waves and S-waves. If, for example, the first P-wave wiggle shows that the ground moved away from the epicenter and toward the instrument, the first P-wave was in a compressional phase; if the ground moved from the instrument toward the epicenter, it would have been in a rarifactive phase. As Figure 11-11A shows, all stations lying in "front" of the hammer motion will receive compactions as first motions, and all those behind will receive rarifactions. Thus, if the orientation of the fault is known, the first motions read at several stations indicate the direction of the offset along it. If the orientation of the fault is not known (perhaps because it lies deep in the earth or under the sea), the first motions at a wide variety of stations indicate the two possible orientations of the fault and of the offset along it (Figure 11-11B). As we shall see, these first-motion solutions are a principal means of interpreting movements in the deep earth.

Earth Movements Causing the 1964 Alaska Earthquake

The fault offset that generated the great Alaska earthquake did not break the land surface, but the epicenter and the two possible fault motions were determined by the methods just described. The actual earth movements were measured by a series of wide-ranging studies—the most thorough ever made after a major earthquake. These studies led to some important scientific discoveries, but their purpose was largely practical: when would the next major quake occur? The results described here are based largely on syntheses made by George Plafker, who participated in many of the studies (Reference 2).

Measuring the 1964 displacements   The main evidence for the earth movements that caused the earthquake came from field studies over a large land and sea area. Measurements of vertical displacements along the irregular coastline were based on sessile (attached) marine organisms that were raised above the sea in some places and submerged in others. Barnacles, for example, live as high as the mean (average) high-tide level, where they form a distinct boundary with lichens that cover coastal rocks in the zone moistened by sea spray (Figure 11-12). The displacement of this line above or below sea level was measured at more than 800 localities. Supplementary data were obtained from (1) tidal gages, which are fixed to the solid earth and automatically record changes in its vertical position relative to sea level; (2) vertical positions of markers (*bench marks*) surveyed before and after the earthquake; (3) changes in the positions of beaches; (4) changes in sea level noted by fishermen and other persons closely familiar with shoreline features; and (5) changes in water depths measured by sounding the sea bottom in places that had been sounded before the earthquake. Assembled on a map, these various measures of displacement showed what a huge area had been

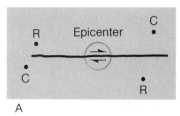

A

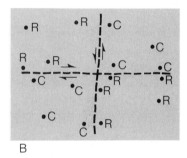

B

FIGURE 11-11

A, when the epicenters and orientation of a fault are known, the sense of offset is indicated by the first ground motions. C = compaction, R = rarifaction. B, two possible fault orientations and senses of offset indicated by P-waves at several stations. C = compaction, R = rarifaction.

FIGURE 11-12

Uplifted shoreline in Prince William Sound marked by a line between dark gray lichens of spray zone (above) and light gray barnacles (with patches of the dark alga called rockweed). U.S. Geological Survey photograph by George Plafker (Reference 2).

affected and how systematically the displacements varied across a region of uplift and an adjacent one of subsidence (Figure 11-13A). Maximum vertical displacements recorded onshore were 38 feet of uplift and 7.5 feet of subsidence.

Bottom soundings were not accurate enough to measure vertical displacements in the large area southeast of Kodiak Island and the Kenai Peninsula (Figure 11-13A), but displacement of the sea bottom was indicated by *tsunamis*, giant sea waves produced by sudden disturbances of the sea or of the sea floor. The arrival times and directions of the tsunamis were recorded at a number of places along the coast. The speed at which such waves travel is also known. The differences between the time of the earthquake and the times of wave arrivals at various places thus made it possible to calculate their source, which is the uplift area labeled in figure 11-13A. The waves also showed that the sea bottom in this area suddenly moved up rather than down, because the first motion recorded on shore was an upsurge rather than a withdrawal of the sea. A detailed record of wave arrivals on Kodiak Island indicated that the maximum offshore uplift was more than 30 feet—about the same as that in the zone of maximum uplift far to the northeast (Figure 11-13A).

The upper part of the earth also shifted horizontally. These displacements were measured by precise surveys of points that had been marked and surveyed before the earthquake. As Figure 11-13B shows, the horizontal offsets were toward the southeast and increased regularly to a maximum of about 65 feet near the most uplifted part of the deformed zone. The water in embayments and lakes commonly surged landward along the northwest shores of these water bodies, indicating that the earth under them snapped suddenly toward the southeast.

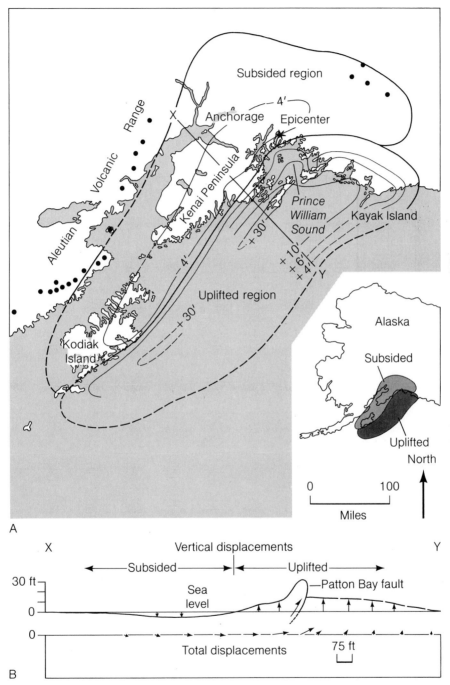

FIGURE 11-13

A, map showing the distribution of uplift and subsidence during the 1964 Alaska earthquake. Note the location of the epicenter (star) and the recently active volcanoes (heavy dots). B, diagrammatic vertical sections across the deformed area at the line marked X-Y on the map. The upper section, which is greatly exaggerated, compares a land surface originally at sea level with the configuration of the same surface after the earthquake. The arrows in the lower diagram show horizontal as well as vertical displacements, the latter less exaggerated than in the upper section. From Reference 2.

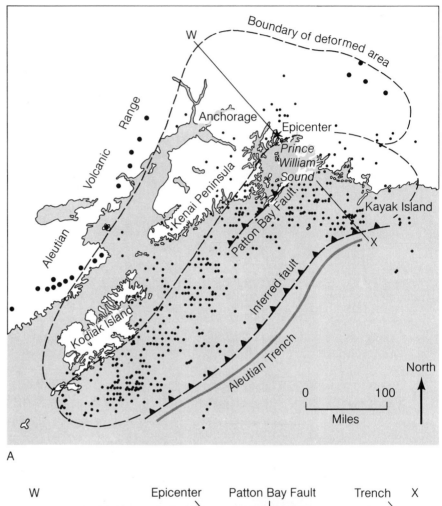

A

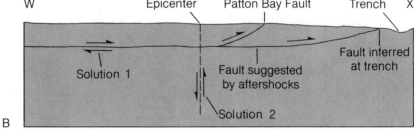

B

FIGURE 11-14

A, map showing the epicenters of aftershocks (dots), the trace of the Patton Bay fault, and the major fault inferred along the edge of the Aleutian trench. B, vertical section along the line W-X on the map, showing the two fault solutions obtained from analyses of first motions of the main shock. From Reference 2.

Another valuable record was that of numerous *aftershocks*—weak to moderately strong earthquakes following the main shock. They decreased rapidly in number and strength, from 120 the day after the earthquake, to 30 on the fifth day, to 15 on the tenth day, and progressively fewer thereafter. The sequence indicates continuing adjustments within the rocks displaced during the main earthquake. The distribution of the epicenters of the aftershocks confirms this idea, for it coincides with the distribution of displaced rocks (Figure 11-14A).

Locating the fault that caused the quake   The fault displacement that caused the earthquake was studied by the first motions recorded on seismograms, as described in the preceding section, and this gave two alternative solutions. One solution indicated a fault that sloped at a low angle from the seafloor near the edge of the Aleutian Trench northwestward under the deformed region (Figure 11-14B, solution 1). As shown in the figure, the rocks above the fault were displaced toward the southeast. The other solution required a fault sloping steeply southeast, with the displacement indicated as solution 2 in Figure 11-14B. Two relations indicate that solution 1 is correct: (1) the epicenters of aftershocks are distributed widely rather than in a line passing close to the main epicenter (Figure 11-14A) and (2) the faults that broke the ground surface are located far from the main epicenter rather than close to it (Figure 11-14A and B).

That movements along the gently sloping fault have caused other major earthquakes in the region is shown by the distribution of epicenters in Figure 11-15. Note that they form an array parallel to the Aleutian Islands and the

FIGURE 11-15
Epicenters of earthquakes of magnitude greater than 7 detected between 1904 and 1952. The dots indicate depths less than 40 miles and the triangles depths of 40 to 120 miles. The 1964 epicenter is marked by a star. The outer dashed line shows the approximate limit at which the 1964 earthquake was felt by people. The lined area was affected by landslides, avalanches, and cracks in the ground. From Reference 2.

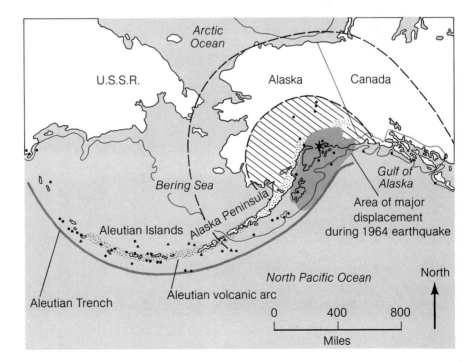

Alaska Peninsula, which were formed by an arcuate chain of volcanoes. None of the epicenters lies south of the Aleutian Trench and few lie north of the volcanic chain—exactly the relations shown by the aftershocks of the 1964 earthquake. Note, too, that the zones of uplift and subsidence in 1964 are elongated parallel to the trench and volcanic chain.

General earth movements associated with the faulting
Altogether, the various lines of evidence indicate the two stages of movement shown diagrammatically in Figure 11-16. The outer part of the earth under the Pacific Ocean is evidently moving northwestward beneath the part of the earth that includes Alaska, the Aleutian Islands, and the Bering Sea. The movement takes place along a gently sloping fault that sticks for long periods during which the rocks above and below it are deformed elastically—similar to the compression of a spring (Figure 11-16A). When some limit is reached, the rocks snap past one another along the fault, allowing a huge slab of the earth to move both horizontally and vertically (Figure 11-16B).

The rock body deformed before the earthquake, however, was too large and probably too unequally affected to snap all at once, such as the one in the simple model of Figure 11-5. Seismograms prove that the rocks broke first under the epicenter but that this triggered the offset of an adjoining "hammer" and then another, producing a series of offsets that propagated outward at an average speed of about 2 miles per second. The succession of offsets took approximately 3 minutes to progress through the entire area, and each one produced a powerful set of earthquake waves. As a result, violent shaking was felt in Anchorage for approximately four minutes. To get some idea of

FIGURE 11-16
A, ground movements (small arrows) resulting from slow elastic deformation of rocks, caused by underthrusting as indicated by the large arrows. B, ground movements caused by a sudden slip on the fault and by displacements of the overlying rocks. All displacements are greatly exaggerated.

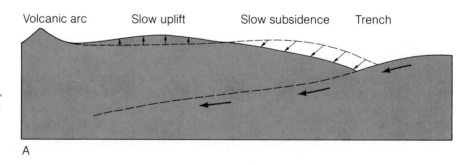

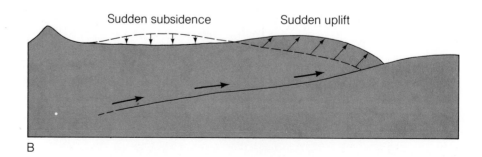

FIGURE 11-17
Beach ridges (the dark tree-covered bands) that have been uplifted one by one during periodic earthquakes to form a stepped series above the present coastline. On the main coast north of Kayak Island, labeled on Figure 11-13A. U.S. Geological Survey photograph by George Plafker (Reference 2).

the total energy released, note that the volume of displaced rock above the fault measures at least 1,300,000 cubic miles! No wonder the Bootlegger Cove Clay flowed under Anchorage, and the large region indicated in Figure 11-15 was affected by landsliding, cracking, and avalanching.

Recurrence of major earthquakes When are such movements likely to occur again? Once the mechanism of the 1964 earthquake was understood, it was possible to answer this question quite directly. Much of the area uplifted during the earthquake had been slowly submerged by the "setting" movements leading up to the quake. Trees and freshwater peat submerged during that period were dated by the radiocarbon method, and they showed that the "setting" lasted about 1,000 years. Additional information came from shorelines that had been uplifted suddenly at the times of previous earthquakes. As Figure 11-17 shows, these ridges are locally preserved in a distinct series, and dated materials in them indicate that past uplifts have taken place at 500 to 1,400 year intervals. The record thus suggests that the same fault is not likely to produce another major earthquake near the epicenter for roughly 1,000 years. Figure 11-15 indicates, however, that the fault may move sooner elsewhere. Other major faults inland may also produce earthquakes. The controversial Alaska oil pipeline crosses part of this seismically active region. Residents must wonder if there is not some way to control the faults or at least to predict more exactly when they will produce additional major earthquakes.

Controlling Earthquakes with Water

A means of predicting and controlling earthquakes was first suggested by an alarming series of events that took place near Denver during the period from 1962 to 1967. The events evidently were started because deep-seated rocks seemed a handy place to get rid of water containing chemical wastes from the Rocky Mountain Arsenal, 10 miles north of Denver. A vertical hole was drilled to a depth of 12,500 feet, and the fluid wastes were injected into the rock at the bottom, a body of ancient granite that occurs widely beneath the sedimentary layers of the region. Between March 1962 and September 1963 fluids averaging 200,000 gallons per day were pumped into the granite at pressures ranging from about 365 to 400 atmospheres. The operation was then halted for a year, after which the fluids were emplaced into the rocks under the force of gravity —the weight of a fluid-column 12,500 feet high (about 360 atmospheres). Approximately 70,000 gallons of fluid per day flowed into the granite between November 1964 and April 1965. Injection was then resumed under pressures as high as 415 atmospheres but was stopped permanently in February 1966 because the operation was evidently causing earthquakes!

The first important evidence that the earthquakes resulted from the water-injection operation was developed by Yung-Liang Wang, a graduate student at the Colorado School of Mines. He plotted the epicenters of local earthquakes recorded by the school's seismograph and found that they clustered distinctly around the injection well. In November 1965, David Evans, a geologist in Denver, compiled data that strongly supported the idea that the fluids were causing the earthquakes. The frequency of earthquakes corresponded with the rates at which fluids were pumped into the rock at the bottom of the well—the more fluids, the more earthquakes.

Most of the earthquakes were mild, but some rattled dishes and were felt widely. Concern mounted among scientists as well as the public, and by January 1966 seismographs had been set up in a network that permitted the precise determination of epicenters as well as first motions. As shown in an analysis by J. H. Healy, W. W. Rubey, D. T. Griggs, and C. B. Raleigh, the epicenters defined an elongate array that extended N60W and S60E from the well site (Figure 11-18) (Reference 3). Moreover, almost all the first-motion studies gave an alternative solution that indicated faulting on vertical surfaces parallel to the array of epicenters. The earthquakes thus appeared to be due to offsets within a zone of vertical faults in the granite deep beneath the arsenal.

The time relations linking the faulting to the fluids are shown in Figure 11-19. Compare the fluid pressure at the bottom of the well to the number of earthquakes per month. The general correspondence between them indicates that the fluids were setting off fault movements, and more of them at high pressures than at low. Note, too, that the peaks representing increased seismic activity lag $\frac{1}{2}$ to 2 months behind the peaks representing increased fluid pressures. This delay must mean that the pulses of fluid pressure causing the faulting took that long to propagate outward from the well. Figure 11-19 shows that the distances involved were typically $\frac{1}{2}$ to 2 miles, so that the speeds of propagation were several hundred feet to many thousands of feet per

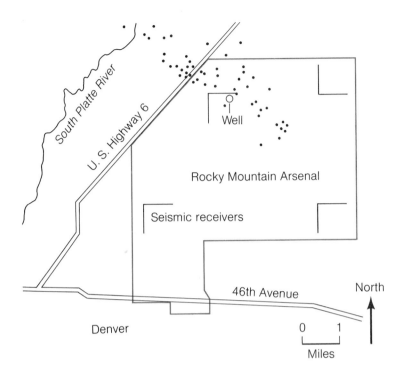

FIGURE 11-18
Epicenters of earthquakes (dots) located in January and February of 1966 by four sets of seismographs that were set out in lines at right angles, as shown. From Reference 3.

FIGURE 11-19
Graph showing correspondence between fluid pressures at the bottom of the Rocky Mountain Arsenal well and the numbers of earthquakes of magnitude greater than 1.5. From Reference 3.

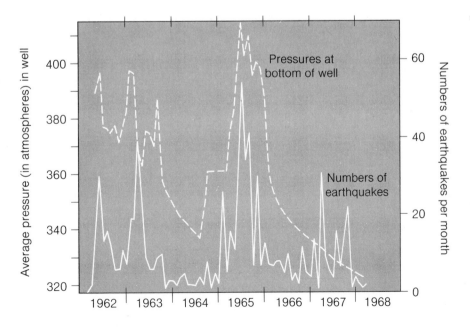

month. Even if we assume that the most distant earthquake, 3 miles from the well, was caused by fluid introduced at the very start of the operation, that fluid would have had to flow at an average rate of 300 feet per month in the ensuing 4 years. This flow is far too rapid to have taken place in solid granite and, thus, indicates that the fluids were moving through a system of inter-connected cracks.

Triggering faults with water at high pressure This, in turn, suggests a cause of the faulting and earthquakes. Evidently, a deeply buried zone of faults had been "set" by slow deformation of the surrounding rocks long before the well was drilled. The deformation had opened fault surfaces and other cracks and pores to some degree, but the faults were locked by irregularities along their surfaces or by mineral cements. When the water pressure was increased, the cracks were forced open enough to separate the irregularities and break through the mineral cements. This part of the fault zone could thus slip, generating an earthquake and pressing water into adjoining parts of the "set" zone. The magnitudes of the earthquakes were typically low (between 1.5 and 3), indicating that most of the offsets involved small fault segments, perhaps 10 feet to several hundred feet long.

The last parts of the record, however, became more alarming. As Figure 11-19 shows, earthquakes continued after the operation was stopped in February 1966, becoming more numerous again in 1967. Three earthquakes in 1967 had magnitudes between 5 and 5.5, together releasing far more energy than all the preceding quakes combined. Water had evidently pressed into parts of the fault zone that either were more strongly set to move or became unlocked over larger areas simultaneously. This relation suggested that water under pressure was moving toward parts of the fault zone that might have still greater potential and might thus produce a truly damaging earthquake. The scientists advising on the project recommended a solution that arose directly from their discovery of how water pressure was affecting the faults. They suggested drilling into the fault zone and pumping water out.

Starting and stopping fault movements The drilling was not done, but fortunately no major earthquake followed. However, the drama of that episode and the knowledge gained from the program of surveillance started many geophysicists thinking that major earthquakes might, in fact, be anticipated and possibly even arrested. Might water be pumped into fault zones at stages of moderate distortion to make the rocks slip in a series of mild events such as those at Denver? Might fault zones set to move disastrously be locked by pumping water out of them?

By 1967 three additional cases suggested a close relation between fluid pressures and faulting. All three involved the impounding of large reservoirs behind dams, which evidently pressed water into fault systems beneath and set off earthquakes, including a disastrous one in India. Then, shortly after the Denver episode, an opportunity arose to inject *and remove* fluids from an earthquake-producing fault, much as in a laboratory experiment. At the Rangely oil field in northwestern Colorado, water had been injected into petroleum-bearing sandstone to increase the productivity of the field, with the result that earthquakes were set off along a fault that crosses the field at depth. A full-scale test was planned in 1969, and from then until 1973 the field was monitored closely by seismographs as (1) water was pumped into the rocks at high pressure, (2) the pressure was reduced and fluid was withdrawn, and (3) water was again pumped in at high pressures (Reference 4). The numbers of earthquakes correlated closely with the fluid pressures, dropping dramatically to near-zero within a day after the withdrawal of fluid. An earthquake-producing fault was thus controlled for the first time!

C. Barry Raleigh, who took part in both the Denver and Rangely studies, has suggested how this knowledge might be used in preventing earthquakes on major faults such as the San Andreas in California, shown on the chapter opening page. Distortions along the fault are known to be accruing constantly toward a major slip event (Reference 4). He has suggested first pumping water out of segments of the fault, leaving unpumped segments 0.5 to 3 miles long between. With the fault locked along the dried segments, water would be pumped into the undried segments, causing them to slip piecemeal as at Denver. Finally, water would be pumped into the dried segments to relieve their elastic distortions by piecemeal movements. The fault would thus be deactivated against major offsets and might be kept in that state by pumping water into the wells occasionally when elastic distortion began to build up again.

Additional studies indicate that drilling and easing might not have to be done along the entire length of the fault, for some segments have been moving quite readily (Reference 5). As Figure 11-20 shows, two large segments of the fault have stuck for long periods and have thus generated major earthquakes, whereas two other large segments have moved frequently by a succession of small but nearly continuous dislocations called *fault creep*. One reason for the two kinds of movement may relate to the faulting mechanism discovered at Denver. Geological surveys suggest that the creeping segments of the fault cut through deeply buried rocks receiving pressurized water released during metamorphism of more deeply buried rocks. The rising water is apparently trapped beneath thick, nearly impermeable layers of rock and is thus forced laterally into the fault zone. Thus, the zone is more or less expanded and therefore able to move easily as distortional forces build up. In the regions that have no capping layers of impermeable rocks, the water released at depth escapes directly to the surface as springs. Water in these segments of the fault zone is thus at low pressures and not likely to trigger displacements until the elastic deformation of the rocks is extreme. The deep-seated movements of water described in Chapter 10 thus have another connection with surface processes!

The practical significance of these connections is that they might make a program of control more feasible, for they reduce the more dangerous parts of the fault by about one-half. A procedure of control would still be expensive because the number of wells needed is large, but the damage caused by a major earthquake in the urbanized regions of California would probably be far more costly. Here is a way to put hard-earned knowledge to use. Major earthquakes take, on the average, about 14,000 lives a year, and one may imagine the losses and suffering inflicted on the hundreds of thousands of persons surviving.

FIGURE 11-20
Map of the San Andreas fault
system, showing by black lines
the parts that have been moving
by frequent small increments,
and by gray lines those that have
stuck for long periods and
therefore caused major
earthquakes. Note that Los
Angeles and San Francisco lie
next to the latter type of
segment.

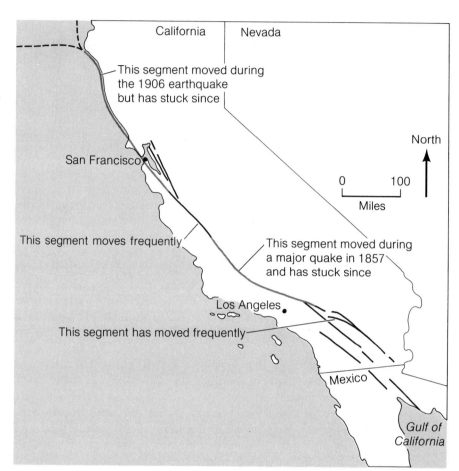

Summary

Prediction and control of earthquakes will probably be possible when we
understand thoroughly how they are generated and propagated. Some important relations discovered so far are:

1. Earthquakes are produced by two related actions: (a) the slow elastic distortion of rocks along faults and (b) the sudden snap and impact when the
rock masses slip.
2. The resulting vibrations travel outward as P-waves, which are compressions
and rarifactions parallel to the direction of propagation, and as S-waves,
which are vibrations at right angles to the direction of propagation.
3. Seismograms of these waves can be used to locate epicenters and to interpret the directions of dislocations on the faults that caused the earthquakes.

4. The magnitude or strength of earthquakes is measured from the amplitude of the waves at a distance of 100 kilometers from the epicenter.
5. The intensity or amount of local ground motion depends on the magnitude, the distance from the epicenter, and local materials under the surface.
6. Water-saturated sediments exaggerate intensities and are a major cause of earthquake damage.
7. The specific causes of major earthquakes can be interpreted from measurements of ground displacement, aftershocks, tsunamis, and the record of past movements and earthquakes.
8. Fluids injected under high pressure tend to expand fault zones and thus set off earthquakes.
9. Fault displacements may be detained by pumping fluids out of fault zones and thus reducing fluid pressures.

REFERENCES CITED

1. John D. Sims. Earthquake-induced structures in sediments of Van Norman Lake, San Fernando, California. *Science*, vol. 182, pp. 161–63, 1973.
2. George Plafker. Tectonic deformation associated with the 1964 Alaska earthquake. *Science*, vol. 148, pp. 1675–87, June 25, 1965. Also, Tectonics of the March 27, 1964 Alaska earthquake. U.S. Geological Survey Professional Paper 543-I, 74 pp., 1969.
3. J. H. Healy, W. W. Rubey, D. T. Griggs, and C. B. Raleigh. The Denver earthquakes. *Science*, vol. 161, pp. 1301–10, 27 September 1968.
4. C. B. Raleigh, J. H. Healy, and J. D. Bredehoeft. An experiment in earthquake control at Rangely, Colorado. *Science*, vol. 191, pp. 1230–37, 26 March 1976.
5. William P. Irwin and Ivan Barnes. Effect of geologic structure and metamorphic fluids on seismic behavior of the San Andreas fault system in central and northern California. *Geology*, vol. 3, pp. 713–16, 1975.

ADDITIONAL IDEAS AND SOURCES

1. Several significant advances in the prediction of major earthquakes have been made in the last few years. The most dramatic was a prediction in Liaoning Province, China, where an earthquake of magnitude 7.3 on February 4, 1975 was anticipated so accurately that three counties were given the final order to evacuate only about 6 hours before the quake. More than a million persons left their homes, and doubtless many lives were saved. In the United States, a 5.2 magnitude earthquake on November 28, 1974 was predicted successfully by an array of instruments 75 miles southeast of San Francisco.

   A number of precursors of earthquakes have been discovered: unusual activity of animals; progressive tilting of the ground and changes in lengths of surveyed lines; progressive changes in water levels in wells; changes in the transmission of electrical currents in the ground; and changes in the earth's magnetic field. Soviet scientists have thoroughly tested changes in velocities of seismic waves, which decrease gradually over a long period while the rocks along a fault are being elastically set and then increase rapidly just before the slip on the fault. With continued research and the installation of many arrays of instruments, it should be possible to predict many major earthquakes by 1985 or 1990.

2. A good idea of how earthquakes affect communities can be obtained from the 33-page booklet *Seismic hazards and land-use planning*, by D. R. Nichols and J. M. Buchanan-Banks (U.S. Geological Survey Circular 690, free on request from the Survey's National Center, Reston, Virginia 22092). More specific and detailed views of seismic hazards and how they can be reduced are given in *Studies for seismic zonation of the San Francisco Bay region*, edited by R. D. Borcherdt (U.S. Geological Survey Professional Paper 941-A, 102 pp., 1975). An additional source, on the effects of the 1964 Alaska earthquake, is the series of U.S. Geological Survey Professional Papers 541 through 546. These papers are available in many separate parts, a list of which can be obtained from the Survey.

3. Some effects of water on the displacement of sediments can be explored in simple experiments using sand and a rubber balloon. (1) Insert a funnel in the neck of the balloon and pour in sand until the balloon is extended to a diameter of 6 inches or so. (2) Add water until the sand is saturated. (3) Thump the mixture lightly but repeatedly on a solid surface, thereby packing the sand grains closer together and causing excess water to gather at the top of the balloon. (4) Squeeze the water out and draw the neck tightly shut—in this state, you will find that the loose sand grains are locked together as tightly as in a solid rock. (5) Finally, open the neck, insert the nozzle of a faucet into it, and inject water into the balloon. You should then be able to deform the sand easily again.

Steam wells at The Geysers, California, at an early stage of development in 1959. This geothermal field was producing 502,000 kilowatts of electricity annually in late 1976 and is being developed toward a capacity of 2 million kilowatts in 1990. Operated by the Pacific Gas and Electric Company, it is the largest commercial geothermal power facility in the world. Photograph by Mary Hill.

12. Molten Rock, Hot Water

Molten Rock, Hot Water

Volcanoes in Action
>   The deposits of Mount St. Helens

Rocks Formed from Melts
>   Growth of minerals in melts
>   Differentiation of basalt melt
>   A suite of rocks from andesite melt

Subjacent Bodies of Melt
>   Emplacement of a pluton
>   Metamorphism by the pluton
>   Crystallization at depth

Transfer of Metals and Energy by Plutons
>   Plutons as geothermal systems
>   The full pluton system

Summary

Besides setting off earthquakes in faulted rock bodies, water circulates to deep-seated, hot parts of the earth, from which it may emerge as a source of thermal energy such as that illustrated on the chapter opening page. These deep-reaching connectives are called *geothermal systems,* and many have their roots in bodies of molten rock or in rocks that have solidified recently from melts. Rocks solidified from melts are called *igneous rocks,* and proof of their origin is found in volcanoes, which erupt lavas heated to temperatures as high as 1,250°C. According to the rates at which temperature increases with depth, described in Chapter 10, a temperature of 1,250°C is normally found at depths of more than 40 kilometers (25 miles). Volcanoes and the molten conduits beneath them thus create an unusual distribution of heat in the outer part of the earth, affording a near-surface source of geothermal energy. The hot water associated with the molten systems also dissolves and transports various metals and other chemicals, locally forming concentrations that are crucial raw materials for human uses. Systems of molten rock and hot water are thus one of our basic resources.

Volcanoes and the molten systems under them are commonly grouped in long chains, such as the one forming the Aleutian Islands and part of the Alaska Peninsula, mentioned in Chapter 11. The volcanoes in Figure 12-1 are part of a similar volcanic chain, the Cascade Range, which extends from British Columbia southward through Washington, Oregon, and northern California. This region has been volcanically active for several long periods during the past 60 million years and is now one of the best exposed and most beautiful volcanic terrains in the world. Mount St. Helens is the youngest of the larger volcanoes of the chain and can be reached by an hour's drive from Portland. As we shall see, a number of recent studies have made it one of the best understood volcanoes anywhere.

FIGURE 12-1
View northward along the Cascade volcanic chain in Washington, with Mount St. Helens in the foreground and Mount Rainier in the distance. Photograph by John S. Shelton.

Volcanoes in Action

The serene beauty of Mount St. Helens is probably only temporary. The entire cone formed in the last 2,500 years, and the last eruption was in 1857. Several other Cascade volcanoes have been active historically: Lassen Peak erupted powerfully in the period from 1914 to 1917, and Mt. Baker has erupted steam intermittently, the latest episode starting in March 1975. Similar volcanoes in Japan and the East Indies are monitored continuously with precise instruments because of the damage inflicted by past eruptions. Increasing human use of the Cascade Range presses some questions. What will the scene be like when one of the big volcanoes comes to life? When is this likely to happen?

We can answer these questions to some degree by examining the volcanoes and their deposits, but this study will have far more meaning if the deposits are compared to those of a similar volcano in actual eruption. As it happens, many volcanoes around the Pacific's perimeter are similar to the Cascade cones, and one especially similar to Mount St. Helens, Mayon Volcano in the Philippines, was well observed in a recent eruption. The description that follows was condensed from an account by James G. Moore and W. G. Melson (Reference 1).

Mayon Volcano ended a 21-year period of quiescence on April 20, 1968, with the emission of glowing lava into the crater at its summit and then with a series of rumbling explosions that increased in intensity during the first two days and continued for three weeks. The explosions blew chunks of incandescent lava as much as 2,000 feet above the volcano and sent dark clouds of fine fragments and gas billowing to heights of 2 to 6 miles. Lava particles were evidently mixed densely into the lower part of the clouds, for huge segments of the glowing mixture fell back immediately on the summit, forming heavy, tonguelike clouds that raced down the mountain in roiling cascades (Color Plate 7). The flow photographed moved with a velocity of 63 meters per second (145 miles per hour) on the steep upper slopes and averaged 31 meters per second (70 miles per hour) from the summit to the outer part of the volcano!

Such flows consist of countless particles, including many large blocks, yet they make little sound. Their soundlessness and great mobility indicate that the fragments are cushioned by air entrapped under the nose of the advancing flow and then heated by the hot fragments so that it expands. Additional hot gases escape from each fragment. The gases and the great surge of such flows have disastrous effects on areas beyond the flows. Moore and Melson found that coconut palms near the flows were uprooted and blown over, indicating a blast of air exceeding 60 meters per second (140 miles per hour). Plants were seared and charred, and all animals were killed in a zone extending 600 feet to more than 6,000 feet from the sides and ends of the flows. These gas-scorched areas are three times as large as the area covered by the fragmental flows themselves.

An accompanying effect of the explosions, typical of many large volcanoes, was the generation of lightning and torrential rains from the eruption clouds. The rapidly rising clouds evidently drew in moist, cool air that nucleated as rain around dust particles. The rains eroded the loose materials just deposited and formed great mudflows that streamed down the mountain and continued 1 to 3 miles beyond the far ends of the hot fragmental flows. Volcanic mudflows are especially destructive because they tend to spread out on river plains. In Java and Sumatra, for example, they are by far the greatest volcanic hazard to humans.

Except for the materials incorporated in the mudflows, the fine particles that fell from the vertically rising clouds were not destructive. They were cold by the time they accumulated on the ground, forming a layer no more than 3 inches thick at the base of the cone. This type of deposit, however, was by far most extensive, forming a layer one millimeter ($\frac{1}{25}$ inch) or more thick over an area of more than one hundred square miles.

The fourth and final kind of activity began on April 27 when an explosion burst the southwest side of the summit crater, and lava began to flow through the gap and down the volcano. The lava was so viscous, however, that it moved at an average velocity of 950 feet per day and took 10 days to reach the base of the cone. Its volume was nearly as great as all the lava erupted explosively, but the flow was nowhere near as destructive because it was confined to a narrow valley on the mountain. As Moore and Melson noted, the flow terminated the 1968 eruption, except for the emission of steam at the summit and additional mudflows generated during periods of heavy rain.

The deposits of Mount St. Helens We can now turn to Mount St. Helens to see if we can find deposits such as those erupted at Mayon. Recognizable at once are recent lava flows that are closely similar to the Mayon flow, though extruded from the sides rather than from the summit of St. Helens (Figure 12-2). They are steep-sided elongate tongues, 100 to 300 feet thick, and their blocky surfaces give some feel for the chaos that occurred when they were moving (Figure 12-3). The angularity of the lava chunks, some over 30 feet across, indicates that the top of the flow cracked and separated repeatedly as the underlying liquid moved slowly downhill.

Abundant deposits on and around the mountain indicate the repeated eruption of hot fragmental flows such as those at Mayon. Figure 12-4 shows vertical exposures of such deposits, 3 miles from the cone. The chaotic mixture of large and small fragments is typical. Tree trunks converted to charcoal attest to the high temperatures of the flows, and some flows are reddened by iron oxides formed by hot gases. The deposit below Goat Rock (Figure 12-2) was a hot fragmental flow that spilled out onto snow and was converted to a mudflow.

Donal E. Mullineaux and Dwight R. Crandell found that the greatest hazards of the volcano are voluminous fragmental flows that are converted partly to mudflows (Reference 2). The mudflows are more numerous and thicker down valley, extending far beyond the hot fragmental flows, as at Mayon. Some hot fragmental flows moved at least 10 miles from Mount St. Helens, and some mudflows at least 33 miles. By comparison, the largest of the hot fragmental flows of the 1968 Mayon eruption flowed about $1\frac{1}{2}$ miles from the base of the cone, and the mudflows about 4 miles.

FIGURE 12-2

Mount St. Helens from the north. Most of the cone is blanketed with hot fragmental deposits (as the one beneath Goat Rock) and with finer fragments that fell from high eruption clouds. The domes (piles of viscous lava) and flows were erupted from subsidiary vents on the sides of the volcano. Photograph by U.S. Geological Survey.

FIGURE 12-3

Upstream view along a recent lava flow on Mount St. Helens that had its source behind the peak in the right-distance. The mountain's summit is partly in view to the left of the clouds.

FIGURE 12-4
Vertical cuts in the deposits of two hot fragmental flows, the one on the left showing a carbonized tree felled by a flow, and the one on the right showing the lack of sorting in an unusually coarse flow.

Rocks Formed from Melts

To predict the volcano's eruptions, it is necessary to understand its inner workings, especially the details of their history. Fortunately, many eruptions at Mount St. Helens have been dated. To use this record, however, we must know more about the erupted materials—what they are as molten lavas and what happens to them as they move up through the volcano. It is intriguing that three kinds of lava have been erupted at Mount St. Helens and that the same three kinds make up many other large volcanoes. Does this mean that volcanoes and lavas evolve in orderly ways? Let's look at the lavas first and then at the succession of their eruptions.

An important lava mentioned in Chapter 6 and erupted twice at Mount St. Helens is basalt. This is the heavy, dark rock that forms the Mid-Atlantic Ridge and occurs in many other volcanoes the world over. When completely molten, basalt is the hottest and most fluid lava, typically erupting with temperatures of around 1250°C and having the consistency of thick syrup. Gases escape from it so easily that they seldom cause explosive eruptions; rather, the lava streams down steep cones and congeals to form tongues only a few feet thick (Figure 12-5). A basalt tongue that moved across flatter ground near the base of the mountain is much thicker but has a delicately wrinkled surface that identifies it as a fluid basalt called *pahoehoe* (Color Plate 8A and B).

Fragmental
dacite pumice

Deposits of
hot fragmental
flows

Thin flows
of basalt lava

FIGURE 12-5
Canyon eroded on the north side of Mount St. Helens, exposing typical volcanic rocks. Note that the flows of basalt slope less steeply than the fragmental deposits. The distant peak is Mt. Adams, another large Cascade volcano.

A very different lava is one called *dacite,* which is so viscous (comparatively rigid) that it piles up in steep-sided domes (Figure 12-6). Occasionally, gases dissolved in parts of the domes explode violently and send a flow of hot fragments down the mountain, forming deposits such as the one below Goat Rock (Figure 12-2). Most dacite, however, contains enough gas to explode totally as it erupts, forming the hot fragmental flows already described, as well as immense eruption clouds. The latter showered particles over huge areas, in one case forming a deposit that extended across Washington and British Columbia to central Alberta, nearly 600 miles from Mount St. Helens. The showers formed thick drifts along the eastern edge of the cone and a

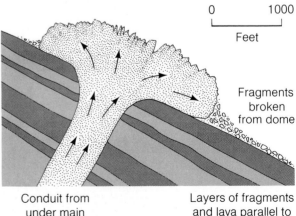

0   1000
Feet

Fragments
broken
from dome

FIGURE 12-6
Lava dome of Goat Rock, cut away to show its feeder. The talus around the dome resulted mainly from cracking of the viscous lava rather than from normal erosion.

Conduit from
under main
volcano

Layers of fragments
and lava parallel to
surface of cone

FIGURE 12-7

Four stages in the vesiculation (bubbling) of a liquid melt as it rises and erupts explosively. Microscope view number 3 is typical pumice, and number 4 shows the fragments resulting from further expansion of the bubbles.

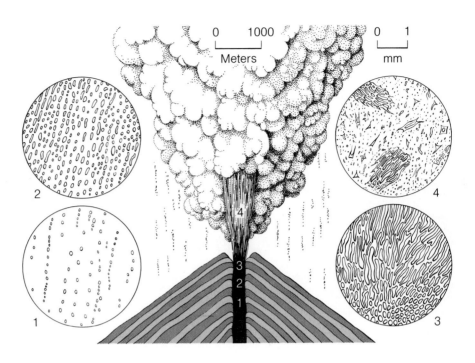

layer 5 to 15 feet thick on the north face (Figure 12-5). Most of the larger particles are of *pumice*, a frothy glass formed when dissolved gases separate from the liquid melt just before it solidifies (Figure 12-7). The gases escape because the confining pressure on the lava is reduced as it rises into the volcano's throat. The bubbles burst with great force because the viscosity of the lava results in an explosion and an expanding cloud of glass chips, pumice lumps, dust, and gas.

The third kind of lava erupted from Mount St. Helens is called *andesite* and has properties between basalt and dacite. Andesite flows are quite viscous but typically move faster and farther than dacite flows. The flows of Figures 12-2 and 12-3 are of andesite, as are many of the hot fragmental deposits. All the 1968 eruptions at Mayon were of andesite.

Many samples of the three lavas have been analyzed chemically in a study made by Clifford A. Hopson. Figure 12-8 shows the amounts of silica (SiO_2) thus determined. Silica is the most abundant component, but other chemical components show the same relations: (1) each kind of rock has a range of compositions and (2) the rocks together show a complete gradation from the least silicic basalt to the most silicic dacite.

Growth of minerals in melts   The mineral grains in the lavas range in size from microscopic to about $\frac{1}{4}$ inch long. They are typically well-formed crystals because they grew unobstructed in a molten liquid, often becoming oriented by its streaming flow (Figure 12-8). The dark glass in the figure is the part of the lava that was molten liquid when the lava was erupted and cooled so fast that the liquid solidified without crystallizing. The different sizes of crystals in Figure 12-8 indicate that some started growing before

263

FIGURE 12-8

Right, compositions of Mount St. Helens rocks with respect to silica. Each dot represents one analyzed sample. Note that the viscosity increases as the silica content increases, probably because silicon and oxygen join in molecular chains and webs within the melt. Below, dacite glass (black) containing feldspar crystals. The small ones are oriented by flow around the large crystals. The view is 2 mm ($\frac{1}{10}$ inch) across.

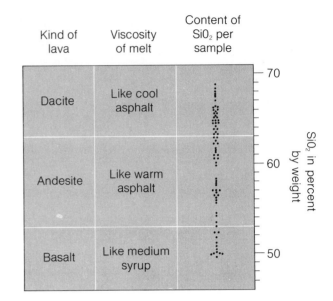

Kind of lava	Viscosity of melt	Content of SiO_2 per sample
Dacite	Like cool asphalt	
Andesite	Like warm asphalt	
Basalt	Like medium syrup	

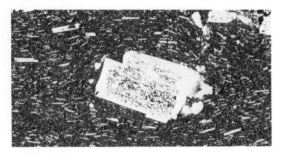

FIGURE 12-9

Thin sections of basalt from the flow shown in Color Plate 8A. Left, the outermost skin of the flow containing much glass (black). Right, the interior of the flow, where all the melt crystallized. The white crystals are feldspar (plagioclase); the large medium-gray ones are olivine (magnesium silicate); and the small dark gray ones are pyroxene (calcium magnesium iron silicate). Both views are 4 mm (0.2 inch) across.

others; all stopped growing when the melt was cooled during eruption. Basalt melt is less viscous than dacite melt, and crystals therefore grow in it more easily; nonetheless, basalts, too, show differences in grain sizes due to rates of cooling. In Figure 12-9, for example, the liquid at the top of the flow was chilled against air and set as black glass, whereas the more slowly cooled parts crystallized completely.

Note that the larger crystals in the basalt, however, are nearly as large in the chilled margin as in the interior of the flow. This similarity in size indicates that the crystals were present in the lava when it was erupted, so that they must have grown when the melt was underground. Most dacite and andesite lavas also carry large grains that grew before eruption. The grains evidently grew gradually and episodically, for when they are examined in polarized light, which brings out their compositional variations, each crystal shows a series of accretional shells, like growth rings on a tree (Figure 12-10A). Similar feldspar crystals have been grown in the laboratory by cooling dacite melt in successive stages (Figure 12-10B).

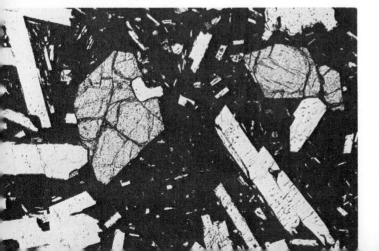

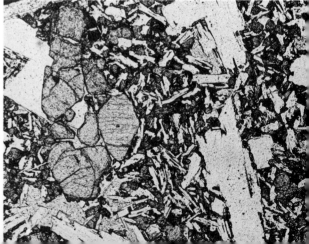

Differentiation of basalt melt If crystals grow to large sizes in melt underground, we may consider some interesting consequences. Rock melts have lower densities than the principal minerals that grow in them. The crystals therefore sink as they grow, like sand grains in thick syrup. The larger crystals in the basalt of Figure 12-9 would sink 100 to 350 feet per year in basalt melt, and crystals one-tenth that size would sink 5 to 10 feet per year. In a matter of 100 years, most crystals formed in the upper 1,000 feet or so of a basalt body would thus have sunk to lower levels in it.

Growth of the crystalline grains, in turn, would affect the composition of the remaining melt. As Figure 12-8 indicates, basalt melt is about 50 percent silica. The mineral *olivine* in Figure 12-9, on the other hand, is about 37 percent silica, and the feldspar is about 46 percent silica. Because both minerals contain less silica than the melt from which they form, the proportion of silica in the melt would gradually increase as the mineral grains grew. If the larger grains in Figure 12-9 were to grow in a basalt melt and then sink from it, we can calculate that the remaining liquid would be about 55 percent silica. As Figure 12-8 shows, such a melt has the composition of andesite!

Indeed, if crystals continued to form and sink from the upper part of the body, the silica content of the melt would continue to increase until it might have the composition of dacite. Is this, then, the way in which the volcano's suite of lavas originated? It is a scheme thought to work at some volcanoes but it is barely plausible for Mount St. Helens. For one thing, if basalt were the parent of all the rocks there, it should have been erupted frequently in the volcano's history, and it has been erupted only twice. For another, the large amounts of andesite and dacite should be balanced by a huge accumulation of crystals, rich in olivine, under the volcano. Melts erupted upward through the accumulation would have occasionally carried chunks of it to the surface, and none have.

A suite of rocks from andesite melt An alternate possibility, however, is that the suite of rocks forms in a similar way from a melt near the middle of the range shown in Figure 12-8. This melt would be an andesite containing about 60 percent silica, and andesites with that silica content have been erupted abundantly at Mount St. Helens and many other Cascade volcanoes. The first minerals formed in these melts are *pyroxene* and the feldspar called *plagioclase*, which have a lesser silica content than the original melt. Sinking of these crystals would leave the upper part of the andesite enriched in silica as just described, *and their accumulation in the lower part of the body would result in rocks with less than 60 percent silica.* The latter rocks would have the compositions shown in the lower half of Figure 12-8. The accumulation of crystals in the lower part of the body of the melt is also supported by the fact that many of the lavas contain large pyroxene and plagioclase grains stuck together in chunks and clots.

The main evidence that the various lavas formed because of the sinking of crystals, however, comes from the eruptive history of the volcano. Because of the recency of the eruptions, many have been dated by the radiocarbon method, using plant materials caught up in the fragmental deposits. The dates show that the main cone was built in three eruptive periods separated by

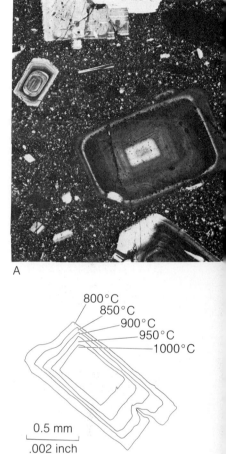

A

B

FIGURE 12-10

A, photograph of dacite lava with large (1 mm) feldspar crystals that show growth layers. The fine-grained matrix (which is partly glass) was molten during eruption but chilled rapidly, forming many small grains. B, drawing of a crystal grown in a capsule such as those described in Chapter 10. The melt was cooled in five stages, being held for 2 days at each of the temperatures shown. Courtesy of William C. Luth.

long periods of quiescence. Each eruptive period commenced with the emission of silica-rich dacite. As eruptions continued, the compositions of the rocks became less silicic, such that the last lavas erupted were silica-poor andesite or, in one case, basalt. As Figure 12-11 illustrates, this order is exactly what would be expected if a column of andesite melt differentiated during a period of quiescence and was then erupted piecemeal from the top of the column. Each eruptive period would begin with the part nearest the surface, the dacite melt, and would progressively produce less silicic melts as eruptions continued.

Also supportive of differentiation during quiescence is the fact that the first eruptions of each cycle have been violently explosive. These eruptions formed the various fragmental deposits, and they were followed in each case by the quiet emission of dacite lava domes or andesite flows. This history would be explained by accumulation of water vapor and other dissolved gases in the upper part of the column of melt (Figure 12-11). Because the dissolved gases are lighter than the other components of the melt, they would thus diffuse upward. The diffusion, however, is very slow, so that the accumulation implies a long period during which the melt stood within the volcano. Possibly much of the dissolved vapor was added from groundwater.

Whether or not this scheme represents what is actually happening under the volcano, the dating shows that the periods of quiescence have lasted for as short a time as 130 years, and probably none have lasted longer than several hundred years. Another eruptive period is thus likely to begin sometime during the next one hundred years. We can, moreover, predict its course. It will start with repeated explosive eruptions of dacite melt, producing scenes likely to be far more violent than those pictured in Color Plate 7. These explosions will be followed by the quiet eruption of domes of dacite lava, by the eruption of andesite, and finally, perhaps, by basalt. It is possible, of course, that the volcano will never erupt again; however, a long life is suggested by the sorts of systems found to lie under volcanoes, as we shall now see.

FIGURE 12-11

Possible differentiation of lava within Mount St. Helens.
1) Intrusion of andesite melt into the volcano, 2) differentiation during quiescence, by sinking of crystals, and 3) eruption—starting with dacite and ending with silica-poor andesite.

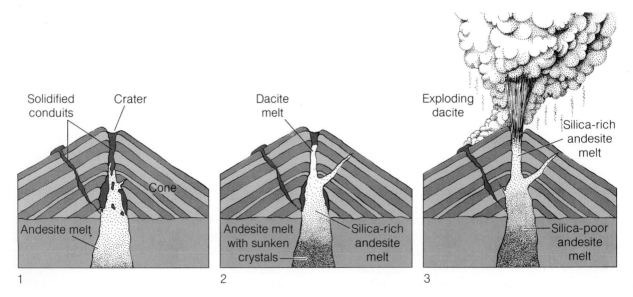

FIGURE 12-12
View northward along the Cascade volcanic chain in central Oregon. On the left is Mt. Washington, in the middle is Three-fingered Jack, and in the distance Mt. Jefferson.

Subjacent Bodies of Melt

Erosion of volcanoes older than Mount St. Helens permits us, in effect, to descend into volcanic feeding systems. In Figure 12-12, for example, the sharp spire on the nearest peak is the volcano's exposed conduit, and the next volcano shows a deeper view of more complex feeders. Feeding systems generally consist of a roughly cylindrical *pipe*, 300 to 3,000 feet in diameter, and branching sheetlike bodies called *dikes*, some of which may feed eruptions on the lower flanks of cones (Figure 12-13). The dikes are especially interesting because their patterns may suggest the presence of larger bodies beneath (Figure 12-13, lower diagrams).

Larger bodies are indeed exposed in many parts of the northern Cascade Range. They fed much older volcanoes, ones active 14 to perhaps 30 million years ago, and erosion has now cut downward as much as a mile or so into them. These large underground bodies of melt (or rock crystallized from melt) are called *plutons* (after Pluto, god of the underworld), and they form a chain parallel to the active volcanic chain (Figure 12-14). One that shows especially clearly how they form is the Tatoosh pluton, named after the Tatoosh Mountains, near Mount Rainier (the more distant volcano in Figure 12-1). The histories of the pluton and the volcano have been worked out by Richard Fiske, Clifford Hopson, and Aaron Waters, who mapped and studied all the rocks in Mount Rainier National Park (Reference 3).

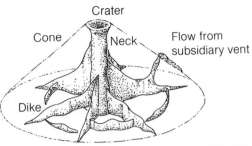

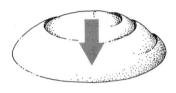

FIGURE 12-13
Feeder system of a cone (top), and two diagrams suggesting how the rise and withdrawal of an underlying body of melt (gray arrows) might crack the overlying cone.

FIGURE 12-14
Plutons (black) and volcanoes of the Cascade Range and vicinity. In addition to the separate cones shown, most of the region is underlain by volcanic rocks of various ages. Chiefly from the Tectonic Map of the United States (U.S. Geological Survey and the American Association of Petroleum Geologists, 1962).

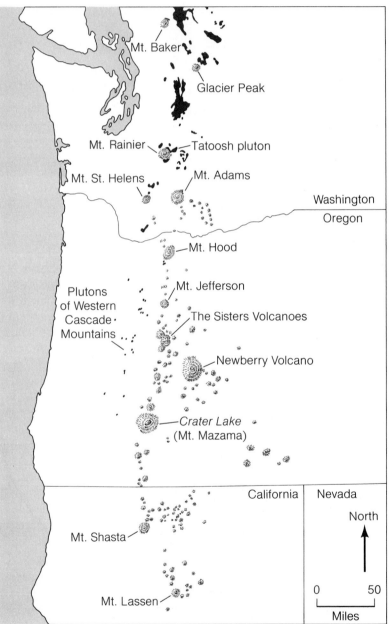

Emplacement of a pluton The Tatoosh pluton is partly concealed by Mount Rainier volcano, which is much younger, and by the older rocks into which the pluton rose. However, one can estimate that it is about 10 miles wide and 16 miles long. The remnants of the pluton's roof are riddled by sheetlike bodies of rocks that look like the dacites and andesites of Mount St. Helens. Fiske, Hopson, and Waters have generalized the sheetlike bodies in a drawing that shows the pluton at a somewhat lower level than the one it finally attained (Figure 12-15). The sheets must be imagined as continuing in and out of the page. Some are dikes, cutting across the layered rocks in a rudely radiating pattern that suggests an upward press by the pluton (Figure 12-13). The upper part of Figure 12-15 shows that some dikes intruded along faults that formed when the pluton lifted segments of its roof.

Most of the sheetlike bodies, however, intruded *between layers of older rocks*—a kind of intrusion called a *sill*. Some sills probably formed when layers of rock collapsed downward into the molten pluton (Figure 12-16A). Most, however, penetrated too far from the pluton to have formed in this way, and the arched roof shows that these sills pressed the overlying rocks upward (Figure 12-16B). This action indicates that the pluton was being pressed upward—presumably because it was less dense than the solid rocks over and around it. The upper part exerted a pressure like that of water in a capped pipe, which drove melt out along cracks developed between rock layers. As each sill solidified, another was typically injected above or below it, so that the overlying rocks were jacked up piecemeal. Eventually, as shown in Figure 12-15, melt burst through the roof and into the open air, forming a new volcano!

The pluton also gained space by eroding its way upward. The evidence can be seen along its margins, where it cuts sharply across sills and older rock layers and carries many fragments of them. Additionally, if the older layered rocks in Figure 12-15 are projected across the space now occupied by the pluton, one can see that a large volume of those rocks is missing—apparently

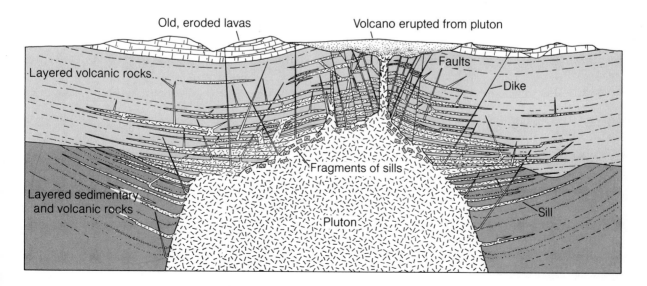

too large a volume to be accounted for by the arching just described. A mystery of this pluton and many others is what happened to these rocks. Did they sink to the depths? Were they melted, or were they slowly changed into granite by diffusion, as mentioned in Chapter 10?

Metamorphism by the pluton   There is no question that the pluton was a long-lived source of heat. It heated the rocks around it so strongly that some were changed to metamorphic rocks comparable to the highest temperature varieties described in Chapter 10. The metamorphic changes can be seen quite clearly by walking toward the pluton from a mile or so away where the surrounding rocks are unaffected. This walk is basically like the one across the tilted rocks of Figure 10-13, only the distance is much less, and the rocks are fragmental volcanic rocks rather than mudstones and sandstones. Figure 12-17A shows a typical volcanic rock in the older sequence, where it is beyond the affects of the pluton. Closer to the pluton, the fragments making up the rock become crisply apparent because of new metamorphic minerals formed in and around them (Figure 12-17B). Note that a black mineral (*hornblende*, described in Supplementary Chapter 1) has grown preferentially in some fragments, and that feldspar (white) has formed around many. Figure 12-17C shows an extreme degree of metamorphism next to the pluton, where hot vapors removed some substances from the rocks and added others.

Crystallization at depth   The heated rocks surrounding the pluton evidently insulated it against further losses of heat, for the main body of the pluton remained molten long enough to grow large crystals. As Figure 12-18 suggests, the various parts of the intrusion show a systematic increase in grain size—from the fine-grained outlying dikes and sills (almost as fine grained as the lavas of Mount St. Helens), to moderately coarse rocks in the outer part of the pluton, to still coarser rocks in its interior. The pluton thus prepared a way for itself by heating the surrounding rocks while it mechanically displaced them.

The deep exposure also provides evidence for what might be happening under the modern Cascade volcanoes, for the rocks making up the Tatoosh pluton are closely similar in composition to those forming Mount St. Helens.

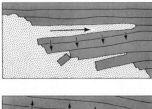

FIGURE 12-16
Two possible mechanisms for the intrusion of sills. Top, sinking of rock-layers into the molten pluton, and bottom, lifting of layers by emplacement of a wedge of melt.

FIGURE 12-17
Fragmental volcanic rocks that were deposited approximately 40 million years ago, when they were like the fragmental deposits of Mount St. Helens. A, rock one mile from the pluton, only slightly metamorphosed; B, rock 500 feet from the pluton, metamorphosed completely to new minerals; and C, rock 100 feet from the pluton, so metamorphosed that fragments not containing black minerals are almost invisible.

A

B
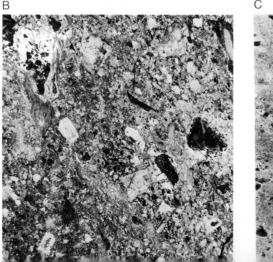

C
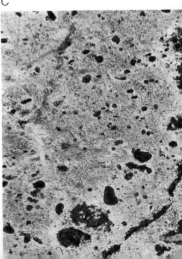

There is no sign of coarsely crystallized basalt (a rock called *gabbro*, described in Supplementary Chapter 2) and no accumulation of olivine crystals. By far the most abundant rocks in the pluton have the compositions of silicic andesite (a coarse-grained rock called *quartz diorite*) and dacite (a rock called *granodiorite*, illustrated in Figure 12-18). Dark rocks composed of silica-poor andesite (called *diorite*) occur locally, and rocks equivalent to silica-rich dacite lie in the upper parts of the pluton. Otherwise, however, there is no compelling evidence for the crystal settling suggested beneath Mount St. Helens. The pluton does, however, show clearcut evidence of repeated activity and intrusion (Figure 12-19). Possibly these disturbances have affected what might once have been a gradation such as that shown in Figure 12-11. Dating of the pluton's rocks by the uranium-lead method (Appendix B) shows that the sills over the main pluton were emplaced about 26 million years ago and that various other melts were intruded and crystallized over the ensuing 12 million years, up to about 14 million years ago. The pluton thus had ample time for crystals to settle and for the results to be rearranged many times.

Transfer of Metals and Energy by Plutons

Several important aspects of the Cascade plutons connect to movements of water in and near them that concentrated metals and heat in certain places. Dissolved water evidently accumulated in the upper part of the Tatoosh pluton much as it did in the upper part of the conduit feeding Mount St. Helens. One indication is the explosive nature of the eruptions through the pluton's roof (Figure 12-15). Additional evidence comes from the effects of water vapor and other gases on the surrounding rocks and on the pluton itself. Fragments of rock in the upper part of the pluton, such as those in Figure 12-19, often contain small bubblelike bodies formed when vapor penetrated the rock and dissolved or melted parts of it (Figure 12-20, left). The rocks surrounding the crystallizing body were also cracked by the intrusion and were locally shattered by explosions of vapor. Vapor then flowed along the cracks, carrying dissolved materials and depositing them locally to form bodies called *veins* (Figure 12-20, right). Water-free minerals in the rock, such as feldspar, were commonly altered to scaly clusters of mica, clay, and other minerals that incorporate water in their compositions. These alterations also

FIGURE 12-18

Rocks of the Tatoosh pluton; left, from a sill; middle, from the outer part of the pluton; and right, from the inner part. All are .4 inch across and were taken with polarized light, such that each crystal is an area of one shade. Most of the grains are plagioclase, the bands being parts of the crystal called twins, each of which has a different orientation from the adjoining parts (see Supplementary Chapter 1, Figure S1-5).

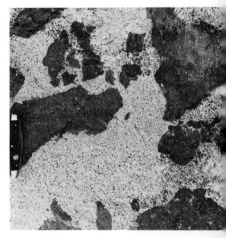

FIGURE 12-19

Fragments formed when a body of diorite was intruded, broken, and partly dissolved by granodiorite melt and by vapor associated with it.

271

FIGURE 12-20
Left, dark diorite intruded by granodiorite melt. The small rounded bodies indicate that vapor dissolved parts of the diorite. The sample is 6 inches across. Right, part of a Tatoosh sill, broken by cracks along which vapor deposited veins of black, iron-rich minerals. The white grains are feldspar, altered by vapor to clays. The sample is 6 inches across. Another vein is shown in Figure 12-17C.

took place along cracks deep in the pluton itself, so that water must have continued to move through the pluton after it solidified. From what we know about the temperatures required to form these various alteration minerals, we can conclude that some formed at high temperatures, perhaps more than 600°C, and some formed at lower temperatures, probably as low as 200°C. This range indicates that some parts of the pluton were affected by vapor and hot water during the entire period of cooling.

Similar alterations formed ore deposits in and near plutons along the western side of the Cascade Range in Oregon (Figure 12-14). Gold, silver, copper, lead, zinc, antimony, iron, and manganese were deposited in a variety of minerals that are disseminated through the rocks or concentrated in veins (mainly mineral fillings in cracks). They have particular significance because Hugh P. Taylor, Jr. made a careful study of the oxygen in the minerals, specifically its composition with respect to the two isotopes ^{18}O and ^{16}O (Reference 4). He found that the isotopic mixture differs greatly from that in water from the deep earth, collected at volcanoes elsewhere, and that it is closely similar to the groundwater in the Cascade Range. This correspondence suggests strongly that the plutons and surrounding rocks were soaked by groundwater as they cooled. The ore metals, however, are only slightly soluble in heated groundwater, even when it is a vapor, so that water must have flowed through the cracks for a long period of time. The metal-rich veins thus imply a long-lived circulation of groundwater as the pluton cooled. As Figure 12-21 suggests, such systems gain surface water mainly from surrounding uplands, from which the water flows downward and laterally to the hot pluton. Heating makes the water less dense, so that hot water tends to be pressed to the surface by the surrounding body of heavier cold water. The water probably dissolves metals where it is hottest, for the solubility of these substances

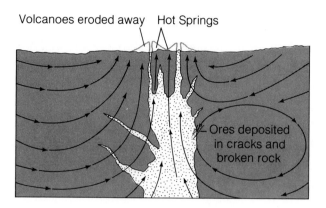

Volcanoes eroded away   Hot Springs

Ores deposited
in cracks and
broken rock

FIGURE 12-21

Vertical section through the upper parts of a solidified pluton showing the water circulation resulting from inflow from surrounding uplands and from heating by the pluton.

typically increases with increasing temperature. The metals are then redeposited where the water is cooled or where it reacts chemically with rocks.

Plutons as geothermal systems   The hot springs formed locally by these circulating systems are only a hint of the vast store of heat that might be tapped and used. The system depicted in Figure 12-21 is an example of a source of heat that has come to be called *geothermal energy*—the heat rising from the earth's interior. Were the example in the figure still active, it would be an especially valuable source because it is large and lies near the surface. Wells could be drilled into the crystallized but still hot parts of the pluton, ideally into natural systems of cracks that would transmit large amounts of vapor and hot water to the wells. If the amounts of vapor and water drawn off in the wells were no greater than the natural flow of groundwater into the system, the resource would last as long as the crystallized pluton remained hot. Use of the vapor and water would be expensive if they contained large amounts of minerals dissolved from the pluton, but the heat itself would cost nothing and might be supplied for periods of thousands of years.

Figure 12-21, however, is a reconstruction of a *past* system, one that is now exposed at the surface and cold. How does one find buried systems that are still hot yet near enough to the surface to be drilled economically? Geologists are currently seeking such systems by examining areas that have recently been volcanically active. An example is afforded by an area just east of the Sierra Nevada, California, around volcanic mountains called the Coso Range (References 5 and 6). The initial clue of a geothermal source came from occurrences of *fumaroles* (vapor springs) associated with a group of extinct volcanoes. The slightly eroded state of the volcanoes indicated their recent age, and the varying degree of erosion among them showed that they were erupted over a considerable period of time (Color Plate 8C). When the ages of the lavas were determined by the potassium-argon method (Appendix B), they were found to range from approximately 40,000 to approximately 1 million years in age. The range of ages thus suggested that the lavas could have been erupted from a subjacent pluton large enough to have remained molten that long and that the pluton might still be partly molten. Mapping (Figure 12-22)

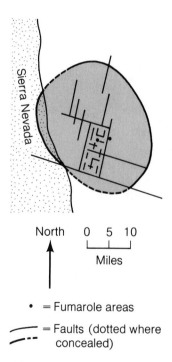

North   0   5   10

Miles

• = Fumarole areas

— = Faults (dotted where concealed)

FIGURE 12-22

Map of the fumarole areas and principal faults of the Coso Range and vicinity. The ring fault suggests subsidence of the shaded area into an underlying pluton.

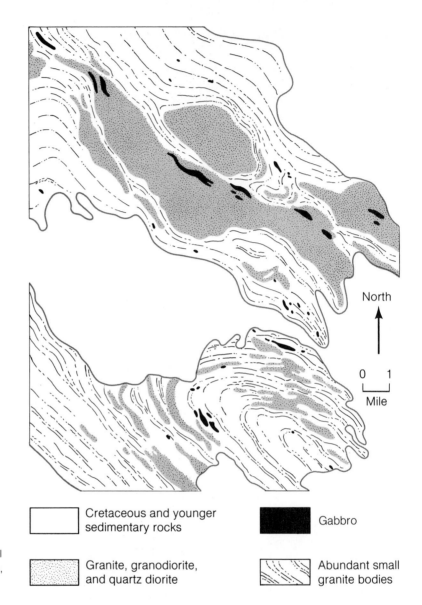

North

0   1

Mile

	Cretaceous and younger sedimentary rocks		Gabbro
	Granite, granodiorite, and quartz diorite		Abundant small granite bodies

FIGURE 12-23

Map showing plutons and associated rocks in the central part of the Santa Lucia Range, California.

disclosed that the main volcanic area lies within a large oval depression surrounded by a zone of faults and broken by additional faults along which some of the lavas were fed to the surface. The arrangement suggests that the depression is underlain by a pluton of about the same size and shape as the oval. It also suggests that removal of melt from this body led to subsidence and faulting of the rocks forming its roof. The Coso Range may thus overlie a large source of geothermal energy. By measuring the rate at which temperature increases downward in shallow drill holes, it should be possible to determine the approximate depth to molten or recently molten rock.

The full pluton system   But what is the ultimate source of the heat; whence the melts? Nowhere in the Cascade Range can we see a bottom to the plutons—their outward-sloping sides only indicate that they get larger with depth (Figure 12-15). Even in more deeply eroded mountains, such as the Coast Range of British Columbia or the Sierra Nevada, the plutons are large, in some cases huge, and have more or less vertical sides. It is only by examining the most deeply eroded parts of the earth that we begin to get a hint of the roots to plutons. These areas are similar to those described in the last part of Chapter 10 and are characterized by the highest temperatures of metamorphism and often by the separation of myriads of small granite bodies (Figure 10-17). Figure 12-23 shows a map of plutons in that region. They are smaller on the average but more numerous than plutons in less eroded regions and commonly are elongated and distorted into arcuate forms. The sides of

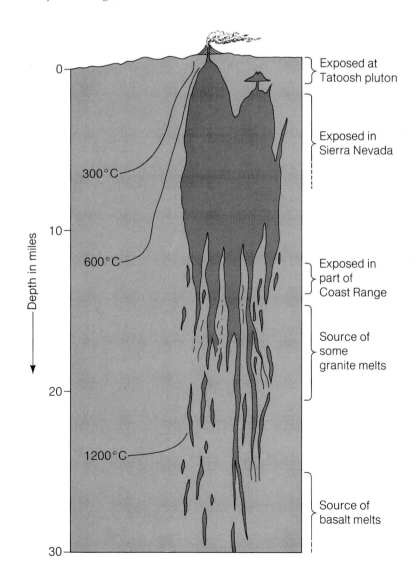

FIGURE 12-24

Vertical section through a pluton system, the roots being placed at depths where temperatures are high enough to produce abundant molten rock. Note the various levels suggested for the different exposures discussed in the text.

275

the plutons shown in the figure typically converge downward. Combining these forms with those already noted at higher levels, one can construct an overall body such as that shown in Figure 12-24.

This towering pluton pierces through all levels of the consolidated sedimentary rocks and metamorphic rocks described in Chapter 10. The temperatures on the left are those expected from depth alone, assuming a downward increase of 30°C per kilometer (9°C per thousand feet). Note that the lines representing temperatures rise sharply near the pluton due to the heat transferred upward in intruded melts. The rocks surrounding the lower part of the pluton, on the other hand, are almost as hot as it is. Note, in fact, that silica-rich melts might be formed from metamorphic rocks at depths of 15 miles or more. This melting gives another way besides crystal settling of developing silica-rich melts.

Occurrences of rocks with the composition of basalt (gabbro) in the deep part of the pluton (Figure 12-23) and the eruptions of basalt from volcanoes require that parts of the feeding system tap deeply enough to obtain melts heated to at least 1,250°C. Possibly some of the heat accompanying these melts is used in melting parts of the metamorphic rocks and thus enlarging the pluton. The basalt melts might also mix with silica-rich melts to produce melts of intermediate composition. Here is an additional way of getting a variety of melts, and thus igneous rocks, within one pluton.

The great central bulb of the pluton suggests that the melts tend to rise because they are lighter than the surrounding solid rock. It also suggests that they generally are arrested below the surface, possibly because of increasingly rapid cooling as they penetrate colder and colder rocks. The pluton thus becomes linked thoroughly with the surrounding rocks, including the water in them. It provides a connection to the deep and truly hot parts of the earth, carrying heat and substances far above the levels normally expected. We will doubtless come to know these giant cauldrons more and more completely as we learn to use them for their energy and for the valuable metal deposits associated with them.

Summary

The parts of the full system and the processes taking place within them may be reviewed as follows:

1. The system is expressed at the surface by volcanoes, which are commonly arranged in chains along mountain belts or arcuate lines of islands.
2. The volcanoes characterizing the margins of the Pacific Ocean periodically erupt lavas with the composition of andesite or dacite, often exploding them to form high-rising clouds and hot fragmental flows and mudflows.
3. The nature of an eruption is determined mainly by the viscosity of the molten lava and its vapor content; the viscosity (and the violence of explosions) increases with the proportion of silica in the melt.
4. Because minerals growing in melts are generally less silicic than the melts,

the remaining liquid tends to become more silicic as the minerals grow and sink, a process that may result in a variety of rocks.

5. Differentiation of a vertical body of melt is also suggested by repeated sequences of eruptions in which melts are most silicic at the outset and become less silicic as eruptions continue.

6. Volcanoes are fed through pipes, dikes, and sills that may be offshoots of deeper lying plutons.

7. Plutons are intruded toward the surface by lifting the pre-existing rocks, by breaking off pieces of the rocks, and by melting or chemically altering them.

8. Plutons also heat and metamorphose the rocks around them, thus permitting the melt within the pluton to cool slowly and crystallize into mineral grains larger than those in lavas.

9. Groundwater and other vapors incorporated into plutons alter the rocks and transport metals that may be deposited in or near the plutons, locally forming valuable ore deposits.

10. The water vapor may also be a potential source of geothermal energy, provided the pluton is near the surface and can be located before it has cooled.

11. The shapes of plutons in deeply eroded regions suggest that they enlarge upward from an intricate system of roots that lie at levels where rocks are partly melted; melts also appear to rise from much deeper levels.

REFERENCES CITED

1. James G. Moore and W. G. Melson. Nuées ardentes of the 1968 eruption of Mayon Volcano, Philippines. *Bulletin Volcanologique*, vol. 33, pp. 600–20, 1969.

2. Donal E. Mullineaux and Dwight R. Crandell. Recent lahars from Mount St. Helens, Washington. *Geological Society of America Bulletin*, vol. 73, pp. 855–70, 1962. Dwight R. Crandell. Preliminary assessment of potential hazards from future volcanic eruptions in Washington. U.S. Geological Survey, Misc. Field Studies, Map MF-774, 1976.

3. Richard S. Fiske, Clifford A. Hopson, and Aaron C. Waters. Geology of Mount Rainier National Park, Washington. U.S. Geological Survey Professional Paper 444, 93 pp., 1963.

4. Hugh P. Taylor, Jr. Oxygen isotope evidence for large-scale interaction between meteoric ground waters and Tertiary granodiorite intrusions, Western Cascade Range, Oregon. *Journal of Geophysical Research*, vol. 76, pp. 7855–74, 1971.

5. Wendell A. Duffield. Late Cenozoic ring faulting and volcanism in the Coso Range area of California. *Geology*, vol. 3, pp. 335–38, 1975.

6. Marvin A. Lanphere, G. Brent Dalrymple, and Robert L. Smith. K-Ar ages of Pleistocene rhyolitic volcanism in the Coso Range, California. *Geology*, vol. 3, pp. 339–41, 1975.

ADDITIONAL IDEAS AND SOURCES

1. You can get a complete view of the crystallization of dacite (granodiorite) melt if you look first at Figure 12-8, then at 12-10A, and finally at the succession in 12-18. Note that 12-8 and 12-10A are enlarged twice as much as the views in 12-18.

2. The top, base, and sides of a basalt flow cool and solidify before the middle does. If the liquid part then bursts out at the lower end of the flow, the core of the flow may be partly drained—leaving a long tunnel called a *lava tube*. Beautiful examples at Mount St. Helens have been described and illustrated by Ronald Greeley and Jack H. Hyde in "Lava tubes of the Cave Basalt, Mount St. Helens, Washington" (*Geological Society of American Bulletin*, vol. 83, pp. 2397–2418, 1972).

3. For further reading on volcanoes, a book by Cliff Ollier (*Volcanoes*. Cambridge; MIT Press, 177 pp., 1969) describes volcanic forms and eruptions. A more advanced book by Gordon A. Macdonald (*Volcanoes*. Englewood Cliffs, N.J.: Prentice-Hall, 510 pp., 1972) describes causes of eruptions as well as the volcanoes themselves. Macdonald's descriptions of the important Hawaiian volcanoes are especially valuable because he has studied them extensively himself.

4. The nature of geothermal energy and its potential value to humans are described by L. J. P. Muffler and D. E. White in an excellent, short article entitled "Geothermal energy" (*The Science Teacher*, vol. 39, no. 3, 4 pp. March 1972).

5. Further evidence of large bodies of melt under volcanoes is provided by the occasional collapse of the volcano into the top of the pluton, typically because voluminous eruptions have emptied the upper part of the pluton and thus left the volcano unsupported. The oval depression of Crater Lake (measuring 5 by 6 miles) formed in this way, as did similar depressions at Newberry Volcano (Figure 12-16). A thorough and beautifully illustrated account of such features is that by Howel Williams: *The Geology of Crater Lake National Park, Oregon* (Carnegie Institution of Washington Publication 540, 162 pp., 1942).

Tilted layers of fine-grained sedimentary rocks that extend along the coast near Santa Barbara, California. The rocks consist largely of remains of microscopic plants and animals and are an important source of petroleum.

13. Sequences of Layered Rocks

Sequences of Layered Rocks

A Formation of Laminated Rocks
    Deducing a former environment
Using Fossils to Identify Stratigraphic Age
    Ages based on assemblages of fossils
    Some stratigraphic principles
Continental Sequences and Unconformities
    Fossils in continental deposits
    The unconformities
A Worldwide Stratigraphic Standard
    The time scale
Effects of Environment on Organic Evolution
    Changing plant communities and horses
    Rates of evolution of the horses
Summary

The science of geology has its main basis in layered rocks such as those shown on the chapter opening page. Layered rocks often contain animal and plant remains that can be used to determine the geologic ages of the rocks as well as the environments in which they accumulated. No part of the earth can be understood fully without this information. The rocks shown on the chapter opening page are particularly valuable because they accumulated slowly and continuously over a considerable period of time. Each foot of rock represents about 1,500 years of accumulation, and the total layers in the photograph represent approximately 30,000 years. We will see shortly how these numbers were determined. The important concept at this point is that layers of rock can be thought of as increments of time. This concept of time, moreover, is not just an abstraction. The rock layers contain shells of specific foraminifers, such as those shown in Figure 13-1. By examining the shells in many layers, we can discover when some species became extinct and others evolved to replace them. The layers thus reveal the evolution of the animals, and once this evolution is known, the fossil species can be used to identify rocks of the same age elsewhere.

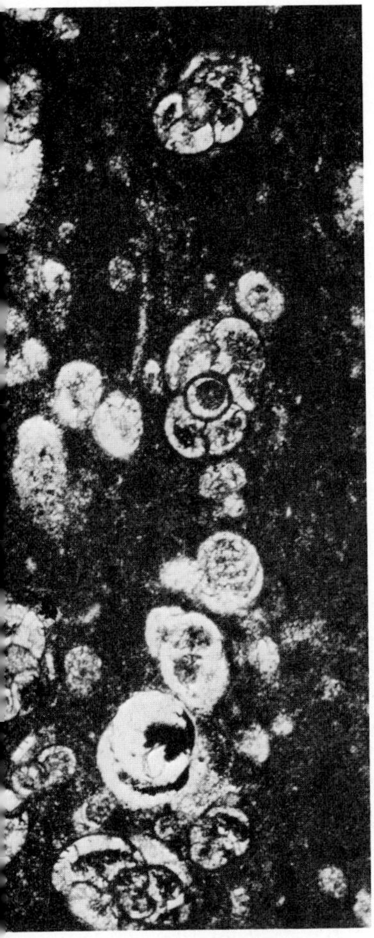

This ordering of fossil-bearing rocks and thereby of events is the branch of geology called *stratigraphy*, the study of layered (stratified) rocks. Its usefulness goes beyond that of putting rocks in a time sequence. We have noted in earlier chapters that certain associations of organisms and sedimentary deposits typify certain environments. By finding similar associations in ancient rocks we can deduce the environments of those times, which may be helpful in understanding the changes that are happening today. The environmental changes, in turn, have evidently caused the evolution of organisms! Stratigraphic studies thus provide a view of the earth's full evolution, both organic and physical. The "wholeness" of the earth, mentioned in Chapter 1, can thereby be carried through many stages in its development.

We should note, too, that much of this history would be difficult to read without the dating methods based on radioactive isotopes, described in Appendix B. These methods enable us to date geologic events in years and thus permit us to put rocks such as granite and basalt into age sequences with fossil-bearing sedimentary rocks. The scope of "seeing" the ancient world then becomes great indeed. We find, for example, that the rocks shown on the chapter opening page were accumulating at the same time that the Tatoosh pluton was cooling under the ancient Cascade chain (Chapter 12), and (as will be related in this chapter) at the time that the first of the great grasslands was forming in the region that is now the High Plains.

A Formation of Laminated Rocks

The layers shown on the chapter opening page are part of a thick sequence of rocks called the Monterey Formation, named in 1856 for its occurrence near the town of Monterey and subsequently recognized and mapped in many other parts of western California. The *Formation* part of the name means that the rocks constitute a distinctive grouping, one that differs from the

FIGURE 13-1
Fossil shells of foraminifers, enlarged about 50 times, in a thin section of one of the layered rocks shown on the chapter opening page.

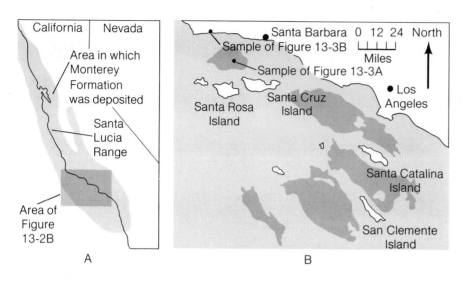

FIGURE 13-2
A, map showing approximate original distribution of the Monterey Formation, now partly onshore and partly offshore. B, map showing basins (gray) in which sediments like those of the Monterey Formation are accumulating off California today. Figure 13-2A indicates location of the map.

rocks above and below it. This particular formation is characterized by clearly defined, thin layers of silica-rich rocks derived from accumulations of diatom shells. These rocks are interlayered with lesser amounts of fine-grained limestone and shale. All the rocks contain organic compounds that color them a deep brown and make the formation a source of petroleum and natural gas. In outcrops, the Monterey Formation is distinctive because the sun's heat drives out the dark organic compounds and thereby bleaches (whitens) the rocks. The rock formations above and below are typically dark gray siltstones that do not become whitened and are less distinctly layered. The top and bottom of the Monterey Formation can thus be located quite easily. The same physical characteristics make it possible to map the extent of the formation on land and to recognize it in drill cores taken at sea (Figure 13-2A).

The boundaries of formations are thus physical surfaces that one may observe and dig into; however, they express a good deal more. The lower boundary of the Monterey Formation identifies when conditions changed such that diatom shells began accumulating abundantly in that part of the ocean. The thickness of the formation represents the duration of those conditions, and the upper boundary marks their culmination. A formation thus expresses certain conditions of origin, a specific environment. The environment is usually not obvious, but it can be interpreted from an inspection of the rocks and a comparison of them with sediments actually forming today. The Monterey Formation has been compared with sediments studied in detail in basins off the coast of southern California (Figure 13-2B).

Deducing a former environment   The more provocative of these studies were made by oceanographers who took cores of the sediments and dated the thinnest layers precisely, using radiometric methods based on thorium and lead isotopes (Reference 1). As Figure 13-3, left, shows, the thin layers record annual seasonal changes. A dark, denser layer was deposited when winter rains carried clay from land to sea, and a light, fluffier layer was

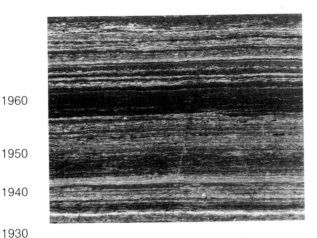

FIGURE 13-1

Left, dated core, one-third actual size, taken in the Santa Barbara basin at a depth of approximately 1,900 feet (located in Figure 13-2B). This is an X-radiograph, in which denser layers appear dark and less dense ones light. From Reference 1. Right, laminated rock of the Monterey formation, twice actual size, from the locality labeled in Figure 13-2B. Courtesy of Caroline M. Isaacs.

— 1960

— 1950

— 1940

— 1930

— 1920

— 1910

— 1900

— 1890

— 1880

— 1870

— 1860

— 1850

— 1840

deposited by summer bottom currents carrying shells of spring-blooming diatoms. The thin layers, called *laminations*, are usually not preserved in sediments because bottom-dwelling organisms destroy them by burrowing in the sediment and ingesting it. In the deeper waters studied, however, dissolved oxygen is scarce because abundant dead diatoms and other plankton particles are oxidized as they sink or are eaten by foraminifers and radiolarians—an additional oxidizing process. Dissolved oxygen is thereby depleted and is minimal at depths of 500 to 3,000 feet—depths that include most of the bottoms of the basins (Figure 13-4). Since few bottom-dwelling animals can live in water containing less than 0.5 part per thousand of oxygen, the sediment laminations are preserved.

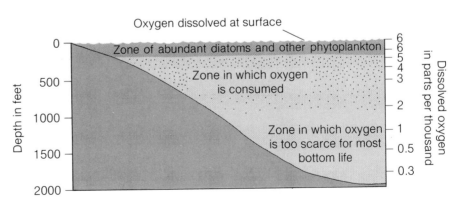

FIGURE 13-4

Distribution of dissolved oxygen in the sea off southern California and the chief processes that govern it. Data chiefly from S. C. Rittenberg, K. O. Emery, and W. L. Orr. Regeneration of nutrients in sediments of marine basins. *Deep-Sea Research,* vol. 3, pp. 23–45, 1955.

As Figure 13-3, right, shows, the Monterey Formation, too, is composed of well-preserved laminations. Although they have been thinned greatly by compaction, they indicate that the formation was deposited in a basin with oxygen-poor bottom waters. This environment is confirmed by the occurrence of a distinctive fossil foraminifer closely related to a modern species that lives only in the oxygen-minimum zone (Figure 13-5). In addition, the laminations are alternately clay rich and silica rich (originally diatom rich), indicating that the formation was deposited near a land mass where rainfall was distinctly seasonal and diatoms grew in abundance. Each pair of laminations must represent one year's increment of sediment. We can therefore conclude that the Monterey Formation provides a continuous (year by year) record of the geological period during which it accumulated.

Using Fossils to Identify Stratigraphic Age

The fossil record in the Monterey Formation must also be continuous, for the remains of organisms that fell from the overlying waters were preserved due to the absence of scavengers. The diatom shells were so thin that they dissolved and recrystallized when buried and compacted, but the larger foraminifer shells are generally well preserved. They can be separated easily from the softer (clay-rich) rocks by pulverizing samples in water and washing the mud through screens that catch the sand-sized shells. The shells are then identified by means of a binocular microscope, either by comparing them to photographs of known species, or, if one is an expert, simply "by eye." Many such examinations have shown a regular order in the appearance and disappearance of foraminifer species with time—the specific events in the evolution of these animals.

Figure 13-6 shows how one important genus (closely related group of species) evolved just before and during the deposition of the Monterey Formation. *Siphogenerina collomi* is the youngest species found in the Monterey Formation, although a single species still lives today in the warm waters of the southwest Pacific. Studies of other formations along the western edge of North America show that the genus was forced to migrate southward by the cooling that preceded the first ice age. The rapid evolution shown in Figure 13-6 is believed to have taken place in comparatively warm waters. Note the short periods during which most of the species existed. Note, too, the distinctive shapes that evolved: the thin, complexly ornamented *mayi*; the lack of longitudinal ribs (costae) on *hughesi*; the peanut shape of *nuciformis*; the few but prominent ribs of *transversa*; the robust form of *collomi*. These features make the fossils easy to identify. The species thus have four attributes that make them exceptionally valuable in recognizing levels (ages) within marine strata: (1) they have short age ranges; (2) they were carried widely in the seas of that time; (3) they were abundant; and (4) they are easily recognized. Such fossils are called *index* or *marker* fossils and can often be used alone to specify the stratigraphic age of a sample.

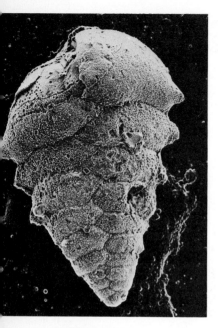

FIGURE 13-5

Matching modern and past environments—the foraminifer *Suggrunda californica*, from the Monterey Formation, a close relative of *Suggrunda eckisi*, which lives in the oxygen-depleted zone shown in Figure 13-4. Courtesy of James C. Ingle, Jr. Electron microscope image by Gerta Keller.

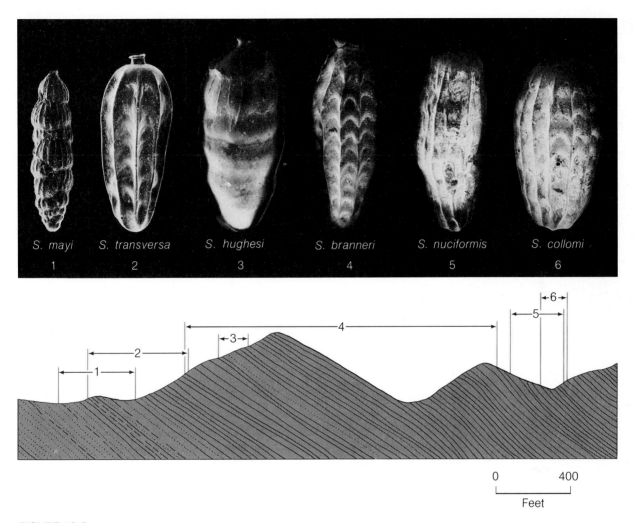

S. mayi 1   S. transversa 2   S. hughesi 3   S. branneri 4   S. nuciformis 5   S. collomi 6

0   400
Feet

FIGURE 13-6

Some of the important species of the foraminifer genus *Siphogenerina*, with a vertical section through tilted strata that contain them. Each numbered line indicates the range of rock layers in which that species occurs. Dotted layers are sandstone, lines are silica-rich rocks or shale. The locality is in Santa Lucia Range, labeled in Figure 13-2A. Data from Reference 2. Specimens from James C. Ingle, Jr., and Ann Tipton Donnelly. Electron microscope images by Gerta Keller. Each test is approximately 1 millimeter long.

Ages based on assemblages of fossils Fossils of only one species, however, are not likely to occur in every part of a sedimentary accumulation. We can think of several reasons: local variations in temperature or salinity, changes in supply of food, unusual turbidity, actions of predators or disease, and so forth. We can identify the stratigraphic position (age) of the fossil-poor layers only if we know the stratigraphic ranges of *many* species. This knowledge requires thorough studies of fossil-rich rocks such as the Monterey Formation. The method used, in its essentials, is to measure the layered sequence and at the same time collect samples at specified positions, for example, at every 5 feet in thickness of rock layers. When all the fossils in each sample have been separated and identified, their occurrences can thus be arranged in stratigraphic order. Figure 13-7 shows such an arrangement for some important species in a part of the Monterey Formation. Each sample is plotted at one horizontal level, its position indicated by the number of feet above the base of the formation (see the left side of the diagram). By marking the lowest and highest occurrences of a species, we can see its total age range (the unshaded bands). Note that many samples within the age range of a species do not contain remains of that species. Their absence confirms our suspicion that we cannot use a single species to identify the stratigraphic position of every sample, even in a fossil-rich formation.

How do we determine the stratigraphic ages of samples without index fossils? Consider the three samples in the interval marked X in Figure 13-7— samples that do not contain the index fossil *Siphogenerina collomi*. Imagine that these samples are from some other, unmeasured sequence and that we want to know their stratigraphic ages. Note that two of the three samples contain two species with age ranges between the upper and middle black lines in the diagram (the species *Bolivina advena* variety *striatella* and *Valvulinaria miocenica*). Note, too, that all three samples contain several species that do not occur above the upper black line and are noticeably abundant just above the middle line (for example, *Valvulinaria californica*). We can thus safely conclude that our "unknown" samples fall somewhere between the upper and middle black lines.

FIGURE 13-7
Abundance of specific foraminifers in 51 samples of the Monterey Formation, collected in the upper part of the sequence shown in Figure 13-6. The numbers on the left are the feet of strata measured above the heavy line shown in Figure 13-6. The dots show the abundance of each species in each sample (see the key). The unshaded bands are the stratigraphic ranges of the species, with the pointed ends indicating continuing (older) ranges. Data from Reference 2.

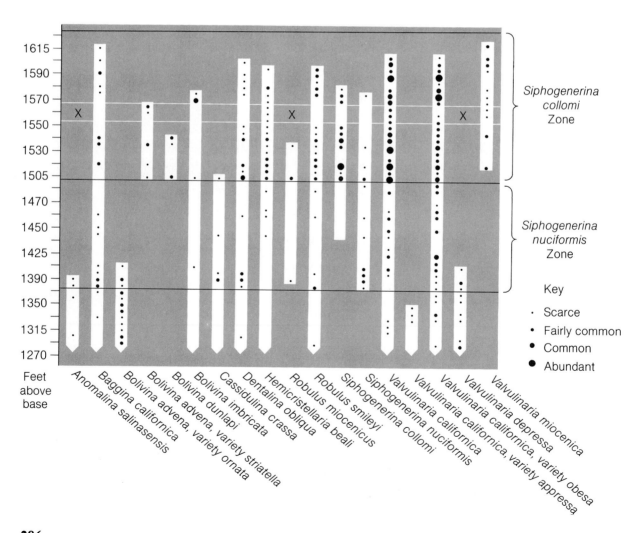

286

As Figure 13-7 shows, this interval has been formalized with a name—the *Siphogenerina collomi Zone*. The horizontal lines or limits, in fact, were fixed by geologists (stratigraphers) who studied the ranges of foraminifers and concluded that these levels can be identified reasonably closely on the basis of several species. Note that Figure 13-7 also shows the base of the next oldest zone, the *Siphogenerina nuciformis Zone*. By looking broadly at the dots and unshaded bands, can you see the *general* contrast in abundances between the two zones? Now find the two species that appeared at about the time the older zone began. Note, too, the four species that became extinct at approximately the same time. Thus, although a number of the species occur in both zones, we can use combined ranges and abundances to delineate the lower zone quite closely.

A *faunal zone* is thus a division of layered rocks based on the evolution of fossil species. Each zone is named after a characteristic species, but *it is based on an assemblage of species* (thus the term *faunal* zone). A sequence of faunal zones provides a standard for recognizing the stratigraphic position of any sample containing a reasonably large number of fossils. A sequence of zones also provides a record of *all* geologic time over the range included. Let us briefly review two reasons for thinking this is so: (1) zones are set up in layered rocks that have accumulated continuously or nearly so, and (2) fossils are so abundant in the rocks that assemblages can be collected at closely-spaced intervals. A third reason is suggested by Figure 13-7: the living ranges of the species overlap quite randomly, indicating that evolution proceeded continuously and that there have been no major pauses in the accumulation of sediments.

Figure 13-8 illustrates the differences between faunal zones and formations and shows how zones may be used to interpret formations. Note, first, that the faunal zones "fill" all the geologic time represented by the vertical dimension of the diagram. Second, note that the top and bottom of the Monterey Formation do not fit zone boundaries at all of the localities. This relation is due to the zones being based on foraminifer evolution whereas the Montery Formation is based on specific kinds of rocks. Finally, note that the top and the bottom of the formation differ in age from place to place, and also that the thickness of the formation differs from place to place. The faunal zones thus provide a means of "seeing" that the basins in which the Monterey Formation was accumulating had somewhat different histories and also received sediment more rapidly in some places than in others.

Some stratigraphic principles   Regardless of these interpretations, our main findings in the Monterey Formation are basic principles that have been tested in many other sequences and may be applied generally to sedimentary rocks:

1. Layered rocks form local sequences deposited over certain periods of geologic time.
2. The layers progress in age from the oldest at the bottom to the youngest at the top; indeed, where the thinnest layers are annual increments, we can count the years between.
3. Remains of organisms in the layers can be set in an evolutionary se-

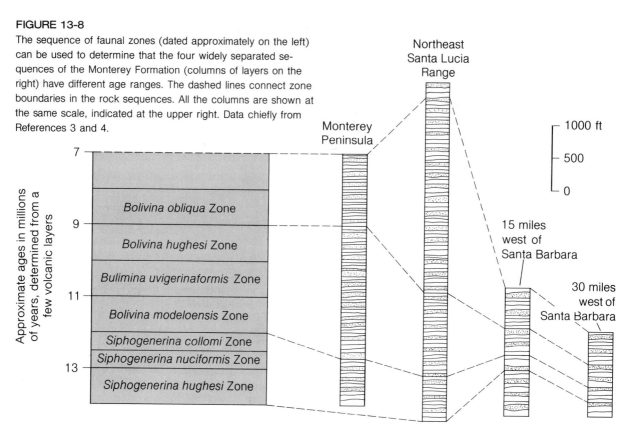

FIGURE 13-8

The sequence of faunal zones (dated approximately on the left) can be used to determine that the four widely separated sequences of the Monterey Formation (columns of layers on the right) have different age ranges. The dashed lines connect zone boundaries in the rock sequences. All the columns are shown at the same scale, indicated at the upper right. Data chiefly from References 3 and 4.

quence, each part of which will be characterized by certain fossil species.

4. The fossil species can be used to recognize rocks of the same age elsewhere; they can also be used to subdivide geologic time into useful periods, such as those equivalent to the faunal zones just described.

5. Where fossils and sedimentary features also indicate the environment in which they accumulated, we can compose an earth history from layered rocks.

Continental Sequences and Unconformities

These principles could be applied readily and widely if all formations were as complete and as rich in fossils as the Monterey Formation, but North America was mainly dry land when the Monterey Formation was accumulating along its western margin. How can we tell what was happening in the middle of the continent? Sediments were accumulating there but were limited to separate river plains and basin-like depressions, much as those of the Rio Grande and the desert basins described in Chapter 9. Besides being laterally discontinuous, some basins were occasionally uplifted and their deposits partly eroded away. In this section we will see how these local, often discontinuous sedimentary sequences can be composed into a more or less complete geologic history. We will be looking at the past 65 million years, a period including the deposition of the Monterey Formation but beginning long before that time.

Features such as that in Figure 13-9 occur here and there in the continental deposits and prove that much of the sediments were deposited by river systems. The rivers often migrated so widely and gradually, however, that they constructed nearly uniform layers of sediment. Some of the finer layers are nearly identical to the loess (windblown silt) described in Chapter 8, and close inspection shows that they consist of minute fragments of volcanic glass carried by the wind from volcanoes far to the west. Alternating silty and sandy layers are especially impressive in river-plain deposits that sloped gently eastward from the Rocky Mountains toward the ancestral Mississippi Valley (Figure 13-10).

Similar formations accumulated along with coarser river sediments and widespread lake deposits in broad basins in the Rocky Mountain region (Figure 13-11). One of these deposits, the Wasatch Formation, underlies a large area in southwestern Wyoming and eastern Utah and commonly has wavy, locally bulging layers that record the lateral migration of the stream channels

FIGURE 13-9

Stream channel exposed in a vertical roadcut in central Utah, with sketches showing its probable development. In stages 1 and 2, the stream migrated laterally by cutting and filling at a bend. In stage 3, the channel was straight and filled symmetrically, perhaps in the way the Rio Grande is filling today. The final fill above the prominent crescent in the photograph probably formed when the channel was abandoned by the stream but received muddy water during floods.

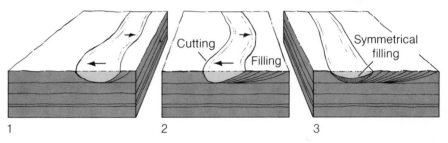

FIGURE 13-10

Part of the Badlands of South Dakota, with the grassy (dark) surface of the High Plains on the horizon. Because the sediment layers are parallel to the modern plains, each probably accumulated on a similar surface. The upper part of the sequence, striped in pale grays, has been named the Brule Formation, and the darker part below has been named the Chadron Formation.

in which the sand accumulated (Color Plate 9A). The red color of the formation is important because it indicates a red soil such as that of Georgia and thus a warm, humid climate. An example of a lake deposit is the Green River Formation, Color Plate 9B, which consists of fine-grained limestone, shale, and fragmental volcanic deposits that formed over thousands of square miles in Wyoming and Colorado. These rocks are typically laminated and whitened as is the Monterey Formation, and in some sequences they contain abundant organic compounds called kerogen that lead to their being termed "oil shale" and make them potential sources of fuel and chemicals.

Fossils in continental deposits   The formations can be measured and traced acrosss each basin by means of surface outcrops and by samples brought up with drills, but fossils must generally be used to match layers in age. Age is determined in much the same way as that used for the Monterey Formation but with greater difficulty because fossils are far scarcer in the continental deposits. Few bones and other hard parts escape the decomposing actions of plants, bacteria, and the weather. Consider, for example, the millions of animals that have lived in and on the Georgia soil described in Chapter 8, yet have left no identifiable remains. Even where bones have been preserved in swamps, ponds, or river channel deposits, they are uncommon and hard to find. A method often used is one adopted from the gold prospector: (1) to seek bits of bone or teeth in dry rill channels; (2) when some are found, to search carefully upstream for more fragments; (3) where the train of fragments ends in the rill, to continue the search upslope to the point where no more appear on the surface; (4) to dig carefully somewhat above this point for the remaining parts of the skeleton.

Such searches, however, tended to favor the discovery of large animals,

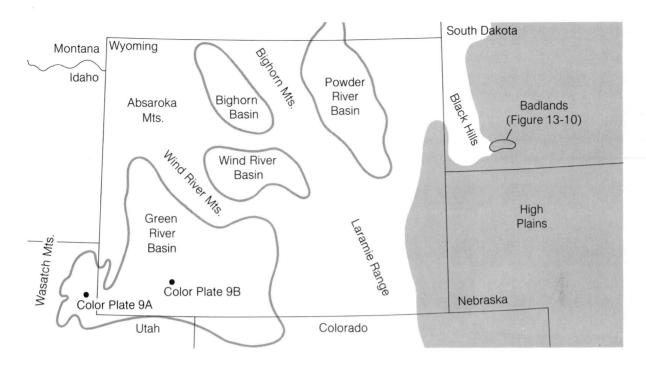

which in some cases biased the interpretation of the fauna and its habitat. Early discoveries of alligators and numerous turtles in the deposits of the Badlands of South Dakota, for example, led some geologists to interpret the environment as one comprised of swampy rivers and lakes. The watery environments of these particular animals, however, often resulted in their burial in water-saturated sediments, beyond the reach of plant roots and the weather and thus more likely to be preserved. Moreover, the river-channel sandstones of the Badlands were searched most thoroughly, so that water dwellers were bound to be discovered preferentially. Fortunately, giant rhinoceros-like animals (titanotheres) and other dry-land herbivores were also found in the river deposits, and when bones of smaller animals were collected (often a difficult task), it became apparent that swamps and stream channels were far less extensive than was thought at first. In an ingenious study by Laurie J. Macdonald, minute bones and teeth were collected in an area near the Badlands from mounds built by ants that gather hard fragments from the silty surface to protect their nest openings (Figure 13-12) (Reference 5). The fossils were the remains of numerous gophers, mice, and other burrowing animals, proving that the fine sediments were not lake deposits or even near-stream deposits, but probably plainsland silts and clays laid down mainly by the wind and consisting largely of fine fragments of volcanic glass. The terrain of 15 million years ago has thus turned out to be much like the loess-covered Great Plains of today!

Terrestrial vertebrates, especially mammals, evolved rapidly, and many migrated widely in North America. The bones of many species are so distinctive that often a foot bone or a few teeth are enough to make an identification. The fossils thus serve as indexes to the stratigraphic sequence, much as the short-lived foraminifer species of the Monterey Formation. The fossils

FIGURE 13-11

Map showing the distribution of sedimentary formations deposited in Rocky Mountain basins and on the ancestral High Plains. Those in the Rocky Mountain basins formed between approximately 60 and 35 million years ago, and those on the High Plains formed between about 32 and 10 million years ago.

FIGURE 13-12
One of the anthills from which large numbers of fossil bones and teeth of small animals were collected near Wounded Knee, South Dakota. Photograph by Laurie J. and James R. Macdonald.

have shown that no continental sequence covers nearly the full period of time between 65 million years ago and the present. The total record must thus be composed by fitting together a number of the sequences using fossils in the parts of the sequences that overlap in age (Figure 3-13). The upper part of the sequence in Utah, for example, contains fossils that show it is of the same age as the lower part of the sequence in central Wyoming (Figure 13-13). Fossils have not proven abundant enough to set up stratigraphic units as brief in duration as the faunal zones of the Monterey Formation, but longer divisions called *stages* have been used, as shown by the shaded bands in Figure 13-13.

The continental stages have been correlated approximately with the faunal zones of the Monterey Formation (and other marine formations) by using the radiometric methods described in Appendix B. The correlations have been checked further at localities in California where fossil-bearing continental formations pass laterally into fossil-bearing marine formations, typically in ancient delta systems. Thus, we have an extremely powerful stratigraphic tool: fossil sequences, dated in years, that can be used to match events in ocean basins with those on continents.

The unconformities   The continental formations emphasize an aspect of layered rocks that must be considered in all stratigraphic studies. The local sequences are interrupted by distinct interfaces that formed when deposition ceased for a certain period and erosion cut away part of the existing deposits before deposition recommenced (Figure 13-14A). Such an interface is called an *unconformity*—in the sense that the layer above it does not fit stratigraphically (by continuous accumulation) on the one below. Because unconformities are commonly parallel to rock layering, they may be difficult to recognize unless one finds evidence of erosion such as that shown in Figure 13-14B. If the older layers are tilted when they are uplifted and eroded, the unconformity will be marked by angular discordance. Perhaps the most useful test is to compare fossils just above an unconformity with those just below it in order to determine how much of the stratigraphic sequence is missing.

An exceedingly important aspect of unconformities is that each must end somewhere in a complete sequence of layers. This assumption is based on the fact that the materials eroded from a given area must be deposited somewhere. By patiently tracing formations laterally, one may thus hope to find age equivalents of the missing layers in an area that was receiving sediments during the same period. This method has been used to fill all major gaps in the continental record except the one labeled in Figure 13-13.

A Worldwide Stratigraphic Standard

Marine rocks can often be matched stratigraphically between continents by means of fossils of pelagic organisms—animals or plants that drift or swim widely in the oceans, as do many foraminifers. Some rocks can also be matched by determining their ages in years by radiometric methods. These means of correlating strata (matching them by age) are especially important to American stratigraphy because the divisions of stratified rocks were first

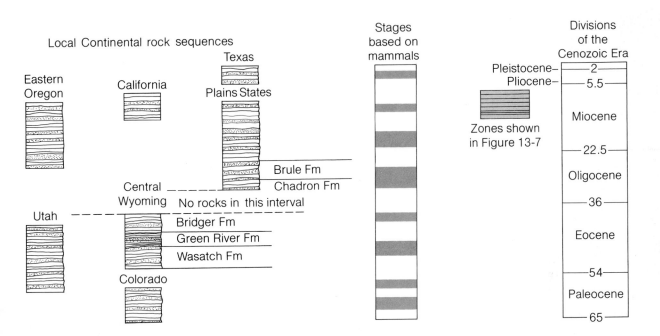

FIGURE 13-13

Chart showing separate continental sequences of the Midwest and West, which have been pieced together into a total record of the period called the Cenozoic Era. The faunal zones from Figure 13-8 are shown for comparison, and on the far right are the principal divisions of the Cenozoic Era. The age in millions of years at each boundary is based on radiometric ages determined on volcanic rocks, and the divisions are from Reference 4.

made in Europe, mainly during the nineteenth century. The divisions shown on the right side of Figure 13-13, for example, were first worked out in the broad sedimentary basin surrounding Paris. These comparatively short divisions make up the Cenozoic Era, which is shown in Figure 13-15 along with the other major divisions of geologic time.

Note that the principal divisions in Figure 13-15 are named in two ways, one pertaining to a period of time and the other to the rocks formed during that time. The Cretaceous Period, for example, is the period of time during which the strata called the Cretaceous System were deposited. The time scale in years in Figure 13-15 was added much later, as we will note shortly. The original basis of each division was a sequence of layered rocks. The parts of the total sequence were well developed in different places, as the map in Figure 13-15 shows. Some sequences clearly lay on top of one another, so that their relative ages could be determined directly, an example being the Cambrian, Ordovician, and Silurian Systems, which were based on three superimposed sequences in Wales. Isolated sequences, on the other hand, had to be fitted into position by the use of fossils, much as we noted for the local sequences shown in Figure 13-13. The science of *paleontology* (the study of ancient life) grew rapidly during the nineteenth century as new fossils were discovered, composed into evolutionary sequences, and used to fit together the stratigraphic record. With a few exceptions (for example, the American use of the Mississippian and Pennsylvanian Systems in place of the Carboniferous System), the named systems became a worldwide standard for naming rocks. The complete assembly of systems, shown in Figure 13-15, is therefore often referred to as the *standard geologic column.*

Although the names of the systems and periods have continued in use, their limits have been changed moderately. Many of the original systems are bounded by unconformities, which make them easy to recognize but mean

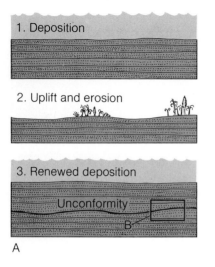

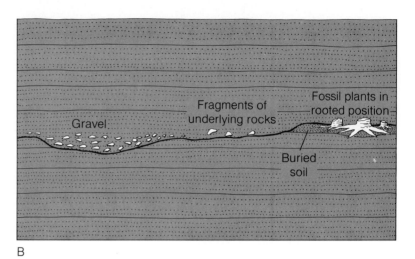

that certain parts of geologic time are not represented. These gaps have been filled one by one as more complete sequences have been discovered, and the boundaries of the systems have been adjusted accordingly. Some original sequences have proved to be undesirable "standards" because they are not nearly as rich in fossils as they might be. The Jurassic System, for example, was originally based on a sequence of rocks in the Jura Mountains of France, but it later became based on a sequence of fossil-rich limestones and shales in southern England, where it could be divided into numerous faunal zones calibrated on ammonites—a group of mollusks that is now extinct but was abundant and evolved rapidly in the seas of that time. Studies of fossils have thus been used widely to sharpen the definitions of the systems and to divide them usefully.

The time scale   An invaluable addition to the geologic column shown in Figure 13-15 is the time scale in years, based on rocks dated by the radiometric methods described in Appendix B. These ages are often called *absolute ages* to distinguish them from *geologic ages*, the ages determined from fossils. You may wonder why all rocks are not dated by radiometric methods rather than by the sometimes difficult procedure of finding and interpreting fossils. The answer is that sedimentary rocks consist largely of minerals eroded from older rocks, so that radiometric dating would not give the age of the sedimentary rock itself. We must thus base the absolute ages on igneous rocks (rocks formed from melts), and even they must be entirely fresh. The slightest amount of metamorphism may drive out some of the substances used in dating and result in incorrect ages. This effect, moreover, cannot be discovered with certainty until several different minerals in a given rock have been dated, and often not until the rock has been dated by more than one method. The total procedure is thus costly and time consuming. Occasionally, however, it results in ages that are closely similar among all the methods used. Such ages give a firm basis for dating that particular part of the earth's history.

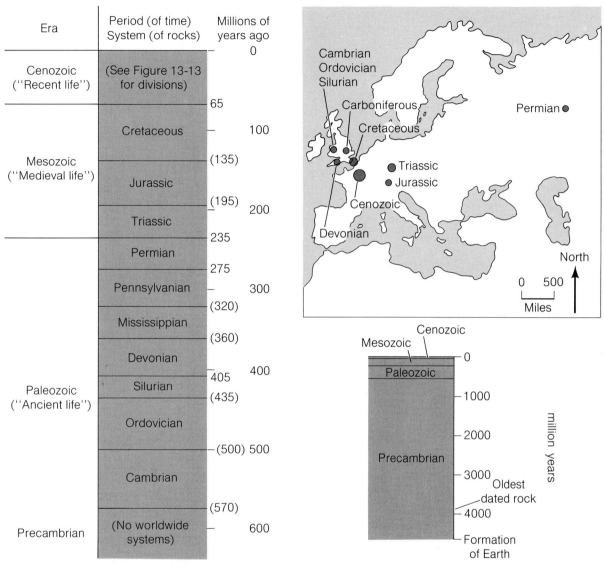

FIGURE 13-15

Major divisions of geologic time (periods) and rocks (systems), first established at the locations shown on the map. The Pennsylvanian and Mississippian Systems were established in the United States for rocks called Carboniferous in Europe. (Ages in parentheses are approximate.) The separate column on the right shows the length of the eras compared to the total age of the earth. The ages in years are chiefly those proposed in ''The Phanerozoic time-scale,'' *Quarterly Journal of the Geological Society of London,* 1964.

To apply absolute ages broadly, however, they must be fitted into the sequence established by the fossil record. The ideal case is one in which the dated rock is a lava or a fragmental volcanic rock within a layered sedimentary sequence. Fossils just above and below the dated volcanic layer will show exactly where the date fits into the sequence of stages or faunal zones. Plutons and other intrusive bodies can also be used for dating sedimentary sequences

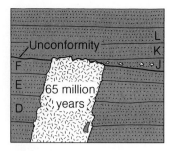

FIGURE 13-16

Relations between a radio-metrically dated pluton and the enclosing sedimentary formations. In the ideal case, layer F will be only a few faunal zones older than J, so that the pluton's age will give a close approximation of the ages of F and J in years.

but only in exceptional cases such as the one illustrated in Figure 13-16. There must be a small difference in age between the youngest fossil-bearing layer intruded by the dated body and the oldest layer deposited over it (the layers labeled F and J, respectively, in the figure). This difference says, in effect, that the intrusive body was exposed by erosion shortly after it solidified and that it was then covered partly by sediments. For example, if we have the remarkable luck to have found a place where the F layer contains fossils indicating the latest Cretaceous age (the youngest stage in the Cretaceous period) and the J layer contains fossils of earliest Cenozoic age, we have dated the boundary between the Mesozoic and Cenozoic Eras! Our 65-million years date may thus be applied firmly to the geologic column of Figure 13-15. As the more approximate "dates" in that column show, however, few major geologic time boundaries have been set so ideally.

The search for ideal cases continues and the time scale will doubtless be improved, but there will still be uncertainties resulting from the radiometric methods themselves. With the methods now in use, the uncertainty typically amounts to about 3 percent of the age determined. Dates determined on even the freshest of Paleozoic lavas and plutons are thus certain only to the nearest 5 to 15 million years, and dates on *unaltered* Precambrian rocks (which are scarce) are certain only to the nearest 15 to 100 million years. At the other end of the scale, dates on Cenozoic rocks have an uncertainty of only 3 million years or less, and Cenozoic volcanic and plutonic rocks are the least likely to be altered. Our time scale for that era is thus so complete and reliable that we can apply it to detailed stratigraphic sequences such as the faunal zones shown in Figure 13-8. Such cases show that individual faunal zones represent between 0.3 and 3 million years of time. Because this time span is about the same as the uncertainty in dating Cenozoic rocks by radiometric methods, fossils and radiometric methods can be used interchangeably to interpret a detailed history of the Cenozoic Era.

Effects of Environment on Organic Evolution

Besides being accurately dated in years, the Cenozoic Era provides the clearest record of the relations between the living and nonliving parts of earth systems. Cenozoic strata are by far the most widely exposed and are least altered by burial. Because many Cenozoic animals and plants are closely related to living species, we can make direct comparisons between the environments of living and extinct species. We can also experiment with the evolutionary capacity of living species, for example, by selective breeding and other genetic studies. The modern theory of organic evolution has grown out of these inter-related kinds of information.

Because evolution is basic to the science of stratigraphy, we should examine a case fairly closely. Consider the effect of Cenozoic climatic changes on the evolution of horses. Horses are one of the groups of mammals that propagated remarkably during that era, filling the environmental "space" left by the extinction of the great variety and numbers of dinosaurs late in the Mesozoic Era. The great abundance and variety of fossil horses in North America shows that the evolution of horses was favored by changes during the

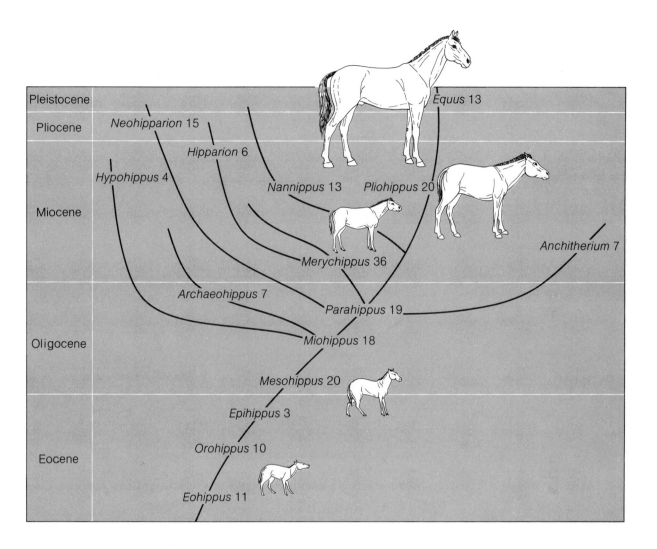

Pleistocene

Pliocene

Miocene

Oligocene

Eocene

Neohipparion 15

Hipparion 6

Hypohippus 4

Nannippus 13   *Pliohippus* 20

Equus 13

Anchitherium 7

Merychippus 36

Archaeohippus 7

Parahippus 19

Miohippus 18

Mesohippus 20

Epihippus 3

Orohippus 10

Eohippus 11

FIGURE 13-17

Evolution of horses, showing the relative sizes of typical species and the geologic range of each genus (the lines extending through the generic names). The numbers after the names indicate the number of species discovered for each genus. Data chiefly from R. A. Stirton. Phylogeny of North American Equidae. University of California Publications, *Bulletin of the Department of Geological Sciences*, vol. 25, pp. 165–98, 1940.

Cenozoic Era. Moreover, the breeding of a variety of domestic forms from a few modern species has shown that horses have a gene pool (hereditary mechanism) capable of producing great variation without sudden (mutational) changes. Finally, the basic activities of modern horses are well known with respect to their bones and other main structures—an important knowledge in view of the fact that only bones and teeth are generally preserved as fossils.

The general sweep of the horses' evolution is depicted in Figure 13-17. The most obvious changes are those of size—from the little eohippus (the genus *Hyracotherium*) that was as big as a good-sized terrier to the Pleistocene species of *Equus* that were as large as a modern range horse. The diagram is not meant to imply, however, that this evolution was one of a steady, "predictable" increase. There were long pauses in the progression. Late Eocene species of *Epihippus*, for example, were as small as were early Eocene species of eohippus, even though these forms represent 20 million years of evolution. There were also distinctly "backward" evolutions to pygmy horses, possibly in isolated environments that were very cool or provided sparse forage. The

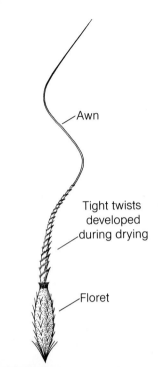

FIGURE 13-18
Top, fallen seed of a modern grass species of the genus *Stipa*, showing the twisted awn and sharply pointed, barbed floret that can emplace the seed into dry ground. Bottom, fossil florets of a similarly adapted grass, *Stipidium minimum*, from the Miocene of western Nebraska. Photograph courtesy of Maxim K. Elias (Reference 6).

causes of the increases in size are not really understood. One possibility is that large individuals were more likely to survive predation by wolves and saber-toothed cats that evolved during Oligocene time. Another is that large size gave an advantage among contending stallions. In either case, offspring would tend to inherit genes favoring larger size.

Changing plant communities and horses   Whatever the cause of the increase in size, there is no question that it resulted in a need for more food. Energy requirements are proportional to weight, and the *Equus* species were at least five times heavier than the large Oligocene species and more than fifty times heavier than eohippus. Of great importance in providing increases in forage were the changing climates and therefore the changing plant communities of the Cenozoic Era. Studies of sediments and the plants contained in them show that the Gulf of Mexico extended up the Mississippi Valley as far as what is now Illinois during Eocene time and that the rest of the continent lay so low that subtropical forest and savannahs covered much of what is now the United States. Eohippus was a browsing (leaf-eating) animal that was evidently well adapted to this environment, for the Eocene genera remained similar.

Near the end of Eocene time, however, the Rocky Mountain region was uplifted broadly (recall the widespread unconformity of that age shown in Figure 13-13). Further uplifts occurred during the Oligocene and Miocene while mountain ranges were forming in California and Oregon. These changes made the climate cooler, drier, and more variable, so that the subtropical forests were replaced by more temperate ones that ranged from upland forests to brushlands and grasslands. Systematic studies by Maxim K. Elias have shown that grasses evolved rapidly, becoming the dominant plants of the ancestral plains and upland valleys (Reference 6). Some of their adaptations to the increasing dryness of the climate are remarkable. One of the main genera (*Stipa*) evolved a long tail on its seed-carrying floret, which twisted as it dried, augering the seed into the soil or into the hair of animals that then propagated it widely (Figure 13-18). A general adaptation of the grasses described in Chapter 8, was their development of annually replenished root systems that stored nutrients for future years and thereby constructed a rich soil.

The horses were among the first large mammals to adapt to this increasing new food resource. This is probably why they could evolve large body sizes and succeed in large numbers. Their most basic adaptation was the development of teeth that could withstand the constant grinding by hard grass tissues (which contain silica) and by gritty silt particles deposited on the grass by the wind. As Figure 3-19 shows, the teeth developed elongated but buried parts that emerged as the grinding surfaces were worn away. Another adaptation was a change toward feet that permitted horses to run more swiftly over the open terrain. Eohippus had three toes and a laterally pliable foot that must have given it the ability to change course quickly among the trees and bushes of the Eocene woodlands. As the terrain changed to a grassy prairie, the middle toe become increasingly dominant until, in *Equus*, it was a powerful unit of the lower limb (Figure 13-20). In addition, the toe was set so as to move back and forth only, giving *Equus* the capability of running

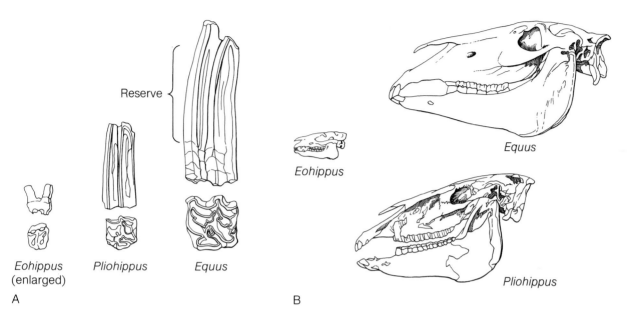

Reserve

Eohippus

Equus

Eohippus
(enlarged)

Pliohippus

Equus

Pliohippus

A

B

FIGURE 13-19

Evolution of horses' teeth. A,
changes from browsing to
grazing tooth forms, with increases in the buried reserve
that was emplaced outward as
the grinding surface wore away.
B, changes in the shape of the
skull, to accommodate the
buried tooth reserve. From G. G.
Simpson, *Horses* (New York:
Oxford University Press, 1951).

surely as well as swiftly across open country. The welding of some of the verte-
bra into a solidly supportive back was another adaptation to a powerful
running gait.

Rates of evolution of the horses   A further suggestion that the
environmental changes caused the horses' evolution is the number of species
and genera discovered in the principal divisions of the Cenozoic Era (Figure
13-17). If the millions of years for each division (Eocene, Oligocene, and so on)
are divided by the numbers of species, we find that an average of 2.5 species
evolved during each 1 million years in the Eocene and Oligocene. The rate
then increased to between 6 and 7 species per 1 million years and continued
at that rate for the remainder of the Cenozoic Era. This increase is coincident
with the changes to extreme environments beginning in the late Oligocene
or early Miocene. Note that the genera (the lines in Figure 13-17) also increased markedly at about that time.

The record is not completely satisfying, however, because the species are
not as graduated as they might be (as much, for example, as are the foraminifers). That the horses' family tree is missing some species is suggested
both by the abrupt changes in the existing record and by the difficulties of obtaining a complete fossil record from continental formations. But perhaps
the fossil record is complete—perhaps the horses did evolve in abrupt steps,
by such means as gene mutations. The latter idea does not fit modern evolutionary theory, but since the fossil record is a major basis of that theory
we must allow the possibility. The answer lies in finding fossils at more stratigraphic levels and at more localities. The same need applies to most groups
of animals and plants. Indeed, the influence of environment on evolution will
be understood only when the total record is more complete, for the environment of a given organism consists in a large share of its contacts with other
kinds of organisms.

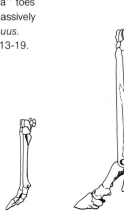

FIGURE 13-20

Changes in the hindfeet of typical horses in the main stem of horse evolution, from four-toed eohippus to three-toed *Mesohippus* to the vestigial "extra" toes of *Merychippus* to the massively one-toed structure of *Equus*. Same source as Figure 13-19.

Eohippus *Mesohippus* *Merychippus* *Equus*

Should you wish to help in filling the missing parts of the record, here are some suggestions about how to proceed:

1. Describe the strata being examined, noting their thickness, their relation to named formations, and any internal features that tell something of the environment in which they were deposited.
2. Measure and describe the location of each fossil-bearing layer in the sequence, so that species can be placed accurately in stratigraphic position.
3. Collect the fossils carefully, preferably embedded in enough rock matrix to protect them. (This can be removed later.)
4. Collect as many kinds of fossils as possible—the total community will tell most about the environment.
5. Using published descriptions or museum specimens, identify the fossils and determine their age and environment; seek help with regard to those you cannot identify.

Summary

These enumerated steps are the basic procedures of stratigraphy, followed in many of the studies mentioned in this chapter. We may briefly review their basis:

1. Sedimentary formations are parcels of rock defined by physical characteristics; these characteristics, in turn, tell us much about how each formation originated.

2. Because sedimentary formations accumulate layer upon layer over long periods of time, the fossils in them can be assembled into evolutionary sequences.

3. Once an evolutionary sequence is defined and tested, the various species can be used to determine the stratigraphic ages of rocks elsewhere.

4. Where the sedimentary rocks are interlayered with volcanic rocks or where intrusions can be related to them exactly, radiometric methods can be used to date the rocks in years.

5. A total "standard column" has thus been constructed from many fossil sequences and many dated rocks; the standard provides a means of correlating geologic events on a worldwide basis.

6. Fossil communities and physical conditions recorded in rocks may indicate how organic evolution has been affected by environmental changes of the past.

REFERENCES CITED

1. Andrew Soutar and John D. Isaacs. Abundance of pelagic fish during the 19th and 20th centuries as recorded in anaerobic sediment off the Californias. *Fishery Bulletin*, vol. 72, pp. 257–73, 1974.

2. Robert M. Kleinpell. *Miocene stratigraphy of California.* American Association of Petroleum Geologists, 450 pp., 1938.

3. M. N. Bramlette. The Monterey Formation of California and the origin of its siliceous rocks. U.S. Geological Survey, Professional Paper 212, 57 pp., 1946.

4. W. A. Berggren. Cenozoic chronostratigraphy, planktonic foraminiferal zonation and the radiometric time scale. *Nature*, vol. 224, pp. 1072–75, December 13, 1969.

5. Laurie J. Macdonald. Munroe Creek (Early Miocene) microfossils from the Wounded Knee area, South Dakota. South Dakota Geological Survey, Report of Investigations, no. 105, 43 pp., 1972.

6. Maxim K. Elias. Tertiary prairie grasses and other herbs from the High Plains. Geological Society of America, Special Paper 41, 176 pp., 1942.

ADDITIONAL IDEAS AND SOURCES

1. Although horses first arose in North America and thrived there for more than 50 million years, they became extinct in both Americas about 20,000 years ago, during the last of the major glaciations described in Chapter 7. Many causes for their extinction have been suggested, including disease and the advent of humans, and some of these are described in a group of interesting papers in *Pleistocene extinctions,* edited by P. S. Martin and H. E. Wright, Jr. (New Haven: Yale University Press, 453 pp., 1967). The fossil record in the deposits of the Great Plains points to an especially interesting cause described by C. Bertrand Schultz in "The stratigraphic distribution of vertebrate fossils in Quaternary eolian deposits in the Midcontinent region of North America" (pp. 115–38 in *Loess and related eolian deposits of the world,* edited by C. Bertrand Schultz and John C. Frye; Lincoln, Neb.: University of Nebraska Press, 1968). The fossils indicate that numerous species of large mammals migrated from Asia across

a land bridge at the site of the Bering Straits. Pleistocene elephants (mammoths), bears, giant bison, elk, moose, caribou, and various ancestral sheep and deer thus began to compete with the native horses and camels for food resources that dwindled during the cooler (glacial) periods. Some of the migrants, such as the elephants, became extinct along with the horses and camels, but others, such as the bison, flourished numerically although they diminished in size. Horses meantime survived on the Eurasian continent and were reintroduced to the Americas by another migration—that of European humans.

2. *Principles of paleontology* by David M. Raup and Steven M. Stanley (San Francisco: W. H. Freeman Co., 388 pp., 1971) is a clearly written and well-illustrated source of ideas on how to collect and study fossils and also on how to interpret the fossil record.

3. George G. Simpson has written several very readable and interesting books pertaining to evolution as determined from the fossil record. *Horses* (New York: Oxford University Press, 1951) adds greatly to the brief account in this chapter, and a broader coverage is given in *The major features of evolution* (New York: Columbia University Press, 434 pp., 1953).

4. A broader view of the nearly horizontal formations of the Great Plains can be obtained from W. H. Raymond and R. U. King's *Geologic map of the Badlands National Monument and vicinity, west-central South Dakota* (U.S. Geological Survey Miscellaneous Investigations Series, Map I-934, 1976), on which the formations are shown in colors. Geologic maps and mapping are described in Appendix A.

The great sandstone cliff at Castlegate, near Price, Utah. The sands were deposited by a river system that overran extensive swamp deposits. (The latter are exposed but in shadow, under the small lower cliff.)

14. Coals and a Sea on the Continent

Coals and a Sea on the Continent

Figure 14-1 shows one of the world's most remarkable rock exposures. The semiarid escarpment, the Book Cliffs, discloses every detail of a 1,000-foot sequence of layered rocks. From the point at which the picture was taken in central Utah, the cliffs can be followed in a broad arc for 50 miles, to the far left corner of the photograph. Behind the camera they extend 150 miles farther —across the rest of Utah and into Colorado!

The name derives from the similarity of the eroded layers to the leaves of a book, one lying nearly flat but tilted a few degrees to the northeast (to the right in the photograph). The lighter faces are the eroded edges of sandstone layers, which are relatively resistant to erosion. The darker slopes mark the outcrop of less resistant shale (mudstone) layers, partly covered by fallen blocks of sandstone. Materials eroded from the cliffs also form a thin cover on the broad alluvial plain sloping gently off to the left. The arroyos that cut into the plain have removed this cover and exposed the lowest shale layer, as in the gullied slopes near the base of the escarpment. This shale extends to the right under the other layers and parallel to them. Drill holes and deeper exposures elsewhere show that the shale is about 3,000 feet thick—three times the height of the Book Cliffs.

This thick shale, a formation called the Mancos Shale, extends from one end of the Book Cliffs to the other. Significantly, however, the other layers do not. Each layer of sandstone becomes gradually thinner eastward and eventually ends. The shale layers between the sandstones generally thicken eastward. Coal beds are associated with some parts of the sequence, and they appear and disappear, even splitting into two beds in some places.

These lateral changes have considerable meaning, and it will be the purpose of this chapter to explore them. They have great practical significance because of the worldwide shortage of fuels. The Book Cliffs sequence contains 12 to 17 billion tons of minable, high-grade bituminous coal—it is one of the world's major reserves. To find additional coals and mine them efficiently,

FIGURE 14-1

The Book Cliffs of central Utah, viewed northwestward from a point near Green River.

we must understand their part in a broader story: the lateral variations in the rocks record the shifts of a vast inland sea that occupied the eastern part of the region during the Cretaceous Period. Here, as in the examples described in Chapter 13, marine deposits must be related to continental deposits of the same age.

Measuring Stratigraphic Variations

We can read these relations clearly because of the bareness and the lateral extent of the cliffs. In fact, the more abrupt lateral changes can be observed in some places by simply approaching the cliffs and looking at the layers. Some more gradual changes can be seen by scanning the cliffs with field glasses from Interstate 70 or Highway 50 (Figure 14-2). It is easy, however, to confuse some of the layers when attempting to match them in these ways. For many years, in fact, a given coal bed at one mine was thought erroneously to be the same as that at other mines miles away, a mistake that led to incorrect estimates of tonnages and of the costs of mining the coal.

More exacting surveys have been made in order to get a complete and precise idea of the lateral changes. In one method, *traverses* (surveyed lines) are measured directly on the escarpment, generally in canyons that permit easy access. The thickness of each layer along the traverses is measured to the nearest foot, and vertical sequences such as those described in Chapter 13 are compiled from the measurements. All the rocks are examined and described in detail, and fossils and rock samples are collected for exact identification in the laboratory. Coal layers are sampled completely.

FIGURE 14-2

Map of the Book Cliffs, the Wasatch Plateau, and related plateaus, ranges, and broad uplifts. Except for the main geographic features, the only ones labeled are those mentioned in the text or figures.

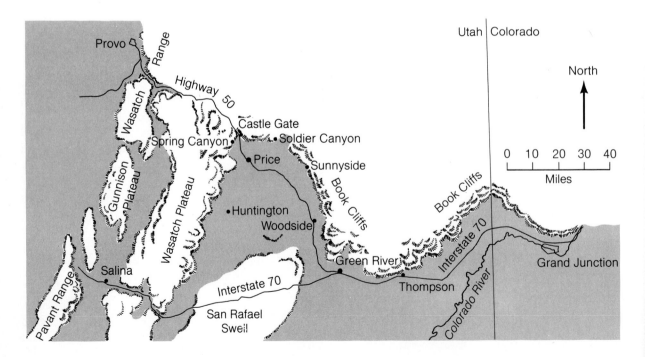

The great length of the cliffs requires that these detailed surveys be miles apart. It is therefore necessary to walk along the main rock layers between the traverses in order to see which layers are indeed continuous and which change laterally into other rocks. This procedure, called *geologic mapping*, is described in Appendix A. Frank R. Clark was the first to map the Book Cliffs systematically and the first to discover many of the lateral changes, including the nonequivalence of many of the coal beds (Reference 1).

Edmund M. Spieker broadened Clark's work greatly by studying the Wasatch Plateau and the smaller plateaus and ranges to the west and southwest of the Book Cliffs (Figure 14-2) (Reference 2). His work proved very informative but went slowly because westward the rocks are increasingly contorted and broken by faults. Moreover, each range to the west is surrounded by alluvial plains that isolate it, so that fossils had to be found in order to match the layers from one range with those of the next, and eventually with the rocks exposed in the Book Cliffs. Diagnostic (marker) fossils were often long in the finding, but Spieker revisited the area many times, often with Ohio State University students, who helped greatly in all aspects of the work. One of his former students, Robert G. Young, made a thorough study of the entire Book Cliffs exposure, which constitutes a major source for this chapter (Reference 3).

Figure 14-3 shows an overview of the findings to date. The diagram represents a vertical slice through the rock layers as they were at the close of the Cretaceous Period. The vertical dimension has been enlarged 60 times the horizontal dimension in order to show the layers distinctly, and as a result the slopes of the layers are greatly exaggerated. The Panther Sandstone on the left, for example, is nearly horizontal. All the layers sloped gently to the east when first deposited.

The figure shows clearly enough that the pages of the Book Cliffs are not the same from west to east. It also shows that the changes are orderly in several ways. For one thing, the accumulation thickens to the west. If the top line (A-B) represents the depositional surface at the close of Cretaceous time, then lines representing older depositional surfaces (such as C-D and E-F) were apparently bent downward at their western ends as more and more sediment was deposited over them.

A more important kind of order relates to the grain sizes of the sediments. The coarser, gravelly rocks are all at the far western end of the array, and the finest rocks (the shales) lie mainly to the east of sandstones of the same age (such as along line C-D). There would, in fact, be a simple west-to-east gradation from gravelly rocks to sandstone to shale were it not for the mixed shale, sandstone, and carbonaceous sediments that include the coal beds. This complication, however, has turned out to be helpful. The coals tell us much about the origin of the entire array.

Interpretations Based on Coals

The significance of the Book Cliffs coal beds derives partly from a broad understanding of how coal forms. This understanding stems from countless

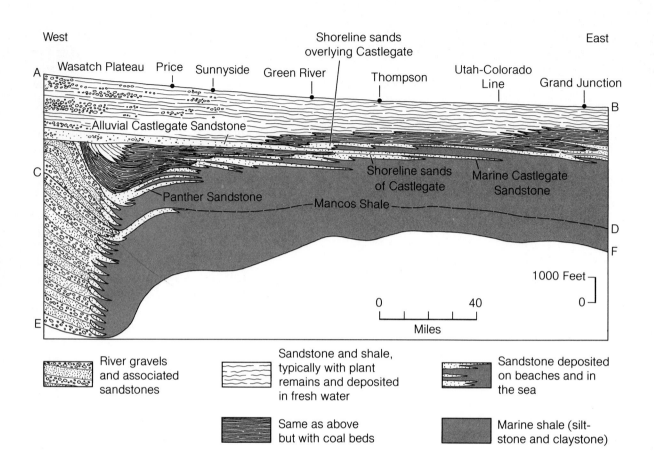

West   Shoreline sands   East
                                    overlying Castlegate

A Wasatch Plateau Price Sunnyside   Green River   Utah-Colorado
                                              Thompson   Line   Grand Junction
                                                                                              B

Alluvial Castlegate Sandstone

C   Shoreline sands   Marine Castlegate
                                                                of Castlegate   Sandstone
                           Panther Sandstone   Mancos Shale

                                                                                              D
                                                                                              F

                                                                        1000 Feet

                                                        0   40   0

                                                              Miles

River gravels   Sandstone and shale,   Sandstone deposited
and associated   typically with plant   on beaches and in
sandstones   remains and deposited   the sea
                           in fresh water

                           Same as above   Marine shale (silt-
                           but with coal beds   stone and claystone)

FIGURE 14-3

Vertical section through the Book Cliffs sedimentary sequence, extended westward to include the Wasatch Plateau. The line A-B represents the surface at the close of the Cretaceous Period, and the lines C-D and E-F show similar surfaces, now bent down by uneven subsidence as the sediments accumulated. The total time represented is about 30 million years (from about 100 million years ago to about 70 million years ago). Modified from References 2 and 3.

observations of coal in various stages of formation. The initial stage in all cases is the accumulation of plant remains, which decompose and compact lightly to form a layer of *peat.* Subsequent changes result from progressive burial. Peat buried to shallow depths is changed to *lignitic* (soft) coal, which may be altered by deeper burial to harder *bituminous* coals such as those of the Book Cliffs. We saw in Chapter 10 that burial increases the temperature as well as the load (pressure) on sediments. The increased temperature causes organic chemical reactions among the carbonaceous (plant) materials, producing water and organic gases. The increased load presses the lightweight materials out, and the coal thus becomes more and more compact. The reduction in volume is large—by the time the Book Cliffs peats had changed to the present coals they were about one-tenth their original thickness.

The minable coal beds are 5 to 22 feet thick, and therefore the original peat layers were 50 to 220 feet thick! What circumstances led to the accumulation of such unusually thick layers? They evidently formed near a sea, for the nearby Mancos Shale and some parts of the sandstones contain fossils of marine animals. We also know that plant litter is destroyed by atmospheric oxygen and soil organisms, so the plant remains must have accumulated in a water-saturated environment. One of more important soil organisms, the fungi, cannot live without free access to oxygen.

Some of the thickest peats evidently accumulated in somewhat salty coastal swamps and estuaries, for they include remains of plants that live in brackish water. Other peats consist entirely of fresh water species and therefore must have accumulated on swampy river floodplains. Well-preserved leaves, seeds, and pollen identify pine trees, redwoods, and other conifers as the main contributors. A large amount of resin in the coals also indicates a conifer source. Coal scientists have studied all the stages in the coalification of conifer litter and found the results identical to the main coal components in the thick beds at Sunnyside (Figure 14-2). Other plants identified in the Book Cliffs rocks include ferns, horsetails, birch, beech, poplar, magnolia, and fig. If they are anything like their modern relatives, these plants and the conifers indicate a wet, temperate climate. Plants such as the horsetails show that conditions were swampy in many places.

Changes in the sediment immediately beneath each coal layer support the idea that the accumulations took place on watery lowlands. Most coals lie directly on sandstone layers that are bleached white just under the coal. The whitening expresses the complete removal of iron and the alteration of feldspar and mica grains to kaolinite-rich clay. Where the coal layers lie on shale, the expandable clays normally found in the shale are changed to kaolinite. These changes are similar to those in strongly developed soils such as those in Georgia (Chapter 8) but with an important exception: the kaolinite under the coal beds is not dissolved and carried downward as it is in the Georgia soils. Thus, groundwater under the peat must have moved mainly horizontally rather than vertically (downward). This movement is consistent with the slow horizontal circulation of water in swampy ground near sea level. Dinosaur footprints are abundant in rocks associated with the coals, but no bones have been discovered, a relation supporting the idea that the water contained strong organic acids.

Significance of the peats   Modern peats accumulate at a rate of about 1 foot in 35 years when conditions are ideal. A 100-foot peat layer thus represents at least 3,500 years and probably about 5,000 years of accumulation. It is therefore significant that the Book Cliffs coals are exceedingly pure. The 10-foot bed at Sunnyside, for example, contains only about 5 percent noncombustible mineral substances, so-called ash. These substances include considerable calcium carbonate and pyrite (iron sulfide) that were added by groundwater after the peat accumulated. Only about 1 percent of the coal is silt and clay deposited with the plant litter.

This purity proves that airborne dust was scarce, which indicates a wet climate and implies further that vegetation covered the terrain for tens to hundreds of miles. Rivers flowing across the forested lowland must have occasionally flooded it, for they deposited the muds and sands that are now the shale and sandstone layers associated with the coal beds. The purity of the thick coals, however, shows that almost no river-borne sediment was introduced to large parts of the swamps for periods of thousands of years. The closely grown plants must have so reduced the velocity and turbulence of flood waters that almost all suspended sediment was deposited near river channels.

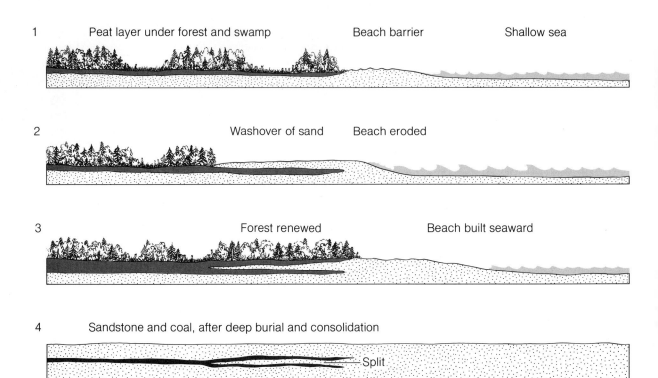

1 Peat layer under forest and swamp   Beach barrier   Shallow sea

2 Washover of sand Beach eroded

3 Forest renewed   Beach built seaward

4 Sandstone and coal, after deep burial and consolidation
                                                        Split

FIGURE 14-4

Probable stages in the generation of a split (a thin wedge of sandstone) in a coal bed.

The sandstone wedges dividing individual coal beds near their seaward margins provide a final indication that the peat accumulated close to sea level. As Figure 14-4 depicts, the sand was probably washed over the peat swamps during unusual storms, interrupting the accumulation of peat until the forest and swamp became reestablished on the washed-over sand. When the deposits finally were buried beneath additional sediments (for reasons we will see shortly), the peat gradually changed to a coal bed with a "split" along one edge.

Cyclic Shifts of Environments

In addition to local incursions of beach sand, such as that just described, the entire area near sea level was periodically overrun completely by the sea. Each event took place decisively, inundating and killing the forests and covering the peat layers with sea water for tens of miles inland. As Robert Young pointed out, the evidence is simple and compelling: many of the coals are terminated along their upper surfaces by shales and sandstones containing fossils of marine and brackish water animals and plants.

Periodic subsidence (sinking) of the entire coastal region and sea bottom caused the inundations. The amount of subsidence differed each time but was about 100 feet for the ten or twelve more pronounced cases. Each subsidence was followed by another long period of stability, during which sediments accumulated along the newly formed shore as well as in the sea. The accumulation resulted in partial filling of the sea, so that the shoreline was slowly pressed seaward (eastward). The new coastal plain was eventually forested and covered with another layer of peat—only to be inundated because of another subsidence.

As Spieker suggested many years ago, this episodic filling and subsidence is recorded in the way the marine sandstone and shale layers interfinger as a series of wedges (Figure 14-3). Let us follow one of the cycles of filling and subsidence to see exactly why the sandstones and shales are wedge-like.

Figure 14-5 illustrates a typical cycle. Sediment brought to the sea by rivers was worked offshore and alongshore to the position of the uppermost diagram.

FIGURE 14-5

Shifts of depositional environments resulting from alternate filling and subsidence of a sea basin. The small numbers and ticks at the ends of the diagrams show the sea levels for each stage. The vertical scale is exaggerated tenfold.

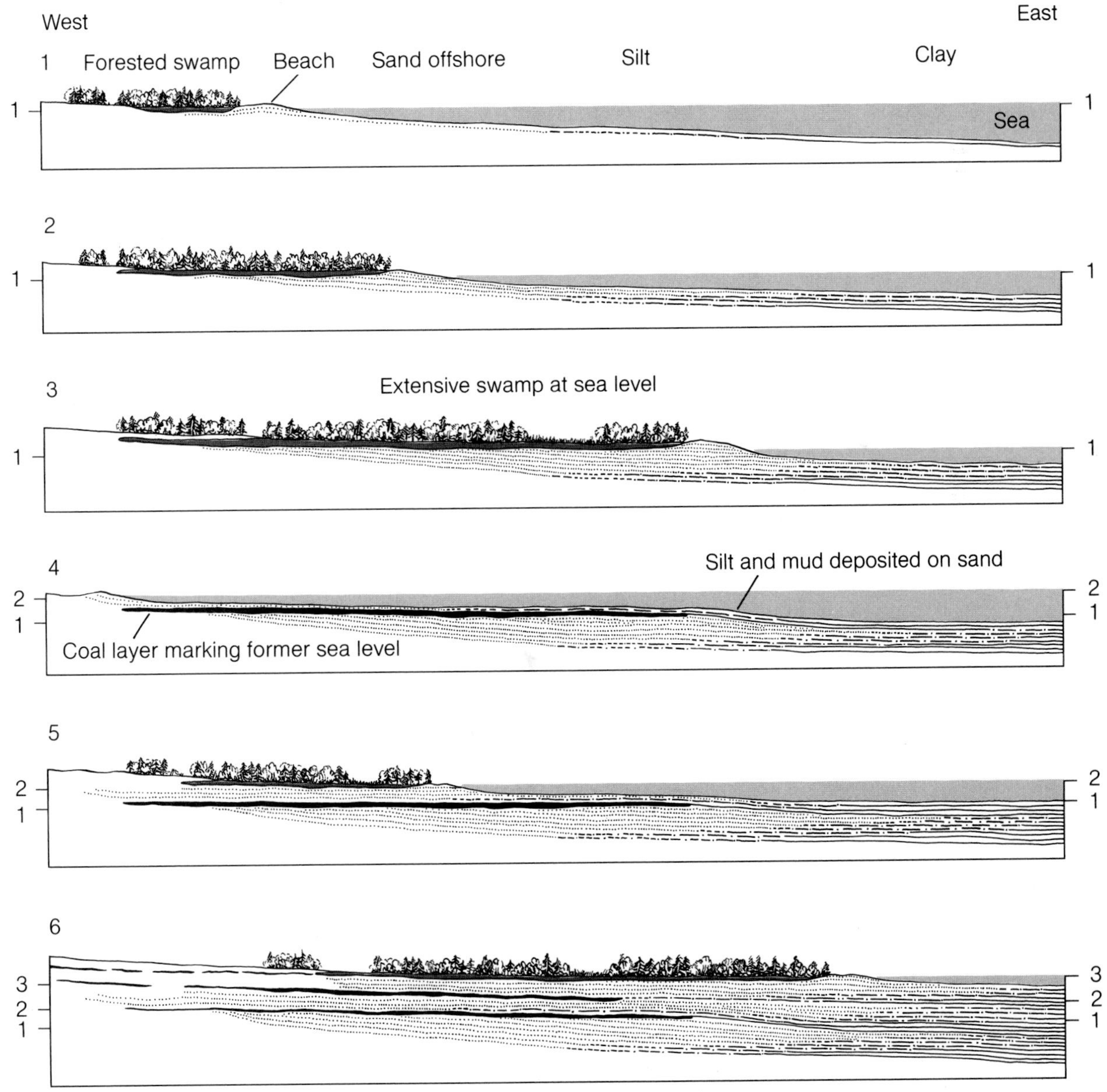

West   East

1   Forested swamp   Beach   Sand offshore   Silt   Clay

Sea

2

3   Extensive swamp at sea level

4   Silt and mud deposited on sand

Coal layer marking former sea level

5

6

Waves and currents sorted it by size, so that clean sand was deposited on the beach and the sea bottom near shore, and silt and clay were deposited farther offshore. The distance from the shore to the silt deposit varied greatly from one situation to another, from $\frac{1}{4}$ mile to about 10 miles.

Gradations from deeper to shallower environments   As shown in diagrams 2 and 3 of Figure 14-5, continued additions of sediment built the beach seaward and filled in the sea bottom. The areas where sand, silt, and clay were being deposited thus shifted seaward to water depths comparable to those in diagram 1. The sediments deposited at any one place became coarser with time, as shown by the upward gradation from clay-rich shales to very fine sandstones and finally to the typical cliff-forming sandstones so widespread in the Book Cliffs (Figure 14-6). Each layer in Figure 14-6 originally sloped very slightly from left to right (west to east) and lay along the actual sea bottom when it accumulated. One may thus imagine this place from the time when the lowest shale was newly deposited mud and seawater extended to the top of the cliff, to the time when the uppermost beach sand filled the last shallow at the shoreline.

The evidence for the gradual construction of a new coastal plain links back to the coals. Most of the sandstones are overlain directly by a coal bed or by swamp and river deposits containing plant remains. Even where the coals have been eroded away, the white tops of the sandstones attest to their former presence (Figure 14-6). Evidently, a swampy forest developed on each newly formed sandy plain, and peat eventually accumulated there.

The record of subsidence and refilling   The thick peats probably subsided somewhat as they formed, but it was a major subsidence that sent the sea landward over the coastal plain (Figure 14-5, diagram 4). As already mentioned, the evidence lies in the sudden appearance of marine fossils in the sediments just above the coals, or above other coastal plain sediments.

Diagram 5 shows an intermediate stage in the second filling of the sea. At first silt and clay were deposited over the more seaward parts of the drowned plain, but as the sediments built up they were overlapped by coarser and coarser detritus, as before.

As this cycle and others like it are completed, the wedges become apparent, as depicted in diagram 6.

We may now understand a broader aspect of Figure 14-3. Note that the positions of the coal-bearing deposits shift to the right (to the east) in the upper half of the sequence. This change must mean that in the long run the sea was slowly filling, the more rapidly as the younger parts of the Book Cliffs sequence were deposited. Seeing the rocks as sedimentary environments thus makes it possible to read the long-term history as well as the many shorter episodes of filling and subsidence.

The Great Wedge of the Castlegate Sandstone

The diagrams and the changes implied, however, are rather abstract. We need to take one rock layer and follow it completely from west to east to see how

FIGURE 14-6

Upward gradation from clay-rich shale (the lower third of the view), into silty and sandy shales (the lower fourth of the vertical cliff), and into sandstone layers that become increasingly thick toward the top of the cliff. In all, 100 feet of strata are shown. Note that the whitening of the top, thick layer extends for 10 to 15 feet into it. The locality is about 10 miles west of Thompson (labeled in Figure 14-2).

FIGURE 14-7
Castlegate Sandstone near the Colorado State line, where it is 1 foot thick (foreground). The same layer shows as a dark line at the top of the vertical cliff in the middle distance and as a thin dark line across the bluffs in the far distance.

its actual parts demonstrate the proposed environments. A formation called the Castlegate Sandstone, labeled in Figure 14-3, ideally serves this purpose. Almost every detail of this great wedge of strata can be seen, from a 400-foot cliff at the western end of Book Cliffs (chapter opening page) to the last foot of sandstone at the thin end of the wedge—100 miles to the east (Figure 14-7). As will be related in this section, the western part was deposited in river channels; the extreme eastern part was deposited in the sea.

Evidence that the western half of the Castlegate Sandstone was deposited by rivers is particularly convincing near the base of the high sandstone cliff. Viewed in certain directions, the individual sandstone layers are discontinuous, one evidently being eroded into the one below it (Figure 14-8). As shown

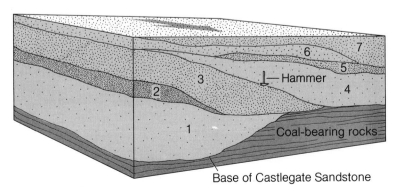

FIGURE 14-8
Close-up view of the lower part of the Castlegate Sandstone near Sunnyside, illustrating the channel-shaped forms of the individual layers. The diagram indicates the numerical order in which the layers were deposited and shows how their shapes appear as simple layers in a surface oriented at right angles to that in the photograph.

313

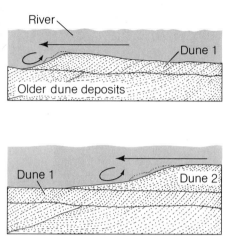

FIGURE 14-9

Outcrop near Castle Gate showing several crosslayered sets of strata in the Castlegate Sandstone. As the diagram suggests, each set represents a river dune, its internal layers slanting in the direction of the currents that deposited it.

FIGURE 14-10

Swamp and associated flood-plain deposits of the Castlegate. The lowest dark layer on the right is a coal bed, which clearly splits near the center of the view. The cliff, 4 miles west of Thompson, is 100 feet high.

in the drawing, the layers have simple, continuous shapes when viewed in a direction at right angles to that in the photograph. Each layer is evidently the result of the filling of a stream channel that was eroded in the underlying layer or layers. The river channels must thus have been oriented approximately parallel to the large arrow on top of the diagram. The total record indicates that a river first eroded the underlying coal-bearing strata and then gradually built up its bed and floodplain with sand, shifting laterally as it did so.

Where the channels are well exposed in three dimensions, they can be seen to be oriented roughly southeast with many variations. The variations indicate moderately sinuous channels, 1 to 20 feet deep, that must have looked much as the braided (parting and rejoining) channel of the Rio Grande in the Middle Valley (Figure 3-2). Even the details are similar. Parts of the sandstones show crosslayered deposits such as those formed by river dunes (Figure 14-9). Recall that comparable features have been formed in flumes (Figures 3-5 and 3-6).

Because these river-deposited parts of the Castlegate Sandstone are widespread on the Wasatch Plateau as well as in the western Book Cliffs, this part of the region must once have been a broad alluvial plain underlain almost entirely by sand.

Thin layers of fine-grained rocks containing abundant plant materials first appear in the Castlegate Sandstone near Woodside (Figure 14-2). The fine-grained layers become thicker and more numerous to the southeast, and in 25 miles the entire Castlegate formation has graded into mixed rocks of a swamp and estuarine environment (Figure 14-10). Fifteen miles farther southeast the deposits begin to show abundant traces of salt water animals (Figure 14-11). Here and there are layers and wedges of well-sorted sandstone with internal layers sloping landward (westward), suggesting that wind dunes and tidal channels carried beach sand landward over salt marshes, much as they do today in the Georgia estuaries (Chapter 4). The whitened tops of many sandstones indicate long periods of leaching, similar perhaps to that now affecting the dune accumulations on the Georgia barrier islands.

Indications of a marine environment Still farther eastward, the rocks of this marsh environment are replaced by exceedingly thick sandstones. Most of them have simple parallel laminations suggestive of beach deposits, and Robert Young has measured extensive cross-laminations in some that indicate alongshore transport of sand from northeast to southwest. He has also found sandstones with cross-laminations such as those of dunes behind modern beaches. Some of the thick sandstones even show channels cut through them and filled laterally, as we found in estuary entrances along the Atlantic coast. Oyster shells are abundant in some places, and sharks' teeth and shells of foraminifers can be found. These fossils indicate a geologic age of late Cretaceous. We have reached the edge of the Cretaceous sea!

Eastward (seaward) from the shoreline, the beach sands thin rapidly to a sand blanket about 50 feet thick. This layer continues for tens of miles but then thins and grades into the Mancos Shale 50 miles east of the shoreline. This part of the Castlegate Sandstone contains marine fossils and nowhere has a whitened top. As can be seen in the foreground of Figure 14-7, the most sea-ward part has delicately rippled laminations that indicate the effects of waves. Relatively shallow water, at most a few hundred feet deep, is also indicated by certain species of benthic (bottom-dwelling) foraminifers found in the shale under the sandstone. Figure 14-7 also shows, in the cliff in the middle distance, that the sandstone grades downward into the Mancos Shale, the same relation we found in other sandstones of the Book Cliffs. The rocks just under the sandstone are silty and cross-laminated, but those lower in the bluff are rich in clay and so thinly (microscopically) layered as to split into paper-like sheets (Figure 14-12). These rocks contrast with the shales of the swamp and marsh environments, which are silty and commonly so mixed by organisms that they show few relics of fine-scale layering (Figure 14-11). In contrast, traces made by organisms in the sea-bottom muds are so widely spaced that probing actions are preserved in detail (Figure 14-13).

The deposition of the Castlegate Sandstone ended when the entire array of deposits subsided and the sea flooded inland (westward). Marine muds started

FIGURE 14-11

Left, mudstone (dark) with sand-filled burrows exactly like those of *Callianassa* in strand deposits of Georgia (Chapter 4). Right, specimens collected by Robert Young that strongly support that origin: a knobby burrow filling (center), a fossil shrimp (left), and a crab pincer.

FIGURE 14-12

Mancos Shale about 30 feet under the Castlegate Sandstone at the same locality as that of Figure 14-7. Note the flaky, thin-layered nature of the rock.

to accumulate over the marine part of the sandstone, the swamp deposits, and even part of the alluvial plain. The environments shifted westward some 35 miles—the distance between the shoreline sands of the Castlegate and those overlying it (labeled in Figure 14-3). The Castlegate Sandstone thus provides evidence for the diagrams of Figure 14-5, both in the array of environments it discloses and in the way they suddenly shifted westward when its deposition was ended.

Reconstruction of a Delta

It is important to see that some of the Book Cliffs sandstones formed somewhat differently from the Castlegate Sandstone. An example is the Panther Sandstone, a wedge exposed in the western part of the Book Cliffs and along the eastern front of the Wasatch Plateau and labeled in Figure 14-3. Its separate layers show an intriguing relationship to the overall form of the body (Figure 14-14). A study of the photograph will show that the individual layers slant downward to the right across the formation; the top layer on the left side of the photograph, for example, lies midway in the formation on the far right.

This relationship in itself is not greatly surprising; we have noted other sandstones that are internally crosslayered on a smaller scale. The oblique layering simply says that the deposit grew toward the right—each layer recording the sloping surface of deposition at that time. The interesting thing, however, is that the layers slope southwestward, whereas the sea in which they were deposited lay generally eastward of the land. A detailed study by James D. Howard has shown that these layers are part of a delta, a lobate mass that extended southwestward along the coast rather than directly eastward into the sea (Reference 4). Let us see how he recognized an ancient delta in what at first glance appears to be an ordinary sandstone body.

When the sloping layers of sandstone are traced to the base of the formation, they can be seen (at the actual exposure) to grade into finer sand and then to silt. Evidently the sediment was sorted to finer sizes in deeper water, much as sediment is being sorted today on the gently sloping shoreface of the Atlantic near Chesapeake Bay (Chapter 5). Howard found, however, that channel-shaped bodies of distinctly coarser sand cut into the uppermost part of the sloping layers (Figure 14-15, top). Crosslayering within the channel-shaped bodies resembles that of river dunes, indicating that the channels were cut and then filled by a river that flowed out over the deltaic mound after it had been built up to sea level.

This interpretation was enhanced by information on the currents that formed the various deposits. The directions of the currents were determined from several kinds of features: (1) the slopes of crosslayering in the channel-filling sandstones, like the ones in Figure 14-9; (2) flutes and grooves eroded in the marine layers by currents that then deposited more sediment in the eroded forms (Figure 14-15, bottom); (3) sediment ripples that have steep sides indicating a downstream direction (as in Figure 3-7); and (4) streamlined forms in the finer bottom sediments, such as those in Figure 14-13. When the current directions derived from these features were plotted on a map, they showed a variety of orientations (Figure 14-16A). Combined with the slopes

FIGURE 14-13
View directly down on the top of a layer of siltstone at the same place as Figure 14-12. The groups of probe marks probably resulted from the systematic feeding of single animals. The curved groove is the trail of an animal that lived on, rather than in, the sediment. Note that a number of small probe mounds were eroded by a current moving from upper right to lower left, before the other marks were made. The view is 6 inches across.

FIGURE 14-14
The formation called the Panther Sandstone, forming the distinctly layered cliff extending left to right through the central part of the photograph. The locality is on Highway 50, about 8 miles north of Price.

and the grain sizes of the deposits, the directions suggest the reconstruction shown in Figure 14-16B. Note how the marine parts of the accumulation slope southward (toward the front of the diagram) but also toward both sides. The sediment carried to the sea by a river was evidently deposited in a submarine mound that sloped outward in all directions. When the initial mound grew up to sea level, the river extended its course in a sinuous pattern, forming channel deposits across the delta. The current directions show that sediment was sometimes transported to the front of the growing lobe and sometimes to its sides—a relation found in the radiating distributaries of modern deltas, for example that of the Mississippi River (Figure 10-1).

Using trace fossils Howard's interpretation was supported further by a study of the burrows, trails, and probe marks left by animals that lived in and on the sediments. These marks are called *trace fossils* because they differ from the fossilized remains of the animals themselves. They are extremely useful because they are often abundant where fossil shells and bones are very scarce. Moreover, fossil bones and shells can be eroded and redeposited in younger sediments—causing gross errors of interpretation—whereas trace fossils are almost always destroyed when eroded.

A systematic search showed that the trace fossils differ greatly among the various parts of the Panther delta. The finest deposits, the silty muds laid down in front of the delta, were so worked and reworked by small animals that they became mottled and spotted by the fillings of burrows and by fecal pellets (Figure 14-17A). Small, undisturbed parts of the original sediments show extremely thin, orderly laminations such as those in the Monterey Formation (Figure 13-3B).

The very fine-grained sandstones that lie above the siltstones and beneath the sloping beds of Figure 14-14 have some mottled zones but generally show fewer and therefore more distinct burrows and trails. Some lie along layer surfaces and some rise at various angles through the sediment (Figure 14-17B). Many consist of a series of cup-shaped units, such as the one labeled in Figure 14-17B.

FIGURE 14-15

Top, cross section through a channel filling in the upper part of the Panther Sandstone in Spring Canyon, about 8 miles northwest of Price. Bottom, bottom of a layer from the Panther Sandstone, showing the fillings of grooves and flutes eroded in the underlying sediment when the sand was deposited. The current moved from left to right. Photographs by James D. Howard (Reference 4).

The slanting sandstone beds of Figure 14-14 have few mottled layers and are otherwise typified by widely spaced and often large trace fossils (Figure 14-17C). Finally, the sands interpreted as river-channel deposits contain almost no trace fossils.

Howard investigated the origins of the trace fossils by observing recent sediments and organisms on and near Sapelo Island. He also experimented with live marine and brackish water animals in tanks containing bottom sediments. In the experiments, thin layers of sediment were added to the tanks from time to time. The movements of the animals and the burrows formed were recorded by taking X-radiographs (photographs taken with X-rays, in order to see through the tank walls and the sediment). They showed that probing and burrowing rapidly destroyed the bedding laminations, especially in layers containing abundant marsh plant detritus (Figure 14-18). This preferential burrowing suggests that the fine muds deposited in front of the Panther Sandstone delta also contained much plant detritus and may have been part of an association similar to the estuarine deltas of today. As Figure 14-16B suggests, deltas provide large areas in which coal-forming plants can grow, and the irregular lobes create shallow bodies of quiet water in which thick peats can accumulate. Recognition of ancient deltas is thus important in seeking and understanding coal deposits.

Rates of sediment accumulation   A provocative discovery is that cup-in-cup structures, such as the one labeled in Figure 14-17B, recorded the upward movements of a single animal, as layer after layer of sediment was deposited over it. A given cupped trace should therefore record the number of sediment layers deposited in one animal's lifetime, or less. Some of the cupped traces in the Panther Sandstone are several inches high, and the animal's life span, although not known, is not likely to be more than 10 years. Thus, such traces record the deposition of sediment at rates greater than 0.1 inch per year and probably around 0.5 inch per year in many cases.

We can easily compare this rate with the average rate of accumulation of the Book Cliffs sequence. The latter has an average thickness of about 7,000 feet, and fossils show that it took about 30 million years to be deposited. These figures suggest an average rate of only 0.0025 inch per year! Clearly, the layers including the cup-like trace fossils accumulated at rates that were far greater than average. This relation is consistent with deltaic growth such as that suggested by Figure 14-16B, which involves first one part of the growing form and then another. A given site might receive as much as an inch of sediment per year for hundreds or thousands of years but then almost nothing for thousands of years. Studies of sediments and fossils in deltaic formations must thus be very thorough to gain a full story of their origin.

Source of the Sediment

The rapid growth of the Panther Sandstone delta and the great volume of the Castlegate Sandstone point up a vital aspect of the Book Cliffs sequence: they represent a vast amount of erosion somewhere. Within the area of Figure

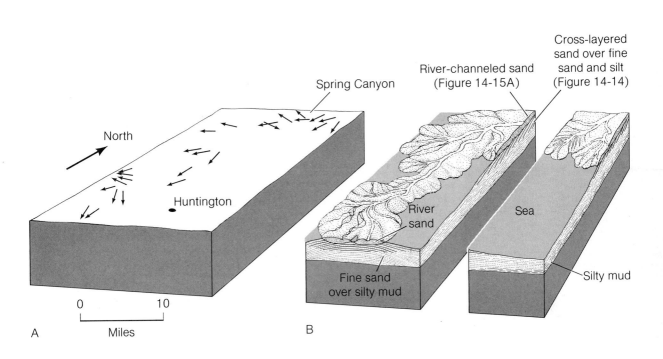

Spring Canyon

River-channeled sand
(Figure 14-15A)

Cross-layered
sand over fine
sand and silt
(Figure 14-14)

North

Huntington

0   10

Miles

A

River
sand

Sea

Fine sand
over silty mud

Silty mud

B

FIGURE 14-16

A, map showing the locations and orientations of current-indicating features in the Panther Sandstone, measured by James D. Howard at the west end of the Book Cliffs and along the east front of the Wasatch Plateau. Data from Reference 4. B, approximate configuration of the delta, based on the current measurements and on the overlapping relations among the different kinds of sediments.

14-2, approximately 50,000 cubic miles of sediment were deposited during the last 50 million years of the Cretaceous Period. What and where were the rocks that yielded these materials?

Figure 14-3 shows the Cretaceous rocks ending abruptly at both ends of the diagram. The cutoff on the right (east) is arbitrary; rocks of the Cretaceous System continue into Kansas and probably once extended far beyond. To the left (west) of the diagram, however, there are no traces of Cretaceous sedimentary rocks for several hundred miles. Spieker, who was especially interested in this point, interpreted it to mean that the region west of the figure was being uplifted during much of the Cretaceous Period and that the erosion of the uplands so formed provided the sediment of the Book Cliffs sequence.

He found considerable evidence for this idea, the most apparent being the coarsening of sediments to the west. In the Gunnison Plateau and at the western edge of the Wasatch Plateau, the Cretaceous rocks are typically gravelly and poorly sorted by size—rocks appropriately called *conglomerate* (Color Plate 10A). These coarse sediments accumulated in lens-shaped bodies and

FIGURE 14-17

Typical trace fossils in the Panther delta, each block being about 8 inches across. From J. D. Howard, "Characteristic trace fossils in Upper Cretaceous sandstones of the Book Cliffs and Wasatch Plateau" (pp. 35–53 in *Central Utah coals*. See Reference 4.

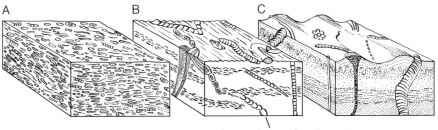

A   B   C

Cupped trace fossil

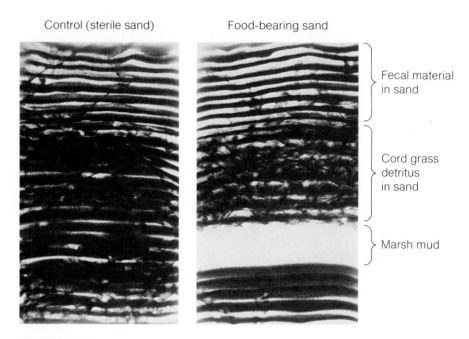

Control (sterile sand) Food-bearing sand

Fecal material in sand

Cord grass detritus in sand

Marsh mud

FIGURE 14-18

X-radiographs of two tanks containing layered sediment, sea water, and equal populations of amphipods. The animals burrowed the sterile sand systematically in their search for food, whereas they quickly destroyed the layering in the part of the right-hand tank that contained favored food. From "Burrowing patterns of haustoriid amphipods from Sapelo Island, Georgia" by J. D. Howard and C. A. Elders (in *Trace fossils*, by T. P. Grimes and J. C. Harper, editors. Liverpool: Seel House Press, 547 pp., 1970).

layers many feet thick, suggesting that swift streams dumped them as channel bars and levee deposits during floods. Their poorly sorted nature implies that they were then covered by additional deposits before the streams could sort and winnow them.

High energy and variability of the streams is also indicated by deep channel forms, cut down sharply into underlying rocks (Color Plate 10B). In modern river systems such as the Rio Grande, these relations are found only in the upper parts of the system, rarely more than a few miles from steep mountain slopes. The gravel fans of Death Valley are probably the closest analog among the modern deposits we have examined. The analogy implies that a mountainous area near the western end of the Book Cliffs area was being uplifted actively by faulting or rapid arching.

Are the older rocks now lying west of the Book Cliffs region appropriate suppliers of the grains that make up the Book Cliffs rocks? They certainly appear to be. Most of these older rocks are sandstones and shales that would readily yield the abundant sand, silt, and clay of the Book Cliffs sequence. As Color Plate 10B illustrates, many of the older sandstones are colored red by iron oxide pigments on the sand grains. Thin relics of these pigments can be seen on sand grains that make up the western part of the Castlegate Sandstone, giving its exposures a pink cast. The most resistant of the older rocks to the west—hard sandstones and the fine-grained, quartz-rich rock called chert—can be found in abundance among the pebbles and cobbles of the conglomerates.

Summary

In the next chapter we will examine additional evidence for uplift and other actions to the west. Those actions have particular significance because they were taking place next to the Book Cliffs accumulation, which lies scarcely disturbed over a large area of the continent. Let's briefly recall the main discoveries and relations in that undisturbed region:

1. Basic sources of information are the bare, widespread cliffs, which afford many places to make stratigraphic studies, such as those described in Chapter 13, as well as to trace and map formations laterally.
2. A lateral succession of environments is thus demonstrated at each stratigraphic level—from an alluvial plain in the west to marshes and lagoons, beaches, and a broad, shallow sea in the east.
3. Fossils show that the environments are in fact the same age; the fossils also help in identifying the environments.
4. When the different kinds of sedimentary deposits are assembled in a vertical as well as a horizontal sequence (Figure 14-3), they show zig-zag junctions expressing lateral shifts of the environments.
5. Deposits of fossil plants, now coals, show especially that these lateral shifts resulted from alternate filling and subsidence of the basin that held the sea.
6. The remarkably thick peats that became the more valuable coal layers require exceptional conditions for plant growth and accumulation, afforded by an extensive swampy terrain and in part by the growth of irregular deltas into the sea.
7. The long-continuing accumulation requires erosion of abundant rock materials somewhere to the west.

REFERENCES CITED

1. Frank R. Clark. Economic geology of the Castlegate, Wellington, and Sunnyside quadrangles, Carbon County, Utah. U.S. Geological Survey, Bulletin 793, 162 pp., 1928.
2. Edmund M. Spieker. Late Mesozoic and early Cenozoic history of central Utah. U.S. Geological Survey, Professional Paper 205-D, pp. 117–61, 1946. See also, Sedimentary facies and associated diastrophism in the Upper Cretaceous of central and eastern Utah. Pp. 55–82 in *Sedimentary facies in geologic history*. Geological Society of America, Memoir 39, 171 pp., 1949.
3. Robert G. Young. Sedimentary facies and intertonguing in the Upper Cretaceous of the Book Cliffs, Utah-Colorado. Geological Society of America Bulletin, vol. 66, pp. 177–202, 1955. See also, Stratigraphy of coal-bearing rocks of Book Cliffs, Utah-Colorado. Pp. 7–21 in *Central Utah coals*. Utah Geological and Mineralogical Survey, Bulletin 80, 164 pp., 1966.
4. James D. Howard. Sedimentation of the Panther Sandstone tongue. Pp. 23–33 in *Central Utah coals*. Utah Geological and Mineralogical Survey, Bulletin 80, 164 pp., 1966.

**ADDITIONAL
IDEAS
AND
SOURCES**

1. Coal-forming situations are well described and illustrated in several articles in *Environments of coal deposition,* edited by Edward C. Dapples and M. E. Hopkins (Geological Society of America Special Paper 114, 204 pp., 1969). Two articles describe the formation of peat in the mangrove estuaries of southern Florida, and one relates its origin in various parts of the Mississippi Delta. Interpretations of ancient environments from coals and associated rocks in Pennsylvania, especially those related to deltas, are presented by Harold R. Wanless, James R. Baroffio, and Peter C. Trescott.

2. The principal rock formations of the Book Cliffs region and the plateaus to the west are shown clearly in the *Geologic map of the northeastern quarter of Utah* (1961) and the *Geologic map of the southeastern quarter of Utah* (1963), edited by William L. Stokes and available from the Utah Geological and Mineralogical Survey (Mines Building, University of Utah, Salt Lake City, Utah 84112).

3. The boundary between Cretaceous and Cenozoic rocks lies within the upper part of the Book Cliffs sequence, within rocks called the North Horn Formation. The sediments below the boundary are identical to those above. Peats continued to accumulate episodically well into the Cenozoic succession, indicating that there were no abrupt environmental changes at the end of the Cretaceous Period (the end of the Mesozoic Era). It is therefore striking that many major animal groups became extinct or changed greatly at that time, notably the dinosaurs and ammonites, both of which were numerous in what is now the Book Cliffs region. These faunal changes and some of their possible causes are reviewed on pages 293–303 in *Principles of paleontology* by David M. Raup and Steven M. Stanley (San Francisco. W. H. Freeman and Company, 388 pp., 1971). An interesting theory relating the extinctions to the transport of water and nutrients to the ocean has been given by M. N. Bramlette in "Massive extinctions in biota at the end of Mesozoic time" (*Science,* vol. 148, pp. 1696–99, 1965).

Layers of sandstone and shale that were folded and contorted when they flowed from left to right during metamorphism. The locality is in the Grouse Creek Mountains, Utah.

15. A Region of Deformed Rocks

A Region of Deformed Rocks

We noted in the last chapter that the vast accumulation of sediments exposed in the Book Cliffs implied an enormous amount of erosion to the west, in a region that was evidently mountainous. This relation raises an intriguing problem. Erosion in most mountainous areas is so rapid that mountains should be reduced to hills in about 5 to 10 million years. Yet the rocks of the Book Cliffs indicate 100 million years of rapid erosion and deposition! What was going on to the west?

Figure 15-1 gives a major clue. The place is Salina Canyon, near the southwestern edge of the Wasatch Plateau and also near the western edge of the Book Cliffs sequence (Figure 14-2). The steeply inclined layers are about 150 million years old (Jurassic) and thus much older than the Book Cliffs rocks. The nearly horizontal layers above them are about 65 million years old (early Cenozoic) and therefore only a few million years younger than the youngest Book Cliffs rocks. The surface between the two sets of layers is an unconformity, the kind of feature we found in Chapter 13 to mark an interruption in the deposition of strata. Unconformities thus express basic changes in earth systems, and this particular one records (1) deposition of the layered Jurassic sediments, (2) tilting of the layers, (3) erosion of part of the layers, and (4) deposition of the Cenozoic sediments on the eroded (unconformable) surface. The final geologic event, the erosion of Salina Canyon, was crucial in exposing the entire relation, but it is otherwise unrelated to the unconformity and its record.

Might the uplifted and eroded parts of the Jurassic rocks have supplied sediment to the Book Cliffs sequence? This one exposure cannot tell us firmly, because the difference in age between the Jurassic and the Book Cliffs rocks is so great (about 60 million years) that the partial erosion of the Jurassic

FIGURE 15-1

Tilted and partly eroded sedimentary layers of Jurassic age, overlain by nearly horizontal Cenozoic strata. The locality is on Interstate 70, 3 miles east of Salina, Utah (located on Figure 14-2).

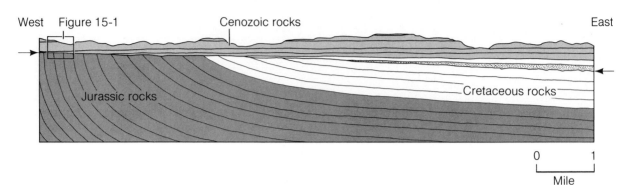

West　Figure 15-1　　　　　　　Cenozoic rocks　　　　　　　　　　　East

Jurassic rocks　　　　　　　　　　　　　　　　　Cretaceous rocks

0　　　　1
Mile

FIGURE 15-2

Vertical cross section along Salina Canyon, showing the unconformity of Figure 15-1 and the variously inclined layers above and below it. The three age groups of rocks are indicated by shading, and the Castlegate Sandstone is dotted.

rocks could have taken place long before the Book Cliffs sequence was deposited. We need to study the unconformity further by following it eastward toward the Book Cliffs sequence. It can in fact, be traced for 10 miles east of the locality shown in the photograph, and the relations shown schematically in Figure 15-2 can be observed. Note how the layers under the unconformity become less and less steeply inclined and younger and younger (stratigraphically higher and higher) to the east. Note also that the rocks just above the unconformity are thin wedges that overlap, so that the rocks lying over the unconformity become younger and younger westward.

An especially important relation is that the Castlegate Sandstone lies on the unconformity in the eastern part of Salina Canyon. Note that the sandstone is parallel to the layers beneath it, so that the unconformity would be difficult to recognize here had we not traced it laterally from the west. Fossils in the rocks show that the Castlegate Sandstone is only a few million years younger than the eroded layer on which it lies. We thus know when erosion took place in this part of the area.

Development of an unconformity　To understand how the western part of the unconformity relates to the eastern part, we must reconstruct the array of layers at various stages during the development of the unconformity. Let's start at the stage at which the Jurassic layers were less bent and the Castlegate Sandstone was first being deposited on slightly eroded Cretaceous strata of the Book Cliffs sequence (Figure 15-3A). We can then proceed in time, reconstructing two intermediate stages in the progressive bending and eroding of the Jurassic and Cretaceous rocks (Figure 15-3B and C). These stages also show the westward enlargement of the wedge of Castlegate sand and the younger sediment wedges. Finally we can reconstruct a stage early in the Cenozoic Era when wedges of sediment had lapped completely across the eroded rocks and the source of sediment had shifted farther west (Figure 15-3D). The last stage is essentially the arrangement now exposed in Salina Canyon.

We have thus determined that erosion of the Jurassic rocks did in fact supply sediment to the deposits of the Book Cliffs, and that this erosion was caused by a progressive upward-bending of the rock layers. The resulting bend

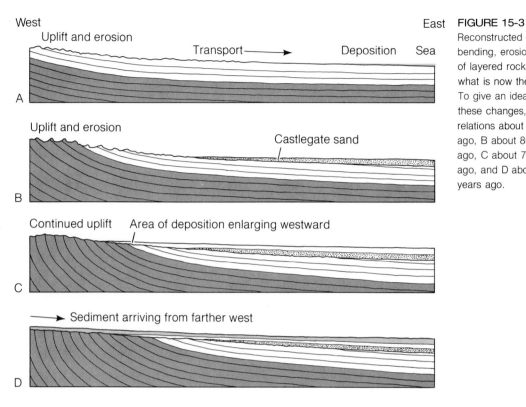

West   East

A — Uplift and erosion / Transport → / Deposition / Sea

B — Uplift and erosion / Castlegate sand

C — Continued uplift / Area of deposition enlarging westward

D — → Sediment arriving from farther west

FIGURE 15-3
Reconstructed stages in the bending, erosion, and deposition of layered rocks to the west of what is now the Book Cliffs area. To give an idea of the rate of these changes, A represents relations about 85 million years ago, B about 80 million years ago, C about 75 million years ago, and D about 65 million years ago.

is part of an important feature called a *fold* (Figure 15-4). Note that folds resemble giant wrinkles in a rug, consisting alternately of arches and troughs that express considerable deformation of the originally horizontal layers. Ranges farther west expose additional folds, and the rocks in many places are broken and displaced along major faults. Indeed, the ranges to the west constitute a region of greatly deformed rocks that extends for hundreds of miles beyond the edge of the nearly flat-lying rocks exposed in the Book Cliffs (Figure 15-5).

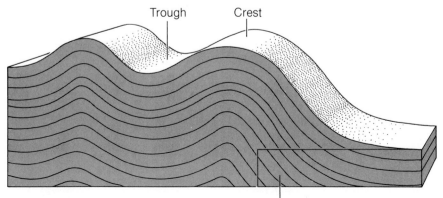

Trough   Crest

Part in Figure 15-2

FIGURE 15-4
Reconstruction of large folds near Salina, greatly idealized because they are now displaced by faults and deeply eroded.

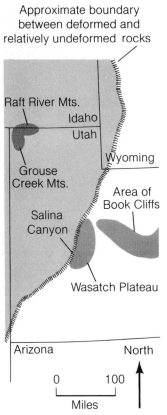

Approximate boundary between deformed and relatively undeformed rocks

Raft River Mts.

Idaho

Utah

Wyoming

Grouse Creek Mts.

Area of Book Cliffs

Salina Canyon

Wasatch Plateau

Arizona

North

0   100

Miles

FIGURE 15-5

Map showing the distribution of strongly deformed rocks (shaded) and weakly deformed rocks (unshaded) in and near Utah.

In this chapter we will examine the folds and faults in order to understand how and why they formed. Our study should answer the question of what was going on to the west. It should also have broad significance, for deformed rocks characterize many other regions of the earth, especially those forming elongate mountain belts. An example is the deformed, earthquake-producing region of the Alaska Peninsula and associated islands described in Chapter 11. Much of the record there, too, is preserved only in folds and faults that resulted from deformation such as that causing earthquakes today. What causes these extensive deformed regions, many lying directly against regions that have long been calm and quite?

Folded Rocks in Northwestern Utah

We cannot hope to study the entire deformed region shown in Figure 15-5. It is essential, in fact, that we stay within a more or less continuous set of exposures in order to be sure of the relations between specific folds and faults. Thus, we must stay largely in one mountainous area, for the deformed region is characterized by small mountain ranges separated by basins in which the deformed rocks are hidden—the general relation of basins and ranges described in Chapter 9. I will concentrate on a basically continuous exposure in and near the Raft River Mountains and Grouse Creek Mountains of northwestern Utah, an area of 1,000 square miles where folds and faults are exposed unusually clearly (Figure 15-5). I have studied much of this area over a period of 10 years, including five summers with geology students from Stanford University (Reference 1). One student, Victoria Todd, went on to map and study most of the southern half of the area outlined in the figure (Reference 2). Additional information has come from others, especially Robert E. Zartman and Charles W. Naeser, who determined a number of important rock ages by radiometric methods (Reference 3).

The rocks of Raft River Mountains were originally thick layers of marine sandstone, shale, and limestone, very similar to the sedimentary layers of the Book Cliffs but much older—ranging from Precambrian to Triassic in age. They were deposited over an extensive body of Precambrian granite that was emplaced about 2,500 million years ago, uplifted and eroded deeply, and finally submerged beneath a shallow sea in which the sediments began to accumulate. The granite and the lower part of the sedimentary sequence were metamorphosed much later, during the deformation I will describe. Recall from Chapter 10 that metamorphism results from heating and slow recrystallization of rocks, without melting. This process changed the sandstone to an exceedingly solid rock called quartzite, the shale to schist, and the limestone to marble (Supplementary Chapter 2). The granite, too, was deformed and metamorphosed to the somewhat flaky rock called gneiss (Figure 10-16).

The layered rocks of Raft River Mountains have been arched upward recently and eroded so deeply that folds formed long ago, during metamorphism, are now exposed (Figure 15-6). The fold may be difficult to see because it is pressed over so strongly on its side (toward the left); however, you can use the diagram to follow the white quartzite and the overlying dark schist from the crest of the range down its flank and into the fold. Such folds are said to

FIGURE 15-6
Large fold exposed on one flank of the arch forming Raft River Mountains. Note the location of the photograph in the vertical cross section, which shows the principal rock layers (formations) and the broad arch of the range.

be *overturned*, in contrast to *upright* folds such as those in Figure 15-4. Only a few folds in the area studied are as large as the one in Figure 15-6, but there are many smaller ones and they, too, are overturned (Figure 15-7).

Interpreting the folds   What can we do to determine how these folds formed? One fruitful approach is to fold layered materials in the laboratory and see how their forms compare with those found in nature. In the example illustrated in Figure 15-8, layers of clay compressed in the apparatus developed folds progressively as shown in the photographs. Such folds are called *buckle folds*, because of the compression that caused them. Note especially that the thickest white layer formed a large fold, whereas the thin layers above it formed much smaller ones. Note also that this difference showed up at the earliest stage of compression (Figure 15-8b), where it is expressed most clearly as the difference in the *wavelength* of the folds—the distance from one trough to the next. This result is the same as that from many other experiments as well as from mathematical analyses: when layered materials are buckled approximately parallel to their layers, the folds have wavelengths proportional to the thicknesses of the layers.

Now examine the folds in Figure 15-7 closely. The folds at 1 are all but invisible, being about 0.5 inch high and involving a layer of quartzite 0.05 inch thick. Perhaps you can just see the folds at 2, where the quartzite layers are

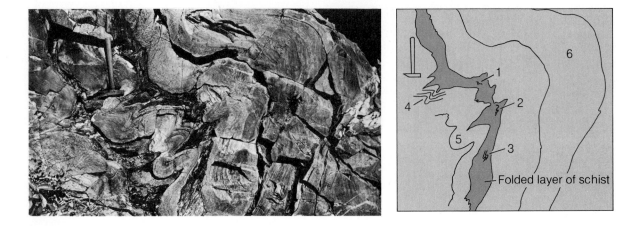

FIGURE 15-7

Folded layers of quartzite (light) and schist (dark), as seen in a vertical cliff. The diagram shows the locations and configurations of six different sizes of folds, each size being typical of layers of that thickness.

0.3 inch thick, and those at 3, where they are 0.5 inch thick. Folds and layer thicknesses are clearly visible at 4 and 5, and fold number 6, involving the thickest layer, is barely contained in the boundaries of the photograph. The fold in Figure 15-7, involving an entire quartzite formation, is 100 times larger. Similar relations can be observed at countless other places in the area studied. We can thus conclude that the folds are buckle folds—they were generated by compression approximately parallel to the original layers.

Experimental buckle folds, however, remain essentially upright, as in Figure 15-8, whereas the folds in the area studied are overturned, many of them extremely so. Moreover, the folds in large parts of the area are overturned in the same direction. Their overturned forms thus imply a systematic flow of material. In the chapter opening photograph, for example, the rocks in the upper part of the view were evidently displaced to the right relative to those beneath. Flow is also implied by the change in thickness of layers that once were the same everywhere. If you examine the photographs carefully you will see that many layers are thickened at crests and troughs of folds and thinned along the more or less straight parts between. Note that the layer of schist labeled in Figure 15-7 appears to have been especially ductile (capable of flowing), for it pinches and swells markedly between the large and small folds in the adjoining quartzite layers.

Folding ductile materials We have already determined, however, that the folds were generated by buckling. Can ductile materials flow laterally and at the same time buckle? Victoria Todd explored this question by a series of experiments in which rock layers were modeled by layers of stitching wax, a solid substance that flows very slowly when warmed slightly. Trays were loaded with layers of wax that were separated in some cases by sheets of tissue paper. When part of a tray was tilted, as shown in Figure 15-9A, the wax flowed slowly down the incline, in some cases forming a bulge in the lower part of the tray and in others forming folds, including strongly overturned ones (Figure 15-9B). Folds were produced when wax in the lower part of the tray did not move readily from in front of the layers flowing down the incline. The moving wax thus pressed laterally against the less mobile wax, causing compression and in turn buckling. The fold forms became strongly overturned in

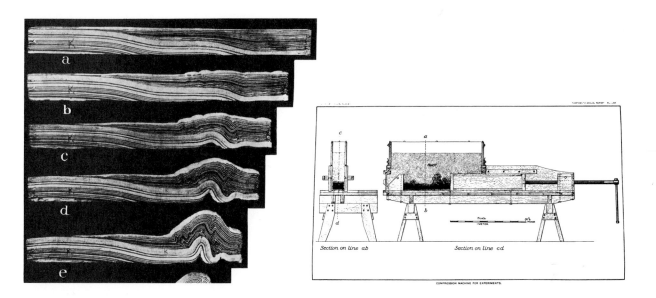

FIGURE 15-8

Apparatus (right) and folds produced by compressing layers of clay laterally, under a load of shot. A side view of the layers before they were deformed is labeled a, and the other photographs show what happened as the right end of the sequence was forced toward the left. The experiments were conducted by Bailey Willis and described in his paper "The mechanics of Appalachian structure" (U.S. Geological Survey, Thirteenth Annual Report, Part II—Geology, pp. 211–82, 1893).

the cases in which the upper parts of the folded wax continued to flow over the lower parts, a movement similar to that suggested by the folds in Figure 15-6 and on the opening page. Two other relations in the models are closely analogous to the natural features studied: (1) separate folds are often over-turned to different degrees (compare the smaller folds in Figure 15-9B with those in Figure 15-7), and (2) extensive parts of the layers are not folded at all (as in the left half of the diagram in Figure 15-6).

How the rocks flowed   The analogy with wax suggests that the rocks studied were molten when they flowed, but melting can be disproven by identifying the mineral grains in the rocks and comparing them with the minerals and textures described in Chapter 10 and Supplementary Chapter 2. The principal minerals are quartz, calcite, and mica, which indicate meta-morphism at temperatures far below those necessary to start melting in the rocks. Moreover, relics of the original sedimentary grains are abundant in the least metamorphosed rocks, and nowhere do the layered rocks cut across others in the manner of molten intrusions.

Relics of the original sand grains, in fact, provide a means of seeing how the rocks flowed as solid materials. Grains in the upper part of the sedimentary sequence, where the rocks were not metamorphosed, can be compared with grains affected by a moderate degree of metamorphism (Figure 15-10). The latter were clearly flattened and elongated by flow, and the fact that many have retained their identity as single crystals (the elongate patches of white and dark gray) must mean that the atom layers making up the crystals shifted systematically as in the deformed ice crystals described in Chapter 7 (Fig-ure 7-2D).

Rocks that were metamorphosed at higher temperatures flowed somewhat differently from those in Figure 15-10. Quartz and calcite recrystallized so easily at high temperatures that the grains could generate interlocking, nearly equant crystal shapes rather than flattened ones (Figure 15-11A). The mica

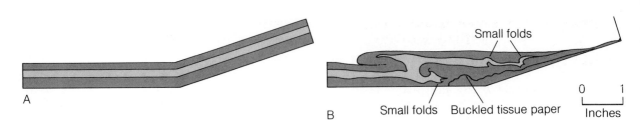

FIGURE 15-9
A, vertical section through tray and wax layers before warming and flow. B, results of heating and flow in a case in which the wax overrode a buckled sheet of tissue paper, forming a large fold much like that in Figure 15-6 as well as smaller, less strongly overturned folds.

FIGURE 15-10
Top, thin section of unmetamorphosed sandstone showing what the grains in the quartzite (bottom) must have looked like before flow. The photos were taken with polarized light, so that each grain (crystal) shows separately. The quartzite is from the fold in Figure 15-6, the flattened grains being parallel to the flattened form of the fold.

flakes in the rocks, however, have become aligned in approximately parallel orientations. As Figure 15-11B suggests, this configuration is a record of the flattening and flow of the rock body, which caused the mica flakes to rotate into a new, common orientation.

Why the rocks flowed Looking at the rocks of the area broadly, we see that both the mica flakes and the flattened quartz grains are oriented in overturned folds as shown in Figure 15-11C. An especially important point is that *all* the metamorphosed rocks have these planar orientations, whether one observes them in folds or in nearly horizontal layers at great distances from folds. Even the gneiss that formed from the ancient granite lying under the sequence has mica oriented in a more or less horizontal plane. It thus appears that all the rocks in the lower, metamorphosed part of the sequence flowed laterally, whether they were folded in doing so or not.

The more or less horizontal orientation of the platy grains and the flattening indicated in Figure 15-11B suggest that the flow was caused by the force of gravity. That is, the rock mass spread (flowed) under its own weight, much like an ice sheet or one of the trays of heated wax. A relation strongly supporting this interpretation is that the folds in one part of the area are overturned in one direction and those in the other part in the opposite direction. As Figure 15-12 suggests, the difference might be explained by the presence of a broad uplift that caused flow in two directions. Note the thinning of rock layers shown at the crest of the uplift.

Our examination and experiments have thus pointed to several tentative but important conclusions concerning the rocks and folds: (1) they were caused by flow in the solid state during heating and metamorphism; (2) flow may have been initiated by a broad uplift; (3) the driving force, gravity, evidently caused spreading and therefore typically thinning of the layers; and (4) folds were generated where the flowing materials became arrested locally and buckled.

Movement of Large Sheets of Rock

In addition to being folded, the rocks in the area studied were displaced along gently inclined surfaces called *low-angle faults*. The faults separate the layered rocks into large sheets, some of them thousands of feet thick and extending

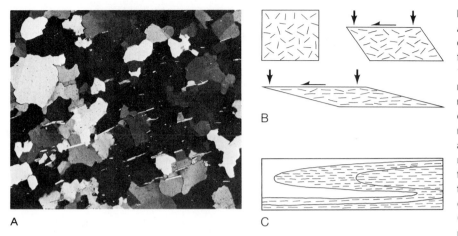

A

B

C

FIGURE 15-11

A, thin section of mica-bearing quartzite from the same formation as that of Figure 15-10, but from a more metamorphosed part of it. The more or less equant grains are quartz and the linear ones are mica flakes that lie at about right angles to the page. The near-parallel orientations of the flakes are probably due to thinning and lateral displacement, as in B. Diagram C shows the orientation of micas in overturned folds.

over areas of hundreds of square miles. The sheets, moreover, have moved along the fault surfaces for miles, even for tens of miles in some cases. As incredible as such features may seem, they are more typical than folds in the deformed region shown in Figure 15-5 and are found in many other deformed regions of the world, often associated with overturned folds such as those already described.

The low-angle faults are not obvious in most places because they typically lie parallel to rock layering or cut across it at small angles. They must be located by tracing and mapping the various rock formations and thereby finding places where the formations have been displaced from their normal stratigraphic position. An example worked out by Victoria Todd is illustrated in Figure 15-13. The dark rocks making up the outcrops on the ridge are formations dated by their fossils as Mississippian and Pennsylvanian. The light rocks in the foregound and along the lower right part of the ridge are less resistant sandstone, conglomerate, and volcanic rocks that are about 11 million years old (Cenozoic). Although the Cenozoic rocks are in their original stratigraphic order, they are overlain by older (Mississippian and Pennsylvanian) rocks—an impossible stratigraphic relation. The surface between the two sets of rocks must therefore be a low-angle fault, a conclusion proven by the fact that the surface locally cuts across layers of Cenozoic rocks (Figure 15-13A), as well as across layers of older rocks (Figure 15-13B). Once the rock formations were recognized here the fault could be traced easily for many miles because the dark, older rocks consistently lie over the light,

Layers greatly thinned

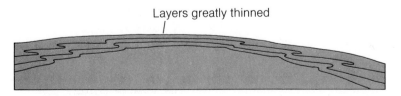

FIGURE 15-12

Overturned folds caused by flow down two sides of an uplift. The thinning of the formations near the crest is substantiated by mica orientations like that shown in Figure 15-11.

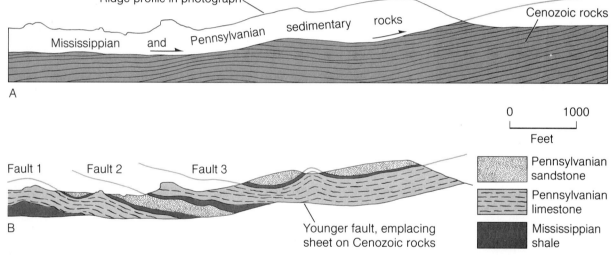

Ridge profile in photograph

Cenozoic rocks

Mississippian   and   Pennsylvanian   sedimentary   rocks

A

0   1000

Feet

Fault 1   Fault 2   Fault 3

B

Younger fault, emplacing
sheet on Cenozoic rocks

Pennsylvanian
sandstone

Pennsylvanian
limestone

Mississippian
shale

FIGURE 15-13

Ridge exposing dark layers of
metamorphosed limestone,
sandstone, and shale of
Mississippian and Pennsylvanian
age, lying above rocks of
Cenozoic age. A is a vertical
cross section showing the rocks
in the ridge as well as under it,
and also the low-angle fault at
the base of the older rocks. B
shows part of the same section
but with the Mississippian and
Pennsylvanian rocks divided
into formations.

younger ones. The ridge in the photograph turned out to be a remnant of
an extensive sheet of older rocks displaced over the Cenozoic rocks and
largely eroded away since.

A detailed study of the older rocks on the ridge revealed additional low-
angle faults as well as folds. The arrangements of the individual Mississippian
and Pennsylvanian formations are shown in Figure 15-13B, and their original
stratigraphic order is indicated to the right of the cross section. Note that
each of the three thin bodies of Mississippian shale lies on younger rocks,
so that the surface under each shale must be a low-angle fault. Three separate
rock sheets apparently rode up and over one another in the order indicated
in the figure. The three faults were later cut off when all the sheets began
to move on the youngest fault, shown separately in Figure 15-13A. Figure
15-13B also shows that the older sheets and faults were folded before moving
into their present position, for the youngest fault and the Cenozoic rocks

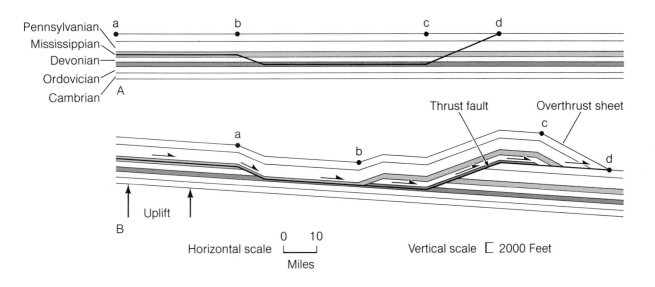

Pennsylvanian
Mississippian
Devonian
Ordovician
Cambrian

A

Thrust fault   Overthrust sheet

Uplift

B

Horizontal scale 0 10 Miles

Vertical scale ⎣ 2000 Feet

FIGURE 15-14

A, vertical section showing broad view of formations before faulting, with a heavy line indicating the position of the low angle fault. B, relations after movement on the low-angle fault, showing how some parts of the formations came to lie on formations older than normal and other parts on formations younger than normal. The letters indicate specific displacements, with the area studied lying mainly between points a and b in diagram B. As the scale bars indicate, the vertical dimension is greatly exaggerated.

under it are not folded parallel with the older rocks. The mapped area thus shows that low-angle faults formed at different times and appear to be linked broadly with the folding.

Similar faults and sheets of rock once extended over the entire area studied and far beyond. The origin of the thickest and most extensive of the sheets is suggested diagrammatically in Figure 15-14 (with folds and subsidiary faults removed so that the main relation can be seen easily). Note that large segments of Mississippian and Pennsylvanian rocks were displaced onto rocks older than normal (Ordovician, rather than Devonian). The part of the system labeled an *overthrust* is the part that emplaced older rocks on younger ones, as in the cases shown in Figure 15-13. Large overthrusts also occur over much of the eastern part of the deformed region of Figure 15-5, suggesting that rock sheets moved for many tens of miles eastward.

Effects of water on low-angle faulting Although the sheets in the area studied did not necessarily connect to the overthrusts in the eastern part of the deformed region, even the 30-mile exposure studied poses a serious problem. How was frictional resistance along the base of the sheets reduced to the point that they could move at all? To appreciate this problem, try pushing 20 bricks together across a rough concrete surface and then imagine those bricks enlarged to a sheet of rock tens of miles across! M. King Hubbert and W. W. Rubey calculated that the force needed to push a horizontal sheet of rock 3,000 feet thick and more than 5 miles across is so great as to crush the sheet where the force was applied, and thus could not move the sheet as a whole (Reference 4). Challenged by this paradox, they considered ways of reducing the frictional resistance and deduced that water in porous rocks beneath a fault might reduce resistance if it could not escape readily into the overlying sheet. They experimented with this idea by placing a 550-pound slab of nonporous concrete onto a horizontal porous contrete base into which water could be injected at various pressures (Figure 15-15). They found that

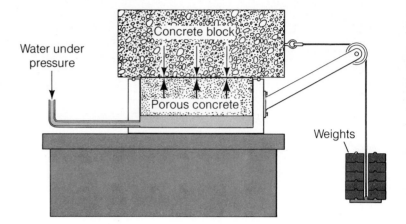

FIGURE 15-15

Diagram of apparatus used by Hubbert and Rubey to measure changes in frictional resistance as water pressure in the porous concrete was increased. The force necessary to move the concrete blocks was measured by adding weights to the carrier in the lower right, and the water pressure was varied by a pumping system outside the intake labeled on the left. The heavy arrows indicate the lifting force of the water counteracting the weight (and therefore the frictional attachment) of the block. From Reference 4.

the force needed to move the slab decreased as the water pressure was increased; at the highest pressure attained (85 percent of the weight of the slab), the slab could be moved easily with one hand. Evidently, the upward press of the water against the impermeable slab greatly reduced the forces that normally held the slab against the base. If the water pressure could have been raised until it was equal to the weight of the slab, the slab would presumably have been cast free so that it could ride buoyantly on the base. Moreover, if the apparatus could then have been tilted even slightly the slab would have glided across it under the force of gravity alone.

These discoveries are exceedingly useful in interpreting the movements of rock sheets in Raft River Mountains. The major low-angle faults commonly lie in rock layers that were originally shales or clay-bearing limestones containing abundant water and hydrocarbons. As described in Chapter 10, fluids are expelled in large quantities when such rocks are buried and metamorphosed. Possibly, the fluids could not escape upward because the overlying formations were already tightly knit quartzite and marble. The fluids thus came to bear the full load of the overlying sheets, permitting them to glide easily even though they were very large.

Several relations show that water was trapped at high pressure under the sheets. Where the sheets moved over the loosely cemented Cenozoic rocks, dikes of sand, mud, and gravel can be seen extending from the fault surface up into the sheets (Figure 15-16, right). Cracks apparently opened in the sheets as they moved, and fluids injected upward from the Cenozoic rocks carried sediments into them. It is especially interesting that the rocks near the dikes disintegrated into masses of angular fragments, many of them becoming only slightly disoriented by the process (Figure 15-16, left). This suggests a process called *fluid-fracturing* that has become well known in the petroleum industry. Fluids are pumped down wells at high pressure and injected into oil-bearing rocks. The pressurized fluids expand and crack the rocks, making it possible to pump petroleum out of them. In the area studied, the fluids were evidently injected upward from the porous rocks beneath the low-angle faults.

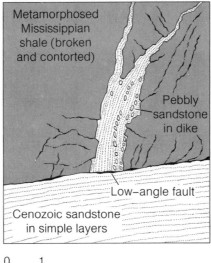

Metamorphosed
Mississippian
shale (broken
and contorted)

Pebbly
sandstone
in dike

Low-angle fault

Cenozoic sandstone
in simple layers

0   1
Feet

FIGURE 15-16
Right, dike of sediments exposed in a vertical cliff, indicating that water was injected from the rocks below the fault. Left, disintegrated rock associated with sediment dikes. Note that the wrinkle lines (small folds) on the surfaces of the larger blocks are still about parallel (for example, the two blocks in the lower left part of the view). The sheet was thus fractured without much rotation of the fragments. The outcrop is near the base of the main rock sheet in Figure 15-13.

The High-angle Faults

Also characteristic of the deformed region of Figure 15-5 are *high-angle* (steeply inclined) *faults* that break across the rock layers. One group of high-angle faults in the area studied formed at about the same time as the folds and low-angle faults, and a second group formed distinctly later. The older group of faults is aligned in the same direction as that in which the horizontal rock sheets moved and in which the folded layers flowed. As Figure 15-17 shows, the faults separate the rock sheets into broad tabular segments, some of which moved farther than others. Because the displacements along these faults are lateral rather than vertical, they are called *lateral faults*.

An important indication that these displacements took place during folding is the fact that the folds in adjoining segments may be overturned to different degrees, much as in two rugs pushed separately across the floor in the same direction. The cause of the segmentation is not known; possibly the frictional resistance under the sheets varied laterally, and the high-angle faults formed because some segments could thus advance more rapidly than others. Another possibility is that some parts of the sheets slid more rapidly because they sloped somewhat more steeply.

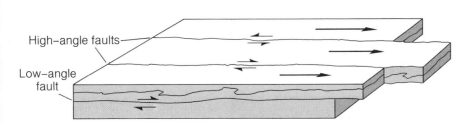

High-angle faults

Low-angle fault

FIGURE 15-17
Segmentation of one of the rock sheets by high-angle faults along which some of the resulting segments have moved farther than others. The half arrows indicate relative movement on the two sides of a fault.

337

FIGURE 15-18
High-angle fault along which
sedimentary layers on·the left
were down-dropped 5 feet. Near
Soldier Summit, Utah.

The second, younger, group of high-angle faults is of great importance in interpreting movements throughout much of the western United States, for they are part of the array of faults that bound the basin-and-range blocks of the desert region described in Chapter 9. These faults cross and displace all the features described so far, and the movement on them is mainly vertical rather than horizontal (Figure 15-18). The displacements range from thousands of feet on the few major faults to a fraction of an inch on the numerous minor ones (as the near-vertical line cutting through Figure 15-10, bottom). The faults can be understood quite readily through the use of models. For example, the layered body of silt and fine sand in Figure 15-19A was broken into fault-bounded segments by stretching the rubber sheet on which it lies. The stretching reduced the laterally directed forces within the body, so that the vertical forces due to gravity became dominant and displaced segments vertically downward (Figure 15-19B). Faults of this kind are called *high-angle gravity faults.* They can be identified (tentatively at least) by their slopes, which are 50 to 75 degrees, and by the simple downslope displacements of features they cut.

Figure 15-20A compares the typical slope and sense of offset on high-angle gravity faults with those of the other kinds of faults we have noted. Lest their names imply that only one kind of fault can be generated by gravity, however, Figure 15-20B shows how all kinds may occur in a single landslide, one not greatly different from the slides in Anchorage formed during the Alaska earthquake of 1964 (Figure 11-2). Indeed, the landslide seems tempting as an analog of all the features described so far, for it may involve flow in clay-rich sediments, and layers in the displaced sediments may buckle into overturned folds!

Interpreting the Causes of Deformation

To help get some ideas as to *why* all these faults and folds should have formed, let's briefly review the observed relations, noting the probable significance of each:

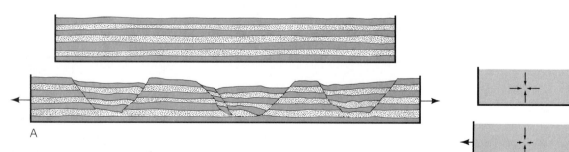

FIGURE 15-19
A, cross sections through rubber-based tray of sand and silt layers, before and after extension of the base to the right and left. B, the arrows in the tray indicate on the top the forces per unit area felt by the sediment due to its weight. When the base is extended (center), the horizontally directed forces are reduced, so that the vertical force (gravity) breaks the sediment along steeply inclined faults and some segments subside (bottom).

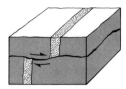

Low angle

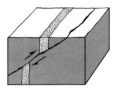

Thrust

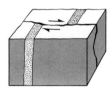

High–angle
lateral

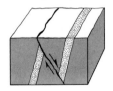

High–angle
gravity

A

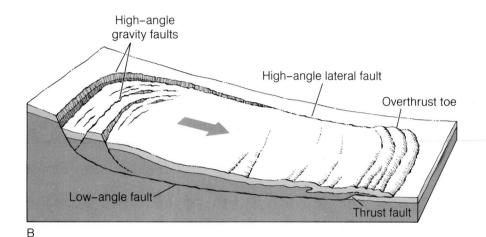

B

FIGURE 15-20

A, diagrams illustrating the typical slopes and movement relations of the four kinds of faults in the area studied. B, the same kinds of faults as they may occur in a single landslide, the front half of which has been cut away. Note that all fault surfaces connect with one another and that all were caused directly or indirectly by gravity.

1. The overturned folds and oriented mineral grains indicate that the rocks flowed laterally.
2. The opposed directions of flow suggest that the materials streamed by gravity down the two sides of a broad arch.
3. The metamorphic rocks are the ones that flowed; therefore, heating evidently made flow possible.
4. The rocks also slid, forming huge segmented sheets that moved in the same direction and during the same period of time as those of flow.
5. The sheets evidently slid because metamorphism produced water beneath them that could not escape through them.
6. High-angle lateral faults developed during the sliding are parallel to directions of sliding, whereas high-angle gravity faults indicate extension of the region after metamorphism and sliding had ceased.

Judging from these relations, one basic cause of the deformation was heating. In Chapters 10 and 12 we learned that heating may be due either to deep burial or to the intrusion of molten plutons. We also learned that metamorphic minerals appear in a progressive series that records increasing temperatures in rocks. This progression takes place through great thicknesses—many miles—of rock where heating is due to deep burial, whereas heating by intrusive plutons produces the same progression in thicknesses of less than a mile. Because the progression of metamorphic changes in the area studied generally spans less than a mile of rocks, it appears that the emplacement of molten bodies beneath the area was the chief cause of heating. This suggestion is supported by the presence of several granite plutons in the most metamorphosed part of the terrain. Morever, sheets of folded and metamorphosed rocks over the intrusions show that the granite was emplaced during the folding and sliding.

Dating the deformation   Radiometric dating of plutons and volcanic rocks has helped greatly in interpreting the heating and deformation. The beginnings of heating have not been dated in the area studied, but Jurassic and Cretaceous plutons in other parts of the deformed region suggest that

339

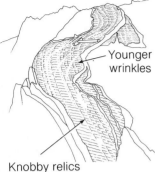

Younger
wrinkles

Knobby relics
of older wrinkles

FIGURE 15-21
Part of a large fold in thinly
layered quartzite
(metamorphosed sandstone),
with small folds (the wrinkles
parallel to the hammer handle)
superimposed on an older set—
the knobby bands that slant
obliquely across the layer
surfaces.

heating started at about the time the Book Cliffs sequence began to accumulate. The oldest pluton and volcanic rocks dated in the area studied are approximately 38 million years old (early Cenozoic). The pluton cuts across some of the folds, indicating that folding was underway by that time. A granite pluton that has been folded along with the surrounding rocks has been dated as 25 million years old by the rubidium-strontium method (Appendix B). Fragmental volcanic rocks were erupted in abundance between about 14 and 11 million years ago, and they are locally overridden by displaced rock sheets. There is thus considerable evidence of prolonged intrusion and eruption of melts during the period in which the rocks were being metamorphosed and deformed.

The relations enumerated at the beginning of this section indicate that uplift of a broad arch was another basic cause of deformation. Evidence for broad uplift during the Cretaceous and the early part of the Cenozoic Era is found in the fact that sedimentary deposits of those ages are missing in almost all of the deformed region. In the area studied, sediments and volcanic rocks accumulated locally about 38 million years ago, but these deposits were then deformed and largely uplifted. Deposition did not recommence until about 14 million years ago, and it took place in far broader basins. The shape of the uplift thus changed during the period of deformation, as substantiated by changes in the directions in which the rocks flowed. Figure 15-21 shows a typical example of the evidence. The wrinkles parallel to the hammer handle are parallel to the large fold that is only partly visible in the photograph. Note the somewhat broader but more subdued wrinkles that curve across the surface under the hammer at oblique angles to the wrinkles just mentioned. These wrinkles are older folds that have been bent around the form of the large fold. Folding (and flow) must thus have taken place first in one direction and later in another. The relation suggests that slopes on the uplifted arch changed during the period of uplift.

The broad basins that formed between about 14 and 11 millions years ago became filled with thick sequences of fragmental volcanic rocks and by sediments transported from the up-arched areas. The sediments are commonly coarse grained, and many are virtually unsorted deposits of debris flows that must have arisen on steep slopes (Figure 15-22). The repeated occurrence of such layers indicates that the uparched areas continued to rise adjacent to the basins. We have already noted that rock sheets were displaced for tens of miles over these same deposits (Figure 15-13). Radiometric dating of mineral grains shows that the rocks cooled shortly after that time, about 10 million years ago. Metamorphism, flow, and lateral movements of large rock sheets thus ended simultaneously. The only deformation that followed was the uplift of arches that became the present ranges, such as the one in Figure 15-6, and the collapse of basins on the high-angle faults already described.

Interrelations of heating and deformation Emplacement of melts, heating, uplift and flow were thus all interrelated, as suggested diagrammatically in Figure 15-23. Diagram A depicts an early stage in the flow, and suggests that it was caused by emplacement of extensive plutons that heated and also up-arched the area. Diagram B shows the sliding of a large rock sheet and the eruption of a volcano from a high-lying pluton. Note that a small basin has formed on the left and that the form of the uplift has changed. Its form has changed more obviously in C, which depicts the broad basins in which abundant sediments and volcanic materials accumulated. Note the younger pluton that has been overrun by a rock sheet. Diagram D shows the sliding and overthrusting of a rock sheet onto the sediment-filled basin and the extensive solidification of the underlying plutons. In E, solidification and cooling are essentially complete and the arches and down-faulted basins of the present landscape have formed.

By examining and dating deformed rocks in some detail we have thus arrived at an interpretation of what was going on to the west of the Book Cliffs region. We discovered one direct connection—that the Castlegate Sandstone resulted in part from the erosion of a major fold near the boundary of the deformed region. In the area including Raft River Mountains and Grouse Creek Mountains, we determined that uplift and erosion continued for a long period after the Cretaceous. Sediments did indeed accumulate as a counterpart in the Book Cliffs region and elsewhere, forming parts of the Cenozoic continental deposits described in Chapter 13. In fact, the uplifts we have been discussing were one cause of the changing climatic patterns of the Cenozoic, which had such profound effects on plant and animal communities. The folded and faulted strata of northwestern Utah thus had a part in the evolution of life as well as the land.

FIGURE 15-22
Sedimentary rocks indicating deformation nearby—a Cenozoic (Miocene) debris flow exposed in a vertical cliff. The largest fragments are 10 inches long. Note how angular and poorly sorted the fragments are compared to the conglomerate in color Plate 10A.

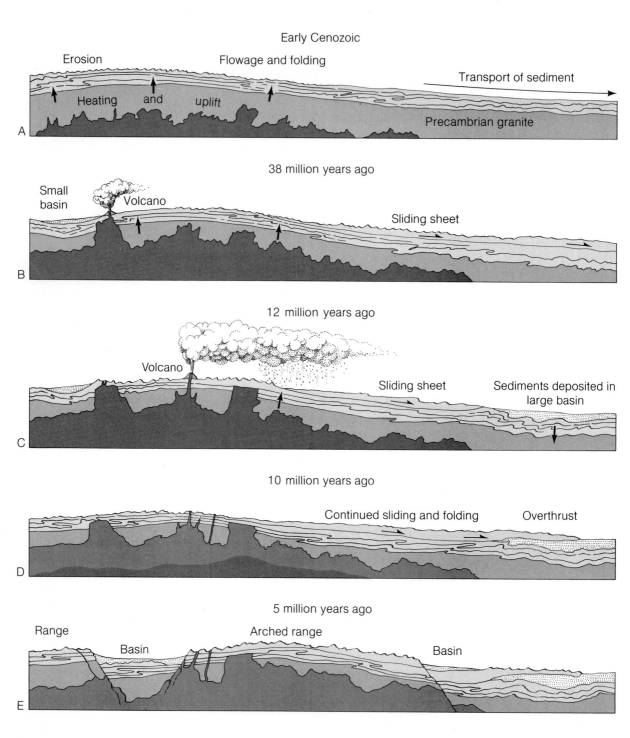

FIGURE 15-23

Interpreted stages in the heating, uplift, and deformation of the rocks now forming Raft River Mountains and Grouse Creek Mountains. The height of each diagram is roughly 3 miles and the width 40 miles, so that the vertical dimension is exaggerated and all features diagrammatic. Dark gray indicates melt; medium gray is crystallized melt.

Summary

The deforming processes and their immediate effects, however, are the most useful discoveries in the chapter, for they may be applied to the study of any deformed region. They may be summarized as follows:

1. Folds are typically produced when rocks are compressed parallel to their layers, such that the layers buckle.
2. Rocks may flow in the solid state, partly by progressive shifts of atoms within crystals and partly by complete reorganization (recrystallization) of the mineral grains.
3. Where flow is arrested, it may result in folds that are overturned in the same direction as that of the flow.
4. Rocks may also be displaced in large sheets that glide laterally and may be thrust upward at a low angle so as to ride over younger rocks.
5. The large frictional resistance to this movement is reduced by water beneath the rock sheets, which may buoy them by a pressure as great as their weight.
6. Besides the low-angle faults at the base of the sheets, high-angle lateral faults form along their sides parallel to the horizontal direction of movement.
7. High-angle gravity faults, on the other hand, involve vertical displacements (collapse) of rock masses.
8. These various processes and effects are often interrelated, as can be determined from the orientations and relative ages of the different features.
9. All or most of the features may result from the force of gravity, although uplift and heating are required to initiate the deforming movements.

REFERENCES CITED

1. Robert R. Compton. Geologic map of the Yost quadrangle, Box Elder County, Utah, and Cassia County, Idaho. U.S. Geological Survey, Miscellaneous Geologic Investigations, Map I-672, 1972. Geologic map of the Park Valley quadrangle, Box Elder County, Utah, and Cassia County, Idaho. *Ibid.*, Map I-873, 1975.
2. Victoria R. Todd. Structure and petrology of metamorphosed rocks in central Grouse Creek Mountains, Box Elder County, Utah. Unpublished dissertation, Stanford University, California, 316 pp., 1973.
3. Robert R. Compton, Victoria R. Todd, Robert E. Zartman, and Charles W. Naeser. Oligocene and Miocene metamorphism, folding and low-angle faulting in northwestern Utah. *Geological Society of America Bulletin*, in press.
4. M. King Hubbert and William W. Rubey. Role of fluid pressure in the mechanics of overthrust faulting. *Geological Society of America Bulletin*, vol. 70, pp. 115–66, 1959.

ADDITIONAL IDEAS AND SOURCES

1. The relations described in this chapter indicate that rocks flow more readily in the solid state when they are heated. Not mentioned is the factor of time, or the rate at which materials are permitted to flow. This factor can be demonstrated dramatically by experiments with silicone putty ("silly putty") or taffy candy bars, which flow plastically when pressed or

bent gradually in the hands but can be snapped sharply in angular pieces if twisted rapidly. Silicone putty will spread and flow slowly under its own weight, yet it will bounce like a rubber ball if dropped on the floor. The latter reaction is analogous to the elastic response of rocks to seismic waves, described in Chapter 11.

2. The effect of scale (size) is also important in visualizing rock deformation, especially in comparing large parts of the earth with table-top models. This factor has been described by M. King Hubbert in "Strength of the earth" (*Bulletin of the American Association of Petroleum Geologists*, vol. 29, pp. 1630–53, 1945). Pages 1635–38 present an especially clear analogy.

3. Another paper by Hubbert, "Mechanical basis for certain familiar geologic structures" (*Geological Society of America Bulletin*, vol. 62, pp. 355–72, 1951), describes and illustrates thrust faults and high-angle gravity faults that can be made in a sand box. Ernst Cloos has described patterns of faults and other fractures made by simple experiments with moist clay in "Experimental analysis of fracture patterns" (*Geological Society of America Bulletin*, vol. 66, pp. 241–46, 1955).

4. The folded and faulted rocks of the Appalachian region are beautifully illustrated in the *Geologic map of Pennsylvania*, which includes the slightly deformed region to the northwest of the mountain ridges as well as a striking belt of folds and thrust faults. Edited by Carlyle Gray, V. C. Shepps, and others, it was published in 1960, in two parts, by the Pennsylvania Geological Survey. Geological maps of other states, which may interest you with respect to major folds and faults or other features, are generally available through the State Geological Survey or the State Geologist.

5. Some of the most splendidly exposed large folds and thrust faults in the world are those in the Canadian Rockies of Alberta. They formed at the same time as the older features in the region described in this chapter and include areas of closely similar metamorphic rocks. The Canadian features have been described by R. A. Price and E. W. Mountjoy in "Geologic structure of the Canadian Rocky Mountains between Bow and Athabasca rivers, a progress report" (Geological Association of Canada, Special Paper 6, pp. 7–26, 1970).

Open vertical fissure in pillow basalt of the Atlantic seafloor, as viewed from the research submersible *Alvin* during the French-American Mid-Ocean Undersea Study (FAMOUS) of 1974. Woods Hole Oceanographic Institution photograph, courtesy of Robert D. Ballard and Tjeerd H. van Andel.

16. Earth's Spreading Crust and Lithosphere

Earth's Spreading Crust and Lithosphere

Figure 16-1, taken on the island of New Guinea, shows some exceedingly important layered rocks. Like the great sheets of rock described in Chapter 15, they have been displaced horizontally for many tens of miles, only in this case somewhat upward rather than downward. They originally lay 6 miles beneath the bottom of the Pacific Ocean but about 43 million years ago they were thrust onto continental rocks that form the main part of New Guinea. The layers in the photograph, however, are not sedimentary rocks. They are a rock called *peridotite*, consisting chiefly of olivine that crystallized in a vast subterranean chamber filled with molten basalt. The olivine evidently settled to the floor of the chamber and accumulated like sand grains that have settled through a tank of water.

These rocks were discovered and studied only recently, at a time when interest in rocks deep beneath oceans and continents had become high and controversy was plentiful. Studies of earthquake waves had shown that the earth is encircled by an outer layer called the *crust*, which lies on a deeper body called the *mantle*. The velocities of seismic waves were found to change abruptly at the boundary between the crust and mantle, thus providing a means of "seeing" that under the oceans the crust is only about 5 miles thick, but under the continents it is thicker and quite irregular (Figure 16-2). The velocities of the seismic waves also indicated that the granites and sedimentary rocks that make up most of the continental crust are scarce or missing in the oceanic crust. Velocities in the mantle were nonetheless found to be about the same under the oceans and the continents. These velocities are also the same as those measured in laboratory samples of peridotite, such as that of Figure 16-1. Solid chunks of peridotite brought to the earth's surface by volcanoes also suggested that this rock occurs widely at depth; however, other kinds of fragments were brought up by volcanoes, so that the evidence was conflicting.

The full significance of the arrangement shown in Figure 16-2 could be learned only by knowing more about the rocks, so scientists suggested drilling

FIGURE 16-1

Layered rocks consisting mainly of olivine in an exposure on the Wele River, Papua New Guinea. Photograph by Hugh L. Davies (Reference 2).

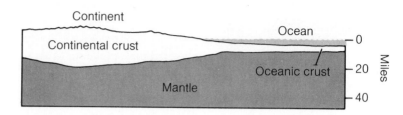

FIGURE 16-2

Vertical section through parts of a continent and ocean basin, showing typical variations in the thickness of the crust as determined by seismic studies made before the late 1950s.

through the crust to obtain samples of the mantle. The continental crust seemed much too thick for this procedure, but it was proposed that a hole might be drilled through the thinner crust under the oceans. Such a project was indeed started in 1958 and was named the Mohole for reasons that will become apparent later. Although the project was bound to be difficult, considerable progress was made by 1965, including the drilling of a shallow test hole at sea (Reference 1). Political difficulties developed, however, and when estimated costs reached 125 million dollars, Congress voted to abandon the project in 1966.

While the Mohole project was being conceived, Jack E. Thompson, a geologist with the Australian Bureau of Mineral Resources, recognized that extensive exposures on the island of New Guinea, including the one in Figure 16-1, are probably oceanic crust and mantle. Hugh L. Davies, a geologist with the same bureau, directed a party that mapped the region in 1963-1964, and his subsequent studies have shown that Thompson's interpretation was correct (Reference 2). The rocks constituted a truly major "sample"—a plate 10 miles thick that extends over a land area measuring 25 by 250 miles. Largely in their original condition, the rocks afford an exceptional view of the layers that lie under all the oceans. In this chapter we will first examine the New Guinea occurrence and then see how volcanic activity at the Mid-Atlantic Ridge helps us understand how the oceanic crust formed—indeed, how it is forming today. We will then explore the continental crust and finally a zone in the upper mantle that profoundly effects the crust as well as many processes at the earth's surface.

A Plate Consisting of Oceanic Crust and Mantle

As Figure 16-3A shows, Papua New Guinea makes up the eastern half of the large island of New Guinea. The backbone of the island is formed by the Owen Stanley Mountains, which rise as much as 12,000 feet above the sea and consist mainly of metamorphosed sedimentary rocks and granites—rocks typical of mountain ranges on continents. The plate of oceanic crust and mantle was emplaced over these rocks along major faults (Figure 16-3B and 3C). Three basic relations indicate that the plate is in fact oceanic crust and mantle: (1) the slope of the faults toward the adjoining sea shows that the plate was emplaced from the ocean basin onto the land; (2) the rocks have seismic velocities comparable to those measured for oceanic crust and mantle

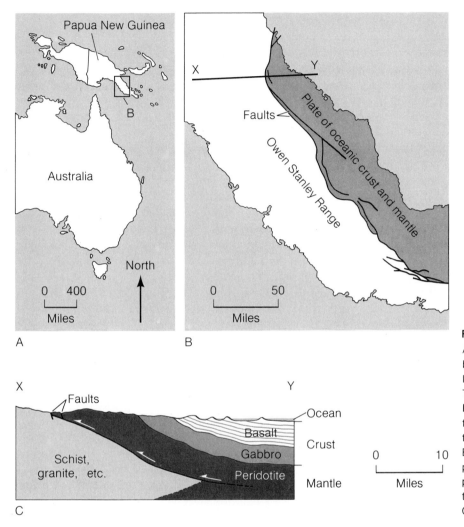

FIGURE 16-3

A, location of the area studied by Davies. B, simplified geological map of the area studied. The main faults, shown by heavy lines, slope northeastward under the adjoining ocean. C, vertical section along the line X-Y in B, showing the layered plate displaced along the faults and the part of the continental crust that makes up the rest of New Guinea. Data from Reference 2.

in place; and (3) the thickness of rock above the peridotite is the same as that measured by seismic methods for oceanic crust in undisturbed ocean regions, as in Figure 16-2.

The main rock layers Figure 16-3C also shows that the oceanic crust is made up of two major layers: an upper one that is 2.5 to 4 miles thick and consists chiefly of basalt, and a lower one that is 2.5 miles thick and consists of *gabbro*. As described in Chapter 12, basalt is a heavy, dark lava that is fine grained because of rapid cooling at the earth's surface, although it commonly contains larger mineral grains that grew slowly at depth before eruption. Its principal constituents are a colorless mineral, plagioclase; a dark mineral, pyroxene; and a pale green mineral, olivine. Gabbro, mentioned only briefly in Chapter 12, consists of these same minerals; indeed, it forms from the same melt as basalt only it does so underground, where it cools slowly, and thus all its grains attain a large size.

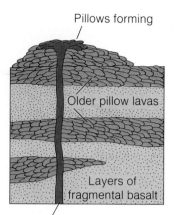

Pillows forming

Older pillow lavas

Layers of fragmental basalt

Fissure carrying basalt melt

FIGURE 16-4

Vertical section through the seafloor showing a vertical dike formed by basaltic melt rising through a fissure to the surface.

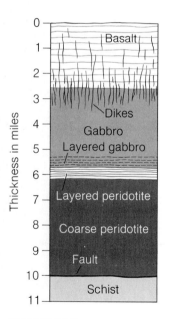

Thickness in miles

0
1
2
3 — Dikes
4 — Gabbro
5 — Layered gabbro
6
7 — Layered peridotite
8 — Coarse peridotite
9 — Fault
10
11 — Schist

Basalt

FIGURE 16-5

Diagrammatic vertical section showing the principal layers composing the plate of oceanic crust and mantle on Papua New Guinea. The thicknesses are approximate averages.

350

The basalt layer on Papua New Guinea is composed of thousands of individual lava flows, many consisting of pillows and elongate forms such as those known to form on the seafloor (Figure 6-16). Mixed with the lavas are layers consisting of small fragments of glassy basalt (a fragmental rock called *basalt tuff*). These layers form when basalt melt explodes or disintegrates when it is erupted underwater. There is thus specific evidence that the rocks on Papua New Guinea were once at the bottom of the ocean. Moreover, they evidently accumulated there one flow on top of another until the sequence was several miles thick.

A close connection between the basalt flows and the underlying gabbro is suggested by the fact that the basalts are cut here and there by dikes of fine-grained gabbro. The dikes apparently formed when melt rose in fissures to feed additional flows on the seafloor (Figure 16-4). The melt in the dikes crystallized to grains that are larger than those found in basalt because it was insulated by the surrounding rock, which is a poor conductor of heat. As Figure 16-5 suggests, the dikes are abundant near the base of the basalt layer and occur locally in the upper part of the main gabbro layer. Some dikes formed after the upper part of the gabbro had crystallized and cooled (Figure 16-6). These relations show that a large body of gabbro remained molten beneath the upper, solidified part.

The layered gabbro and peridotite   Other features indicating protracted solidification of the main body of melt are the mineral layers at the base of the gabbro body (Figure 16-5). Several kinds of evidence show that these layers resulted from the accumulation of mineral grains that settled to the floor of the body of melt. For one, the pyroxene and olivine grains look like sand grains that fell and accumulated loosely, finally being cemented by minerals that grew from the melt between the grains (Figure 16-7). Another indication of the sedimentation of the grains comes from the layers themselves, many of which have denser (heavier) minerals (olivine or pyroxene) concentrated near the base. This relation suggests that the melt in the upper part of the body sometimes crystallized episodically, forming showers of falling crystals in which the denser grains fell faster than the lighter ones.

Perhaps the most convincing evidence of the settling of grains, however, is the order in which the principal minerals appear within the layered sequence. Olivine forms almost all the lower layers; pyroxene then appears in increasing abundance; plagioclase joins olivine and pyroxene at somewhat higher levels. This order is the same as that in which the three minerals form in basalt melts. Hatten S. Yoder, Jr., for example, has melted basalts of about the same composition as the basalts in Papua New Guinea and then let them crystallize in a number of stages, noting which mineral species formed at each temperature (Reference 3). He found that olivine crystals appeared first, at temperatures between 1240 and 1210°C, that pyroxene crystals began to form when the melts had cooled to about 1180°C, and that plagioclase started to form soon thereafter, at around 1165°C. This sequence thus explains why the lowest layers in the body consist almost entirely of olivine (which crystallized and accumulated first) and why pyroxene appears before plagioclase in the overlying layers.

Experiments have also shown that only about 10 percent of a body of basalt melt will crystallize as olivine. The layered accumulation of olivine (the

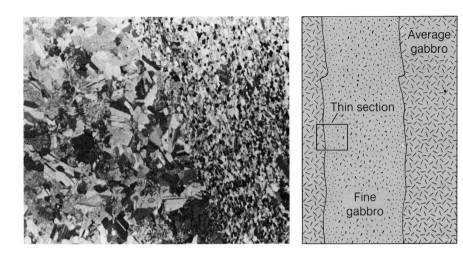

FIGURE 16-6

Thin section, 1 centimeter across, of a dike of fine-grained gabbro and the coarser gabbro into which it was intruded. The difference in grain size implies that the coarser gabbro had solidified and cooled before a vertical fissure formed and was filled with melt. Note that the somewhat elongate crystals, which are plagioclase, are aligned vertically, as though oriented by flow in that direction. Photograph by Hugh L. Davies (Reference 2).

layered peridotite) is 0.3 mile thick (Figure 16-5), so that the body of melt from which these grains accumulated would have to have been about 3 miles thick. The body of melt would thus have extended from the base of the layered peridotite nearly to the top of the gabbro layer. We can therefore conclude that the oceanic crust exposed on Papua New Guinea must have formed largely from a molten pluton of gabbro.

The metamorphic peridotite   The pluton, however, had a floor of peridotite, and it is important to see that this underlying peridotite differs considerably from the layered peridotite lying on it. Its olivine grains are larger than the sand-like grains in the layered rock, and their irregular shapes indicate crystallization in the solid state (Figure 16-8, left). This metamorphic-appearing rock contains no plagioclase, and its pyroxene grains differ in composition from those in the overlying rocks. Another difference is that the layers in the underlying peridotite are not simple sediment-like features but are streaked out and folded as the quartzites described in Chapter 15. Laboratory studies indicate that olivine must be heated to at least 900°C to flow in this way. Because temperatures ordinarily increase downward at rates no greater than 9°C per 1,000 feet (30° per kilometer), the peridotite must once have lain deep in the mantle or temperatures in the upper part of the mantle were once unusually high. In either case, the lower, metamorphic peridotite has had a longer and more complex history than the layered peridotite and gabbro lying on it. Like a major unconformity in sedimentary rocks, the interface of the two kinds of peridotite expresses extensive changes in this part of the earth.

Can we be sure that all these things happened under an ocean basin, before the plate was thrust onto the continental rocks of New Guinea? Dating of the rocks by radiometric methods shows that the gabbro body crystallized approximately 145 to 150 million years ago and was emplaced onto the continental rocks about 43 million years ago. The emplacement produced new features in the metamorphic peridotite near the base of the plate, deforming the olivine grains into lenticular forms and reducing many of them to groups of small grains (Figure 16-8, right). The peridotite is even finer grained where

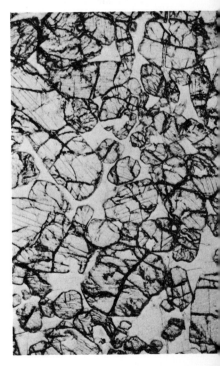

FIGURE 16-7

Thin section of gabbro showing grains of pyroxene (gray, with many cracks) in plagioclase (white) that crystallized around them after they accumulated. View is 2 millimeters across. Photograph by Hugh L. Davies (Reference 2).

it lies against the fault, having been deformed there to a tough, flinty rock
called *mylonite* (literally, milled-out rock). The quartz-bearing (continental)
rocks under the fault were also metamorphosed and deformed by the emplacement, and among the new minerals is one called *glaucophane,* which has been
formed in the laboratory at exceptionally high pressures but only at comparatively low temperatures (Supplementary Chapter 1). The presence of this
mineral, along with the deformed, fine-grained nature of the rocks, supports
the idea that the oceanic plate was cool when emplaced onto the continental
rocks 43 million years ago. All the processes relating to the basalt melts and
the crystallization of the peridotite and gabbro, then, must have taken place
when the plate still lay beneath an ocean basin.

Production of Crust at Oceanic Rises

A remarkable recent discovery is that basalt and gabbro like that exposed on
Papua New Guinea are forming today along linear rises such as the Mid-
Atlantic Ridge and are gradually being spread outward to form the entire
oceanic crust. An important part of the discovery is that linear rises occur in
each major ocean basin, and photographs and samples of the seafloor show that
they consist of basalt covered partly with pelagic sediment. Heat probes
lowered into bottom sediment have detected higher temperatures along the
axes of the rises than along their flanks, suggesting that recently erupted basalt
and subterranean bodies of melt are concentrated along the axes. The rises
appear to be offset along gash-like valleys that have floors 2 to 3 miles below
the nearby volcanic peaks (Figure 16-9, left). The valleys must be zones of
active faulting, for many of them generate numerous mild earthquakes,
especially where they intersect the axes of the rises. Fragments dredged from
the upper parts of the valley walls are basalt, whereas those dredged from near
the valley floors may be gabbro and, in some places, peridotite. These relations
suggest the same vertical succession as seen in Papua New Guinea: a basalt
layer underlain by gabbro and then peridotite.

An undersea study of the Mid-Atlantic Ridge   The rises had so
intrigued scientists by the early 1970s that a direct exploration was made in

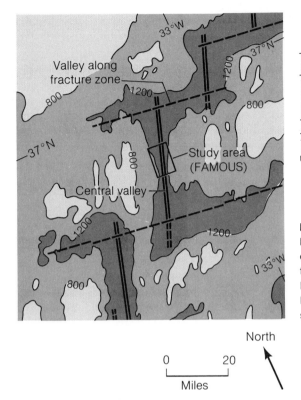

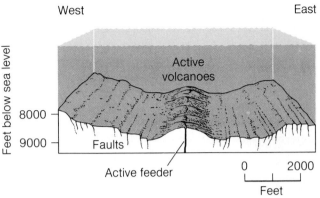

FIGURE 16-9
Left, map of the Mid-Atlantic Ridge 200 miles southwest of the Azores, showing the central valley and the deep transverse valleys that cross it. Contours in fathoms. Right, part of the central valley of the Mid-Atlantic Ridge based on soundings and direct observations from the submersible *Alvin*. From data in Reference 4.

1974 using a combined fleet of French and American research submersibles (submarines) and other vessels. Approximately 278 hours were spent along the bottom in the submersibles, examining, mapping, and sampling the central part of the area shown in Figure 16-9, left (Reference 4). Studies from the American submersible *Alvin* disclosed the features in the part of the central valley shown in Figure 16-9, right. The valley is bordered by a series of steep slopes that locally expose crushed rock, indicating they are the surface expression of high-angle faults. Dozens of open vertical fissures trend parallel to the central valley and its fault system, some gaping as much as 25 feet (see the chapter opening page). The fissures and faults indicate that the Mid-Atlantic Ridge is gradually spreading apart along its axis, with its flanks moving outward toward the east and west.

No volcanic eruptions were seen in progress, but linear mounds of pillow lava and volcanic mountains had evidently formed by eruptions through some fissures, especially those along the central part of the valley (Figure 16-9). The eruptions would explain the high temperatures measured near the central parts of oceanic rises, for the melt retained in fissures beneath the surface would have cooled slowly and warmed the rocks above and around the fissure systems. The bodies would become dikes, like those discovered in Papua New Guinea, when they solidified.

The formation of several dikes and flows is depicted in Figure 16-10. Note that the opening of fissures may be independent of the intrusion of melt. Only the fissures that open deeply enough, or in just the right places, tap the underground reservoir of molten basalt, which then rises through the fissures

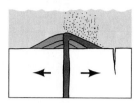

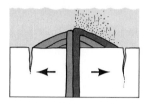

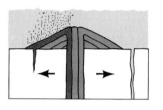

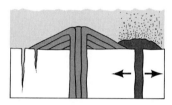

FIGURE 16-10

Steps in opening of fissures and eruption of the basalt that forms the upper layer of the oceanic crust. Note that each dike adds a foot or more to the width of the crust beneath the lavas. The crust is thus growing and spreading episodically toward the right and left (east and west).

to the surface. Each dike thus adds a flow to the growing basalt layer *and enlarges the underlying crust by an amount equal to the dike's width.* The entire Atlantic Ocean is enlarged eastward and westward by the same amount. Oceanic crust generated at the midocean rise thus comes to underlie the entire ocean.

A pluton beneath the central valley The growth of gabbro and peridotite bodies beneath the medial valley could be determined only indirectly at the Mid-Atlantic Ridge. The most useful information came from dating many lava flows and from studying their minerals and chemical compositions (Reference 5). All the flows along the axial line of volcanoes shown in Figure 16-9, right, were found to be younger than 1,000 years. Lavas elsewhere in the valley are *on the average* progressively older away from the axial line; however, lavas as young as those along the axis were found here and there elsewhere in the valley. Evidently an underground reservoir of melt extends under the entire valley, but it comes closest to the surface at the center, thus being most often tapped by fissures near the valley's axis.

An important discovery, then, was that the compositions of the younger lavas change systematically outward from the axial line. Basalts at the axis contain numerous olivine crystals and have a silica content of around 49 percent. Eastward and westward from the center, the olivine diminishes and the silica content of the lavas increases gradually to somewhat more than 52 percent. These relations are like those discussed in Chapter 12 with regard to the Mount St. Helens lavas; the increase in silica and reduced amount of olivine suggest that olivine (a mineral with a low silica content) has crystallized and sunk from the melt. This interpretation is supported by variations in chemical components that are not contained in olivine, for example, potassium oxide (Figure 16-11). The increases of potassium oxide away from the center of the valley suggest crystallizing and sinking of mineral grains that contain little if any potassium.

Figure 16-12, left, shows how these various data might be explained by movements in a pluton that would result in rock layers like those on Papua New Guinea. As suggested at the bottom of the diagram, the molten system is fed through fissures that open in the upper part of the mantle and introduce melt from deeper sources. Because the newly arrived melt is hotter than that already in the pluton, it is less dense and therefore rises to the top. Here it cools and begins to crystallize as olivine; part of this olivine-bearing melt is erupted to form the flows in the center of the valley. Along the sides of the pluton, however, cooler, crystal-laden melt sinks in a movement contrary to the rise of new melt up the center, generating the convection shown by the lines of arrows in the diagram. The crystallizing melt at the top thus flows slowly outward along the roof, showering down olivine crystals and thereby becoming enriched in silica and potassium, as noted. The olivine crystals fall to the floor of the chamber, constructing a series of layers. The cooler melt flowing slowly down the walls is largely depleted of olivine, and thus it contributes to pyroxene and plagioclase grains growing there, forming unlayered gabbro like that making up most of the gabbro layer exposed in Papua New Guinea.

Figure 16-12, right, summarizes these various additions to the central part of the ridge and to the oceanic crust beneath it. Note that the rocks at each level are added piecemeal as the crust and underlying mantle spread outward from the medial valley. The walls of the pluton never grow together because they are separated as fast as melt crystallizes along them. A layered oceanic crust is thus constructed by processes proceeding regularly under an oceanic rise.

Sounding the Continental Crust with Seismic Waves

The complete continental crust has not been found in an exposure such as the one in Papua New Guinea, and it is unlikely that it ever will be. As Figure 16-2 suggests, it is so thick that it would probably be greatly broken and distorted were it turned up to our view by some gigantic upheaval. We have already examined rock bodies typical of the upper third or so of the continental crust—down to a depth possibly as great as 15 miles (the exceptionally uplifted rocks described near the end of Chapter 10). To "see" below that level we must resort to seismic waves, and this section describes some of the exceedingly important studies using them.

Variations in velocities of seismic waves   All studies using seismic waves are based on the fact that they travel at different velocities through different materials. The more rigid and less compressible the material, the faster the waves travel through it. Rigidity and compressibility depend largely on the minerals making up the rocks. Olivine, for example, responds to seismic

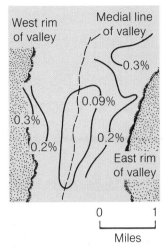

FIGURE 16-11

Map showing amounts of K_2O (potassium oxide) in young basalt of the Mid-Atlantic Ridge. From W. B. Bryan and James G. Moore (Reference 5).

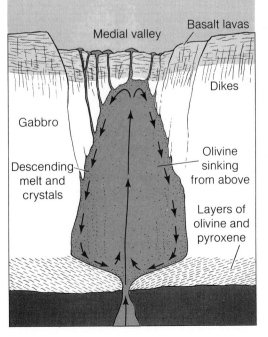

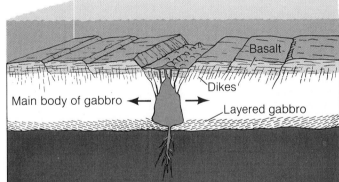

FIGURE 16-12

Left, schematic vertical section through the crust and upper mantle under the Mid-Atlantic Ridge, showing how the circulation of basaltic melt in a pluton explains the variations in lavas erupted in the valley. Note that the accumulating crystals also explain the layered rocks exposed on Papua New Guinea. Based on data from James G. Moore. Right, cross section suggesting how spreading at a medial rise, accompanied by intrusion and crystallization of melt, gradually forms all parts of the oceanic crust.

355

TABLE 16-1

Velocities of P-waves in Common Sediments and Rocks, from the Earth's Surface to a Depth of 20 Miles*

Materials, Depths	Velocities (km/sec)
Water, near the surface	1.5
Sand and gravel, near the surface	0.8–1.5
Clay, near the surface	1.0–2.0
Moderately consolidated shale and sandstone, 1 mile deep	2.5–3.5
Well-consolidated shale and sandstone, 2 miles deep	4.0–5.0
Quartzite and schist, 2 miles deep	5.5–6.5
Granite, 2 miles deep	5.9–6.3
Granite, 12 miles deep	6.0–6.5
Gabbro, 2 miles deep	6.4–7.1
Gabbro, 12 miles deep	6.7–7.2
Peridotite, 12 miles deep	7.9–8.3
Peridotite, 20 miles deep	8.0–8.4

*Generalized from data in *Handbook of Physical Constants*. Sydney P. Clark, Jr., editor. Geological Society of America, Memoir 97, 587 pp., 1966.

waves more rigidly than pyroxene and plagioclase, so waves travel faster through peridotite than through basalt and gabbro. Pyroxene and plagioclase respond more rigidly than quartz, and thus seismic waves travel faster through gabbro than through granite, quartzite, and sandstone.

Velocities of seismic waves are also affected by the general increase in rigidity of rocks with depth. As we noted in Chapter 10, the increasing weight of overlying rock and sediment presses cracks and other small openings more and more tightly together as rocks are buried. Consolidation thus reduces the sizes and numbers of water-filled openings that reduce the velocity of seismic waves. At great depths, recrystallization and metamorphism knit grains together so tightly that only very small water-filled openings remain, and velocities of seismic waves in these rocks increase accordingly.

Table 16-1 shows how velocities of P-waves are affected by consolidation as well as by the kinds of minerals making up the chief rocks of the crust and upper mantle. S-waves travel more slowly but show similar variations. The vertical order in the table is the same as the typical distribution of sediments and rocks in the outer part of the earth. The data thus show that the velocities of seismic waves typically increase with depth. The important exception is where rocks are molten or partly molten. Small pores and films of melt reduce velocities of seismic waves much as do small bodies of water in rocks near the surface. In addition, minerals heated so greatly as to be near the point of melting are less rigid than when they are cooler. Recall, too, that S-waves (Chapter 11) cannot be transmitted through fluids, so that large bodies of melt stop them completely.

Refraction of seismic waves with depth   To see the effects of the general increase of velocity with depth, let us trace a seismic wave that has been generated by faulting high in the crust and does not pass through molten rocks. Looking down in "map view," the first wave will expand evenly outward like a ripple formed by dropping a pebble in a quiet pool of water (Figure 16-13A). In vertical section, however, the shape of the expanding wave changes progressively because its velocity increases with depth. As Figure 16-13B shows, increase in velocity causes the wave to bulge increasingly downward. By following one point in the expanding curve, such as the one numbered in sequence in the figure, we can see that the path of the point curves back upward. We can construct any number of *wave paths* (the white lines in the figure) by drawing lines perpendicular to the wave front in each position. The array of lines shows clearly that the direction of the wave is bent (*refracted*) back toward the surface, the only exception being the part of the wave moving straight downward.

Let us now see what happens when the velocity increases abruptly at a more or less horizontal surface, such as the interface between gabbro and peridotite at the base of the oceanic crust. Figure 16-14 shows this case, simplified by using a constant velocity in the crust and a greater velocity in the mantle. Note that the increase in velocity causes the wave to bulge downward in the mantle, resulting in the angular refraction of wave paths, such as OAB. The wave path OP is especially important because it makes just that angle with the top of the mantle so that it is refracted along path PQW in the top of the mantle. The important angle OPX can be calculated from the relation Sin $(90° - \angle OPX = V_1/V_2)$, where V_1 is the seismic velocity in the crust and V_2 that in the mantle.

As you can see, the wave in the top of the mantle moves increasingly ahead of the same wave in the top of the crust. Moreover, the wave in the top of the mantle disturbs (impacts) the base of the crust and thereby sends a wave obliquely up through the crust (the gray lines in the figure). As the figure indicates, this wave is comparable to the bow wave of a ship plowing through quiet water. All wave paths within this part of the array, such as QR, extend upward at the same angle as the descending path OP.

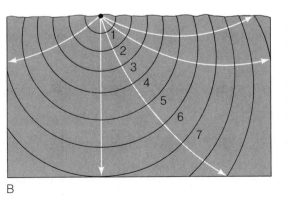

A   B

FIGURE 16-13
A, horizontal surface showing a seismic wave at equal time increments after it was generated. B, vertical section showing that the successive positions of the wave become spaced farther and farther apart due to increase of velocity with depth. The wave thus bulges downward and outward, and wave paths (the white lines) curve back toward the surface.

357

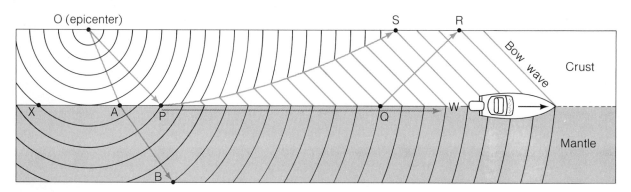

FIGURE 16-14

Progress of a seismic wave at equal time increments. The effects of the abrupt increase in velocity at the boundary between the crust and mantle are shown. The part of the wave traveling along the top of the mantle (*PQW*) generates a wave in the adjoining crust much as a ship generates a bow wave. This wave travels obliquely upward in the direction *QR* and is the first wave recorded at the earth's surface at points beyond *S*. At points between *S* and the epicenter, on the other hand, the wave traveling directly through the crust is recorded first.

Now imagine a series of seismographs set up at various positions across the earth's surface. All those between *O* and *S* will receive the wave traveling through the crust first, and all those beyond *S* will receive the "bow" wave first. If we now assemble the seismograms in the same order and with the same relative spacing as the stations at which they were received, we will find that their first wiggles form two linear arrays (Figure 16-15). By drawing lines along these two sets of wiggles, we can locate *S*, the point at which the direct wave and the "bow" wave should have arrived simultaneously. We thus obtain the important distance *OS* in Figure 16-14. We have already noted that we can calculate the critical angles *OPX* and *RQW* from the velocities in the two layers, and we now have enough information to calculate the thick-

FIGURE 16-15

Nine seismograms that recorded the earthquake depicted in Figure 16-14, placed in vertical order based on their distances from the epicenter. When the two lines are drawn along the first wiggles, their intersection locates the point *S* in Figure 16-14.

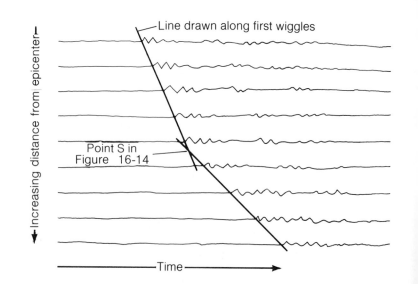

ness of the crust. The equation is: thickness $= \frac{1}{2}\sqrt{(V_2 - V_1)/(V_2 + V_1)}\,d$, where V_1 is the seismic velocity in the crust, V_2 in the mantle, and d the distance OS (Reference 6).

Measurements of crustal thickness   The thickness of the continental crust was first calculated in essentially this way in 1909 by A. Mohorovicic, who used data from a strong earthquake in Yugoslavia. In his honor, the seismic discontinuity at the base of the crust has been named the *Mohorovicic discontinuity*, often shortened to the *Moho*; thus, the name Mohole for the project whose goal was to drill down to the discontinuity.

Studies using similar calculations have shown that the thickness of the continental crust averages about 18 miles but varies considerably. Variations in seismic velocities due to different rock bodies in the continental crust are also large, so that seismic studies must be especially thorough to give accurate results. One problem has been that the epicenters and times of natural earthquakes can rarely be fixed as exactly as needed to resolve the variations; however, this problem has been overcome in recent studies by using seismic waves generated by powerful explosions. The place and time of origin of the waves are thus known exactly, and seismographs can be set in advance in groupings ideally suited to "look" at specific parts of the lower crust and even to determine local slopes in the Mohorovicic discontinuity. These methods were used extensively in 1965 to probe the crust and mantle of the Middle Atlantic States (Reference 7). Over 100 large explosions were fired on land and at sea, and the resulting seismic waves were recorded by nearly 1,000 setups of seismographs. The setups were arranged to look at the crust in many directions and thus take into account major rock masses within it. The positions of the explosion and the seismographs were also reversed for some of the lines studied to provide a check on refraction paths such as those shown in Figure 16-14. In addition, considerable data from drill holes, surface outcrops of rocks, and offshore surveys were used in interpreting the seismic data.

The Moho was thereby mapped over a large area, and Figure 16-16 shows one vertical section through it. The section lies across the James River system and the nearshore part of the Atlantic Ocean described in Chapters 2, 4, and 5. It shows several relations typical of regions that have not been disturbed by recent earth movements: (1) the crust thickens distinctly under the highest part of the terrain to as much as 35 miles; (2) the thickness of the crust elsewhere is around 18–25 miles, which is about average for all continents; (3) the

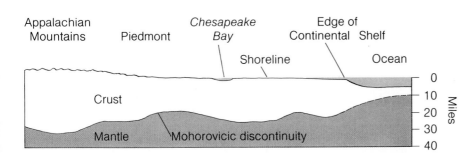

FIGURE 16-16

Vertical west-to-east section through the crust and upper mantle in Virginia, with vertical scale (shown right) somewhat greater than horizontal scale. Data from James and others (Reference 7).

continental crust extends out under the shelf; and (4) the crust then thins to the 3–5 miles typical of the ocean basins. The study showed quite conclusively that the mantle under the continental crust has the properties of peridotite and is therefore similar to the mantle under the oceans. The data of this survey (and others) also suggest a possible connection between the oceanic and continental crusts, for seismic velocities in the lower one-quarter or so of the continental crust are comparable to that of a rock called *amphibolite*, formed by the metamorphism of gabbro and basalt under conditions suitable to the depths involved. This relation suggests that an oceanic crust may have been slowly buried beneath the lighter weight rocks that now form the bulk of the continental crust and that the basalt and gabbro were metamorphosed in the process.

The Asthenosphere, A Ductile Zone in the Upper Mantle

Soundings by means of seismic waves have indicated another layer-like body in the outer part of the earth, one that has great influence on horizontal and vertical movements of the crust as well as on the generation of rock melts. The body is called a zone rather than a layer because its boundaries are gradational. Its essential characteristic is that rocks in it are more ductile (less rigid) than those above and below it. S-waves pass through the zone and some earthquakes are generated within it, so that the zone cannot be molten as a whole; nonetheless, rock melts are erupted from it and must therefore form in it locally or perhaps temporarily. Slow movements within the zone indicate that the rocks there can flow readily in the solid state, presumably like the rocks described in Chapter 15. The streaky character of the metamorphic peridotite forming most of the displaced mantle on Papua New Guinea indicates that this rock was once part of the ductile zone. The gradational nature of the zone makes it difficult to locate, but it is typically 100 to 200 miles thick and its upper boundary lies about 30 miles beneath oceans and perhaps twice as deep under most continents. The zone thus lies well within the mantle in most places; however, we shall find that it rises locally to the base of the crust.

Discovery of a low velocity zone   The zone was discovered in the 1950s by Beno Gutenberg, a seismologist who studied the upper mantle by measuring the strength of seismic waves originating at various depths (Reference 8). By carefully comparing seismograms recorded at a large number of stations, he found that most earthquakes are felt weakly between 60 and 600 miles from their epicenters. He explained this phenomenon by the relations shown in Figure 16-17A, which depicts paths of seismic waves generated near the surface (lines like the white lines in Figure 16-13). Note that paths *a*, *b*, and *c* are bent toward the surface as already described but that path *d* is bent downward because of a decrease in wave velocity in the ductile zone. Note that path *d* is then gradually bent back upward because of increasing velocity in the lower part of the ductile zone and beneath it. This change is an expression of the gradational nature of the boundary. The effect of the ductile zone is seen again as the wave travels upward from point *y* to *z* and

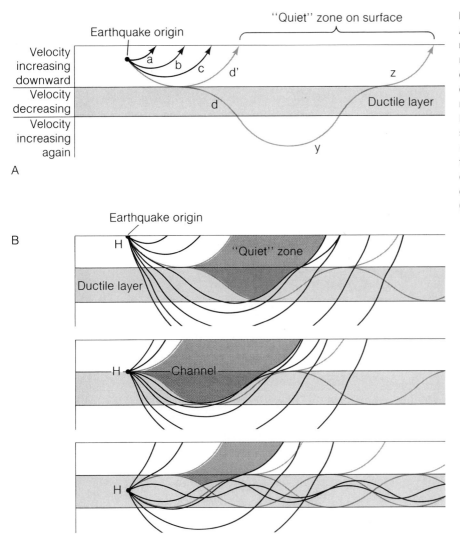

Velocity increasing downward

Velocity decreasing

Velocity increasing again

A

Earthquake origin

"Quiet" zone on surface

a   b   c   d'   z

d   Ductile layer

y

B

Earthquake origin

H

"Quiet" zone

Ductile layer

H   Channel

H

FIGURE 16-17

A, cross section extending 300 miles into the earth, showing refraction of wave paths by the ductile zone in the mantle, creating a seismically quiet region on the earth's surface. B, sections such as those of A, showing how the depth of the origin (H) affects the extent of the quiet region and the degree of channeling of waves in the ductile zone. After Beno Gutenberg (Reference 8).

finally curves to the surface in a normal course. Now note the gray path *d'*, the course the wave would have taken had there been no ductile zone. The zone has thus displaced waves hundreds of miles beyond their normal points of arrival, creating a seismically quiet region on the surface and concentrating seismic energy just beyond it.

By studying earthquakes originating at various depths, it was found that the quiet region widens and then narrows systematically as earthquake sources become deeper (Figure 16-17B). An important discovery was that earthquakes originating deep within the ductile zone have their wave paths and most of their energy channeled within the zone (the bottom diagram in Figure 16-17B). This channeling was found to extend from under ocean basins to under continents, indicating that the ductile zone is worldwide. By studying the travel times of the various waves shown in Figure 16-17, it was also possible to determine the approximate depth and thickness of the zone.

Seismologists often call the ductile zone the *low-velocity zone,* because its discovery depended on the decrease in velocity of seismic waves that enter it. More generally, though, it is called the *asthenosphere*—literally, the weak (ductile) zone. The overlying mantle and crust, which are comparatively rigid, are together called the *lithosphere* (the more brittle, "rock-like" layer). The basic contrast between ductile and rigid behaviors is proven by the distribution of earthquake sources, for most are in the lithosphere. Moreover, the earthquakes generated in the asthenosphere are so mild that they account for less than 1 percent of the earth's total output of seismic energy. The lithosphere's dominant role in generating earthquakes follows from what we learned of earthquake mechanisms in Chapter 11: (1) rocks must be rigid enough to deform elastically before they are "set" to generate an earthquake, and (2) triggering of fault displacements is commonly due to incursion of water at high pressure. Rigidity of rocks and the presence of abundant water are characteristic of the outer, cool parts of the earth. Rocks in the asthenosphere, on the other hand, are evidently so hot that they generally flow in the solid state before forces can "set" them elastically.

Melting in the asthenosphere   The high temperatures in the asthenosphere, in fact, bring us to the problem of why melting does *not* take place more generally. If, for example, we assume that temperatures increase downward at a rate such as that commonly measured near the surface, about 8°C per thousand feet (25° per kilometer), rocks deeper than about 30 miles (50 km) should be hot enough to yield melts of basalt. The reasons basalt is not melted so rapidly are shown graphically in Figure 16-18. The dark gray band expresses one reason: temperature increases less rapidly at depth than suggested by measurements near the surface. This relation is due in part to the high rate of cooling at the surface and in part to the greater concentration of radioactive heat-producing elements in the crust.

The other reason basalt does not arise abundantly at great depths is expressed by the slope of the line on the right side of the graph. The line shows that the temperatures at which basalt melts from water-free peridotite increase with depth. The cause of this relation lies in the fact that rocks have smaller volumes when solid than when molten. Increased pressure thus raises the melting temperature, "forcing" a rock to remain solid at higher temperatures than those we measure in lavas near the surface. If you now compare the line with the dark gray band on the left, you will see that melting should not take place at any depth but that *it is most closely approached in the depth range of the asthenosphere.*

Origin of basalt   Molten basalt is nonetheless erupted widely and repeatedly, and its temperatures, around 1,250°C, suggest that it should arise from depths of 50 miles or more (the vertical dashed line in Figure 16-18). Chemical studies of many freshly erupted basalts on Hawaii show that they arrive from the mantle with a small but consistent amount of water dissolved in them—about 0.4 percent by weight. This water is very significant, for laboratory studies have proven that even as little as 0.1 percent lowers the melting temperature of the mantle by 100–200°C. The lowering is indicated in Figure 16-18 by the pale gray band to the left of the melting line for dry mantle.

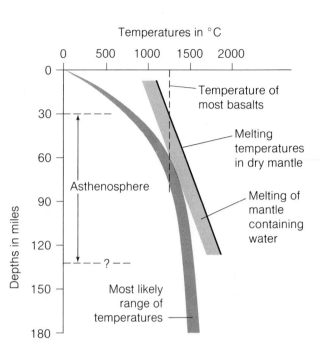

Temperatures in °C

FIGURE 16-18

Graph showing temperatures at which melting begins in dry mantle rocks (solid line) as compared to likely temperatures at depth (dark gray band). Note that addition of water (the pale gray band) brings rocks in the asthenosphere just to the melting range. Depth relations are for an ocean basin.

The relation shows that the asthenosphere in the 50- to 100-mile depth range should begin to melt where small amounts of water are present.

A remaining problem is to bring enough basalt melt together in the partly melted peridotite to form the large volumes of lava that are occasionally erupted to the surface. The metamorphic peridotite normally seen at the earth's surface, for example, in Papua New Guinea, consists entirely of olivine (essentially Mg_2SiO_4) and a pyroxene that is also a magnesium silicate. Such rocks could supply very little of the other chemical components of basalt; indeed, less than 1 percent of this kind of peridotite could be converted to molten basalt under any conditions. An important discovery, therefore, is that some basalts carry to the surface chunks of a different kind of metamorphic peridotite—one containing minerals other than olivine. Included are small amounts of mica and amphibole (water-bearing minerals) and quite large amounts of pyroxene and garnet—mineral varieties that carry many of the chemical components present in basalt. When these peridotites are melted under the conditions shown in Figure 16-18, they yield 10 to 20 percent basalt melt. This liquid evidently then gathers from rather large bodies of peridotite and moves rapidly through fissures to the surface or to a subterranean chamber such as that described beneath the Mid-Atlantic Ridge. Indeed, spreading and fissuring such as that under the Mid-Atlantic Ridge may be essential to moving batches of melt from the asthenosphere to the crust.

We may complete this exploration of the asthenosphere and its partial melting by considering the materials left behind. Experiments show that they should be refractory (difficult to melt) minerals such as magnesium-rich olivine and pyroxene. We have just noted that these minerals are the very ones that make up most metamorphic peridotites displaced from the mantle to the surface. They are also the minerals in which seismic waves move at

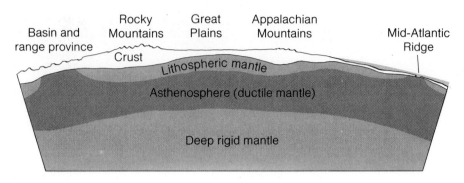

FIGURE 16-19
Diagrammatic cross section through part of America and the western half of the Atlantic Ocean, showing that the asthenosphere is much closer to the surface in some places than others. Vertical scale greatly exaggerated.

velocities typical of the rigid part of the mantle above the asthenosphere. Very likely, then, these upper parts of the mantle had their original basaltic components melted long ago. That they were once part of the asthenosphere is proven by exposures such as that on Papua New Guinea—the streaked, metamorphic character of the rocks shows that they were once ductile, that they flowed just as the present asthenosphere.

Summary

We can get some feel for recent advances in knowledge of the outer earth by comparing Figures 16-2 and 16-19. Since the 1950s, the crust has been measured quite exactly in many places, and the mantle has been found to contain the thick though somewhat hazy zone called the asthenosphere. Although the asthenosphere has not been measured as exactly as Figure 16-19 may suggest, it is almost certainly thinner under the Atlantic Ocean than under most of the continent, and it rises to the base of the crust at the Mid-Atlantic Ridge and under the basin and range region of the western states. Its position in the latter place correlates well with the extensive spreading and collapse of that region and with the evidence of recent heating described in Chapter 15.

Let's summarize the basic attributes of the various layers and zones:

1. The crust is a worldwide outer layer on the earth that can be measured because the velocities of seismic waves increase abruptly when they pass into the underlying mantle.
2. Displaced plates of oceanic crust and mantle show that the upper mantle is peridotite (essentially olivine rock) and that the oceanic crust consists of a lower gabbro layer and an upper basalt layer.
3. Velocities of seismic waves indicate that the same rocks form the oceanic crust and mantle elsewhere.
4. Oceanic crust is forming at oceanic rises as a consequence of spreading and of erupting of basalt melt, part of the melt accumulating at the surface and part filling fissures and huge chambers at depth.
5. The continental crust is much thicker than the oceanic crust and is thickest under mountain ranges and thinnest under continental shelves.

The rocks forming at least the upper two-thirds of the crust are granites, metamorphic rocks, and sedimentary rocks such as those described in earlier chapters.

6. The crust and the relatively rigid part of the upper mantle are joined in the continuous body called the lithosphere. Had the term crust not been preempted in the way already noted, the lithosphere would be the natural candidate, for it is the earth's outer, rigid layer, lying on the ductile asthenosphere.

7. The asthenosphere lies mainly in the mantle and is mainly composed of peridotite, but it is defined basically as a ductile zone. We can detect its present position by the way it slows seismic waves and by the nature of molten materials that rise from it. Rocks that were formerly in the asthenosphere can be identified by their streaky, ductile appearance and by the fact that they have been depleted of minerals that can be utilized in rock melts. As we shall see, the somewhat irregular and hazy asthenosphere takes a major part in most earth processes.

REFERENCES CITED

1. Willard Bascom. *A hole in the bottom of the sea; the story of the Mohole project.* Garden City, New York: Doubleday, 325 pp., 1961.
2. Hugh L. Davies. Peridotite-gabbro-basalt complex in eastern Papua: an overthrust plate of oceanic mantle and crust. Bureau of Mineral Resources, Geology and Geophysics (Australia), Bulletin 128, 48 p., 1971.
3. H. S. Yoder, Jr., and C. E. Tilley. Origin of basalt magmas: an experimental study of natural and synthetic rock systems. *Journal of Petrology*, vol. 3, p. 342–532, 1962.
4. Robert D. Ballard and Tjeerd H. van Andel. Project FAMOUS: morphology and tectonics of the inner valley at 36° 50′ N on the Mid-Atlantic Ridge. *Geological Society of America Bulletin*, in press, 1977.
5. W. B. Bryan and James G. Moore. Compositional variations of young basalts in the Mid-Atlantic rift at 36° 49′ North. *Geological Society of America Bulletin*, in press, 1977.
6. The complete derivation of this equation and other interesting relations derived from seismic records are given by Milton B. Dobrin in *Introduction to geophysical prospecting*, New York: 2nd ed. McGraw-Hill Book Company, 446 p. 1960.
7. David E. James, T. Jefferson Smith, and John S. Steinhart. Crustal structure of the middle Atlantic states. *Journal of Geophysical Research*, vol. 73, pp. 1983–2007, 1968.
8. Beno Gutenberg. Low-velocity layers in the earth, ocean, and atmosphere. *Science*, vol. 131, p. 959–65, April 1, 1960.

ADDITIONAL IDEAS AND SOURCES

1. A general account of the French-American Mid-Ocean Undersea Study (FAMOUS), part of which was described in this chapter, is given in the article "The floor of the Mid-Atlantic rift" by J. R. Heirtzler and W. B. Bryan (*Scientific American*, vol. 233, pp. 78–90, August 1975). In addition to some excellent photographs and diagrams, the article includes interesting descriptions of the surveying methods.

2. Additional seismic evidence of the asthenosphere (low-velocity zone) is described by Don L. Anderson in "The plastic layer of the earth's mantle" (*Scientific American*, vol. 207, pp. 52–59, July 1962). An important part of the evidence comes from seismic waves of very long wave length that are described by Jack Oliver in "Long earthquake waves" (*Scientific American*, vol. 200, pp. 131–43, March 1959). Both articles are available as offprints from W. H. Freeman Co., 650 Market Street, San Francisco, CA.

3. In addition to the occurrence in Papua New Guinea, plates of oceanic crust and mantle have been displaced onto continents and islands in a number of places around the margin of the Pacific, in the West Indies, and in the Mediterranean region. Places in California where these rocks may be seen are described by Edgar H. Bailey, M. C. Blake, Jr., and David L. Jones in "On-land Mesozoic oceanic crust in California Coast Ranges" (pp. C70–C81 in *Geological Survey Research 1970*. U.S. Geological Survey Professional Paper 700-C, 1970).

4. Earth scientists in the U.S.S.R. have reportedly drilled to depths of more than 30,000 feet in the continental crust, and further efforts for deep drilling to explore the crust are being planned there as well as in the United States and elsewhere. These are some of the interesting studies within what is called the International Geodynamics Project, an intensive program aimed at understanding the basic mechanisms of earth deformation. Ongoing information on these studies, as well as many other important projects, is reported frequently in *Geotimes*, a monthly news magazine of the earth sciences published by the American Geological Institute (5205 Leesburg Pike, Falls Church, VA 22041).

Evidence of uplift following retreat of the last ice sheet from eastern Canada—a shoreline 840 feet above sea level (the horizontal terrace in the distance) that was once the edge of ancestral Hudson Bay. The shoreline is also marked by boulders dropped by the ice sheet (foreground); those above water level have remained in place and those below were removed by waves. Photograph by George M. Stanley.

17. A Balance in the Outer Earth

A Balance in the Outer Earth

Although the ductile rocks of the asthenosphere may seem remotely deep, they affect the earth's surface in many ways. As we shall find in this chapter, they yield under the weight of water bodies and ice sheets, so that the earth sinks when these loads are added and rises when they are removed. The sinking and rising affect slopes and vertical positions on the surface and therefore modify all the gravity-driven processes examined in earlier chapters. Our purpose at this point is to use our knowledge of surface processes to detect and interpret the movements beneath the surface. We will need to do some sleuthing, for neither the movements nor their effects are necessarily obvious. The terrain in Figure 17-1, for example, must have seemed just another desert to immigrants crossing it in wagon trains bound westward through Utah. The cliff in the right foreground, however, was cut by powerful water waves only 20,000 years ago, and a lake nearly 1,000 feet deep extended from the foreground to the horizon and 150 miles beyond!

This was Lake Bonneville, the largest of several lakes that filled basins in the western deserts during the Pleistocene glaciations (Figure 17-2). The lakes formed as a result of the increased rainfall and the reduced evaporation of the glacial periods rather than from the effects of the ice sheets themselves. They are therefore called *pluvial* (rain-caused) lakes rather than glacial lakes. Their levels rose and fell with the changes in temperature and precipitation that caused the cyclic glaciations described in Chapter 7. Lake Bonneville was thus brimful during the period of the last glacier advance, around 20,000 years ago, and stayed at that level long enough to erode cliffs in solid rock. The cliff in the foreground of Figure 17-1, for example, was eroded into a hill of lava, and the hill in the middle distance shows a similar cliff at the same level, eroded into the same hard rock. The mountain ranges in the distance, too, have wave-cut cliffs and benches high on their sides, showing that the mountains must have been islands in the deeper parts of the lake.

FIGURE 17-1
View southward from the Grouse Creek Mountains into the Great Salt Desert of western Utah. The gravel at the base of the cliff on the right was a beach, and another beach at the same level crosses the butte in the middle distance, at the foot of the shadowed cliff. Other shorelines cross the butte below that level, and still lower ones formed low beach ridges and lagoons (the white clay flats) in the center of the photograph.

Ancient Beaches and Lake Sediments

The lake overflowed about 16,000 years ago at the point shown in the upper right part of Figure 17-2. The outlet river eroded downward so rapidly that the lake's surface dropped 350 feet in the ensuing 1,000 years, leaving the lake's wave-cut cliffs as distinct high-water marks. Then, about 13,000 years ago, the climate became drier and warmer and the lake evaporated to levels hundreds of feet below the highest shoreline. The water surface rose again moderately during another pluvial period about 11,000 years ago, but it then fell further and underwent additional cyclical changes. The modern remnant, Great Salt Lake, is thus bordered by slopes that were once underwater and show a series of terraces developed by waves as the lake fell from one level to another.

Dating this history in years is especially important in order to understand Lake Bonneville's effects on the deep earth. The dating has been done by the radiocarbon method, using plants and mollusk shells buried in sedimentary

FIGURE 17-2

Map showing maximum extent of Lake Bonneville and other large pluvial lakes of the basin-and-range region. Based on *Glacial map of North America* (Geological Society of America, 1945).

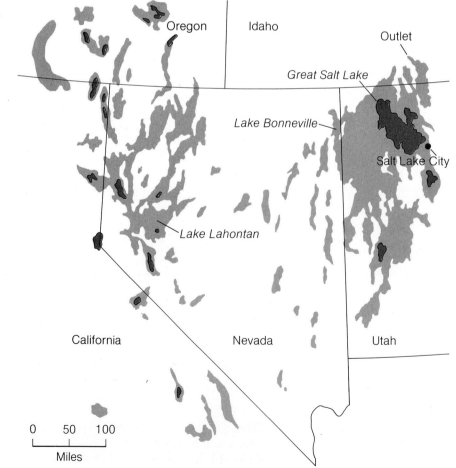

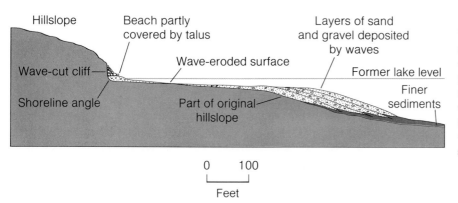

Hillslope
Beach partly
covered by talus
Layers of sand
and gravel deposited
by waves

Wave-eroded surface
Former lake level

Wave-cut cliff
Finer
sediments

Shoreline angle
Part of original
hillslope

0   100
Feet

FIGURE 17-3

Vertical section through a cliff
and an offshore shelf, which
were eroded by Lake Bonneville
from a hillslope of rock. The
resulting deposits have a greater
volume than the eroded rock
because much of the sediment
was brought to the lake by
streams.

deposits during the erosion of the successive shorelines. How does one use sedimentary deposits to date erosional features such as the cliffs? As Figure 17-3 shows, the cliff lies next to a gravel beach that passes laterally into patchy, thin deposits of sand and gravel on a gently inclined rock surface. Waves apparently eroded this surface downward as they cut the cliff landward, and the angular junction between the two surfaces, called the *shoreline angle*, marks the lake level at that time. Note that most of the sediment was transported to deeper water, where it accumulated in thick layers that slope into the lake basin. The tops of the terraces are thus partly cut in rock and partly constructed of detritus. Fossils found in the terrace deposits can thus be used to date the shoreline and cliff associated with those deposits.

Distinct cliffs and terraces, however, formed only where the lake lay against steep slopes, and much of it lapped onto broad desert plains. The high lake levels can be identified there, but only by searching for low beaches that are not nearly as obvious as the cliffs. The search entails walking over the desert and looking for low ridges of pebbles that are distinctly rounded and well sorted by size. To show that the beaches can be recognized quite easily, Figure 17-4 illustrates the surface materials in an actual case—starting on a hillslope above the highest lake level, proceeding to a typical stream bed near the shoreline, and finally ending on the ancient beach. The three photographs also illustrate the actual history of most of the lake sediment, for more was brought to the lake by streams than was eroded at the shoreline.

The sediment was then transported by waves, which often carried it alongshore because the waves arrived obliquely against the beach, a phenomenon described for ocean waves in Chapter 5. In Figure 17-1, for example, gravel bars were built downwind (toward the left) of the beach at the foot of the hill in the middle distance. The bars enclosed lagoons that filled partly with clay to form the dry whitened flats in the middle distance. Each bar and clay-floored lagoon represents a temporary level of the lake. The highest formed first, and the lake then fell about 50 feet before the next bar and lagoon were formed. These deposits and their contained fossils can thus be used to date the lake's history, just as can the deposits associated with the cliffs formed at steeper lake margins.

A final part of the record, and an important one, lies in the finer sediment winnowed (sorted) from gravels near shore and transported into deeper parts of the lake. The grains in this sediment generally become finer as the water

FIGURE 17-4

Gravels near the edge of Lake Bonneville. Top, rubbly soil on a hillslope of 10°. Middle, wash 0.5 mile downstream. Bottom, highest Bonneville beach.

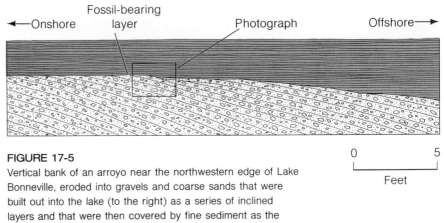

←Onshore Fossil-bearing layer Photograph Offshore→

FIGURE 17-5

Vertical bank of an arroyo near the northwestern edge of Lake Bonneville, eroded into gravels and coarse sands that were built out into the lake (to the right) as a series of inclined layers and that were then covered by fine sediment as the lake deepened. Courtesy of Victoria Todd.

0   5

Feet

becomes deeper, so that vertical gradations in grain size at a given site can be used to judge the filling or emptying of the lake, much as in the vertical sequences exposed in the Book Cliffs (Figure 14-6). In Figure 17-5, for example, a gravel bar sloping gently offshore and grading laterally to sandy sediment was covered by pebbly sand and then by silty sand as the lake deepened. Snail shells were found in the lowest layer of silty sand, and they were dated at approximately 15,000 years by the radiocarbon method. The deposits thus record a partial refilling of the lake at that time. In such ways, then, the history of the lake has been determined with considerable precision.

Reactions of Lake Bonneville with the Asthenosphere

It is important that this history be convincing because of another, unexpected discovery: the shorelines that can be dated and walked-out so exactly are no longer level! This relation is especially striking with respect to the highest shoreline, which can be identified positively in countless places and must originally have been as level as the surface of the lake. At present, however, its traces on islands near the deepest parts of the lake are fully 200 feet higher than its traces in the lake's outermost bays. Even a difference this large might seem reasonable if the shoreline were tens of millions of years old, but it is only 20,000 years old—as young as the youngest parts of the estuaries and bays of the Atlantic coast. Moreover, the upbowed shape of the shoreline cannot be an arch caused by regional changes such as those described in Chapter 15, for it rises symmetrically over the lake basin and is confined to it (Figure 17-6). The deformation must somehow relate to the lake itself.

It is important to note that these discoveries did not require intricate, modern instruments. The upbowing was noted in the late 1800s by G. K. Gilbert, whose explanation of it is considered generally correct today (Reference 1). The weight of the lake, Gilbert suggested, was sufficient to press deep-seated materials outward from under the basin. The basin thus became downbowed, and the highest shoreline was formed at that time (Figures 17-7A

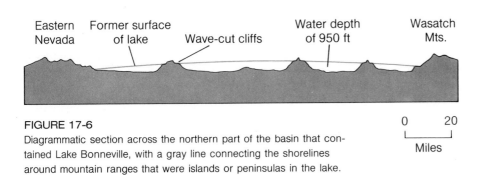

Eastern Nevada Former surface of lake Wave-cut cliffs Water depth of 950 ft Wasatch Mts.

0   20
Miles

FIGURE 17-6
Diagrammatic section across the northern part of the basin that contained Lake Bonneville, with a gray line connecting the shorelines around mountain ranges that were islands or peninsulas in the lake.

and B). When the lake evaporated, the deep-seated materials flowed back under the basin, reinstating its original shape and thus upbowing the shoreline to its present configuration (Figure 17-7C).

Measuring the degree of buoyancy Such evidence of flow in the deep earth indicated the presence of the asthenosphere long before it was detected by the seismological studies described in Chapter 16. The early investigations did not, however, establish much about its nature. How much of a load is needed to initiate flow? Does the asthenosphere respond in the same way in all parts of the earth? Seeking answers to such questions, Max D. Crittenden, Jr., restudied Lake Bonneville's shorelines (Reference 2). He was especially interested in seeing how closely the details of the upbowed surface correspond to the detailed distribution of the lake's weight. To do this, he first determined the water depths at many localities by finding the difference in altitude between the lake bottom (the present land surface) and the closest trace of the high-level shoreline. The specific water depths were then averaged over circles 25 miles in radius, for it was desirable to simplify the resulting map to that degree of detail that the asthenosphere might "feel" through the lithosphere. The contours in Figure 17-8A are based on this set of depth values.

He then plotted another map, showing the upbowing of the highest shoreline (Figure 17-8B), by using the elevations measured at many places along the shoreline, data that in no way depend on the first map. The correspondence

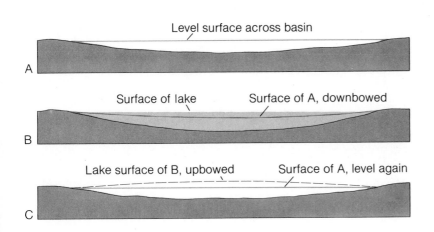

Level surface across basin

A

Surface of lake   Surface of A, downbowed

B

Lake surface of B, upbowed   Surface of A, level again

C

FIGURE 17-7
Changes in the basin of Lake Bonneville caused by the filling and evaporation of the lake. A, the basin before the lake formed; B, the basin downbowed by the lake, when the highest shoreline features formed; C, upbowing of the shoreline when the lake evaporated and the earth returned to its former state.

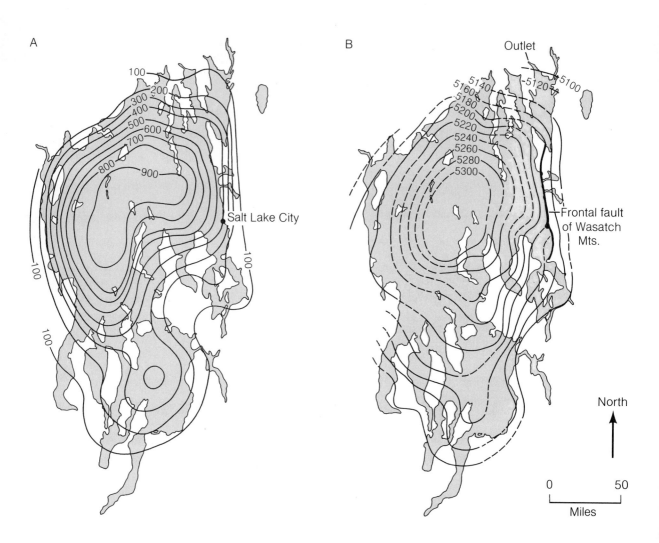

A

B

Outlet

Salt Lake City

Frontal fault
of Wasatch
Mts.

North

0   50

Miles

FIGURE 17-8

A, map showing depths of Lake Bonneville, averaged over circles 25 miles in radius and depicted by lines through points of equal water depth. B, map showing present elevations of the up-bowed Lake Bonneville surface, with solid lines showing where the highest shoreline was measured accurately. From Max D. Crittenden (Reference 2).

between the two maps is complicated somewhat by the downfaulting of the basin near Salt Lake City, but the general relation they show is striking: the deeper the water, the more elevated the shoreline. Note especially that the southern end of the lake, which constituted an irregular water body only 50 miles across and mainly less than 300 feet deep (Figure 17-8A), corresponds to an uplifted arch of the same shape (Figure 17-9B). As the contours in Figure 17-8B indicate, the lake in that southern lobe pressed the basin some 60 to 75 feet below its pre-lake level. We can conclude that the material in the asthenosphere is caused to flow outward by loads that are very small compared to the mass of the lithosphere.

This conclusion implies a very important relation: the various segments of the lithosphere act as though they were floating on the asthenosphere. Like wooden rafts in a pond, they sink when loaded and rise again when unloaded. Moreover, in the case of Lake Bonneville they did so under comparatively small loads (equivalent to sprinkling a few sand grains on a wooden raft!). Crittenden saw a way of determining if this buoyancy had indeed been perfect. He reasoned that if it had been perfect the weight of the asthenosphere pressed out by the lake should equal the weight of the lake, just as the weight of water displaced by a floating block of wood equals the block's weight. He was able to determine the weight of the lake accurately by measuring its volume from the depths shown in Figure 17-8A and multiplying that figure by the density of water. He then determined the volume of displaced asthenosphere by measuring the lens-shaped space between the elevation of the outermost shoreline and the uplifted form shown in Figure 17-8B (Figure 17-7C shows this space in cross section). This volume was then multiplied by the density of the asthenosphere, as measured from peridotite such as that exposed in Papua New Guinea. Crittenden's results showed that the weight of the displaced asthenosphere was at least 75 percent of the weight of the lake and probably nearer 90 percent of it. If we consider the probability that the depression extended beyond the lake's margins and the likelihood that the basin is still rising due to return flow of asthenosphere, we see that the result is impressive. We can conclude that the lithosphere in that region floats with nearly ideal buoyancy on the asthenosphere.

Loading and Unloading by Ice Sheets

Have other regions been affected similarly? Is the asthenosphere indeed worldwide? To answer these questions, we must find other places affected by temporary loads; those covered by the last great ice sheet are ideal because we have a detailed history of its growth and decay. Moreover, lakes that formerly occupied depressions next to the ice sheet can be used to date the subsequent upbowing of the ground. The basins of the Great Lakes, for example, are bordered by shorelines that are now tilted due to uplift after the melting of the ice sheet. Most of these shorelines are less obvious than those of Lake Bonneville because the moister climate of the Great Lakes region has covered them with plants and soil (Figure 17-9). They are, however, basically similar to the Bonneville beach lines and gravel bars, and they have been examined and mapped over much of the region. They have also been dated by radiocarbon determinations made on pieces of wood included in the shoreline gravels and other lake sediments. This information is especially valuable because the former positions of the ice sheet have been mapped and dated, as described in Chapter 7. From these two kinds of data maps can be constructed showing the extent of the lakes and the ice sheet at various stages (Figure 17-10).

An important difference between these lakes and Lake Bonneville is that they always received enough water to overflow through outlet rivers—they never shrank because of evaporation. The principal changes in the shapes of the lakes in Figure 17-10 are due to the melting back of the ice sheet, which had dammed the former outlets of the lake basins and then reopened them as it retreated. At the stage shown in map A, the lakes drained into the ancestral Illinois River at the site of Chicago and into a seaway that occupied the St. Lawrence River Valley. Retreat of the ice to the positions shown in map B freed an outlet into what is now the St. Croix River. The central lakes were reduced dramatically at a later stage when a low outlet that had been covered by ice was exposed at North Bay (map C). As map D shows, however, the lakes then filled again even though they continued to drain through that outlet!

FIGURE 17-9
Left, grassy terraces (in the distance) marking two former levels of a glacial lake near the eastern edge of modern Lake Huron, which lies 105 feet vertically below the upper terrace. Right, excavation in the lower terrace, in a bouldery deposit that was a beach sloping toward the right foreground. The finer gravel beneath was deposited in deeper water, so that the sequence records the shallowing of the lake at this site. The patterned board was used in surveying the terrace elevations. Photographs by George M. Stanley, from "Lower Algonquin beaches of Cape Rich, Georgian Bay" (*Geological Society of America Bulletin,* vol. 48, pp. 1665–86, 1937).

To understand the latter change, we must turn to elevation surveys of the shorelines, which record the vertical movements since the ice retreated. The diagrams in Figure 17-11 are vertical sections aligned about north-south through ancestral Lake Michigan, and each represents a stage depicted in the maps of Figure 17-10. The vertical scale has been exaggerated enormously to show the changes. The basic evidence for the reconstructions is shown in the lowest diagram, which depicts the tilted shorelines about as they are today. Diagram C was made by bringing the youngest shoreline back to level to reconstruct the lake surface of that age. Diagrams A and B were made similarly, using the shoreline of appropriate age. The reconstructions have the effect of reloading the earth with the ice sheet, step by step. Two important relations can thus be seen: (1) the ice sheet pressed the earth down so that the deepest part of each lake lay next to it, and (2) the depression followed the ice sheet northward as it retreated. Note how these relations explain the refilling of the central lakes illustrated in Figure 17-10D. The deep outlet opened by the ice at North Bay was held down when the ice front was near it, so that the lakes drained through it to reach an all-time low level (Figure 17-10C). When the ice wasted back far to the north, that part of the lake basin rose more than the southern part and the water level in the lakes thus rose again.

A dynamic model of a load on the earth   Perhaps the best way to visualize the "dimple" caused by these loads, as well as the broader changes caused by the ice sheet, is to construct a table-sized model of the lithosphere, the asthenosphere, and an ice sheet (Figure 17-12). The one described here

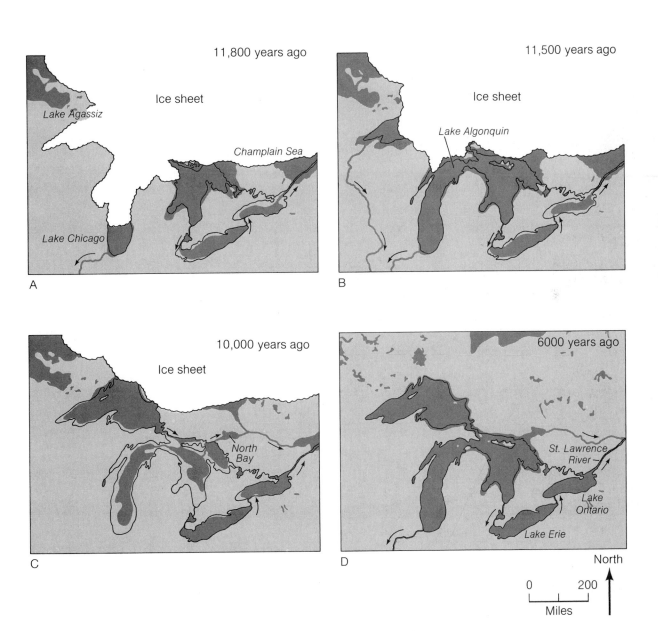

11,800 years ago

Ice sheet

Lake Agassiz

Champlain Sea

Lake Chicago

A

11,500 years ago

Ice sheet

Lake Algonquin

B

10,000 years ago

Ice sheet

North Bay

C

6000 years ago

St. Lawrence River

Lake Ontario

Lake Erie

D

North

0   200

Miles

FIGURE 17-10
The Great Lakes at several stages of wasting-back of the last ice sheet, with outlines of modern lakes for comparison. After V. K. Prest, ''Quaternary geology of Canada,'' Chapter XII in *Geology and economic minerals of Canada* (Department of Energy, Mines, and Resources, Ottawa, Canada, 838 pp., 1970).

was made from a rectangular glass tank filled partly with layers of gelatin, a substance that is elastic to short-term forces (it wiggles when thumped) but flows very slowly when pressed over long periods. It is, of course, much weaker than rocks, but if we scale the earth down to this size we should also scale its strength down, and the gelatin is more or less appropriate. To make the asthenosphere of the model less rigid than the lithosphere, the gelatin in it was mixed with more water than that in the lithosphere (3 cups of water to 1 tablespoon of dry gelatin compared to $1\frac{1}{2}$ cups per tablespoon). The difference became apparent when a body of gelatin was added to proxy for an ice sheet (Figure 17-12B and C). The lithosphere bent without thinning, whereas the asthenosphere flowed laterally from under the load. The sinuous

FIGURE 17-11
Diagrammatic sections through ancestral Lake Michigan and the ice sheet, with vertical scale greatly exaggerated. The stages correspond to those in Figure 17-10.

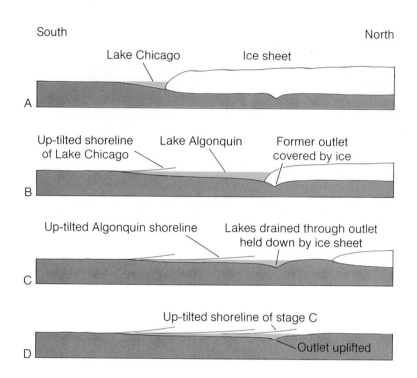

FIGURE 17-12
A, model constructed of relatively rigid gelatin ("lithosphere") lying on less rigid gelatin ("asthenosphere") that is layered to make deformation more visible. B, same model $\frac{1}{2}$ minute after a gelatin load ("ice sheet") was added. C, same model 25 minutes later, when "asthenosphere" had flowed from under the load and distinct depressions and low bulges had formed. The ice sheet, too, flowed (spread).

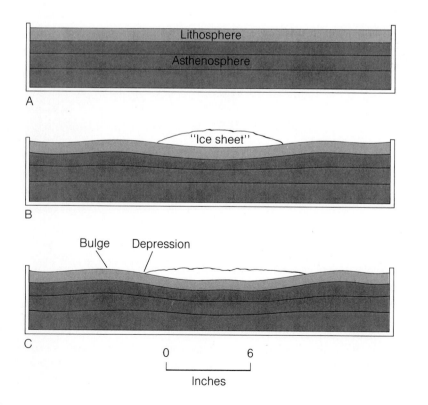

378

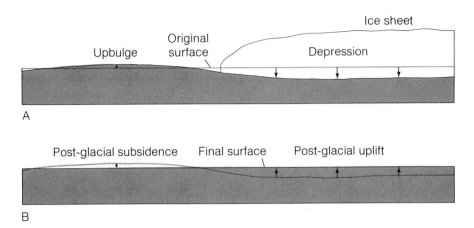

A

B

FIGURE 17-13
Vertical sections through a depression and upbulge, with arrows indicating the movements that caused the bulge (A) and those during return to normal after the ice sheet melted (B).

shape of the lithosphere shows that the layer was stretched slightly, but this does not effect the main result: the "ice sheet" formed a depression (dimple) that extended beyond its edges and would thus hold the glacial lakes on the actual earth.

The "ice sheet" also formed the low bulge that lies just outside the depression and creates extra depth in the lake basins. Is there evidence of a similar bulge in the earth? Let us first consider the likely nature of this evidence. If there had been an upbulge, the earth would have recently returned to normal by sinking, as in the lower diagram of Figure 17-13. One indication of sinking is that the sea has encroached inland along the eastern seaboard in areas well beyond the margins of the ice sheet. Evidently the sinking is one cause for the development of large estuaries along the eastern coast of the United States. Figure 17-14 shows the approximate distribution of the subsided areas compared to that of areas that have risen since the ice sheet melted. Note that the areas that have risen are those that were covered by the main ice sheet or near it, and those that have subsided are a long distance from it. Note, too, that the comparative amounts of subsidence and uplift indicate that the upbulge was of much lesser magnitude than the depression, just as in the gelatin model. The lithosphere of the real earth thus appears to have acted as a comparatively rigid layer that was pressed down bodily under the ice sheet and rose in a low bulge outward from the loaded region. The material in the asthenosphere must have flowed laterally to accommodate these changes.

We can conclude that buoyant interactions such as those under Lake Bonneville have affected most of North America. Similar effects are associated with an ice sheet that once lay over northern Europe. These relations strongly support the results of seismic studies described in Chapter 16: (1) the asthenosphere is worldwide; (2) the rocks in it are ductile; and (3) the lithosphere lying on it typically acts as a coherent, rigid layer.

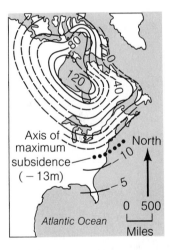

FIGURE 17-14
Generalized uplift and subsidence of eastern North America since approximately 6,000 years ago, as compiled from various sources by Richard I. Walcott. Contours are in meters. From "Late Quarternary vertical movements in eastern North America: quantitative evidence of glacio-isostatic rebound" (*Reviews of Geophysics and Space Physics*, vol. 10, pp., 849–84, 1972).

Continents and Oceans: Evidence from Gravity

The sensitivity of the asthenosphere to loading by ice sheets and pluvial lakes suggests that the various segments of the lithosphere should themselves be balanced in a floating equilibrium. We saw in Chapter 16 that the oceanic crust differs from the continental crust in composition. The basalt and gabbro of the oceanic crust give it a density of around 3 grams per cubic centimeter, whereas the abundant granite and other quartz-bearing rocks of the continental crust give it an average density of around 2.8 grams per cubic centimeter. If we imagine two blocks of these materials floating in ductile peridotite with a density of 3.3 grams per cubic centimeter, we realize that one will be buoyed only slightly higher than the other (Figure 17-15A). If, however, we use the dimensions of the crust and lithospheric mantle determined by the seismic studies described in Chapter 16, we find that a block of continental lithosphere should float distinctly higher than one of oceanic lithosphere (Figure 17-15B). The differences in elevation between the tops of continents and the floors of oceans is thus explained partly by the greater thickness of continental crust and partly by the lesser densities of the rocks composing it.

We may thus imagine a model such as that of Figure 17-12 but with gelatin layers having densities and thicknesses appropriate to the real earth (Figure 17-16). We think of it as gelatin rather than rock because we need to scale down the strength of the material in a model this size. We must imagine large masses of rocks slowly bending and flowing even though they are elastic (rigid) to impacts caused by sudden movements, such as those on faults. The model also gets us away from thinking of the outer earth as a series of gigantic blocks. The curving depressions under Lake Bonneville and the ice sheets show that the rigid lithosphere yields by bending rather than by moving up and down as separate chunks. As the model suggests, all parts of the earth may deform readily enough to settle into a continuous, internally balanced configuration.

We could test the gelatin model for ideal buoyancy by inserting pressure gages at various places on any one level in the asthenosphere. The pressures should all be the same. Or we could freeze the gelatin and cut it into vertical segments of equal cross-sectional area and extending to the same depth below sea level in the asthenosphere. These pieces should also be the same in weight. This relation is true not only in the model's ocean basins and continents but also in mountain ranges, plateaus, and broad continental basins, each of which is underlain by a particular thickness of crustal material (Figure 17-16). All these relations are due to the simple fluid balance achieved by flotation on the asthenosphere. We now know that this balance in the real earth is so nearly perfect that if we could place an ice sheet or a Lake Bonneville on our diagram (at this scale they would be so thin as to be imperceptible), asthenosphere material would flow out from under them until the various segments would again have the same weight!

Isostasy and gravity Earth scientists have given the name *isostasy* to this concept of buoyant equilibrium. Vertical changes such as those described are called *isostatic adjustments*, and regions that have attained equilibrium

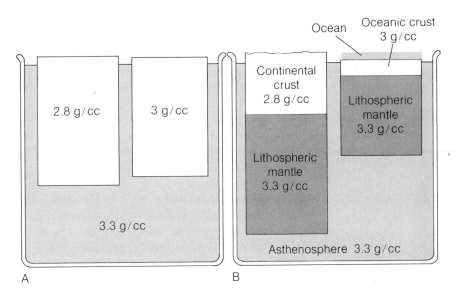

FIGURE 17-15

A, blocks of equal weight but different densities, floating in a somewhat denser liquid. B, blocks of continental and oceanic lithosphere, with the top of the continent about 3 miles above the bottom of the ocean basin, and the surface of the ocean about 1,000 feet below the average continental surface (these are the average differences in elevation between the actual features).

are termed *isostatically balanced*. But how does one test the actual earth to see which regions are isostatically balanced and which are not? Ideally, we would like to weigh segments of the earth, as we imagined doing with the model. This procedure is impractical, but we can compare the weights of segments by measuring their gravitational pull on sensitive instruments placed on the surface. The method is based on the law of gravitation—any two bodies attract one another with a force proportional to the product of their masses (weights) and inversely proportional to the square of the distance between them.

The earth's gravitational force is called gravity and is generally measured in units called gals, named after Galileo. One gal is defined as the force that will accelerate a falling body by 1 centimeter per second during each second of fall. Gravity has an average strength of 981 gals or 981 cm/sec^2 (32 ft/sec^2) on the earth's surface but varies moderately (by a few gals) from place to place. The chief cause of these variations is elevation—greater or lesser distances from the earth's main mass. This effect can be calculated readily if the local elevation is known (to within a foot or so, for precise determinations of gravity). The earth's rotation also effects measurements of gravity, for the centrifugal (outward pressing) force due to rotation acts in the opposite sense (direction) to the force of gravity. As Figure 17-17A indicates, the effect of

FIGURE 17-16

Imaginary gelatin model of the earth with average densities indicated. The heights of mountains are exaggerated, but the parts of the lithosphere are shown at approximately natural scale.

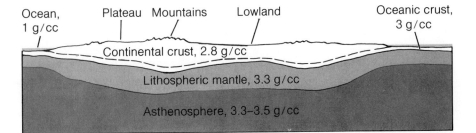

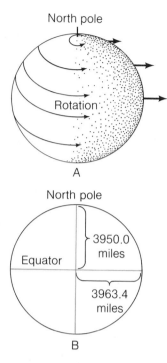

FIGURE 17-17
A, schematic view of the earth, with heavy arrows indicating how centrifugal forces (forces due to rotation) vary with latitude. B, section through the earth showing an exaggerated flattening due to centrifugal forces, and the actual lengths of the polar and equatorial radii.

rotation decreases from a maximum at the equator to nil at the poles. Because the effect is regular, it can be accounted for by calculations. The centrifugal force also makes the earth bulge at the equator, such that sea level is about 13 miles farther from the earth's center at the equator than at the poles (Figure 17-17B). If we assume that the resulting shape is regular, we can also take this effect into account by calculations.

Results of gravity measurements   Gravity can be measured with great precision (to the nearest 0.000,01 gal) and has been so measured at thousands of localities the world over. When corrected for the variations due to elevation and latitude, the measured values are typically within about 0.01 gal of being the same. The remaining differences are caused mainly by unusually heavy or light bodies of rock close to the instrument, which have an appreciable effect because of the distance-squared relation in the law of gravitation. When the differences due to these rock bodies are accounted for, the corrected values of gravity indicate that most large segments of the earth have very nearly the same weight. Most parts of the lithosphere must thus be isostatically balanced, as in the model in Figure 17-16.

The region around Hudson Bay, however, which was covered by the last remnant of the ice sheet, is a segment that is still undergoing isostatic adjustment. This region is rising at a rate of 0.4 to 0.7 inch per year near the center of the uplifted area shown in Figure 17-14. Figure 17-18 shows the departures from normal gravity for this region, which were determined by correcting thousands of measured values for latitude and elevation. The resulting values were then averaged over areas measuring about 70 by 90 miles in order to smooth local variations due to relatively small rock bodies. The results show that the lithosphere and upper asthenosphere under Hudson Bay are lighter than they should be for the surface elevations—the earth here is not in isostatic balance. R. I. Walcott has calculated the amount of disbalance from the gravity values and suggests that the center of this area will continue to rise until it is between 500 and 1,200 feet higher than it is now. The size and shape of Hudson Bay would thus be changed greatly, as would those of all other coastal water bodies in eastern Canada. The movements would also tilt the Great Lakes basins increasingly to the south, causing retreat of the northern shorelines and increased drainage through outlets in the southern parts of the lakes.

Conclusions   Because isostatic adjustments are driven by gravity, we must allow that all types of loading and unloading may cause vertical movements. Sedimentary deposits should start to depress their basins when the deposits are little more than 100 feet thick. Might this depression have been a cause of the continued deepening of the basin under the Book Cliffs accumulation? The results of the Lake Bonneville study indicate that it should have been. And what of eroded regions—should they not rise slowly as rocks are stripped from them? The evidence of continued elevation in eastern Canada suggests that they should. Isostasy is thus a universal concept in interpreting all earth systems affected by shifts of material on the surface.

Isostatic adjustments also give us a view of the intricate connections between the earth's depths and the processes going on at its surface. We must,

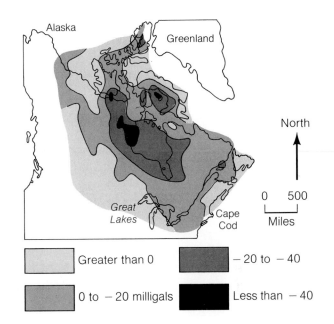

FIGURE 17-18

Map of eastern Canada, showing areas with gravity values less than expected at these latitudes. The lines connect all points of equal gravity, and the numbers indicate the departures, in gals, from average gravity. From R. I. Walcott, "Late Quaternary vertical movements, in eastern North America: quantitative evidence of glacio-isostatic rebound" (*Reviews of Geophysics and Space Physics,* vol. 10, pp., 849–84, 1972).

in fact, recall our investigation of the ice ages to see the full extent of the connections described in this chapter: a change of a few degrees in the average temperature of the atmosphere led to the pluvial lakes and ice sheets that generated the flow of solid rock 100 miles beneath the surface! And we may imagine still further connections. Would not, for example, the shift of 300 feet of water from all the oceans have led to a disbalance causing flow of asthenosphere from under the continents toward the ocean basins? Crittenden's results suggest that the loads involved were sufficient to cause such a rebalancing.

Summary

Our overall findings may be summarized as follows:

1. The asthenosphere is ductile enough to flow from under loads imposed on the earth's surface, even though it is rigid to short-term forces such as those associated with seismic waves.
2. The study of Lake Bonneville shows that small loads may be effective—for example, a few hundred feet of water covering an area 50 miles square.
3. The depressions beneath ice sheets extended for tens of miles beyond the outer edges of the sheets, and the surface rose in a low bulge just beyond each depression. These relations indicate that the lithosphere is a rigid, elastic layer floating buoyantly on the asthenosphere.
4. Measurements of gravity support this concept of buoyancy (isostasy), for they show that segments of the earth with equal surface areas have very nearly equal weights.
5. Where measurements of gravity are either too large or too small for the rocks that form that part of the earth, isostatic balance has evidently not been achieved, and we may expect such a region to rise slowly or to sink.

REFERENCES CITED

1. G. K. Gilbert. *Lake Bonneville.* U.S. Geological Survey, Monograph 1, 438 pp., 1890.
2. Max D. Crittenden, Jr. New data on the isostatic deformation of Lake Bonneville. U.S. Geological Survey, Professional Paper 454-E, 31 pp., 1963.

ADDITIONAL IDEAS AND SOURCES

1. As mentioned for G. K. Gilbert's interpretation of the Lake Bonneville basin (Reference 1), some of the early surveys and studies of the West resulted in some truly remarkable advances in understanding the earth. A general description of Gilbert's various studies is given in Chapter XXI of *Giants of geology* by Carroll Lane Fenton and Mildred Adams Fenton (Garden City, N.Y.: Dolphin Books, Doubleday & Co., 318 pp., 1952, paperback). Chapter XX likewise describes the explorations and discoveries by John Wesley Powell in the great canyons of the Colorado Plateau.
2. *Glacial isostasy* (Stroudsburg, Pa: Dowden, Hutchinson & Ross, 491 pp., 1974) is an exceedingly useful collection of 28 papers edited and annotated by John T. Andrews. Recent work and ideas are well represented, and the concept of glacier loading and isostatic adjustment is traced historically by presenting a number of earlier papers.
3. An additional analysis by Crittenden showed how ductile the asthenosphere was during the uplift of the Lake Bonneville basin. Recall that radiocarbon dates on shells and wood fragments buried in the lake sediments have given a history of filling and emptying of the lake. The dates made it possible to calculate the *rate* of uplift, and thus the rate at which the peridotite in the asthenosphere flowed back under the basin. This value gives a measure of the ductility or viscosity of the asthenosphere. It is interesting that the ductility of the asthenosphere under Scandinavia and eastern Canada is 10 to 100 times less (that is, the asthenosphere there is more rigid). Measurements made in Glacier Bay, Alaska, indicate a ductility similar to that under Lake Bonneville, as does a recent study in Greenland by Norman W. Ten Brink entitled "Glacio-isostasy: new data from west Greenland and geophysical implications" (*Geological Society of America Bulletin*, vol. 85, pp. 219–28, 1974). The greater ductilities suggest that the asthenosphere lies closer to the surface in those places, as supported by seismic studies in the basin-and-range region (Figure 16-19).
4. The idea that the earth is bowed downward under large loads is supported by data from greenland. Note in the cross section in Figure 7-4, for example, that the island under the central part of the ice sheet is partly below sea level. Assuming a density of 3.3 grams per cubic centimeter for the asthenosphere, we can calculate that the central part of the island would rise approximately 2,600 feet if the ice sheet were to melt.

The crater Lambert, approximately 18 miles (30 km) across, blasted by the impact of a meteorite into the lava plain of Mare Imbrium, one of the large basins on the side of the moon facing earth. The view is turned so that the moon's north pole is to the left (in order that the shadows will make the vertical dimension more apparent). Apollo 15, mapping photo 1153. Courtesy of Don E. Wilhelms, U.S. Geological Survey.

18. A History Derived from the Moon and Planets

A History Derived from the Moon and Planets

The view in Figure 18-1 shows one of the most heavily bombarded, cratered regions of the moon. It is part of the lunar highlands, a terrain that makes up most of the far side of the moon and stands a mile or so above the moon's main basins; thus, it is somewhat comparable to a continental surface on the earth. Similar though less extensive highlands on the moon's near side were visited by the astronauts during the Apollo missions. It was found that impacts have disintegrated the rocks to rubble and dust, even melting them in part, but that the solid rocks beneath are evidently gabbros similar to some rocks in the earth's crust. It was therefore amazing to discover, by radiometric dating, that the main constituents of the moon's crust formed approximately 4,500 million years ago—within a comparatively short time (perhaps 100 million years) after the moon and planets were assembled from orbiting particles of the primitive solar system.

The chaotic materials on the moon's surface have thus brought new light to our view of the earth's history. Before the lunar and planetary programs, many scientists believed that the earth accreted as a cool body and only slowly heated to the point that it could melt and segregate a crust, mantle, and core.

FIGURE 18-1

Part of the highlands on the far side of the moon (the side always turned away from the earth). The large depression in the foreground is the crater Mendeleev. Part of an instrument boom projecting from the spacecraft is visible at the right. NASA photograph AS16 metric #1305. Courtesy of Don E. Wilhelms, U.S. Geological Survey.

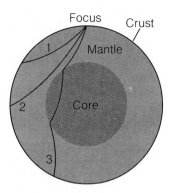

FIGURE 18-2

Refraction of P-waves by the deep mantle and core, such that they are received weakly at all stations between 6,800 and 9,500 miles from epicenters. After Beno Gutenberg.

The melting was thought to be due to heating by radioactive uranium, thorium, and potassium, a process that would take several billions of years to cause a major share of the melting and segregation. Studies of the moon and of Mars and Mercury now show this interpretation to be unlikely. A crust evidently was segregated almost at once and then cooled rapidly. As we shall see, recent dating of the earth's oldest known rocks supports the new interpretation. In this chapter we will first examine the deeper of the earth's segregated layers and then see how the moon and nearby planets help us interpret the development of the entire earth.

Soundings of the Deep Mantle and Core

Long before the lunar studies, seismic waves had been used to determine the position and likely constitution of the earth's deep mantle and core. The first determinations were based on seismograms of earthquakes with epicenters more than a third of the way around the earth from the receiving station. A striking discovery was that P-waves are received strongly up to about 6,800 miles from their epicenters, fall off sharply (almost to nothing) between 6,800 and about 9,500 miles, and at greater distances come through strongly again. The total travel times at various distances from the epicenters show that this phenomenon is due to a sudden reduction in the velocity of the waves at a depth of about 1,800 miles (2,900 km), resulting in the strong refraction of the deep-traveling waves (Figure 18-2). Wave paths 1 and 2 in the figure reflect the usual downward increase in seismic velocity, for they are refracted in curving lines back to the surface as explained in Chapter 16. Wave path 3, on the other hand, is refracted drastically at 1,800 miles depth as it enters the inner body and arrives at the earth's surface many minutes later than predicted for refraction through a homogeneous earth. The inner body is called the *core*. The great reduction in the velocity of P-waves at the boundary of the core—from about 14 kilometers per second to only 8 kilometers per second—suggests that it consists of molten material. This molten state is corroborated by the fact that S-waves, which cannot travel through liquids, are not received at all beyond about 6,800 miles from their epicenters.

Another important aspect of the core is indicated by P-waves that travel nearly straight downward and are reflected from deep interfaces such that they return almost vertically to the earth's surface (Figure 18-3). In the case illustrated, P-waves generated by a nuclear explosion at the test site in southern Nevada were reflected from the outer boundary of the core and also from an inner boundary lying 1,370 miles (2,200 km) deeper (Reference 1). The reflections are so weak that they can be detected only on sensitive arrays of seismographs designed so that all interfering seismic "noise" is filtered out. Such seismic records show that the inner core is about 1,500 miles (2,430 km) in diameter. Seismic waves that travel through it speed up so much as to indicate it is solid.

Figure 18-4 shows how the seismic wave velocities calculated through all depths in the earth reflect the nature of substances at various depths. The broad increase downward in the mantle is due to the increased rigidity of

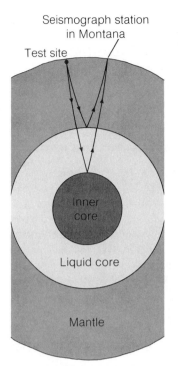

FIGURE 18-3

Reflections of downward traveling P-waves from the outer boundary of the core and also from a boundary within the core. From Bruce A. Bolt, "The fine structure of the earth's interior" (*Scientific American*, vol. 228, pp. 24–33, March 1973).

rocks under increasing pressure. This effect is similar to that of compaction at shallow depths, described in Chapter 16, only here atoms rather than mineral grains are being squeezed closer together. Studies in the laboratory, using very high pressures, indicate that at depths of around 250 miles (400 km), pyroxene is changed to garnet (a more compact mineral), and the atoms in olivine are forced into a more tightly packed arrangement to form a mineral with cubic symmetry. This cubic mineral then changes into another with still more tightly packed atoms at pressures equivalent to 420 miles (650 km) depth, and garnet begins to break down to more tightly packed metal oxides. Temperatures also increase downward, to perhaps 3,700°C at the base of the mantle, but S-waves traveling through the mantle show that little if any of it is molten. The very high pressures evidently maintain the solid state of the minerals.

Nature of the earth's core The sudden appearance of melt at the core boundary therefore implies a major change in the composition of the earth at that level. The core, in fact, is mainly metallic iron, which has a distinctly lower melting temperature than the silicate minerals of the lower mantle. Evidence that the core is mainly iron comes from several sources. The strength of the earth's gravity field indicates a far denser planet than one containing only the silicate minerals of the crust and mantle. The earth's average density is about 5.5 grams per cubic centimeter, whereas the average density of the mantle and crust is about 4.4 grams per cubic centimeter. A core with an average density of approximately 11 grams per cubic centimeter would resolve the difference. This density is too light for pure iron at the great pressures

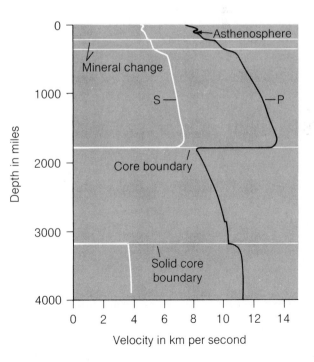

FIGURE 18-4

Variations in the velocities of seismic waves with depth. Note the great change at the boundary between the mantle and the core and the lesser ones within the mantle. From Bruce A. Bolt (see Figure 18-3).

FIGURE 18-5

Iron-stony meteorites (pallasites) cut and polished to show the relations between the metallic and silicate parts. In the left meteorite, the iron matrix is white and appears to have been intruded among the angular silicate fragments. In the right one, the iron matrix has been etched with acid to bring out its internal crystalline units—the elongate patches that intersect at angles of 60°. The stony grains on the right include some well-formed crystals (as the large dark grain). Photographs from The Center for Meteorite Studies, Arizona State University; courtesy of Carleton B. Moore.

in the core but would be suitable for iron with moderate amounts of lighter elements. Another kind of evidence comes from the sun, which emits energy (photons) showing that roughly 30 percent of its solid core is iron—a far greater abundance than is apparent in the earth's crust and mantle. Iron appears also to be abundant in parts of the solar system farther from the sun, for about 7 percent of the meteorites that have fallen to the earth consist largely of free iron, and the trajectories (travel directions) of most of these small bodies indicate an origin in the asteroid belt, which is 115 million miles farther from the sun than the earth is.

The meteorites, in fact, provide some suggestions about the earth's core and the planetary cores in general. For one thing, the iron-rich meteorites contain considerable nickel and other elements lighter than iron, resulting in a mixture with a density appropriate to that calculated for the earth's core. A second kind of evidence comes from the proportions of iron and stony (silicate) meteorites collected from actually observed meteorite falls (about 1,000 have been so collected). The meteorites that are metallic iron are essentially *all* metal, and the great proportion of the meteorites that are stony are essentially all silicates. This is in agreement with the seismic evidence from earth that the core and mantle have a sharply defined boundary. The few mixed iron-stony meteorites include an interesting type called *pallasites,* in which iron forms a continuous body around fragments and single crystals of silicate minerals (Figure 18-5). The arrangements suggest that liquid iron locally intruded the solid silicate mantle of an asteroid (or its parent body), much as plutons intrude solid rocks on earth to form mixtures of liquid melt and solid fragments (Figure 12-19). The small numbers of pallasites are suitable to a process going on at a sharp core–mantle boundary. These meteorites thus provide a suggestion of what the earth's core–mantle boundary might look like.

Origin of the Main Planetary Layers

The earth is thus partitioned into three sharply defined compositional layers: a thin crust of lightweight rocks, a thick mantle of heavier magnesium-rich silicates, and a core consisting mainly of metallic iron. One of the most provocative findings of the space program is that Mars, Mercury, and perhaps the moon also have iron-rich cores and thick silicate mantles and that the moon and probably the two planets have crusts of lighter silicate rocks. Another major discovery is that core size relative to planet size diminishes away from the sun (Figure 18-6). The substances in the primitive solar system must thus have been sorted systematically when the planets formed. Note, however, that the moon has a smaller core (if any) than it should have for its position in the solar system. Besides being deficient in iron, it is also deficient in volatile (easily vaporized) elements and relatively enriched in refractory elements, for example, uranium, thorium, and titanium. Perhaps it accreted from a belt of particles orbiting around the accreting, primitive earth, and the earth's gravitational field attracted a great share of the iron and volatile materials that would otherwise have accreted to the moon. The early relations between the moon and the earth, however, remain highly speculative.

A history based on meteorite impacts The moon's history after the main stage of accretion is far better known, for it can be based on maps of surface features and on dated rock samples collected by the astronauts. The mapping is a special kind called *remote sensing*, whereby features are studied and delineated on photographs or other kinds of images, without actually walking over the area. The photographs of the moon were taken from the earth or from spacecraft, and many were studied decades before the Apollo program. Figure 18-7, for example, is an early composite photograph that shows the main features of the moon clearly. The dark, smooth areas are called *maria* (plural for *mare*, the Latin word for sea). The Apollo missions showed the maria to be nearly flat plains consisting of basalt lavas. The lighter areas are mainly uplands roughened by countless meteorite impacts, such as those shown in Figure 18-1. Meteorite craters are clearly apparent in Figure 18-7, and the circular shapes of many of the maria suggest that they, too, are meteorite craters—giant ones that have been partly filled with lava.

FIGURE 18-6

Iron-rich cores of the inner planets sensed by space probes and shown here at the same scale. Outside of some indications of a former magnetic field, there is little evidence of a core in the moon. Distances from the sun are indicated by the arrows.

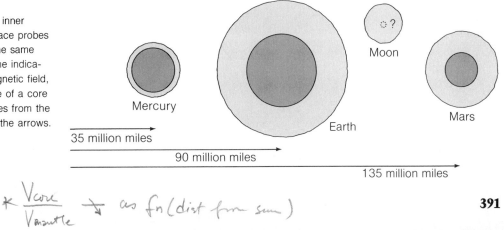

Mercury

Moon

Earth

Mars

35 million miles

90 million miles

135 million miles

The moon photographed from the earth. The image is a composite of two halves (note that the illumination of the right side is from the right, that of the left side from the left). This gives features near the central meridian an appearance of greater relief than features in the right and left parts of the image. Lick Observatory photograph L-9.

Cratering and other effects of meteorite impacts are thus important in deciphering the moon's history, and these topics have been studied in detail on earth, partly at well-preserved craters, such as Meteor Crater in Arizona, and partly by firing projectiles into various kinds of materials. It was found that craters result from sudden expansion after the intense compaction caused by the impact and penetration of the meteorite—a compression amounting to hundreds of thousands of atmospheres of pressure. A great amount of fragmental debris is ejected. The most compressed materials melt as they recoil, and they are jetted outward in incandescent streams, mixing with the solid fragments erupted along with them. The resulting crater is much larger than the meteorite, and the ejected materials cover a vastly larger area. The fragments settle in a blanket-like deposit that becomes thinner away from the crater and commonly has a pattern of rays radiating from the crater—the result of the ejection of debris in concentrated jets.

Figure 18-8 illustrates how craters and their surrounding deposits have been mapped and ordered into an age sequence (Reference 2). Of the large craters, Copernicus, Aristillus, Autolycus, and Timocharis are interpreted as the youngest because the fragmental deposits formed from them are light in tone and show the radiating patterns just mentioned. Older deposits have

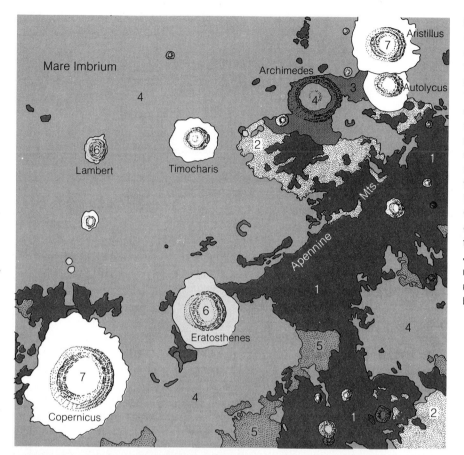

FIGURE 18-8

Photograph and map of the craters and surface deposits in the region lying just to the left and above center in Figure 18-7. The surface materials delineated on the map are numbered in order of their formation. The map and photograph are projected differently with respect to the moon's spherical surface and thus do not correspond exactly. NASA photograph from Lunar Orbiter IV, frame M-126 (courtesy of Don. E. Wilhelms) and map from Don E. Wilhelms and John F. McCauley, Geologic map of the near side of the moon (U.S. Geological Survey, Map 1-703, 1971).

Mare Imbrium

Aristillus

Archimedes

Autolycus

Lambert

Timocharis

Apennine Mts

Eratosthenes

Copernicus

0   100

Miles

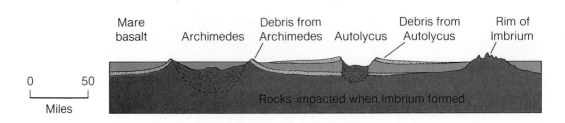

Mare basalt · Archimedes · Debris from Archimedes · Autolycus · Debris from Autolycus · Rim of Imbrium

0   50
Miles

Rocks impacted when Imbrium formed

FIGURE 18-9

Diagrammatic cross section showing probable interrelations of the deposits in Figure 18-8, along a line extending from Archimedes through the rim of Imbrium. The thicknesses of the fragmental deposits are greatly exaggerated.

been darkened mainly by dark glass formed by melting during meteorite impacts. The craters Lambert and Eratosthenes must be older than the large craters just mentioned because the deposits surrounding them have been darkened, and the pale rays from Copernicus lap across them. A closer view of the debris deposit around Lambert is shown on the chapter opening page.

All these craters were impacted from the smooth dark mare surface that covers most of the area in Figure 18-8, so that the surface must be older than they are. The crater Archimedes, on the other hand, was partly flooded by mare lava, and thus must be still older. The crater's great age is verified by its relatively flattened (eroded) rim, caused by the impacts of countless small meteorites. This method of distinguishing age is important, for some lavas have been added to the maria more recently. Flows with well-preserved steep edges, for example, lap up against Lambert and must thus be younger than that crater. These lavas are not clearly visible in the figures here but can be seen in the original photographs.

By using our knowledge of impact deposits and their age sequence, we can draw a vertical section showing their probable relations in the third dimension (Figure 18-9). Note that the mare basalts lap up against the Apenine Mountains, which Figure 18-7 shows to be part of the upraised rim of the giant crater Imbrium, measuring 750 miles across. Many isolated mountains within the rim of Imbrium are also relics of that impact, for example, the sharp, lighted peak in the upper part of the chapter opening photograph. The worn aspect of the main rim of Imbrium must have been caused by the countless impacts of small meteorites, for the moon has neither surface water nor an atmosphere that might cause erosion. The extent of the meteorite erosion of the rim of Imbrium supports the conclusion that this feature is the oldest shown in Figure 18-8.

The spacing of craters in each deposit shows that the impacts dwindled rapidly after the moon formed. The maria, for example, have far fewer craters than the highlands near the top and bottom of Figure 18-7. The earlier impacts, moreover, include all the giants—those of Imbrium and the other more or less circular maria in Figure 18-7. These impacts evidently heated large areas of the moon's surface about 4,000 million years ago, for radioactive isotopes and their decay products were redistributed ("reset") in most rocks at that time (Reference 3). The record before that time is therefore difficult to date. We do not know whether the giant impacts were the last part of a long period of rapid accumulation or whether they record a special event—a sudden accretion of exceptionally large meteorites. In either case, the spacing of craters in mare basalts younger than 4,000 million years shows that meteorite impacts decreased 100-fold during the period from 4,000 to 3,100 million years ago.

FIGURE 18-10
Closeup of fragmental ma-
terials next to a small crater
visited by the astronauts during
the Apollo 15 mission. NASA
photograph AS–25–82–11140;
courtesy of Johnson Space
Center, Houston.

Interpreting the moon rocks   One might suppose from the foregoing that the surface of the moon would consist largely of meteorite fragments, but this is far from true. Meteorites, as we have noted, are much smaller than the craters they produce. Most of the rocks on the moon's surface are fragments impacted outward from the solid basalt and gabbro beneath. The origins of these rocks must be determined from small chunks and bits of rocks and minerals, for the surficial layer is greatly pulverized and so thick that the astronauts could not dig or core through it to solid rock (Figure 18-10). Much of the deposit consists of spherules (globules) and chips of glass formed by melting and chilling during meteorite impacts (Figure 18-11). Impact-melted glass also cements fragments together in the surficial layer. Even the younger basalts are covered by the fragmental layer, some of it representing material exploded locally by small impacts and some of it thrown from distances of tens or hundreds of miles by major impacts. Most of the fragments lie about where impacts have left them, for although they creep gradually downslope— due to impacts of small meteorites, to ground tremors (moonquakes), and to expansion and contraction due to day-night temperature changes—there are neither water nor wind to erode and transport them long distances.

FIGURE 18-11

Thin section of part of the
moon's fragmental surface deposit, collected during the Apollo
15 mission. The view is 0.1
inch (2 mm) across and was
taken with polarized light so that
the glass particles, including
much of the finer materials,
are dark, whereas crystalline
grains (mainly plagioclase) are
white or gray. The sample is
number 15265, and the photograph and data were supplied by
Odette B. James.

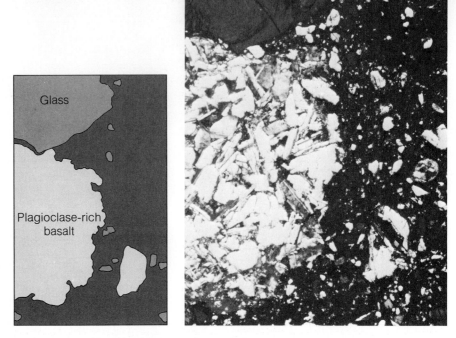

Lack of an atmosphere has kept clays and iron oxides from forming in the surficial layer, in great contrast to soils on the earth. The lunar rocks also are free of water-bearing and hydrogen-bearing minerals, such as mica, indicating that water was exceedingly scarce at depth when the rocks formed. Minerals and bits of glass have thus remained unchanged for billions of years, making it possible to determine precisely their original nature and to measure their ages by radiometric methods.

Segregation of a crust   Another great value of the fragmental layer is that one may find in it rocks ejected from all parts of impact craters, some materials from the larger craters being thrown from depths as great as approximately 6 miles (10 km). A great preponderance of plagioclase feldspar fragments at the surface in and near the highlands thus indicates that the outer 6 miles or so of the moon in those areas consist of plagioclase-rich gabbro (Figure 18-12). Plagioclase-rich gabbros form on the earth in the upper parts of large bodies of molten gabbro due to the sinking of the denser grains, pyroxene and olivine. Scarce fragments of average gabbro (about half plagioclase) and of peridotite (nearly all olivine) support the idea that the moon's plagioclase-rich gabbro is the upper part of a thick body of molten gabbro in which pyroxene and olivine settled and accumulated at greater depth. This segregation formed the moon's crust, for the plagioclase-rich gabbro is now a layer of comparatively light rock lying on heavier olivine-rich rocks that form part of the moon's mantle. Measurements of seismic waves and of the moon's gravitational forces indicate the crust is approximately 36 miles (60 km) thick on the near side of the moon and perhaps 90 miles (150 km) thick on the far side. These thicknesses imply that the body of gabbro melt from which the crust was segregated must have been at least 100 miles (160 km) thick and that it must once have covered the moon as a great molten sea.

A fragment of gabbro (a variety consisting of plagioclase and olivine) and one of peridotite, both of which escaped the effects of heating by the giant impacts 4,000 million years ago, have given radiometric ages of approximately 4,500 to 4,600 million years by the rubidium-strontium method, and determinations on other materials by the uranium-lead method have given somewhat younger ages, around 4,420 million years (Reference 3). These ages indicate that the sea of gabbro melt formed at the time the moon accreted, and it crystallized soon thereafter. The melting may have been caused by the rapid succession of impacts during accretion and by the heating due to compaction of the particles. It may also have been caused by the radioactive decay of short-lived isotopes. Analysis of one meteorite has indicated that a rapidly decaying radioactive isotope of aluminum (^{26}Al) may have been abundant enough in the primitive solar system to have caused the melting of all accreted bodies more than a few miles in diameter within a few tens of millions of years after they accreted (Reference 4).

Continued eruption of basalt All the basalts dated so far give ages younger than the giant impacts that occurred 4,000 million years ago. As already noted, basalts fill the basins made by these impacts, forming the smooth dark maria in Figure 18-7. The mare basalts could not have been melted and emplaced by the meteorite impacts, for smaller craters were impacted in the solid floors of the giant basins long before the basins were filled with lava (as Archimedes in Figure 18-8). Basalts dated so far have given a wide range of ages, from approximately 3,900 million to 3,100 million years, and we have noted that younger flows have been recognized in photographs such as that on the chapter opening page. Eruptions of basalt thus continued intermittently for more than 800 million years, probably becoming less abundant with time. Their total volume is very small compared to that of the earlier sea of melt, but they show that the moon was partially melted in its solid interior somewhat as the earth is partially melted today.

Early histories of the planets The moon's early melting and meteorite impacts are tremendously significant because of similarities among the moon and inner planets. Images transmitted from the Mariner space probes show that Mercury was also impacted by countless meteorites, including very large ones. X-rays activated by the sun indicate that Mercury's highland surfaces include abundant light (plagioclase?) fragments and that the lower areas were flooded by basalts after a crust had formed. Mars has relics of ancient cratered terrains that were probably once similar to the highlands of the moon and Mercury, suggesting a similar early history. The earth's early history has been essentially erased by its atmosphere, by surface water, and by broad movements in the crust and lithosphere such as those described in Chapters 15 and 16. Rocks recently dated as more than 3,850 million years old, however, show that the outer part of the earth had cooled to "normal" before that time, for some of these ancient rocks include metamorphosed sands, gravels, and limestones similar to those forming today (Reference 5). The segregation of the earth's core and initial crust thus probably took place at the same early stage as for the moon.

FIGURE 18-12
Thin section of plagioclase-rich gabbro showing partial granulation due to impacts. The dark and light bands are multiple twins in two plagioclase crystals. View is 0.5 mm across. U.S. Geological Survey photograph courtesy of Odette B. James.

Planetary Lithospheres and Asthenospheres

The moon and inner planets thus appear to have crusts and, with the possible exception of the moon, iron-rich cores. These similarities indicate strongly that the bodies became segregated in the same way. What of their later development? Do they have asthenospheres and lithospheres similar to those of the earth?

Although the moon's youngest lavas remain to be dated, the surface appears now to be volcanically inactive. What measurements have been made indicate that this inactivity is due to an exceptionally thick lithosphere, beneath which the moon remains hot. The astronauts installed instruments that detected seismic events (moonquakes), the flow of heat from the moon's interior, and the strength of the magnetic field, and the data obtained suggest that the boundary between the lithosphere and asthenosphere is about 600 miles (1,000 km) beneath the surface, the depth at which S-waves are markedly attenuated. Temperatures below that level are probably near the melting point of basalt, as in the earth's asthenosphere. Temperatures increase with depth at a lesser rate than on the earth because of the moon's smaller size; however, heat from radioactivity may compensate for part of this difference because of the greater concentration of uranium and thorium in the moon. Detailed studies of the minerals in lunar basalts indicate that most were melted from olivine-rich rocks at a depth of around 300 miles (500 km), thus suggesting that the lithosphere was about half as thick during the period from 3,900 million to 3,100 million years ago as it is now. The present-day lithosphere is evidently far more stable than the one on earth, for the only moonquakes recorded by the Apollo seismographs have been mild ones. Recalling that water under pressure sets off earthquake-generating movements in the earth, one wonders if the absence of water in the moon may contribute to the scarcity of moonquakes.

Mercury is so similar to the moon that we can probably infer a thick lithosphere there, too, but Mars has had a long volcanic history, suggesting a thinner lithosphere and long-continued melting in the asthenosphere. Martian volcanoes differ from those on earth in that many of them are much larger, several measuring between 250 and 400 miles across and as much as 16 miles high! As Figure 18-13 illustrates, the larger have the form of broad lava shields (convex lava mounds) and have an elliptical basin at the summit formed by the collapse of the shield into the body of melt that fed the volcano—a feature called a *caldera*. Basalt shields on the earth, such as those forming the Hawaiian Islands, are smaller but have nearly identical slopes and similar summit calderas; it thus seems almost certain that the Martian volcanoes are made of basalt. Their large size compared to that of the earth's lava shields is probably due to the lithosphere on Mars remaining fixed in position over sources of melt in the asthenosphere. The earth's lithosphere, on the other hand, moves laterally over the asthenosphere, so that volcanoes tend to become disconnected from their feeding systems. The lithospheres of Mars, Mercury, and the moon do not appear to have moved laterally, although

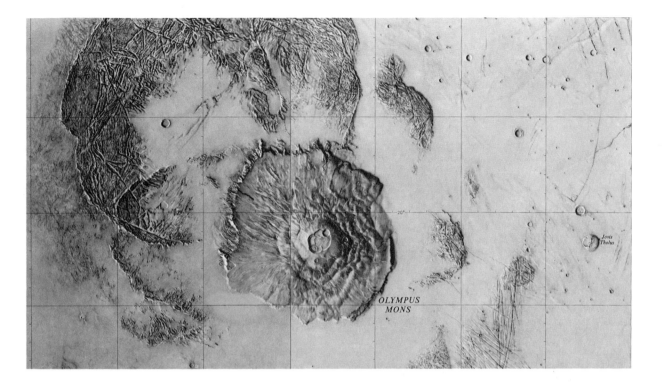

broadly arcuate fractures and ridges on Mars indicate vertical movements associated with heating and volcanic activity, and local wrinkle systems on the moon and Mercury suggest shrinkage of lava and impact-heated deposits.

Suggestions from planetary atmospheres   Atmospheres probably formed on the moon and all the inner planets by outward working of gases soon after accretion, but they were lost because of the light weight of hydrogen and the erosive sweep of the *solar wind* (elemental particles streaming from the sun). Younger atmospheres then formed by gradual degassing of the planetary interiors, especially by volcanic activity. The near absence of atmospheres on the moon and Mercury thus support other indications of little if any volcanism during the latter half of their histories. In great contrast, the earth's heavy atmosphere and abundant hydrosphere are suitable to the geologic evidence of volcanic activity continuing on a large scale. The atmosphere on Mars, measured by the Viking Missions, is about $\frac{1}{50}$ as concentrated as the earth's atmosphere but dense enough to suggest young volcanic activity. Its composition (95 percent carbon dioxide, with traces of oxygen and other gases) might support some form of life, although the Viking instruments had not detected any organic substances as of October 1976 (the detection limit was a few parts of organic substances in a billion parts of the sediment tested).

Some of the physical effects of the atmosphere and of running water were

FIGURE 18-13

Shaded relief map of Olympus Mons, one of the largest of the volcanoes on Mars, and some of the surrounding areas of fractured and "wrinkled" terrain. The vertical grid lines are about 1,500 miles apart. From the Amazonis and Tharsis quadrangle maps (MC-8 and 9) of the U.S. Geological Survey, published, respectively, in 1976 and 1975.

FIGURE 18-14
The surface of Mars, a view composed from two photographs taken from Viking 2 lander on September 7, 1976. The horizon is approximately 2 miles away, and the largest rock near the center is 2 feet long. What appears to be a small stream channel, partly filled with wind-blown sediment, crosses the view from middle left to lower right. The slope of the horizon is due to the 8° tilt of the spacecraft. NASA photograph courtesy of Viking News Center, Pasadena, California.

superbly documented by the Viking cameras. Winds evidently transport dust and sand across large areas of the planet, forming dunes and scouring rounded pits in surface stones (Figure 18-14). The small linear course in the photograph appears to be a stream channel, and photographs taken from the orbiters show distinct stream networks over much of the planet's surface, the larger of the main channels being tens of miles wide and hundreds of miles long. Floods hundreds of feet deep must once have coursed down these main channels, but all the channels are now dry and the atmosphere contains so little water that rains could not have produced the streams—at least not recently. A possible source of the water is ice (or some water-rich compound) that saturates the fragmental rocks under the surface and produces abundant water when disturbed by volcanism or meteorite impacts. This theory is supported by the deposits around impact craters, which commonly have lobate extensions that look like huge water-charged debris flows and thus suggest the sudden mixing of impacted debris and abundant water. Figure 18-14 shows the eroded surface of one of these unsorted flows.

Isostatic balance and asymmetry Mercury and Mars have relatively low-lying basinal areas that are partly filled by basalt and relatively high-lying areas that are heavily cratered and are evidently relics of the original light-weight crust. Comparable areas already noted on the moon are the maria and the highlands, underlain, respectively, by basalt (density of approximately 3.1 grams per cubic centimeter) and plagioclase-rich gabbro (density of approximately 2.9 grams per cubic centimeter). Orbiting satellites have been used to measure the gravitational attraction over these two kinds of terrain, and the results indicate that the highlands and some basalt-filled basins are isostatically adjusted but that some basalt-filled basins must sink farther to become adjusted. Similarly, the highlands and basinal regions on Mars are apparently isostatically adjusted, but the large basalt shields just described are not—the asthenosphere should flow out from beneath them and the lithosphere should subside accordingly. The general relation, then, is that the older features of the moon and Mars are isostatically adjusted, but the younger features are not. This relation suggests that the asthenospheres in these bodies have receded inward with time, presumably because the moon and planets have cooled.

It is probably an important fact that in spite of the indications of isostatic adjustment between many of their higher and lower parts, the moon and nearby planets are distinctly lopsided. The maria of the moon all lie on the side that always faces the earth and are within 60° of the lunar equator. The moon is thus divided into a highlands hemisphere (the far side) and a maria hemisphere (the one shown in Figure 18-7). Mars is also divided into a highlands hemisphere (the ancient cratered terrain) and a hemisphere consisting mainly of less cratered plains (Figure 18-15). Note that the big volcanoes, too, are not distributed evenly on Mars. Mercury has not been mapped completely, but its lava-filled basins appear to lie mainly in one hemisphere. On the earth, 80 percent of the continental areas are in a hemisphere that centers on western Europe, and the opposite hemisphere, centering on the Pacific Ocean, is 90 percent water. The causes of these asymmetric distributions are not understood, but one possibility is suggested by studies of the earth's magnetic field, as we shall now see.

Planets as Magnets

An interesting similarity among the earth, Mars, and Mercury is that they all have magnetic fields—they act as though they were giant but weak magnets. The moon does not have a magnetic field now, but some of its surface rocks are magnetized as though they had solidified in a former magnetic field. The earth's field is much stronger than those of the other planets, suggesting that its magnetism may connect with its larger size and longer volcanic history.

The magnetic fields of the planets may be compared in their essentials to the field of a bar magnet. Although invisible, the field can be "seen" by its effects on grains of iron (which are induced to become tiny magnets) or the mineral magnetite, an iron oxide that is magnetic as found in nature (Figure 18-16A). The curving lines of grains are linear traces of the magnetic force field, which in three dimensions would look something like Figure 18-16B. This field is called a *dipole field* because of its symmetry with respect to the two ends (poles) of the magnet. The earth's field acts as though a large bar magnet in the core were oriented about parallel to the axis of rotation. The dipole fields of Mars and Mercury are also about coincident with their axes of rotation, implying that the rotation of the planets has much to do

North

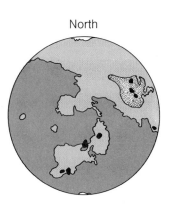

North

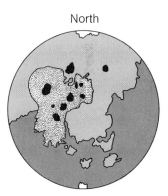

FIGURE 18-15
Map of Mars, aligned to the north-south rotational pole, showing the hemisphere of ancient cratered terrain (dark gray), the sparsely cratered plains (light gray), the super-imposed (younger) volcanic fields (dotted), the large volcanoes (black), and the polar ice caps (white). From James B. Pollack, "Mars," (*Scientific American*, vol. 233, pp. 106–17, September 1975).

with the generation of magnetic fields. Further evidence is that Venus, which has about the same size and density as the earth, rotates very slowly (once in 8 months) and has, at most, a very weak magnetic field.

Magnetic measurements and the nondipole field Other indications of where and how magnetic fields arise come from irregularities in the earth's magnetic field. These are departures from the simple dipole field and can be discovered by measurements taken on the earth's surface. A magnetized needle that rotates on a vertical pivot (as in a compass) is used to measure the horizontal orientation of the force lines, and another needle, mounted on a horizontal pivot, is used to measure their inclination (Figure 18-17). The local strength of the field is measured with an instrument called a *magnetometer*. When such measurements are made widely and plotted on maps, it is found that the actual magnetic field departs appreciably from an ideal dipole field (Figure 18-18). The components causing the irregularities in the total field are called the *nondipole field.*

A striking aspect of the nondipole field is that its oblique patterns, such as those in Figure 18-18, shift distinctly within periods of a few years, some parts of them moving systematically westward as though affected in some way by the earth's rotation. The rates of shift average around 0.2° per year (about 14 miles per year at the equator). The orientations of the force lines also change, typically at rates of 1° in every 10 years or so. The dipole field, too, shifts its orientation relative to the earth's axis of rotation, for it now makes an angle of 11.4° with the axis but in the geologic past has often been coincident with the axis. Finally, the magnetic field has decreased in intensity by about 6 percent since it was first measured in about 1830.

Cause of the magnetic field These changes are certainly too rapid to result from the solid-state flow of rocks or from any known fault movements. They are so extensive geographically and so unaffected by most surface features as to suggest that they are generated at great depths in the earth. It

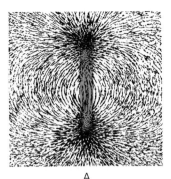
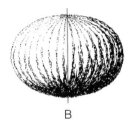

FIGURE 18-16
A, lines of force in magnetic field generated by a bar magnet, as shown by aligned trains of iron filings placed on a sheet of plastic over the magnet. B, sketch showing three-dimensional form of one "layer" of the magnetic field of a bar magnet.

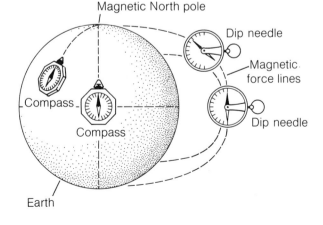

FIGURE 18-17
Measuring the local orientation of the earth's magnetic field with a compass and dip needle, shown with respect to a three-dimensional earth.

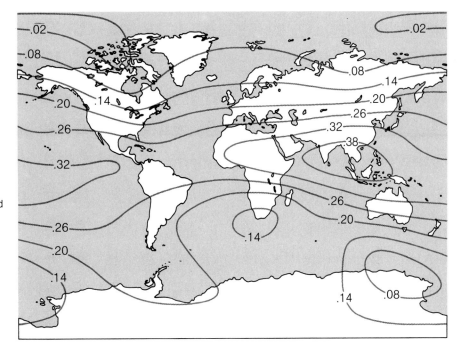

thus seems likely that both the dipole and nondipole parts of the magnetic field are generated in the liquid part of the core.

The actual mechanism must result in some way from a coupling of the earth's rotation with fluid motion in the core. Theoretical analyses of electric dynamos suggest that if the molten liquid in the core were moving in currents oblique to the earth's rotation, the added affect of rotation would be sufficient to generate the magnetic field. One possible cause of oblique stirring in the core is unequally distributed heating at the base of the liquid core by crystallization of the solid core. Another possibility is that the boundary between the liquid core and mantle is slightly undulatory, thus diverting the currents in patterns that are asymmetric with respect to the earth's rotation.

Asymmetry of the nondipole field   One strong indication of asymmetry in the core or lower mantle is an asymmetry in the nondipole field measured at the earth's surface. The asymmetry was discovered by Richard R. Doell and Allan Cox, who examined records of the present and past magnetic fields and found the nondipole field markedly weaker in the Pacific Ocean region than elsewhere. The magnetic field for 1965, in fact, divides the earth into two hemispheres. The one centering on the Pacific Ocean includes all the areas in which the nondipole field is unusually weak and the opposite hemisphere includes all the areas in which it is unusually strong (Figure 18-19). These specific areas have changed in size and strength compared to the earliest reliable measurements, made in 1829, but the general hemispheric division and the large area of weak values in the Pacific region

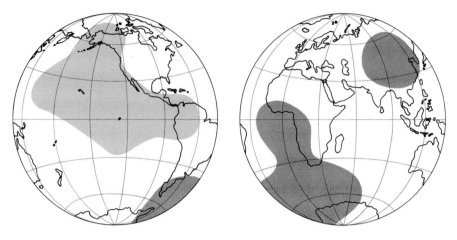

FIGURE 18-19

Regions where the earth's nondipole field is consistently weak (light gray) and strong (gray), thus dividing the earth into two opposite hemispheres. Data for 1965, from Doell and Cox, "The Pacific geomagnetic secular variation anomaly and the question of lateral uniformity in the lower mantle," pp. 245–84 in *The nature of the solid earth,* E. C. Robertson, ed. (McGraw-Hill, 677 pp., 1972).

were similar then and have remained so. Doell and Cox found evidence in magnetized rocks in the Pacific region that the general pattern extends back in time for at least 2,000 years.

Besides the possibility of asymmetry within the core, speculations on the cause of the asymmetric field include the possibilities that temperature varies laterally at the boundary between the core and mantle and that the lower mantle differs laterally with respect to composition or physical state. These ideas are supported by the fact that seismic waves traveling through the lower-most mantle are affected differently under the Pacific Ocean than under the Atlantic.

It is intriguing that the two magnetic hemispheres coincide roughly with the two hemispheres dominated by ocean basins and continents, mentioned in the last section. This relation suggests that the deep mantle and core are somehow coupled with the asthenosphere and thus with the positioning of broad surface features. Possibly the deep-seated differences already mentioned affect the distribution of heat throughout the mantle. The fact that Mercury, Mars, and the moon preserve asymmetries produced at an early stage in their history suggests that the earth's asymmetry also arose at that time. Perhaps the planets became segregated slightly off-center because melting was initiated off-center. Perhaps the accreted bodies included large masses of slightly different makeup. As we shall see in the next chapter, earth's asymmetry is also reflected in broad movements of the lithosphere, which affect most surface processes.

Summary

Whatever the cause of their asymmetries, the moon and the planets probed by the space programs have provided indirect evidence of the earth's early history. If the similarities are indeed as close as they seem to be, the earth's principal stages and events may be summarized as follows:

1. Condensation of solid particles from the primitive solar system, which accreted into planets so rapidly (over a period of 10 to perhaps 200 million years) that the earth was nearly its present size by 4,500 million years ago.
2. Melting of a large fraction of the newly accreted planet, the heat being supplied by accretional impacts, by gravitational compaction of the accreted particles, and by decay of short-lived radioactive isotopes.
3. Segregation of molten iron inward to form the core and of molten gabbro outward to form a liquid body at or near the surface at least 100 miles thick.
4. Solidification of all or most of the gabbro by about 4,400 million years ago, the outer parts being formed of more feldspar-rich and thus lighter rocks.
5. Cooling of the surface before 4,000 million years ago, so that large water bodies could form and surface water could alter, erode, and deposit sediments.
6. Periodic eruptions of basalt and more silica-rich melts, due to fissuring of the lithosphere and slow build up of heat by radioactivity.
7. Continued modification of the crust by these processes, up to this day.

REFERENCES CITED

1. Bruce A. Bolt. The fine structure of the earth's interior. *Scientific American*, vol. 228, pp. 24–33, March 1973.
2. Don E. Wilhelms and John F. McCauley. Geologic map of the near side of the moon (with descriptive text). U.S. Geological Survey, Map I–703, 1971.
3. G. J. Wasserburg, D. A. Papanastassiou, F. Tera, and J. C. Huneke. Outline of lunar chronology. *Proceedings of the Royal Society* (in press), 1976.
4. Typhoon Lee, D. A. Papanastassiou, and G. J. Wasserburg. Demonstration of ^{26}Mg excess in Allende and evidence for ^{26}Al. *Geophysical Research Letters*, vol. 3, pp. 109–12, 1976.
5. Stanley R. Hart and Samuel S. Goldich. Most-ancient known rocks may be found in all Earth's Precambrian shields. *Geotimes*, vol. 20, pp. 22–24, March 1975.

ADDITIONAL IDEAS AND SOURCES

1. Superb descriptions of planetary atmospheres and evidence relating to them and their possible effects on life are included in the September 1975 issue of *Scientific American* (vol. 233, no. 3), which deals entirely with the solar system.

2. The earth's nearest neighbor, Venus, is similar to the earth in size and probably has a similar core, but one can say little about the planet's history because its surface is hidden by a cloudy CO_2-rich atmosphere. A first look at the surface was provided in October 1975 when two Soviet spacecraft, Venera 9 and 10, landed on Venus and transmitted photographs showing a terrain of large rock outcrops or partly buried boulders (see page 14 of the January 1976 issue of *Geotimes*; vol. 21, no. 1). Another point of interest, described by Andrew and Louise Young on page 73 of the *Scientific American* issue just mentioned, is a possible asymmetry in the planet's mass, as suggested by the orientation of the planet when it passes closest to the earth.

3. Michael H. Carr's article, "The volcanoes of Mars" (*Scientific American*, vol. 234, pp. 32–43, January 1976), presents a well-illustrated and clearly written view of the planet and its interpreted history.

4. Besides the specific data mentioned in the chapter, Bruce Bolt's article (Reference 1) presents excellent figures and descriptions of how seismograms are used to interpret the travel histories of seismic waves that pass through the core or come close to it.

5. A very readable description of the geophysical soundings of the moon is given in "The interior of the moon" by Don L. Anderson, in the March 1974 issue of *Physics Today*.

View from the *Gemini II* mission satellite when it was over the south end of the Red Sea (left foreground). The Gulf of Aden is in the middle distance and the Indian Ocean in the distance. The land on the right is Africa and that on the left Saudi Arabia. NASA photograph courtesy of Johnson Space Center, Houston.

19. Earth's Mobile Lithosphere

Earth's Mobile Lithosphere

The photograph on the chapter opening page, taken by the astronauts from an earth-circling satellite in 1966, records a remarkable on-going event. The Gulf of Aden and the Red Sea look like shallow estuaries flooded onto the continent from the Indian Ocean (in the distance), but that is far from the case. Each is as deep as 7,000 feet near its medial axis, along which a series of narrow basins defines a spreading system comparable to the axial valley of the Mid-Atlantic Ridge. Note that the opposite coasts of the Gulf of Aden would fit nearly together if Saudi Arabia were to lie against Africa. Dating of rocks has shown that the coasts were indeed joined some 30 to 40 million years ago. The continental mass probably began to separate at about that time and has done so episodically since, the average rate during the past 4 million years being nearly 1 inch per year. If this rate of separation continues, the Gulf and the Red Sea will be a major arm of the Indian Ocean in 10 million years and in 100 million years will have expanded into a new ocean 2,000 miles across!

A special significance of this spreading system is that it continues as a rise along the axis of the Indian Ocean. The lithosphere under the Indian Ocean in the far right side of the photograph is thus continuous with the lithosphere under Africa. The lithosphere forming Arabia and the lithosphere under the left-hand side of the Indian Ocean constitute a second continuous plate. Thus, spreading lithospheric plates may be partly continental and partly oceanic—a discovery that links the oceanic spreading discussed in Chapter 16 with the fragmentation and "drift" of continents.

Continental drift The movements of these lithospheric plates were not well documented until the 1960s, but the idea of continental drift was a lively scientific topic between 1910 and 1935. The concept stemmed from the apparent fit of the eastern coastlines of the Americas with the western coastlines of Africa and Europe. Similar correspondence was found between the coastlines of some other continents, and considerable evidence was marshalled to support the idea that the continents were once joined: (1) groups of fossil plants and animals on the two sides of the present day oceans were found to be similar; (2) mountain ranges and rock bodies could be matched from one continent to another; and (3) the positions of glacial features dating from a widespread glaciation 250 million years ago were found to correspond closely between continents. The concept of drift was nonetheless put down by geophysicists because they could see no way in which continental masses might slide over the rocks forming the ocean basins. That the continents might be moving *along with* parts of the ocean basins was not considered, perhaps because little was known of the ocean floor and because earth scientists tended to deal with the oceans and the continents separately. Some of the crucial relations recognized later have been described in Chapters 16 and 17: (1) the oceans are spreading along medial rises; (2) new lithosphere is being formed at the rises; and (3) the ductile asthenosphere provides a worldwide base on which the lithosphere can move easily.

In this chapter we will look at the more complete evidence for these movements. We will find that all parts of the outer earth are linked in a worldwide system of moving plates of lithosphere, a system that is remarkably balanced by flow in the asthenosphere. Most earth processes are now thought to be connected to this system, notably the forming of mountain chains and ocean deeps, the activity of plutons and volcanoes, and the generation of earthquakes. Recent studies suggest that the cyclic movements of metals, fuels, and groundwater are also related to the plate system, an understanding of which should help in our exploration for new resources. Despite the importance of the plate movements, however, their causes are only beginning to be understood.

Discoveries Based on Rock Magnetism

Our present knowledge of the lithospheric plates and their movements stems largely from studies of magnetized rocks. Most rocks have weak magnetic fields due to numerous but minute grains of magnetic minerals in them. The minerals all contain considerable iron, an element that is induced to form tiny magnetic domains (parts) parallel to the earth's magnetic field when the minerals form. When basalt crystallizes, for example, many small grains of the black iron oxide called magnetite commonly form in it. The iron atoms in the grains have variously ordered magnetic fields at high temperatures, but a majority become aligned with the earth's magnetic field when the rock cools below about 500°C. The grains then act as tiny permanent magnets that give the rock a magnetic field parallel to the earth's field at the time the grains formed. Sediments, too, become magnetized, because magnetic grains become oriented parallel to the earth's field as they settle. These magnetic fields are permanent as long as the rocks are not heated above 500°C or altered chemically. Many rocks thus retain their original magnetic fields even if they are moved relative to the earth's field—they "remember" the orientation of the earth's field that affected them initially.

We can therefore use the magnetic fields measured in rocks to decipher subsequent movements of continents. Collecting and measuring proceeds as follows: (1) a number of rock samples of about the same age are collected from each of several widely spaced localities on one continent, a north arrow and horizontal line being marked precisely on each sample before breaking it from its outcrop; (2) the orientation of the magnetic field of each sample is measured with a sensitive magnetometer in the laboratory; (3) the field localities are also mapped and studied to determine how much, if at all, the rocks have been tilted by folding or faulting since they were deposited, and each sample is reoriented to correct for these changes; and (4) by using the orienting marks made at the outcrops, the measured magnetic fields are plotted in their actual orientations on a world map. Figure 19-1 shows such a map diagrammatically and indicates how it may be interpreted. Each arrow in A shows the horizontal orientation of the field measured at that locality, and the convergence of the arrows indicates the former orientation of the earth's magnetic field. Note the inclinations of the measured magnetic fields in A and their correspondence to the inclined lines in B. These relations show that the continent has shifted in latitude as well as rotated with respect to the geographic poles, as illustrated in C. Such results are typical, and their indication of continental drift is strengthened by the fact that suites of older rocks indicate lesser amounts of displacement than those of younger rocks.

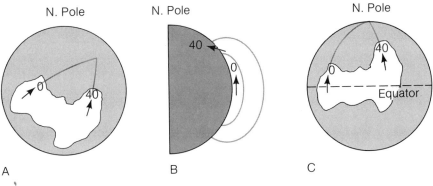

A

B

C

FIGURE 19-1

A, orientations of magnetic fields (arrows) in rocks of the same age. The numbers show the
local inclination of the field in degrees from the horizontal, and the gray lines show
the convergence toward what must have been the north magnetic pole. B, cross section
through the earth's magnetic field, showing how the inclinations in A indicate the original
latitude of each sample. C, position and orientation of the continent when the rocks formed,
based on the relations in A and B.

Another overall test is illustrated by Figure 19-2. The diagram represents
the assemblage of continents suggested by various kinds of evidence dating
back to the Triassic (about 220 million years ago). It is based on the correspondence of coastlines and the matching of various fossil assemblages, glacial
features, and mountain belts, as mentioned earlier. The arrangement is about
the same as that suggested early in this century by Alfred Wegener in the first
major proposal of continental drift (Reference 1). Carl Seyfert, however, has

FIGURE 19-2

Approximate arrangement of continents during the Triassic
Period (220 million years ago) with dots showing the position
of the south magnetic pole derived from suites of samples
on each continent. From Carl K. Seyfert and Leslie A. Sirkin,
Earth history and plate tectonics (New York: Harper & Row,
504 pp., 1973).

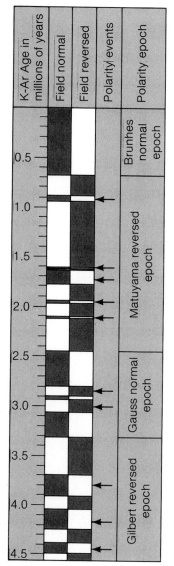

FIGURE 19-3

Time scale for periods of normal and reversed magnetic fields, back to the time when potassium-argon ages are too uncertain to be useful in dating such short-term variations. From Allan Cox, ''Geomagnetic reversals'' (*Science*, vol. 163, pp. 237–45, January 17, 1969).

used additional information to improve it and has added the data of particular interest here—the South Poles of the various continents determined in the manner just described. The close grouping of these points indicates strongly that the continents were arranged this way 220 million years ago and, therefore, must have moved long distances to attain their present positions.

Polarity reversals and rigid plates Just how they moved apart has been determined from another, surprising discovery: the earth's magnetic field has reversed its polarity many times in the past—that is, the dipole described in Chapter 18 has switched its north and south poles. Many magnetic readings in closely spaced rock layers indicate that the field weakens slowly at first, then more rapidly for a thousand years or so, and then reverses polarity and intensifies in the new orientation. We do not know why this happens, but we can guess that the dynamic patterns in the liquid core change. The reversals have been detected systematically in many places and dated quite precisely in rocks younger than 4.5 million years. As Figure 19-3 shows, this record is composed of periods called *polarity epochs* that last from $\frac{1}{2}$ to $1\frac{1}{2}$ million years and may include shorter *polarity events*. The ages of polarity epochs older than 4.5 million years are known only generally because the uncertainty of radiometric dating becomes greater than the duration of most of the magnetic epochs.

The polarity scale led to a highly provocative interpretation of the spreading of ocean basins as well as continental drift. Instruments carried on ships and in aircraft had detected variations in magnetic intensity that form broad bands parallel to nearby ocean rises (Figure 19-4A). Note the symmetry of the pattern: a single band straddles the rise, and bands of approximately equal width lie equidistant from the rise crest on each side. F. J. Vine and D. V. Mathews suggested in 1963 that the more intensely magnetic bands (gray) consist of lavas magnetized parallel to the earth's present field and therefore additive to it (Reference 2). The more weakly magnetic bands are lavas erupted during reversed polarity epochs, so that their magnetization subtracts from the earth's measured field. The bands thus record the spreading of the oceanic crust during millions of years (Figure 19-4B).

We can identify the bands in Figure 19-4 by using the dated reversals shown in Figure 19-3. The central black band would represent the present (Brunhes normal) epoch. The broad white bands on each side of it would have formed during the relatively long Matuyama reversed epoch, and the next black band would represent the rather brief Gauss normal epoch. By measuring the widths of the bands, we can thus determine that the approximate rate of separation along this part of the rise is 0.7 inch per year. Each plate of oceanic crust is thus growing at a rate of approximately 0.35 inch per year.

Similar magnetic bands have now been mapped across most of the oceans, and their patterns indicate strongly that all ocean basins have grown by spreading at rises. The nearly straight, parallel nature of the successive bands, even those over 100 million years old, indicates that the oceanic lithosphere was not deformed as it spread. Each oceanic plate must therefore be coherent and rigid.

In view of the nonrigid behavior of many parts of the continental litho-

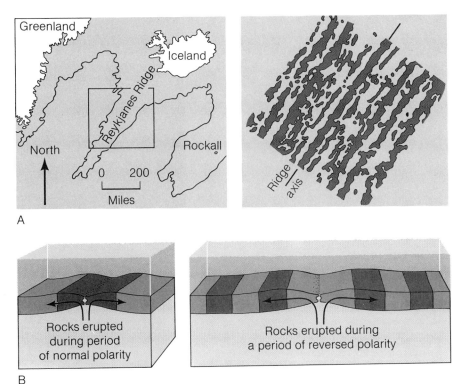

A

B

FIGURE 19-4

A, map showing unusually strong (gray) and weak (white) magnetic intensities over the Reykjanes Ridge, south of Iceland. From J. R. Heirtzler, X. Le Pichon, and J. G. Baron, "Magnetic anomalies over the Reykjanes Ridge" (*Deep-Sea Research*, vol. 13, pp. 427–43, 1966). B, diagrams suggesting how the magnetic bands are explained by spreading of the oceanic crust and lithosphere.

sphere, such as that described in Chapter 15, this discovery was so amazing that it was disbelieved by many scientists. How might it be checked? Oceanographers reasoned that the first sediments deposited on the seafloor basalts should have progressively greater ages away from rises, so that drill cores of the sediments might be used to date the spreading even if the underlying basalts could not be sampled. The experience gained from the Mohole Project, described in Chapter 16, made it possible to undertake the drilling and coring of sediments in the deep sea. Figure 19-5 shows the ship specially designed for the drilling, the *Glomar Challenger*, which has the capacity for recovering cores through sedimentary sequences several thousand feet thick and for drilling and coring basalt and other hard rocks as well as unconsolidated sediments. During the initial drilling program, from 1968 to 1975, holes were drilled and cored at 392 sites distributed widely through the oceans. Most were not drilled deeply enough to give a measured age of the basal sediments (those lying on the layer of basalt), but enough dates were obtained to prove that the magnetic bands do, in fact, record a regular spreading of the oceanic lithosphere. The results in the Pacific Ocean are especially striking, for they show that lithospheric plates have been spreading from the Pacific rise system at rates of 1.5 to 5 inches per year for more than 160 million years (Figure 19-6). As the arrows that cross the figure suggest, the plate of oceanic lithosphere on Papua New Guinea traveled some 6,600 miles from the rise where it was formed!

FIGURE 19-5

The drilling research vessel *Glomar Challenger* in the Atlantic. The 10,500 ton ship is 400 feet long, and the top of her drilling derrick is 194 feet above the water. Photograph by Deep Sea Drilling Project, Scripps Institution of Oceanography; courtesy of Tom Wiley.

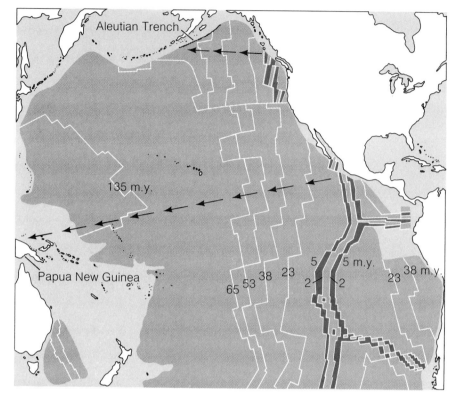

Seismic Zones and Sinking Plates

By following the spreading lithosphere outward from rises, we come eventually
to places where it sinks and is consumed. If we start at the rise just west of
British Columbia, for example, we find that lithosphere spreads westward and
finally sinks at the Aleutian trench (the arrows in the upper part of Figure
19-6). The direct evidence for sinking was presented in Chapter 11: the deformation and aftershocks associated with the 1964 Alaska earthquake indicate that a plate of oceanic lithosphere is descending beneath the adjoining
plate, dragging the ocean basin downward to form a trench and thereby setting
off major earthquakes. Indeed, the intense seismic activity and the presence of
the trench permit us to locate other places where lithospheric plates are sinking (Figure 19-7). Note that earthquake epicenters are spinkled loosely along
rises but form broad, dense bands along seismic zones such as that of the
Aleutian Islands. The average intensity of the earthquakes generated in the
seismic zones is also much greater than that of earthquakes associated with
rises. The volcanoes associated with the main seismic zones, too, differ from
those erupting basalt along the oceanic rises. They are typically towering peaks
such as Mount St. Helens and Mayon, and most of the rocks erupted are
andesite and dacite rather than basalt.

A major earthquake in Chile has been used to determine the actual motion
of the sinking plate, and the data show that the plate bends downward quite
steeply to an angle of about 45° (Reference 3). Records of deeper earthquakes

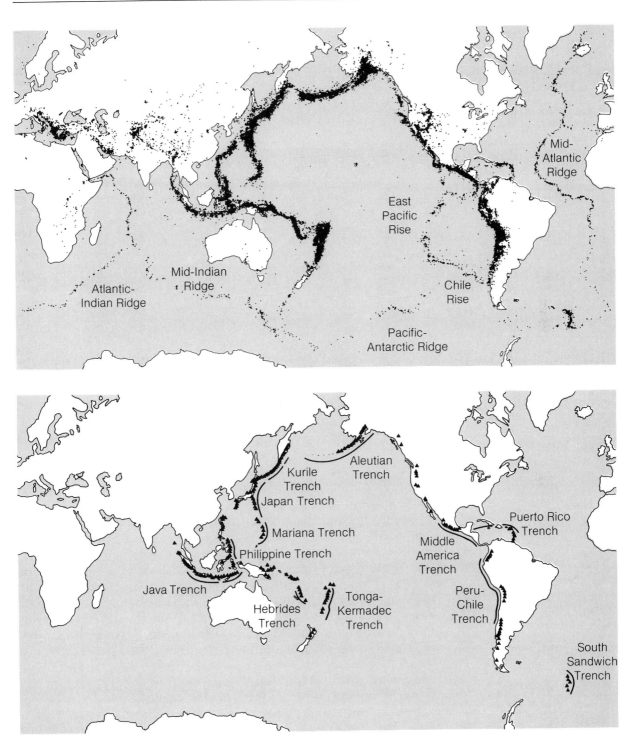

FIGURE 19-7

World maps showing the epicenters of earthquakes of magnitude greater than 4, for the period 1961–1967 (upper) and the distribution of oceanic trenches (water depths of more than 20,000 feet) and active or recently active volcanoes (triangles) that typically erupt andesite or dacite (lower). Upper map from Muawia Barazangi and James Dorman, ''World seismicity maps compiled from ESSA, Coast and Geodetic Survey, epicenter data, 1961–1967'' (*Bulletin of the Seismological Society of America,* vol. 59, pp. 369–80, 1969).

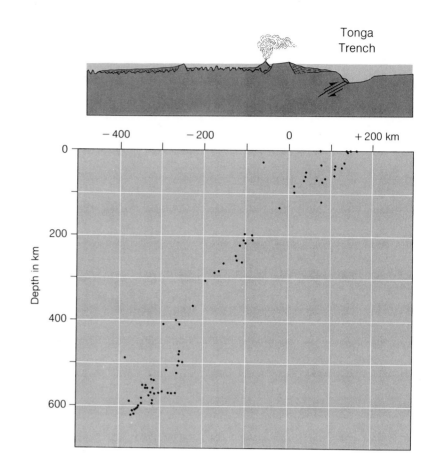

FIGURE 19-8
Vertical section through the Tonga trench (located on Figure 19-7) showing depths to sources of earthquakes occurring during 1965. The topographic relief (above) is exaggerated, but the vertical and horizontal scales for the seismic diagram are the same. From Bryan Isacks, Jack Oliver, and Lynn R. Sykes, "Seismology and the new global tectonics" (*Journal of Geophysical Research*, vol. 73, pp. 5855–99, 1968). The profile (top) is from Daniel E. Karig, "Ridges and basins of the Tonga–Kermadic island arc system," *Journal of Geophysical Research*, vol. 75, pp. 239–54, 1970.

in the Aleutian region show that that plate, too, descends more steeply with depth, and seismic studies indicate that most plates slant downward at angles from 40° to 60°. Where seismographs have been set up near the seismic zones in order to measure accurately the depths and geographic positions of all earthquake sources, the sinking plates have been delineated to depths as great as 400 miles (Figure 19-8). Earthquakes are evidently generated in the sinking plates as long as they remain more rigid than the surrounding asthenosphere. The broad seismic zones in Figure 19-7 are thus the surface expressions of comparatively thin plates of rigid lithosphere that slant downward from the trenches to depths of 200 miles or more.

Trenches as subduction zones   The floors of the trenches have not been sampled extensively because of their great depths (generally 20,000 feet and more), but the sediments recovered so far are mainly muds and sands deposited by turbidity currents that evidently flow down the sides of the trenches and then along their axes. Trenches lying far from major sources of sediment contain little mud, whereas those near land may fill so rapidly that they become relatively shallow troughs and in some cases fill completely. Some of the seismic zones in Figure 19-7 are thus associated with a filled trough rather than a trench, an example being the one parallel to the coasts of British Columbia, Washington, and Oregon.

FIGURE 19-9

Section through the landward margin of the Shikoku trough showing folds, faults, and slumps caused by westward transport and subduction of the sediments. From J. Casey Moore and Daniel E. Karig, "Sedimentology, structural geology, and tectonics of the Shikoku subduction zone, southwestern Japan" (*Geological Society of America Bulletin*, vol. 87, pp. 1259–68, 1976).

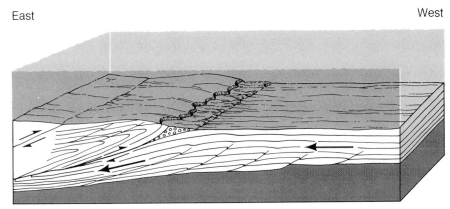

Seismic waves reflected from the sediment layers in the trenches and troughs show that the layers are typically horizontal rather than distorted, a surprising relation in view of the forceful convergence proven by events such as the 1964 Alaska earthquake. Drilling from the *Glomar Challenger*, however, has shown that sediment accumulates so rapidly in some trenches that they would have filled long ago had not sediments at the bottom of the accumulation been carried downward and away by the sinking lithosphere plate. This process is called *subduction*, literally the carrying of materials to the earth's depths. A case in which subduction has distorted the sediment layers was discovered recently by J. Casey Moore and Daniel E. Karig, during a study of the Shikoku trough, off Japan. As Figure 19-9 shows, the layers have been progressively buckled and folded against the landward side of the trough, indicating that they have been carried toward and then under it. This movement has evidently pressed the sediments strongly, for the muds are far more compacted near the landward side of the trough than they would be by normal burial loads such as those described in Chapter 10.

Evidence of deep subduction has come from measurements of gravity over trenches and seismic zones. As Figure 19-10 indicates, the force of gravity over the trenches is weaker than over the earth generally, continues to be weaker over the part of the seismic zones near the trench, and then increases to values greater than average. The lower than average values of gravity indicate a deficiency of mass appropriate to the empty part of the trench and to the lightweight (water-bearing) sediments being subducted beneath its landward edge. The increase of gravity values away from the trench suggest that the subducted sediments are compacted with depth and eventually metamorphosed to unusually dense minerals. Gravity values also increase away from the trench because the plate descends more rapidly than it is heated by the surrounding rocks. Because it is cooler than normal at each depth, the plate remains less expanded and therefore denser than its surroundings.

The low gravity values associated with the subducted rocks tell us that this part of the lithosphere is not isostatically balanced—that it normally would be elevated by the buoyant action of the underlying asthenosphere. A force must thus be pulling that part of the earth downward, and the force is evidently the downward thrust of the sinking plate.

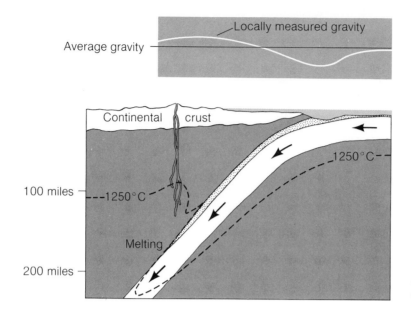

Average gravity

Locally measured gravity

Continental crust

1250°C

100 miles —

---1250°C

Melting

200 miles —

FIGURE 19-10
Diagrammatic section showing variations in gravity over a sinking plate and its subducted sedimentary rocks. The relatively low temperatures in and near the plate are suggested by the dashed line showing depths at a temperature of 1250°C.

Volcanic chains and the growth of continental crust   The subducted sediments and the plate of oceanic lithosphere heat up slowly as they sink and parts of them finally melt, giving rise to the chains of volcanoes associated with the trenches and seismic zones (Figures 19-7 and 19-10). Because the volcanic chains typically lie 100 to 200 miles from the trenches and because the sinking plates slant downward at angles averaging 40° to 50°, melting in most cases takes place at depths of around 80 to 200 miles. We found in Chapter 16 that temperatures high enough to produce basalt melts (around 1,250°C) should normally occur no deeper than 65 miles (100 km) under ocean basins (Figure 16-18). The lithosphere plate must thus sink so rapidly that it is not heated to the melting point until it reaches abnormally great depths. Figure 19-10 shows this relation by the dashed line indicating a temperature of 1,250° in each part of the system. The upward jog of the line near the upper surface of the sinking plate is based on the idea that the rocks there will be heated by metamorphic changes and by frictional heating along the surface of the moving plate. Melting in the upper part of the plate is indicated by the andesite and dacite typically erupted in the volcanic chains, for laboratory experiments have shown that the metamorphosed oceanic crust and associated sediments should yield abundant melts with the composition of andesite or dacite.

A study suggesting how these melts slowly construct a continental crust has been made by David E. James, who examined age relations among the rocks of the central Andes and used seismic waves to measure thicknesses of the crust and the lithosphere (Figure 19-11). Note the great thicknesses of the modern crust under the Andes and also the exceptional thickness of the lithosphere under that part of the continent (the lower diagram). James believes these thicknesses were attained gradually, for fossil-bearing sedimentary rocks and dated igneous rocks show that subduction started about 250 million years ago and that the volcanic chain has migrated inland since that time (compare the upper diagram in the figure). Parts of the Andes are eroded deeply enough to expose the large plutons as well as the volcanic rocks that gradually enlarged the crust. Sediments derived from these rocks were also added to the growing crust. When James calculated the total volume of these rocks, he found it could be produced by melting approximately one third of the oceanic crust and subducted sediments that have moved under the edge of the continent since subduction began. Major parts of the continental crust thus appear to have been distilled from the oceanic crust by subduction and partial melting.

The Circulation Overall

By mapping magnetic bands, seismic zones, and physiographic features, the boundaries of the lithospheric plates have been delineated over much of the world (Figure 19-12). Some small plates are not shown accurately because of the scale of the map and the question marks indicate unproven boundaries, but most of the boundaries have been delineated reliably. Note that mountain

FIGURE 19-11

Cross sections showing interpreted stages in the development of the Andes and the associated subduction zone. Wiggly arrows indicate rising melts, which form molten plutons (dark gray) and volcanoes. The thickened, unpatterned part of the Andean crust is mainly crystallized plutons. From David E. James, "Plate tectonic model for the evolution of the central Andes" (*Geological Society of America Bulletin,* vol. 82, pp. 3325–46, 1971).

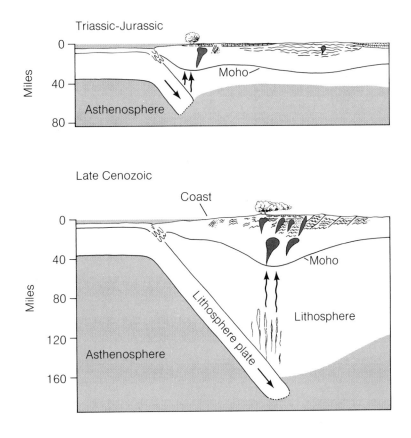

ranges mark boundaries where one or both plates consist of continental lithosphere, such as along the western edge of the America plate or between the Eurasia and Africa plates. The continental lithosphere is evidently too light and too thick to be subducted and is therefore buckled and faulted into elongate welts that are eroded to form mountain chains. Note, too, the many short boundaries called *transforms*, which are vertical faults along which plates move past each other horizontally. Many transforms in the oceans are marked by scarps or deep gash-like valleys, as already noted for the Mid-Atlantic Ridge (Figure 16-9). Transforms produce many minor and moderate earthquakes but at depths no greater than the thickness of the lithosphere—important evidence that the moving plates are no thicker than the lithosphere.

Transforms are also valuable in determining the direction of plate movement, because that direction must be parallel to the trace of a transform on a map. The curving transforms of the South Atlantic, for example, show that the ocean opened in a manner suggested diagrammatically in Figure 19-13. Evidently the transforms initially were offsets along the original spearation in the seafloor. Note that the opening resembles that between two spherical shutters pinioned at one end. The spreading rates thus increase away from the pinion, in the case of the Atlantic from about 0.3 inch per year near Iceland to nearly 2 inches per year near the equator.

By determing the opening rates and directions of all the spreading rises, it should be possible to calculate how their motions interact on a worldwide basis. Xavier Le Pichon has done this by using the five principal rises and

FIGURE 19-12

The principal lithospheric plates, delineated by three kinds of boundaries (see the key under the map).

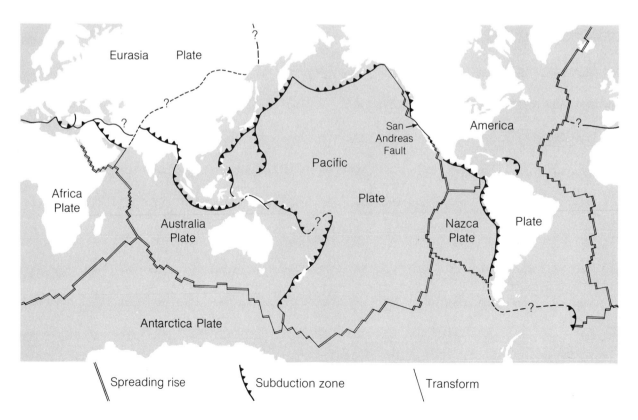

has determined that if the Antarctica plate has remained in position over the South Pole, all the other plates, as well as their rises and subduction zones, have been moving westward (counter to the earth's rotation) (Reference 4). This motion increases toward the equator, where Le Pichon calculated that the average westward shift has been about 300 miles during the past 10 million years, a rate of nearly 2 inches per year. Spreading rises and subduction zones, then, would result because some plates are moving faster than others or in somewhat different directions.

Rates of subduction Subduction of the lithosphere is not related in a simple way to the production of the lithosphere at the rises, for there are only about half as many miles of subduction zones as there are miles of rises. As a case in point, we may take the historically measured amounts of convergence and subduction at the Peru-Chile trench and attempt to identify the sources of the converging material. Subduction associated with the 1960 Chile earthquake was approximately 65 feet—the actual offset between the Nazca and America plates (Figure 19-12) (Reference 3). The historical record of major earthquakes in Chile, which goes back to the sixteenth century, indicates that such events recur about once in 100 years. The rate of convergence during recent times has thus been approximately 8 inches per year. Where does this large amount of material arise? It has been determined from the magnetic bands that the Nazca plate is spreading eastward with respect to the Pacific rise at about 3 inches per year and that the America plate is spreading westward with respect to the Mid-Atlantic Ridge at about 0.8 inch per year (one-half the total spreading rate at the ridge). These rises thus generate only about half the lithosphere converging at the Peru-Chile trench.

Other rises that may be contributing to the convergence are suggested in Figure 19-14. Note that there is no subduction zone west or east of the Africa plate. Therefore, the lithosphere produced at the rise in the Indian Ocean as well as *all* the lithosphere produced at the Mid-Atlantic Ridge may be shifting westward toward the Peru-Chile trench. As the numbers in Figure 19-14 indicate, these sources would increase the input from the east to about 2 inches per year. We may obtain additional lithosphere by noting that the site of the 1960 earthquake lies east of the Chile Rise as well as the East Pacific Rise. Adding these contributions from the west, we get 6 inches and can thus obtain the total rate of 8 inches per year indicated by the historical earthquakes. The result by no means proves the specific motions noted, for the relations west of the trench are not well known, and the historical earthquakes may have been more or less frequent than the average over millions of years. The general approach is nonetheless important; plate movements must be accounted for across entire hemispheres of the earth in order to balance productions and consumptions of the lithosphere.

Cooling and heating of plates Material must then flow back in the asthenosphere to complete the circulation, and Figure 19-15 indicates this return flow as well as the transfers of heat connected with the circulation. Measured flows of heat near the earth's surface indicate that the lithosphere is warmest at rises and volcanic chains. Heating at rises is caused by the intrusion of basaltic melt from the underlying mantle and by the recent crystal-

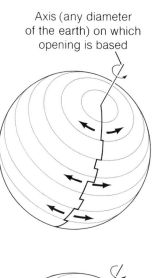

Axis (any diameter of the earth) on which opening is based

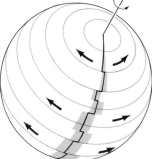

FIGURE 19-13
Diagrammatic views of the opening of an ocean such as the Atlantic, based on an axis or pinion point at the north end of the ocean. Top, the spreading rise at its inception. Bottom, two stages in the spreading and growth of the ocean basin.

421

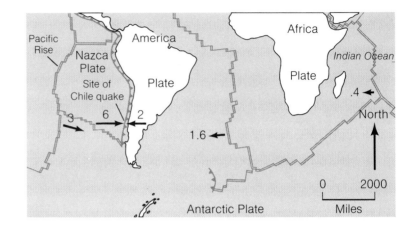

FIGURE 19-14

Map showing approximate directions and rates of spreading in inches per year that would account for the rapid rate of convergence at the site of the 1960 Chile earthquake.

lization of the oceanic crust on both sides of the rise from melts, as described in Chapter 16. The lithosphere then slowly cools as it moves away from the rises and, as a result, shrinks, causing a reduction in volume that accounts for the broad shape of the rises—the gentle slopes from their crests to the abyssal plains hundreds of miles away on each side.

We have already noted that the cool lithosphere is then slowly heated again as it sinks in subduction zones, even to the point of partly melting. Most of the heat, however, is obtained from the surrounding asthenosphere, which must therefore be cooled by the descending plate. The cooling may even transform the asthenosphere to rigid lithosphere, a change suggested, for example, by the exceptional thickening of the lithosphere under the Andes (Figure 19-11). The cooling of the asthenosphere should thus act against the continuing descent of the plates, so that plate movements over hundreds of millions of years are not as "easy" as they may seem in short-term view. Real forces are evidently needed to sustain the movements. How might these forces arise?

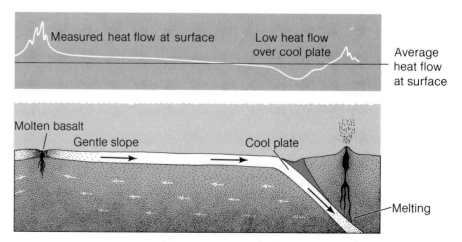

FIGURE 19-15

Vertical cross section suggesting flow of material at depth (small arrows) needed to compensate for materials spreading from an oceanic rise. Also indicated by spacing of dots are changes in temperature of the plate at various stages of spreading and subduction.

What Moves the Plates?

One early theory explaining plate systems is illustrated in Figure 19-16A. The basic cause is local heating deep in the mantle, perhaps by heat escaping from the core or perhaps by decay of radioactive elements. Because the heating expands and thereby lightens the rocks, they are displaced upward by gravity, flowing slowly in the solid state. Divergence of the current near the top of the mantle pulls the lithosphere apart, and melts arising in the hot mantle flow into the cracks thus formed. The lithospheric plates are carried laterally until the mantle cools, becomes dense, and sinks, thus dragging the plates downward. Circulation is completed by the mantle materials flowing laterally at depth, where they are slowly heated and eventually replace the material rising toward the surface.

This theory is attractive because it explains how the plates can spread as rigid bodies yet set off few earthquakes except at rises and in subduction zones. The horizontal current also gives the plates the active push suggested by the deformation and intense seismic activity at boundaries of convergence. The theory, however, has serious drawbacks. For one thing, we have seen that the most ductile part of the mantle, the asthenosphere, is a layer just under the lithosphere. This layer should act to decouple the lithosphere from movements in the deeper mantle. The deeper horizontal currents would thus have to flow faster than the plates move, as suggested by the lengths of the arrows in Figure 19-16A. The deep mantle, however, is more rigid than the asthenosphere, so that the swifter currents there seem very unlikely.

Another problem is that there is no evidence of an upward thrust by the rising column. Measurements of gravity give no suggestion that an excess of material has here been pressed to the surface. As we have noted, the slightly arched shapes of the rises are explained by cooling of the lithosphere away from the axes of the rises.

Another explanation of plate systems is based almost entirely on reactions in the asthenosphere and lithosphere (Figure 19-16B). The main driving force is the weight of the sinking lithospheric plate, which has become heavy because of cooling and contraction and because of the formation of dense metamorphic minerals in it as it sinks. The sinking part of the slab thus drags the rest of the plate behind it, causing a simple pulling apart at the rise. The upper part of the asthenosphere is dragged along somewhat by the moving plates,

FIGURE 19-16

A, hypothesis of convective overturn of the mantle, resulting in spreading and subduction of lithospheric plates. B, hypothesis of rifting along a mid-ocean rise caused by the pull of heavy (cool) sinking plates.

A

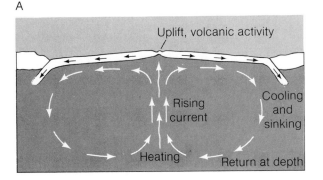

B

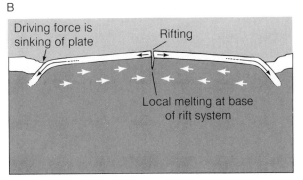

but it mainly provides a decoupling between the rapid movement of the plates and the slow return in the deeper part of the asthenosphere.

This scheme gets around most of the problems just mentioned but has some of its own. For one, it would seem that plates should cool and sink at a characteristic distance from their rises, but Figure 19-12 shows that this is far from the case. Note that the plates on the east side of the Pacific may sink after a few hundred miles of spreading (off British Columbia and Mexico), whereas the plate on the west side spreads thousands of miles before sinking. Even stronger contrary evidence is found in the fact that several plates do not sink at all in subduction zones and thus would have no cause to spread from their rises. The America and Africa plates are examples.

Seismicity and earth tides   The distribution of subduction zones with respect to latitude suggest another cause of plate movements. Figure 19-17 shows especially impressive data, which are compilations by Robert C. Bostrom of the numbers of earthquake sources per unit area of the earth's surface at different latitudes. The sources shallower than 70 kilometers (42 miles) include earthquakes generated at spreading rises as well as the relatively shallow earthquakes caused by subduction, whereas the deeper earthquakes are almost entirely the results of subduction. Note that the deeper sources are concentrated within 30° of the equator and that even shallow sources are scarce more than 60° from the equator. If you examine either Figure 19-12 or the seismic map of 19-7, you can see the reason for these differences: subduction zones are concentrated within 30° of the equator and are absent at more than 60° from the equator.

The correlation between subduction and latitude suggests that plate movements are connected with the earth's rotation and the gravitational pull of the moon. These forces result in what are called *earth tides*—broad bulges on the solid earth that are a few inches high at the equator and decrease systematically toward the poles (Figure 19-18). Unlike ocean tides, the equatorial bulges of the earth tides can sweep freely westward as the earth rotates with respect to the moon. The progress of the moon in its orbit, however,

FIGURE 19-17

Variation of seismicity with latitude, shown as the number of earthquake sources per 10,000,000 square kilometers, for each degree of latitude. A, sources shallower than 70 kms (42 miles); B, sources deeper than 70 kms. From R. C. Bostrom, ''Arrangement of convection in the Earth by lunar gravity'' (*Philosophical Transactions of the Royal Society of London*, A, vol. 273, pp. 397–407, 1973).

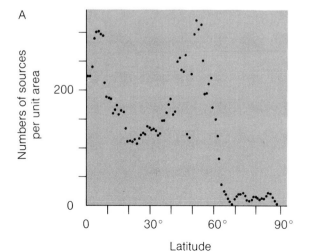

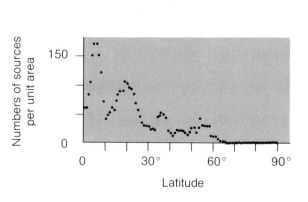

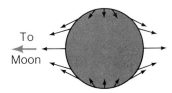

FIGURE 19-18
Diagrammatic view of the earth and the variation of lunar tidal forces near its surface.

causes the bulges to move about 55 minutes per day more slowly than the earth rotates. If this tidal drag were to retard the equatorial lithosphere enough to shift it westward relative to the inner earth, we would have a simple explanation of plate movements, especially the westward shift near the equator discovered by Le Pichon. A tidal control of plate movements would also explain why Le Pichon's analysis should have worked when the Antarctica plate was held steady over the South Pole, for the tidal forces within 20° of the poles are small and are directed mainly downward (Figure 19-18). As compelling as these relations are, however, most geophysicists believe the moon's retarding forces are too small to move the plates. Possibly the tidal forces will be found to order other movements that are the more direct cause of the plate movements.

Asymmetry of subduction A final relation that has not been explained is the great concentration of subduction around the margins of the Pacific (Figure 19-7). The concentration gives the earth an asymmetry, a sort of two-hemisphere aspect. If we compare this asymmetry with that determined for the earth's magnetic field, shown in Figure 18-19, we can see that the two sets of hemispheres are roughly coincident. We may recall, too, that the earth's continental and oceanic areas are distributed in somewhat the same asymmetric way and that Mercury, Mars, and the moon can also be divided into hemispheres dominated by highlands and lava-plain terrains, respectively. The asymmetries of the moon and nearby planets, in fact, suggest strongly that the earth's asymmetry arose during accretion and initial melting; it cannot, therefore, be due to plate movements (which have evidently never taken place on the moon and nearby planets). The asymmetric distribution of subduction thus suggests that plate movements are ordered to some degree by small differences deep within the planet. The differences might, for example, control the flow of heat to the asthenosphere, but such suggestions are almost entirely speculation.

Summary

We thus have no satisfying explanation of plate movements, but the relations discovered since 1960 are exceedingly impressive. The principal facts and ideas may be summarized as follows:

1. The oceanic lithosphere is being generated at a series of rises and is spreading outward at rates of 0.3 inch to perhaps 6 inches per year; it has been doing so for at least 160 million years and probably much longer.
2. The plates include continental as well as oceanic lithosphere, and they move as coherent bodies on the ductile asthenosphere.
3. Some of the horizontally moving plates sooner or later bend abruptly downward and follow a slanting course into the asthenosphere to depths as great as 400 miles.
4. This subduction forces the ocean floor and its sediments downward in trenches and troughs, causing the world's principal earthquakes and creating chains of plutons and volcanoes.

5. Continental lithosphere, which is too thick and light to be subducted, is buckled and thickened into mountain ranges where plates converge.

6. The lithosphere tends to shift westward in lower latitudes so that convergence and subduction are concentrated within 30° of the equator, a fact suggesting that earth tides order the plate movements.

REFERENCES CITED

1. A. Hallam. Alfred Wegener and the hypothesis of continental drift. *Scientific American*, vol. 232, pp. 88–97, February 1975.
2. Fred J. Vine and Drummond H. Matthews. Magnetic anomalies over oceanic ridges. *Nature*, vol. 199, pp. 947–49, 1963.
3. George Plafker. Alaskan earthquake of 1964 and Chilean earthquake of 1960: implications for arc tectonics. *Journal of Geophysical Research*, vol. 77, pp. 901–25, 1972.
4. Xavier Le Pichon. Sea-floor spreading and continental drift. *Journal of Geophysical Research*, vol. 73, pp. 3661–97, 1968.

ADDITIONAL IDEAS AND SOURCES

1. Although data and ideas on plate movements, commonly called "plate tectonics," are comparatively recent, the literature is already large. Articles appearing in *Scientific American* are especially valuable because of their superb illustrations and availability. John F. Dewey's article, "Plate tectonics," (vol. 226, pp. 56–68, May 1972) gives a lucid description of the geometric relations among plates, including the interesting places where three plate boundaries meet in what are called triple junctions. "The subduction of the lithosphere," by M. Nafi Toksöz (vol. 233, pp. 88–98, November 1975) presents dimensional relations of seismicity and heating during subduction. Bruce C. Heezen and Ian D. MacGregor's article, "The evolution of the Pacific," (vol. 229, pp. 102–12, November 1973) describes some connections between the world's largest ocean and plate movements.

   An interesting sign of the rapid and diverse evolution of data and ideas on plate systems are the differences in interpretations that the reader can discover among these articles and between them and Chapter 19. It must never be assumed that the latest report is always the most accurate, for some ideas have become somewhat obsolete without being disproven!

2. The satellite *LAGEOS* (*Laser Geodynamics Satellite*), launched in May 1976, provides a means of measuring rates of movement of specific parts of the lithosphere by means of a reflected laser beam. The results will be important in interpreting the cause of plate movements and will probably appear from time to time in *Geotimes* and *Science*.

Alluvial fans and steep mountain front along the eastern edge of the Sierra Nevada, California, resulting from rapid uplift during the past 5 million years. The high-angle faults along which the mountain block rose and the adjoining basin subsided define the margin of the basin and range province. The highest peak on the horizon is Mt. Whitney, approximately 14,500 feet above sea level. Photograph by U.S. Geological Survey.

20. Movements within the Continent

Movements within the Continent

Do interactions of lithospheric plates such as those described in Chapter 19 fully explain the earth's recent history? As you probably recall, plate movements were detected mainly in ocean basins or at plate boundaries near oceans. What has happened within the continents, where rocks and surface features can be visited and studied in detail? Figure 20-1 introduces an important case in point. This is the San Juan River, shown where it crosses the Colorado Plateau 120 miles south of the Book Cliffs. Complex loops such as these (meanders) are formed where rivers can migrate laterally over nearly horizontal floodplains, yet these loops form a canyon in solid rock. The photograph thus indicates that the San Juan River once flowed on a broad floodplain that was later uplifted, giving the river the potential to incise its meanders into the underlying rocks. Additional river canyons and other landforms indicate that this uplift extended over 250,000 square miles, a region including much of the Rocky Mountains as well as the Colorado Plateau. The region, however, is isostatically balanced, and therefore the uplift requires the addition of lightweight materials to the lower part of the crust to increase the buoyancy of the lithosphere. The uplift thus implies a major transfer of lightweight materials at depth. Can this transfer be explained by the simple westward movement of the America plate over the asthenosphere, or must we seek other explanations for it?

The purpose of this chapter is to see if the geologic record on the continent can provide answers to such questions. Our method will be to explore the Cenozoic history of major parts of North America—the continental events of the past 65 million years. We have already noted that this period of time

FIGURE 20-1
Goosenecks of the San Juan, in southeastern Utah, incised vertically downward when the Colorado Plateau was uplifted. Note that the two tributaries in the upper half of the view have also eroded canyons into the plateau surface. Photograph by John S. Shelton.

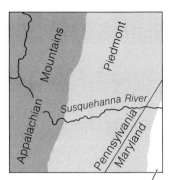

Coastal Plain

is recorded in relatively well-preserved and accurately dated rocks and fossils.
The period should also be long enough to show what order, if any, has typified
the continent's development. Many topics examined in earlier chapters relate
to this period and thus provide us with important details, such as the part of
the geologic column we examined closely in Chapter 13. By bringing together
a variety of information we may hope to see some broad connections in time
as well as place.

The chapter is organized on a geographic basis, dealing with three regions
that have certain distinctive characteristics but include a variety of physiographic provinces. The first region, the largest and most typically continental,
extends from the western edge of the Colorado Plateau and Rocky Mountains
to the Atlantic basin (see the map inside the back cover). The second and
next largest region is composed of the volcanic provinces west of the Rocky
Mountains and Colorado Plateau, and the third region consists of the greatly
deformed and presently mobile provinces along the continent's Pacific margin.

The Main Part of the Continent

The physiographic provinces shown inside the back cover are based on the landforms shown inside the front cover. The maps relate directly to geologic history. Each province is typified by certain kinds of rocks that have become arranged by certain earth movements. The simplest arrangement forms the broad band of plains and low hills bordering the Atlantic Ocean and the Gulf of Mexico. This coastal province is underlain by young sedimentary formations that have been tilted gently (less than 1°) seaward. In contrast, the Appalachian Mountains and the Piedmont province are developed on ancient sedimentary rocks that were buckled into folds and locally metamorphosed and intruded by granite plutons more than 180 million years ago. The Appalachian terrain is eroded deeply, and the ridges and valleys developed on individual formations express the folds by their loops and zig-zags (Figure 20-2).

The folds die out on the west flank of the Appalachian Mountains, and from there the formations slope gently westward under the irregular mountains and plateaus carved by tributaries of the Mississippi River (Figure 20-3). Formations of the same age are buried by the young marine and alluvial deposits of the lower Mississippi Valley but are exposed again in the lower parts of the Great Plains, although they are partly hidden there by thin deposits of glacial till, loess, and sand. In the High Plains, the formations are covered more deeply by the Cenozoic deposits described in Chapter 13, but they finally rise abruptly to form the main parts of the Rocky Mountains and Colorado Plateau (Figure 20-3).

The coastal province   The fuller meaning of the coastal province is suggested by a thickening of the sedimentary deposits under the shelf (Figure 20-4). Thickening is even more dramatic along the Gulf coast, where the base of the Cenozoic formations now lies as much as 5 miles below sea level. Evidently the outer parts of the continent along the Atlantic and Gulf coasts —including the shelf and upper continental slope—were downbowed or downfaulted during the Mesozoic and Cenozoic Eras.

Drill cores show that many of the sediments were deposited in shallow water and that some are nonmarine, indicating that the coastal plain and shelf lay at approximately their present vertical positions (near sea level) throughout the Cenozoic Era. This relation implies that the shelf has always

FIGURE 20-3

Generalized vertical cross section from the center of the Colorado Plateau to the Atlantic basin. The dots indicate Cenozoic deposits and the lines older (chiefly Paleozoic) formations. The vertical scale is greatly exaggerated.

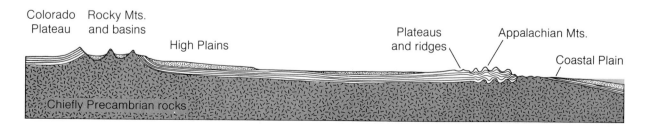

Colorado Plateau   Rocky Mts. and basins   High Plains   Plateaus and ridges   Appalachian Mts.   Coastal Plain

Chiefly Precambrian rocks

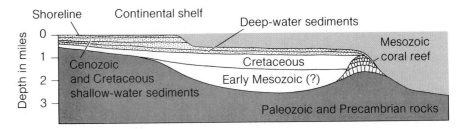

FIGURE 20-4

Vertical section from eastern Florida into the Atlantic basin, showing three indications of subsidence of the continental margin: (1) shallow-water Cenozoic and upper Cretaceous rocks now as much as a mile below sea level; (2) a coral reef that must once have been in shallow water; and (3) downbowed Mesozoic deposits. From Elazar Uchupi, ''Atlantic continental shelf and slope of the United States—shallow structure'' (U.S. Geological Survey Professional Paper 529-1, 44 pp., 1970).

been underlain by a continental (lightweight) crust. The progressive subsidence and thickening of the deposits therefore required that the underlying crust was progressively thinned, or this part of the lithosphere would become so light that it would bob up isostatically and form a plateau rather than a sea-level plain. This thinning of the crust evidently was gradual and orderly, for the Cenozoic deposits along the coastal plain are deformed only slightly.

Interpretations of uplift   While the eastern and southeastern edges of the continent were subsiding, the Appalachian region and adjoining uplands were evidently being uplifted. Deposits of late Mesozoic and Cenozoic age are missing there, and the older, deformed rocks are eroded deeply. The upper cross section in Figure 20-5 indicates diagrammatically how the unconformity under the coastal plain sediments was tilted upward in a broad arch and how deeply the mountain province has been eroded. The cross section shows two other relations that have led to differing interpretations of the uplift: (1) the summits of the high Appalachian ridges are *accordant*, that is, at approximately the same level, and (2) the summits of hills in the Piedmont are also accordant at a lower level, as are the narrower uplands bordering some of the broader valleys in the Appalachian Mountains. One interpretation of these relations is that the entire region stood at one level long enough to be eroded to an extensive surface of low relief, an undulating plain. The region then uplifted, and streams cut valleys into the undulating surface, leaving the accordant summits as remnants of the surface (the upper dashed line in the figure). The region then again lay undisturbed while another plain was eroded at a lower level (the lower dotted line in the figure), but this plain, too, was uplifted and dissected. Still lower surfaces have been interpreted as the result of additional uplifts followed by periods of standstill.

Detailed studies of several areas have shown, however, that the various sets of summits include many that are not accordant. John T. Hack has pointed out that roughly equal summit heights should be expected in well-watered regions where streams are distributed evenly (Reference 1). Hack has also

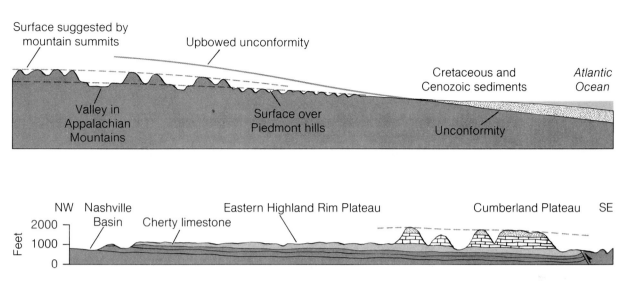

FIGURE 20-5

Indications of uplift and erosion of the Appalachian province. Above, generalized vertical section showing the upbowed unconformity beneath the Cretaceous and Cenozoic sedimentary formations and the approximately accordant summits that have been interpreted as eroded uplifted plains (the dashed lines). Below, detailed cross section showing an extensive surface of low relief developed on the top of a resistant cherty limestone. Note that the Cumberland Plateau is also delimited by a nearly horizontal formation. Lower diagram from John T. Hack (Reference 1).

shown that some of the "relict" surfaces are modern surfaces eroded on nearly horizontal formations that are more resistant than the formations overlying them (see the lower diagram in Figure 20-5). Hack thus suggests that there may be no need to call on brief periods of uplift alternating with long periods without uplift; the array of surfaces may have developed during a long period of continuous uplift.

The resolution of the two interpretations will be an important step toward understanding the cause of uplift. An indication that uplift may be continuing is found in the unusually thick continental crust under the Appalachian Mountains (Figure 16-16). This downward projection is as great as that under the Rocky Mountains, yet the Appalachian Mountains are not as high as the Rockies. The relation implies that the Appalachian terrain should be rising in order to attain isostatic balance. Direct measurements are difficult to interpret because of the continuing adjustments to removal of the ice sheets and to changes in the amount of water in the ocean; however, precise surveys of elevations indicate that the province may now be rising in a broad arch (Reference 2).

Uplift of the Rocky Mountain region   Similar but more dramatic movements elevated the Rocky Mountains and the Colorado Plateau, leading to the transport of large amounts of detritus to the Gulf of Mexico and the Pacific Ocean. The uplifts began during a period of folding and faulting that commenced late in the Cretaceous Period and lasted until about 40 million years ago (in the Eocene). The widespread occurrence of subtropical plant

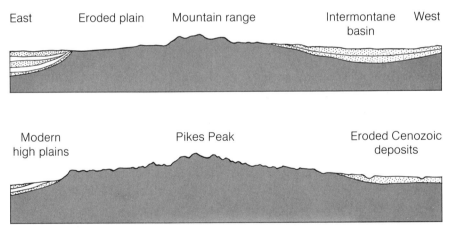

FIGURE 20-6

Above, view southward over the High Plains (on the left) and the level-topped upland of the Front Range, Colorado, with Pikes Peak standing above it. Photograph by John S. Shelton. Right, vertical sections suggesting how the upland surface looked before and after late Cenozoic uplift and erosion. The deposits in the intermontane basin are those described in Chapter 13.

East   Eroded plain   Mountain range   Intermontane   West
                                              basin

Modern   Pikes Peak   Eroded Cenozoic
high plains   deposits

communities during that period indicates that the climate was not diversified by mountain barriers. Uplift must thus have been so gradual that erosion kept the uplifted areas at moderate elevation. We may recall the evidence from Chapter 13 that rapid uplift then took place during the late Eocene: (1) erosion produced a widespread unconformity on the earlier Cenozoic deposits; (2) climates and plant communities became deversified; and (3) alluvial and air-borne sediments were deposited in great sheets over the ancestral Great Plains. This late Eocene uplift, which amounted to several thousand feet, was followed by lesser uplifts in the Miocene. The peaks of the Rocky Mountains were thus elevated higher and higher. The ranges were eroded at the same time, and the high peaks became surrounded by gently sloping plains such as the pediments described in Chapter 9.

Figure 20-6 shows the evidence of the final uplift of the mountains and pediments—towering peaks of the former mountains stand above a widely accordant surface that was dissected to form the many lower peaks. This most recent uplift must have taken place late in the Cenozoic, for the uplifted deposits of the High Plains include formations as young as the Pliocene. The Colorado Plateau was uplifted at about the same time, causing the rivers of that province to incise deep valleys (Figure 20-1). In addition to the canyon of the San Juan River, the dramatic canyons of the Green River and the Colorado River (the Grand Canyon) formed at that time.

In summary, the interior of the main continent was uplifted and eroded during the Cenozoic Era, and shallow seas lapping over the Atlantic and Gulf shelves received a large part of the eroded materials. Erosion was greatest in two mountainous provinces within the continent, and the sedimentary accumulations became thickest where the shelf gradually subsided. It is difficult to explain these changes by the lateral movement of the America plate, for the uplifted areas are remote from the plate's boundaries and the subsided margin faces in many directions. The uplifted and subsided areas were evidently parts of a system in which lightweight materials were gradually transferred from one part of the lithosphere to another. Part of the uplift might have been caused by lightweight materials that rose bit by bit from the deep mantle and were caught beneath the heart of the continent, thus slowly buoying it up. Possibly the sediments shed from the uplifted tracts pressed down on the continent's margins and caused them to spread toward the ocean basin. This spreading would help explain the thinning of the crust around the continent's margin. In this view, the driving force in each part of the system would have been gravity.

A Region of Vulcanism and Broad Extension

Of the volcanic provinces of the western states, the Cascade Range is the smallest but the easiest to interpret, for it appears to be related to the plate movements described in Chapter 19. It lies about 170 miles from the subduction zone off Oregon, Washington, and British Columbia, a distance typical of other volcanic chains around the Pacific (Figure 19-7). The chain is parallel to the subduction zone and consists of lavas like those being erupted from similar chains around the Pacific basin. The cascade chain differs from most of the others, however, in that earthquakes are not being generated under and behind the chain. One possible explanation, as we shall see, is that the rocks there are so ductile that they do not stick and become elastically deformed.

The full history of Cascade vulcanism, however, is not easily explained by simple spreading and subduction of oceanic lithosphere. The principal difficulty is that the bands of normal and reversed magnetic polarity west of the subduction zone indicate fairly continuous spreading and subduction during the Cenozoic Era, whereas Cascade volcanic activity has been distinctly episodic. Vulcanism during the first half of the Cenozoic Era (during Eocene and Oligocene times) occurred not on the site of the present range but, rather, near what is now the coastline of Oregon and Washington and also over large areas of what is now eastern Oregon. The first eruptions along the range took place in the Miocene and were remarkably voluminous during the period lasting from about 16 to 14 million years ago (Reference 3). After a period of quiescence, the chain was active again in the late Miocene (12 to 9 million years ago), in the Pliocene (6 to 4 million years ago), and finally in the Pleistocene and Recent. One explanation for the episodes of activity is that they may correspond to periods when the plates were moving somewhat faster than otherwise. Another possibility is that the periods of quiescence represent the time required for new melt to accumulate at depth and be intruded upward as plutons. Recall from Chapter 12 that at least some of the Miocene volcanoes were fed from plutons.

The Columbia Plateau An intriguingly different volcanic province is the Columbia Plateau, which lies just east of the Cascade chain in Washington and north of similar lava plateaus in Oregon (see the map inside the back cover). The Columbia Plateau is one of the world's largest volcanic accumulations, consisting of basaltic lavas that aggregated as much as a mile in thickness and originally covered an area of about 50,000 square miles. This great extent is due partly to the fluid nature of basaltic liquid but is also due to the nature of the volcanic conduits, which were vertical fissures tens of miles long that opened first here, then there, and thus distributed flows widely.

When each flow solidified and cooled, it shrank just enough to generate cracks dividing the flows into elongate prisms oriented at right angles to the cooling surfaces (Figure 20-7). The sizes and patterns of shrinkage cracks commonly differ in different flows, and individual flows can also be identified by the effects of weathering, the details of their chemical composition, and the proportions of phenocrysts (larger mineral grains) in them. Single flows have thus been traced over large areas, and many spread for distances of 100 miles and had volumes as great as 150 cubic miles (Reference 4). In comparison, *all* the volcanic rocks erupted from Mount St. Helens add up to less than 15 cubic miles! The Columbia lavas thus imply that unusually large bodies of melt in the asthenosphere were tapped quite suddenly by fissures that extended through the entire lithosphere. These relations suggest either that the region was unusually hot or that the lithosphere was rifted (fissured) unusually strongly. These conditions were rather shortlived, for the first floods of basalt appeared 16.5 million years ago and the great bulk of the lavas had been erupted by 13 million years ago. This is also the period during which the Cascade chain erupted most voluminously.

Vulcanism in the basin and range province The largest of the volcanic provinces of the West is the basin and range province, which includes the desert region described in Chapter 9 and much of the deformed region mentioned in Chapter 15. The rocks exposed in the ranges are of many kinds and many ages, but the principal formations are Paleozoic sedimentary rocks that were folded and displaced laterally as giant horizontal sheets, chiefly during the Mesozoic Era and part of the Cenozoic. This deformation ended 40 million years ago (in the Eocene) in many places but continued episodically until 10 million years ago in some areas (such as in the specific area described in Chapter 15).

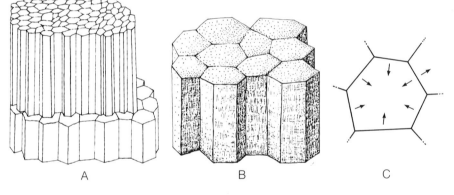

FIGURE 20-7
Sequence of basalt flows exposed by the Columbia River at Vantage, Washington, with sketches illustrating (A) a typical arrangement of the columnar joints (shrinkage cracks) in a single flow; (B) the polygonal forms of the columns where well developed; and (C) the tendency for shrinkage cracks to join in three-crack intersections—the cause of the polygonal forms.

A   B   C

High-angle faulting evidently started locally about 16 million years ago, but the terrain did not break up into the present arrangement of faulted basins and ranges until around 6 million years ago, a deformation that continues today. The faulting was evidently preceded by broad arching within the province, for even after collapse the basins are typically 3,500 to 5,000 feet above sea level and the summits of the ranges are 7,000 to 11,000 feet above sea level. We have already noted that the high-angle faulting implies the extension of the province toward the east and west, and the even spacing of the sunken blocks in Figure 20-8 indicates that the extension must have been orderly and evenly distributed. Note in the figure, too, that the basins and ranges near the center of the province are oriented north-south, whereas those near the eastern and western margins are parallel to the province's curving boundaries, which are convex outward.

The volcanic activity took place mainly before the present ranges formed. A variety of lavas was erupted, but basalt is scarce compared to silica-rich rocks such as dacite and rhyolite. The latter were commonly erupted as fragmental flows that were so voluminous and moved so swiftly that they remained incandescent for distances as much as 100 miles from their sources.

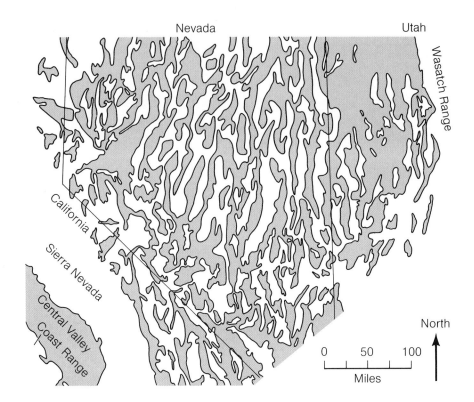

The evidence lies in the fact that the fragments in the lower parts of the
deposits were welded into compact rock, like lava (Color Plate 11A). Further
evidence of the great volumes and rapid emission of the flows is the collapse
of the upper parts of the volcanoes into the plutons beneath, forming calderas
that measure miles across (Figure 20-9).

A possible cause of the broad uplift and volcanic activity is suggested by
the distribution of the volcanic rocks according to their ages (Figure 20-10A).
Note that the oldest are in the central and eastern part of the province, that
rocks of intermediate age lie to the south and west of them, and that the
youngest lie still farther south and west. A possible explanation for this arrangement is shown in Figure 20-10B: (1) a heated body within the asthenosphere arched and fissured the lithosphere, leading to the initial volcanic
activity; (2) as the heated body spread westward and southward, the lithosphere over it was fissured and thinned and volcanic activity spread accordingly; and (3) as the asthenosphere cooled and the arch gradually collapsed,
westward drift of the lithosphere led to extension within the province and
thus to the present basins and ranges. The westward drift (the large arrows
shown in Figure 20-10B) may have been caused by whatever forces normally
propel the plate (Chapter 19). The movement may have been augmented by
the sliding of the western part of the plate toward the Pacific margin, under
the force of gravity.

Conclusions The deep fissuring and remarkably voluminous eruptions
in the Columbia Plateau province suggest that that part of the region, too, was

FIGURE 20-9

Upper part of Valles Volcano, New Mexico, soon after its summit subsided into an underlying pluton. The mounds on the caldera floor are domes of viscous silicic lava erupted after subsidence.

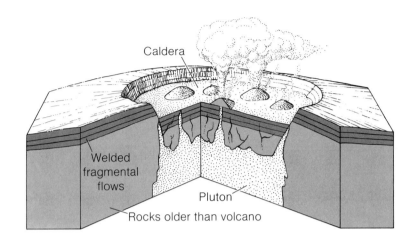

Caldera

Welded fragmental flows

Pluton

Rocks older than volcano

0   5

Miles

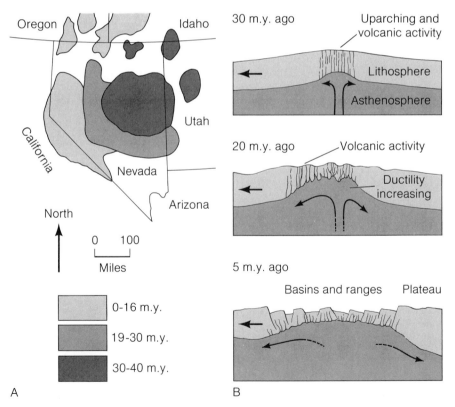

Oregon

Idaho

California

Utah

Nevada

Arizona

North

0   100

Miles

0-16 m.y.

19-30 m.y.

30-40 m.y.

A

30 m.y. ago

Uparching and volcanic activity

Lithosphere

Asthenosphere

20 m.y. ago

Volcanic activity

Ductility increasing

5 m.y. ago

Basins and ranges   Plateau

B

FIGURE 20-10

A, map of the central part of the basin and range province, showing the distribution of volcanic rocks and shallow plutons of three age groups (m.y. = millions of years). Data chiefly from E. H. McKee, "Tertiary igneous chronology of the Great Basin of Western United States—implications for tectonic models" (*Geological Society of America Bulletin,* vol. 82, pp. 3497–3502, 1971). B, three possible stages in the volcanic history, here explained by progressive arching and thinning of the lithosphere by heating and flow in the underlying asthenosphere.

affected by abnormal heating and by extension. Similar relations are found in the Oregon plateaus to the south of the Columbia Plateau, and we have noted that the Cascade chain showed increased activity when the Columbia lavas were erupted. A broadly heated asthenosphere and lithosphere may also explain the apparent ductility beneath the Cascade chain, that is, why earthquakes are not being generated there frequently.

The volcanic region of the West thus seems to be a part of the America plate that was so heated and thinned from beneath that it lost much of its rigidity and locally even its coherence. Perhaps the increased ductility of the rocks under the plate led to increases in the rate of westward drift of this part of the continent. The lithosphere was thus stretched, thinned, and locally caused to collapse in a series of separate blocks.

Subduction at the Pacific Margin

The last of the three continental regions is comparatively small but active and intriguing. Through much of the Mesozoic Era and the first half of the Cenozoic, the entire west coast was the site of convergence between the America plate and plates spreading from rises in the Pacific Ocean. We have already noted that the edge of the ocean basin off British Columbia, Washington, and Oregon is still an active part of this system. Formerly active parts must be interpreted by the rocks and structural relations now seen onshore from western British Columbia south to the islands off northwestern Mexico. These can be reconstructed into a system closely analogous to the active subduction systems described in Chapter 19.

A simplified view of the formerly active system is shown in Figure 20-11. Starting with the left half of the figure, the evidence for the long-continuing convergence of oceanic lithosphere and the America plate lies in deep-sea deposits now exposed on the continent. These include abundant pillow basalts, dikes, and fragmental basaltic rocks such as those forming along modern oceanic rises. Further evidence comes from pelagic sediments associated with the volcanic rocks: (1) layered cherts consisting of fine-grained quartz formed by the accumulation and crystallization of shells of radio-

FIGURE 20-11

Reconstructed vertical section through the plate system that paralleled the Pacific margin of central California from about 130 million years to about 30 million years ago. The arrows express relative motion between the plates.

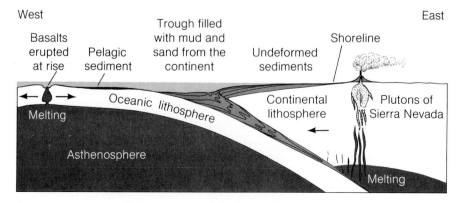

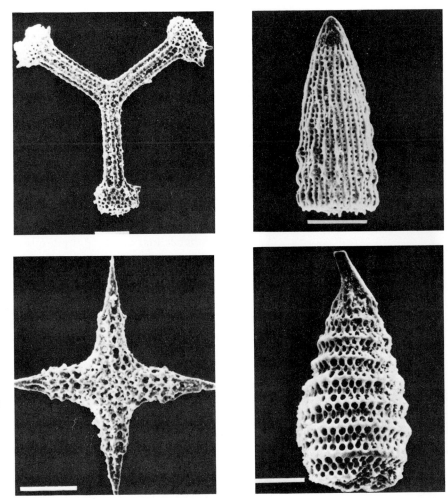

FIGURE 20-12
Above left, chert of the Franciscan Formation, near San Francisco, showing typical layers and folds. Photograph by Arvid M. Johnson. Right, shells of radiolarians separated from Franciscan chert and photographed with a scanning electron microscope. From Emile A. Pessagno, Jr., "Age and geologic significance of radiolarian cherts in the California Coast Ranges" (*Geology,* vol. 1, pp. 153–56, December 1973).

larians (Figure 20-12); (2) fine clay carried into the ocean basin beyond the trough; and (3) sparse limestones consisting largely of pelagic foraminifers and coccoliths (Figure 20-13). The scarcity of limestone compared to siliceous sediments suggests that most of the pelagic materials accumulated in the deep sea, at depths where the $CaCO_3$ was dissolved.

As the oceanic plate conveyed the basalts and pelagic sediments into the trough, they evidently became overrun by floods of muddy sand eroded from the continent, for the latter materials make up most of the marine sequences now exposed onshore. Size grading of the sandstones and mudstones indicates most were deposited from turbidity currents (Figure 20-14). Linear forms produced as a result of cutting and filling by the currents show that the currents swept into the trough system and then along its axis. The formation shown in Figure 20-14 is from trough deposits in Oregon that are comparatively undeformed and thus can be measured quite accurately. Approximately 10,000 feet of these sediments accumulated in 5 million years in the deepest

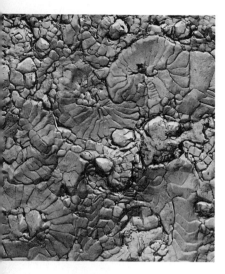

FIGURE 20-13

Greatly enlarged electron microscope image of a limestone consisting of coccoliths (compare with the modern coccosphere shown in Figure 6-17B). From Robert E. Garrison and Edgar H. Bailey, "Electron microscopy of limestones in the Franciscan Formation of California" (U.S. Geological Survey Professional Paper 575-B, pp. B94–B100, 1967).

part of that trough, giving an average rate of 2 feet of sediment per 1,000 years (Reference 5). This rate is at least 20 times greater than the average rate of sedimentation in the Atlantic Ocean, even when the unusually rapid accumulations of the glacial periods are included. The layers deposited in the Pacific troughs are also individually thicker than the turbidity current layers in the Atlantic, with the important exception of the layer generated by the Grand Banks earthquake. It is thus likely that each graded layer of sandstone in the Pacific troughs was caused by a major earthquake and that the earthquakes were generated by the convergence of the lithospheric plates.

Deep subduction and melting Subduction of a large proportion of the accumulated rocks is indicated by their metamorphism. The great thickness and the compression (buckling) of the partly subducted sediments shown in Figure 20-11 led to unusually strong compaction and to the recrystallization of the clay minerals that made up the muddy matrix of the sandstones. The rocks thus became unusually tough and also dark gray in color. The sandstones are called *graywacke*—the name used for dark sandstone made up of poorly sorted angular grains surrounded by a recrystallized matrix of clay. Rocks that were subducted more deeply and pressed more powerfully were metamorphosed to new minerals of high density. Calcite, which has a density of 2.7 grams per cubic centimeter, was changed to *aragonite*, another form of $CaCo_3$ with a density of 2.9 grams per cubic centimeter (Supplementary Chapter 1). Plagioclase and iron-rich clays were metamorphosed to denser minerals such as pyroxene, garnet, and the blue amphibole called glaucophane. These changes are induced by low temperatures as well as high pressures, and the metamorphism thus implies that subduction was so rapid that the rocks remained relatively cool while they were being carried to great depth.

Buckling of the layered rocks is further evidence of plate convergence (Figure 20-12). Subduction also produced countless faults, along which the

FIGURE 20-14

Sharply defined bases of sandstone layers (light toned) that grade upward into siltstone. The enlarged part of a thin section, which is 0.3 inch across, shows the angular and poorly sorted nature of the sand grains (compare with the sandstone in Figure 15-10, top). The rocks are in the Tyee Formation of western Oregon.

FIGURE 20-15

Part of the southern Sierra
Nevada, showing granite
plutons (light-toned) and the
dark metamorphic rocks into
which they were intruded.
Photograph by John S. Shelton.

rocks were ground into fragments of all sizes, forming a chaotic mixture called *mélange*. Single exposures of mélange commonly include every variety of rock from the trench and the oceanic plate, metamorphosed as well as unmetamorphosed (Color Plate 11B). Mélange thus indicates a tremendous amount of faulting and mixing during subduction. Still more deformation was caused where slabs of oceanic crust and mantle were thrust locally onto the edge of the continent, just as in Papua New Guinea, or where peridotite of the mantle was intruded into the overlying rocks.

On the east (right) side of the plate-tectonic system in Figure 20-11 are the roots of the volcanic chain—a great complex of plutons that now forms the cores of mountains from northwestern Mexico through the Sierra Nevada to northwestern Nevada and Idaho (Figure 20-15). The many radiometric dates determined on rocks from the Sierra Nevada indicate that most of those plutons were emplaced between 150 and 70 million years ago (Reference 6). Because subduction under that part of the volcanic chain ended only 25

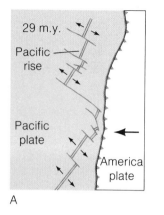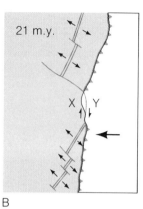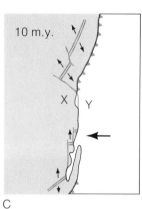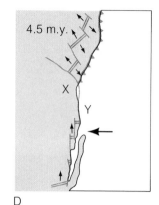

A   B   C   D

FIGURE 20-16

Evolution of the plate system near the western margin of North America during the latter half of the Cenozoic Era, with numbers showing ages in millions of years. A, America plate moving westward toward the Pacific rise, with subduction in progress all along the boundary between the plates. B, C, and D, America plate progressively contacting and deactivating the rise, after which the motion between the plates becomes one of lateral offset, such that points X and Y become increasingly separated. E, the present coastal region south of Oregon is thus dominated by transforms and the San Andreas fault.

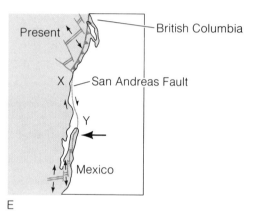

E

million years ago, no melts were produced during the last 45 million years of subduction or else melts did not rise as high as the level now exposed.

The late record of erosion of the volcanic chain is beautifully documented in the wedge of undeformed sediments lying between the trough and the plutons in Figure 20-11. These sediments are fossil-bearing sandstones, shales, and conglomerates that can be dated in detail. Fragments of granite in the sediments show that the first plutons were exposed by erosion in Mesozoic time, soon after they cooled, and that the last of the volcanic chain was eroded away by the start of the Cenozoic Era. The erosion of the plutons continued to pour granite fragments into the trough during the Cenozoic Era. The trench thus came to be filled by materials once emplaced deep in the subducting system and melted under the volcanic chain!

The Mobile Border During the Late Cenozoic

About 25 million years ago, the subduction system became inactive along the part of the continental margin that is now western California. The rocks in the upper part of the system were uplifted and eroded to form a broad new addition to the edge of the continent. This history is shown conclusively by shallow-water sandstone deposited 25 million years ago across the eroded edges of the subducted rocks.

These changes were due basically to the America plate moving westward against the spreading Pacific rise, as shown by studies of the magnetic bands in the Pacific Ocean (Reference 7). The results are summarized diagrammatically in Figure 20-16, which shows several interpreted stages in the movement of the America plate and its progressive deactivation of the rise. Note that the changes initiated lateral movement between the Pacific and America plates, generating an early version of the San Andreas fault (Figure 20-16C). The fault grew in length and developed several branches as more of the America plate was carried against the rise. Parts of the former continental margin also became attached to the Pacific plate and have thus been carried hundreds of miles north of their original positions, leading to the major offset shown in Figure 20-16E.

Some earth scientists have suggested that lateral movements such as those along the San Andreas fault have caused all the subsequent deformations in the region, but, as we shall now see, much of the deforming movement has been at right angles to the San Andreas fault and has often included large vertical displacements. Considerable evidence comes from basins that started to form about 20 million years ago in western California and deepened rapidly to the point that the sediments of the Monterey Formation and other deep-water deposits began accumulating in them. The rate of deepening of some basins increased about 7 million years ago, in some cases remarkably. Nearly 18,000 feet of turbidity-current deposits and other deep-water sediments accumulated in the deepest part of a basin northwest of Los Angeles between 6 million and 1 million years ago. This accumulation gives an average rate of nearly 4 feet of sediment per 1,000 years—greater than the rate in many active trenches. Even where basins were subsiding more slowly, newly deposited sediments periodically slid and mixed on the basinal slopes, suggesting episodic shaking by strong earthquakes (Figure 20-17 left). Earthquakes also fluidized unconsolidated sand and injected it into overlying impermeable sediments to form dikes of sandstone in shale (Figure 20-17 right).

There is also reason to believe that the land areas between the basins were being uplifted rapidly, for most of the sediments contain abundant grains of feldspar and mica that are nearly unaltered by soil-forming processes, showing that erosion was cutting rapidly into fresh rock and thus implying steep slopes on land. The most remarkable thing, however, is that many of the basinal deposits were themselves folded and uplifted in the last 3 or 4 million years, so that they now form parts of mountain ranges! Figure 20-18A shows a section through one of the ranges, with arrows at the two ends of the section indicating the amount of shortening due to buckling. When the amount of shortening is divided by the duration of buckling (a few million years), the rate of shortening turns out to be 0.5 to 1 inch per year—as rapid as the converging movements at some subduction zones. Note in Figure 20-18B, moreover, that this movement was at right angles to that expressed by the offset along the San Andreas fault.

FIGURE 20-17

Left, layers of disrupted sediment in a laminated sequence deposited in deep water. Because each disrupted layer is underlain and overlain by undisturbed laminations, the disruptions must have occurred periodically and many years apart. Sisquoc Formation near Point Conception, California. Right, wave-cut cliff exposing dikes of sandstone that were injected upward into a shale formation when an unconsolidated sand beneath was fluidized. Near Santa Cruz, California.

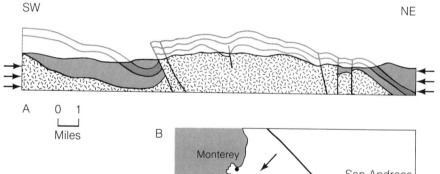

SW   NE

A   0   1
         ⊢——⊣
          Miles   B

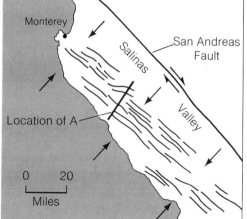

FIGURE 20-18

A, vertical section through the Santa Lucia Range, California, showing folds and faults that formed only a few million years ago, partly reconstructed above the present surface (black line) to show the full distribution of the formations. The formation shaded dark gray was deposited in a deep marine basin, and was uplifted several miles when the range was formed. B, map showing location of A and the alignments of the wrinkle-like folds (lines) in the range. Note that the range is separated from the San Andreas fault by the broad Salinas Valley and that the direction of compression (the full arrows) is nearly at right angles to the fault.

Evidence from coastal terraces Are similar movements going on today? Studies of coastal ranges and of islands near shore have shown that some have been rising persistently during the past 1 million years. Most of the evidence comes from old wave-cut cliffs and former submarine surfaces that now form terraces much like those around Lake Bonneville (Figure 20-19). The youngest terrace, dated at several localities on the California coast, formed during the last interglacial period, when sea level was high. When sea level fell at the beginning of the last period of glaciation, the wave-cut cliff was left high and dry and part of the seafloor was exposed. Earth movements then gradually uplifted the coastal ranges and islands during the ensuing 100,000 years, so that the sea cliff was well above sea level when the glaciers melted and sea level rose at the start of the present interglacial period. Dates determined for terraces above the lowest one confirm this interpretation, for some correspond to older interglacial periods. Sequences of terraces such as that in Figure 20-19 are thus due to more or less continuous uplift, each cliff being cut during a high stand of the sea.

An especially pertinent case has been studied just southwest of San Francisco Bay, where a set of terraces on the seaward side of a coastal range evidently formed in the manner just described (Reference 8). The youngest (lowest) is approximately 90 feet above sea level at the southern end of the coastal range and rises gradually to a maximum of 120 feet where the western flank of the range rises most steeply from the sea. This relation indicates that the range has risen as an elongate dome since the terrace formed. The dome has been arched even more notably at right angles to its elongate form, for the older (upper) terraces slope more steeply toward the sea than do the lower terraces. Radiometric dating indicates a tentative age of 125,000 years for mollusk shells uplifted with the lowest terrace, an age which is compatible with the emergence of the terrace at the start of the last glacial period. None of the other terraces has been dated, but the highest, 800 feet above sea level,

FIGURE 20-19

Marine terraces consisting of uplifted segments of the seafloor (the gently inclined surfaces) and former sea cliffs (the steep slopes), each one modified by stream erosion. The locality is on San Clemente Island, off southern California (see Figure 13-2B for its location). Photograph by John S. Shelton.

would be approximately 1 million years old if the rate of uplift has been more or less the same as that for the lowest terrace. The elongate form of the arch is parallel to the San Andreas fault and thus suggests compression (buckling) at right angles to the faults. The bay region is thus being deformed in ways other than those due solely to lateral movement along the San Andreas fault.

Conclusions

It thus appears that the main part of the America plate, described in the first section of the chapter, has been moving westward more or less quietly and regularly as new oceanic lithosphere is generated along the Mid-Atlantic Ridge but that powerful interactions have taken place along the western edge of the plate. Two interactions, those involving subduction and horizontal offsets along the San Andreas fault system, can be explained by the movement of the Pacific plate relative to the America plate. Other interactions, however, which cannot be explained so readily, appear to be as effective. The western parts of California, for example, have been more strongly compressed since subduction ceased than have the western parts of Oregon, Washington, and British Columbia, where subduction has remained active! The direction of the compression, moreover, has been at about right angles to the San Andreas fault system. Evidently, the America plate has continued to move westward against the Pacific plate.

An intriguing connection is that these deformations may be related to the volcanic provinces of the western interior. We saw that the volcanic activity there indicated province-wide heating and also indicated fissuring due to the westward extension of parts of the America plate. The heating was associated with doming in what is now the basin and range province, suggesting that westward movement was quickened by increased ductility of the asthenosphere and possibly by flow and sliding under gravity. We may now note that the position of the westward bulge of the America plate across the Pacific rise correlates exactly with the region of greatest extension within the continent. The two actions also fit well in time, for the greatest compression at the Pacific margin took place within the last 6 million years, the period of greatest extension in the basin and range province.

As compelling as these relations may sound, however, the connections are highly conjectural. We have not accounted for movements north and south of the regions described in this chapter, nor have we explained the thickening of the crust that occurred at the same time east of the basin and range province—the cause of the uplift of the Colorado Plateau and Rocky Mountains. Indeed, we have proposed no reasonable cause for the heating that led to the doming and separations! Such is the present state of the science. A worldwide system of moving plates seems well founded, and one can identify local systems that "work." A truly comprehensive theory, however, will require far more thorough knowledge of continental and oceanic regions. As I have mentioned before, any thoughtful, honest study will contribute toward the whole, often more fully than anticipated.

A return to San Francisco Bay Additional studies, moreover, may have more significance than simply testing theories of earth movements, and we might use our return to the San Francisco Bay area to consider that significance. Let's think about our progress since we first became aware of problems concerning the bay and of our need for understanding the earth. If you were to visit a remnant of the bay marshes now, would you see it with an improved understanding? Besides recognizing some problems and some interesting details, would you see the marsh as part of the whole estuary and the estuary as one of a family of estuaries? The effects of the bay's tides as well as its river-driven circulation would thus be predictable, and you could probably guess how each part of the system might have evolved during the cyclic glaciations that affected our environment so greatly. Do you see now, too, that this particular estuary is affected by worldwide movements of the outer parts of the earth, that its basin was formed by complex deformation in a mobile region?

If you want to plan a major structure—a beautiful building—next to the marsh, you know a variety of questions that must be answered. Is this part of the bay area rising or sinking because of the movements described in this chapter? Is it also sinking because of compaction of the bay mud? What of isostatic adjustments due to the weight of the bay and its recent deposits? Will the pumping of groundwater from deep-seated gravels cause subsidence of the building site? Does the bay mud hide layers of sand that might liquify during an earthquake?

If you were to ask local experts these questions, you would probably get useful answers but only on a one-by-one basis. Experts live in fragments of their science and are cautious about crossing boundaries! You know now that the real subject (the earth) has no such boundaries, but that won't help you in exacting a whole answer to your questions. The significance of earth studies is that each adds to all the others, that it is a whole system that is being explored. We are all part of that system. Do you see that your understanding is needed badly? There is still so much to be learned and so much to be done.

REFERENCES CITED

1. John T. Hack. Interpretation of Cumberland Escarpment and Highland Rim, south-central Tennessee and northeast Alabama. U.S. Geological Survey Professional Paper 524-C, 16 pp., 1966.
2. Larry D. Brown and Jack E. Oliver. Vertical crustal movements from leveling data and their relation to geologic structure in the eastern United States. *Reviews of Geophysics and Space Physics*, vol. 14, pp. 13–35, 1976.
3. J. F. Sutter and A. R. McBirney. Periods of Cenozoic volcanism in the Cascade Province of Oregon (abstract). Geological Society of America Abstracts, vol. 6, no. 3, pp. 264–65, 1974.
4. D. A. Swanson and T. L. Wright. Linear vent systems and estimated rates of magma production and eruption for the Yakima Basalt on the Columbia Plateau. *American Journal of Science*, vol. 275, pp. 877–905, 1975.
5. Parke D. Snavely, Jr. and Holly C. Wagner. Tertiary geologic history of western Oregon and Washington. Washington Division of Mines and Geology, Report on Investigation 22, 25 pp., 1963.
6. Marvin A. Lanphere and Bruce L. Reed. Timing of Mesozoic and Cenozoic plutonic events in Circum-Pacific North America. *Geological Society of America Bulletin*, vol. 84, pp. 3773–82, 1973.

7. Tanya Atwater. Implications of plate tectonics for the Cenozoic tectonic evolution of western North America. *Geological Society of America Bulletin*, vol. 81, pp. 3513–36, 1970.
8. W. C. Bradley and G. B. Griggs. Form, genesis, and deformation of central California wave-cut platforms. *Geological Society of America Bulletin*, vol. 87, pp. 433–49, 1976.

ADDITIONAL IDEAS AND SOURCES

1. A particularly well written and interesting history of the continent is given by Philip B. King in *Evolution of North America* (Princeton, N.J.: Princeton University Press, 189 pp., 1959). The book was written before the discovery of plate movements ("plate tectonics"), so that its interpretations are outdated in that respect, but the descriptions and most interpretations are accurate and well organized on a geographic basis. Another useful source is *Geological evolution of North America*, 2nd ed., by Thomas H. Clark and Colin W. Stearn (N.Y.: Ronald Press Co., 570 pp., 1968).

2. The interpretations of dissected plains and accordant mountain summits, mentioned for the Appalachian province, are important in many other parts of the world. The contending theories of uplift and erosion are well described by William D. Thornbury in *Principles of geomorphology*, 2nd ed., (New York: John Wiley and Sons, 594 pp., 1969). An annotated collection of the original papers presenting the theories and evidence has been published in *Planation surfaces*, George F. Adams, ed. (Stroudsburg, Pa.: Dowden, Hutchinson & Ross, 476 pp., 1975).

3. Much of the history of the volcanic provinces mentioned in the chapter, as well as the coastal provinces of Oregon and Washington, are described by Bates McKee in *Cascadia: the geologic evolution of the Pacific Northwest* (N.Y.: McGraw-Hill, 394 pp., 1972).

4. The semitropical Eocene climate mentioned for the western United States evidently had a counterpart in Canada at least as far north as 78° latitude. Fossil turtles, alligatorids, and other animals that now live in temperate to warm climates have been reported by Mary R. Dawson and others in "Paleogene terrestrial vertebrates: northernmost occurrence, Ellesmere Island, Canada" (*Science*, vol. 192, pp. 781–82, May 21, 1976). Eocene phytoplankton in sediment cores from the bottom of the Arctic Ocean have shown that that ocean was free of ice and comparatively warm at that time, as reported by David L. Clark in "Late Mesozoic and early Cenozoic sediment cores from the Arctic Ocean" (*Geology*, vol. 2, pp. 41–44, January 1974). The plankton record also indicates that the Arctic Ocean had frozen over by about 5 million years ago, but the cores are not complete enough to show when the initial freezing took place.

Open pit mine at Bingham Canyon, Utah, an excavation in a granite pluton that is so pervasively enriched in copper-bearing minerals that nearly the entire intrusion and its marginal rocks are mined as ore. Started in 1904, the mine has produced nearly 11 million tons of copper metal as well as important amounts of gold, silver, molybdenum, platinum, palladium, selenium, and rhenium. Photograph by John S. Shelton; data courtesy of Kennecott Copper Corporation.

21. Materials and Energy from the Earth

Materials and Energy from the Earth

Our dependency on the earth's supply of fuels, metals, and building materials gives rise to many problems. Uses of these things have increased so rapidly in the past three decades that there is now a serious question about the adequacy of the supply. The impending shortages and the resulting social problems have four basic causes, each of great importance:

1. Deposits of minerals and fuels form so slowly that they are for all practical purposes nonrenewable. They are finite resources and can be mined or pumped from the earth but once.
2. Most of the richer deposits have already been found and have largely been depleted.
3. The deposits are distributed unevenly among nations, which has led to political problems, even wars that have consumed immense quantities of the resources.
4. The human population is increasing at a rate that leads to a doubling about every 35 years. The increases are greatest in nations with relatively undeveloped industries.

To understand the resulting problems, we need to look at a specific example, and we would do well to consider the basic data for copper. This metal is essential to the electrical industry and is used so widely in buildings, machinery, vehicles, and ships that it epitomizes the needs of the modern industrialized world. The history of copper production, shown in Figure 21-1, is thus an indicator of modern industrial expansion and its dependency on mining. Note that prior to 1950 the United States typically used more newly mined copper than all the other nations combined. Then note the rapidly increasing uses *outside* the United States after about 1950 (the vertical distance between the curves for the United States and the world). Finally, contemplate the ever steepening shape of the world curve and its projection into the near future. Is there enough copper to supply the 16 million tons that are the projected need for the year 1985, let alone the 40 million tons that may be the annual need in the year 2000?

How much copper is there in the outer part of the earth? Measurements of copper in many samples of rocks collected near the surface indicate that roughly three quadrillion (3×10^{15}) tons lies within the outer mile of the crust—apparently enough to supply the most affluent society imaginable for millions of years. However, an unfortunate reality of earth resources is that almost all this copper occurs in mere traces (less than 0.005 percent of the rocks). The great volume of the rocks rather than the concentration of copper in them makes copper seem so abundant. It would be far too costly in energy and materials to use ordinary rocks for our copper supply. For humans, then, an available resource is not the total amount of a material in the earth, *but the amount that can be recovered economically.*

Fortunately, most substances have been concentrated locally in rock that can be mined economically. For metals, the naturally enriched rock is called *ore*, and the body of ore is called an *ore deposit*. Such deposits are few and far between; all of them together would make up a very small fraction of the

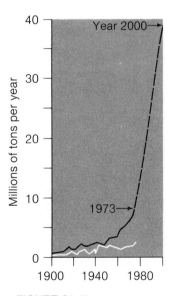

FIGURE 21-1
Quantities of copper mined and produced each year in the entire world (upper curve) and in the U.S. The dashed line is a prediction of future needs, based on the shape of the curve since 1940. Production data from U.S. Bureau of Mines.

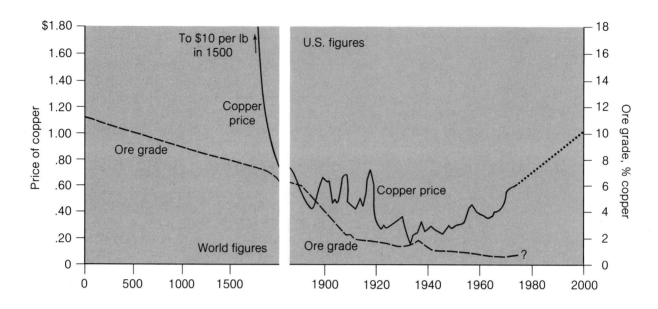

FIGURE 21-2

Curves showing the average copper content in mined ores (the ore grade) and the price of a pound of copper metal (all calculated with 1970 dollar values). Prices since 1960 are those in the United States (prices elsewhere, as quoted in London, have been about 17 percent above those shown). Modified slightly from a figure by J. David Lowell (Reference 1).

earth's crust. There is, moreover, no certain, preordained number of ore deposits. Rocks that contain almost enough metal to be ore will become economical deposits when less expensive methods of mining or milling are developed or when demand sends the price of the metal upward faster than the costs of obtaining it.

The amounts of *available* earth resources are thus tied strongly to human economics and human ingenuity. J. David Lowell has illustrated these relations for copper ores by the two curves reproduced in Figure 21-2. If you compare these data up to the year 1932, you can see that the mining industry was able to produce copper at broadly diminishing prices from ores with steadily decreasing concentrations of metal—a tribute to its ingenuity. Note, however, that the price has been increasing quite rapidly since 1932. Note, too, that the average grade of ore being mined has probably risen slightly since 1974—the first reversal in the long-term downward trend. The two reversals are due mainly to the increasing costs of the energy and the materials needed to mine and produce the copper and thus express the *general* scarcity of basic resources.

The post-1932 part of Figure 21-2 suggests that it will be a great challenge to meet the needs expressed in Figure 21-1. Moreover, the costs of a large number of commodities and energy sources will be involved, for all are related to some degree. Approximately 80 of these commodities are sufficiently important to be researched and accounted for regularly by the U.S. Bureau of Mines. The United States annual consumption of the twenty-three most valuable is summarized in Table 21-1. Note that the United States imports a large proportion of its needs. Note, too, the tremendous values of the fuels needed by an industrialized society and the surprisingly large values of "ordinary" materials such as clays, stone, and sand and gravel.

In this chapter, we will consider the supply of a few key substances and the earth's potential for meeting the increasing demand for them. We will first continue with our examination of copper, to see how one uses geologic

TABLE 21-1

Most Valuable Mineral Products Put in Use in the United States in 1973[1]

Product	Annual value in millions of dollars	Principal uses	Principal sources	Percent from U.S. deposits
Aluminum	3,400	Construction; transportation	Bauxite (aluminum-enriched soil	7
Clays	1,300	Construction; refractories; paper	Sedimentary deposits	103[4]
Coal[2]	4,800	Utilities; industrial fuel; coke	Altered deposits of plant materials	107
Copper	2,800	Electrical wire and parts; alloys	Sulfides in altered rocks, veins	88
Gem stones	420	Jewelry; art objects	Peridotite (diamond); stream sediments	1
Gold	670	Jewelry; arts; industry	Sedimentary deposits; quartz veins	19
Gypsum	630	Construction; agriculture	Sedimentary layers	66
Iron and Steel[3]	25,000	Construction; vehicles	Sedimentary deposits, chiefly Precambrian	65
Lead	520	Batteries; gasoline	Sulfide in veins and altered rocks	54
Lime	370	Chemicals; industry	Limestone	99
Natural gas[2]	5,000	Industrial, household fuel	Permeable sedimentary formations	98
Nickel	790	Stainless steel; other alloys	Sulfide in igneous rocks	7
Nitrogen (and ammonia)	960	Fertilizer; plastics	Atmospheric nitrogen	99
Petroleum[2]	22,000	Gasoline; fuel oil; kerosene	Permeable sedimentary formations	59
Phosphate	2,300	Fertilizer; detergents	Marine sedimentary deposits	135
Sand and gravel[2]	1,400	Construction; glass; ceramics	Recent sedimentary deposits	100
Silver	500	Photography; silverware	Veins; igneous bodies	23
Sodium salt products	1,500	Chlorine; soda ash; other chemicals	Sedimentary formations	96
Stone[2]	2,000	Construction	Various rock formations	100
Sulfur	320	Agriculture; industry	Sedimentary formations	107
Tin	340	Plating metals; solder	Stream deposits; veins	0
Titanium oxide	670	Paints; paper products	Igneous, metamorphic rocks; sands	44
Zinc	680	Alloys; galvanizing steel	Sulfide in veins and in altered rocks	31

[1]Calculated by Caroline M. Isaacs from data in *Minerals yearbook 1973* by Bureau of Mines Staff (U.S. Government Printing Office, Washington, 1975) and in *Minerals in the U.S. economy: ten-year supply-demand profiles for mineral and fuel commodities* by Bureau of Mines Staff (U.S. Government Printing Office, Washington, 1975).

[2]Values are for unrefined (crude) materials.

[3]Steel contains other materials beside iron, including approximately 500 million dollars worth of manganese and chromium, almost all of which are imported.

[4]Amounts over 100 percent are exports.

knowledge to look for ore deposits and also to judge the total supply. Most of the other metals in Table 21-1 (gold, lead, nickel, silver, tin, and zinc) would be similar cases, as would many crucial metals not listed, such as cadmium, molybdenum, and tungsten. For the fuels, we will be especially interested in the domestic supplies still available and how they are estimated. We can then judge the need to develop other sources of energy. Finally, we will consider the need to balance the acquisition of industrial commodities against our needs for still more basic resources: water and soil.

Finding a Crucial Metal

To find more copper, we need to know how it occurs and how it forms. In economically minable deposits, copper occurs primarily as a compound with sulfur and iron, a shiny yellow mineral called *chalcopyrite* (Color Plate 12D). Other metal-sulfur compounds commonly occur with it, for example, *galena* and *sphalerite* (sulfides of lead and zinc, respectively) and *pyrite* (iron sulfide) (Color Plates 12E and 14A). These minerals are concentrated locally in thin veins or small irregular bodies of ore that may contain as much as 5 to 15 percent metal; these were the economic source of most copper mined prior to 1850. These relatively small but rich ore bodies occur near the margins of granite plutons and related intrusive bodies, as noted in Chapter 12, and this relation has been a guide in the search for these deposits. From 1850 to the early 1900s, a principal ore was native (metallic) copper that makes up only a few percent of the altered basalts of the Keweenaw Peninsula of Michigan. These relatively diffuse deposits of copper could be used because of improvements in mining and milling. Today we supply most of our needs from large deposits containing less than 1 percent (in some cases as little as 0.35 percent) of copper.

Plutons as ore deposits   Some of these large deposits (about 25 percent) are sedimentary formations in which copper-bearing sulfides were deposited in unusual concentrations, and some (about 10 percent) are basalts and gabbros with unusual amounts of copper, such as in the Keweenaw deposits. Most (65 percent), however, are plutons of granite peppered with small grains of chalcopyrite and other ore sulfides. The granite ore bodies conform roughly to the shapes of the small plutons in which they occur, but they also include veins and other small bodies outside the intrusive contacts, such as those just mentioned (Figure 21-3). The figure also illustrates places where downward percolating rainwater dissolved copper from the upper part of the ore body and redeposited it near the top of the water table, thus increasing the copper content of this part of the ore from about 0.5 to 0.8 percent. Local concentrations may also occur in vertical columns of rock that were exploded into fragments by water vapor escaping after most of the pluton had solidified (Figure 21-3). The effects of vapor are seen also in the alteration of feldspars to mica, clays, and carbonate minerals, as described in Chapter 12. These relations suggest that copper was concentrated by vapors that passed through the pluton at a late stage in its solidification and cooling.

FIGURE 21-3
Vertical and horizontal sections through a pluton ore deposit, showing typical concentrations of copper.

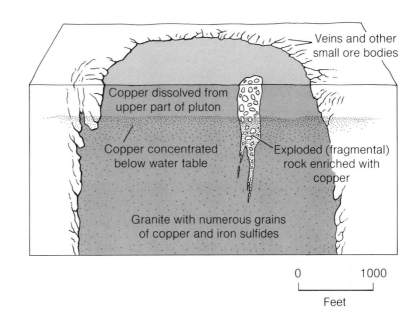

Plutons thus affected, however, are scarce, and it is important to know why some carry large quantities of metal sulfides while most do not. One hint comes from the location of the discovered ore plutons relative to the subduction zones that were active when the plutons formed (Figure 21-4). The rocks and sediments of oceanic regions provide accessory evidence: basalt, gabbro, and oceanic sediments commonly contain moderate concentrations of copper, and volcanic vapors rising along spreading rises have locally added large amounts of copper to these deposits. Evidently, then, copper is added to the oceanic crust at spreading rises, and additional amounts are dissolved and eroded from the continents and deposited in oceanic sediments. When the oceanic crust is subducted and heated, vapors and melts carry copper upward and concentrate it in the upper parts of some granite plutons.

A search for additional deposits This hypothesis suggests that plutons arising from the melting of the oceanic crust would be especially rewarding to explore for copper deposits. The many possibilities may be reduced further by two related strategies: (1) to explore regions where valuable deposits have been found, but (2) to concentrate on places that have not been explored so thoroughly that the chances of a new discovery would be small. As Figure 21-4 indicates, many deposits have been found in the southwestern United States and in Peru and Chile; however, surface exposures of rock have been examined closely in those places. Many deposits probably still occur there, but they are hidden beneath alluvial deposits and other sediments and thus would be costly to find. We might thus turn our efforts first to western Canada, Alaska, and regions adjacent to subduction zones in the southwest Pacific, where valuable deposits occur but exploration is incomplete. Indeed, this strategy has recently led to discoveries of major copper-rich plutons in

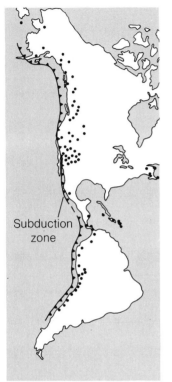

FIGURE 21-4

Locations of pluton ore deposits of copper in the Americas and the subduction zones (toothed lines) active when most of the plutons formed. Note that many deposits in the United States and northern Mexico seem too far inland to have formed from subducted oceanic crust. Possibly they are related to the widespread volcanic activity described in Chapter 20. Locations courtesy of S. C. Creasey, U.S. Geological Survey.

British Columbia, the Philippines, Papua New Guinea, and the Solomon Islands.

Having chosen the most likely region, how does one find a deposit? The most direct "indicator" is a body of reddened rock that forms at the top of the ore deposit when iron sulfide is altered to iron oxides by oxygen-bearing rainwater. The "rusty" rocks may also contain blue or green copper carbonates and silicates formed by the alteration of copper-bearing sulfides, although most of the copper is carried downward in solution as already noted. A prospecting method used where soil and vegetation hide the reddened rocks is to analyze samples of plants, soil, or water for unusual concentrations of dissolved copper and related metals.

The next step is to drill holes through the surface rocks to see if there is indeed ore beneath and, if so, to drill additional holes to determine the size of the ore body. Thousands of *assays* (analyses of metal content) are made from the drill cores thus obtained, so that mining and milling can be planned ideally for each deposit. Here is where pluton ore deposits prove uniquely valuable, for their large, simple forms can be mined optimally with large scale machinery. The huge pit shown on the chapter opening page is an example. The economy of open pit mining is the main reason copper prices have increased only moderately in spite of rapidly increasing demand.

The future of relatively low-cost copper thus depends on constant improvements in mining as well as on finding additional large deposits of ore. Considering the areas prospected and the numbers of deposits found so far, J. David Lowell has estimated that roughly 170 deposits remain to be found in British Columbia and the southwestern United States (Reference 1). If copper resources elsewhere in the world could be increased similarly (which seems reasonable), it appears that world needs through 1985 can be met. New kinds of resources may help considerably. The manganese nodules of the seafloor, for example, which commonly contain 1 to 2 percent of copper as well as valuable amounts of nickel, cobalt, molybdenum, vanadium, and manganese, may soon become economic resources (Figure 21-5).

The amount of copper needed by the year 2000, however, seems unattainable. Even if sufficient low-grade deposits are found (perhaps down to a grade of 0.15 percent copper), it seems unlikely that the mining industry can grow rapidly enough to handle the vast quantities of ore needed. Besides an inadequate supply of trained personnel, there will be a general increase in the costs of exploration and production. As Lowell noted, "In 50 years the price of copper (not considering inflation) has increased by $3\frac{1}{2}$ times, but the cost of building a plant to produce a unit amount of copper has increased by over 12 times." (Reference 1). Thus, we might have the theoretical capacity to produce enough metal but be unable to supply the other resources needed.

Energy from Oil and Natural Gas

A key factor in obtaining all earth materials is the availability of large quantities of cheap energy. As Figure 21-6 suggests, almost all this energy has been obtained from *mineral fuels*, (also called *fossil fuels*) mainly from oil and natural gas. Oil and gas are exceptionally easy to handle and until recently were available in large quantities at low prices. Although they have been criticized widely for atmospheric pollution, their *average* production of pollutants is lower than that of coal, the only other mineral fuel now available in large quantities.

In spite of increasing uses and prices, however, domestic production of oil and natural gas has lagged steadily behind demand. Imports amounted to one third of the United States consumption in 1974, and unless domestic production can be increased substantially, half our needs in 1980 and roughly two thirds of them in 1985 will have to be supplied from imports. These projections suggest that our main source of energy will be increasingly costly and less and less reliable—a serious situation for a society in which agriculture, industry, space heating, and transportation rely mainly on these fuels. Can we count on increasing domestic production of oil and gas, or should we try to convert to other sources of energy? The decision will have exceedingly far-reaching effects on our economy and on our way of life.

How much oil and gas will be discovered? The decision rests largely on knowing how much domestic oil and gas can be found and produced in the next two decades. Several estimates of this figure have been made, but they differ greatly. To understand our energy situation, we must see how these estimates were made and which seems the most accurate. The largest estimate results from a series of calculations made by the U.S. Geological Survey, starting in 1956 and continuing through 1971 (Reference 2). These estimates were based on two assumptions: (1) that wells drilled in the future to find new fuel reservoirs (called *exploratory wells*) would succeed in the same proportion as the average for all such wells drilled in the past, and (2) that all rock bodies still to be drilled contain fuel reservoirs in the same proportion as similar rocks already drilled. The success of past drilling was converted into a simple factor that has been used in other estimates: the number of barrels of oil (or cubic feet of gas) that have been discovered per foot of exploratory wells drilled. When the U.S. Geological Survey multiplied this number by the total amount of sedimentary rocks still to be explored by drilling, it obtained figures of 450 billion barrels of oil and 2,100 trillion cubic feet of gas as the fuels still to be discovered. The 450 billion barrels was based on the belief that 40 percent of the oil discovered underground could be brought to the surface, but because 32 percent is the proportion that can be recovered currently, I have reduced the number in Table 21-2 accordingly (all the other estimates are based on the same percentage of recovery).

The consumption of oil in the United States was 6.3 billion barrels in 1973, so that the estimate indicates that we could increase domestic production considerably and still have enough oil for 30 to 40 years. Also listed were 2,100 billion barrels of what were termed "undiscovered submarginal oil resources," implying that improvements in technology or increased prices would put gigantic amounts of petroleum "in hand."

The next two sets of numbers in Table 21-2 come from thorough studies of regions that have produced oil and gas or have the potential for doing so. The estimates were made by teams of geologists who were best qualified to analyze each region. Many types of data were used: drilling records, records

FIGURE 21-5

Manganese nodules from the Pacific Ocean, one being broken to show its hollow interior. Similar nodules on the seafloor can be seen on the opening page of Chapter 6. The metal content of nodules varies geographically, but most have their greatest values in nickel and copper.

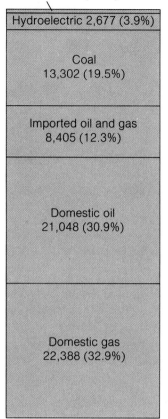

Geothermal 7 (0.017%)
Nuclear 240 (0.4%)

Hydroelectric 2,677 (3.9%)

Coal
13,302 (19.5%)

Imported oil and gas
8,405 (12.3%)

Domestic oil
21,048 (30.9%)

Domestic gas
22,388 (32.9%)

FIGURE 21-6

Amounts and sources of energy used in the United States in 1970. A B.T.U. (British thermal unit) is the amount of heat that will raise a pound of water from 63 to 64 °F.; it is equal to 252 calories. Data from the National Petroleum Council (Reference 3).

of production, knowledge of specific sedimentary sequences, and indications of favorable arrangements of strata based on geophysical methods (seismic studies, measurements of gravity, and measurements of the local magnetic field). The estimate submitted by the National Petroleum Council (industry's advisory group to the Secretary of the Interior) was based on regional estimates made by the petroleum industry (Reference 3). The U.S. Geological Survey estimate (of 1975) was made under the press of time (between September 1974 and June 1975) and could not use much of the data available to petroleum companies (Reference 4). If one considers that many industrial geologists may be somewhat too optimistic in judging oil in the ground, however, the two estimates are in reasonable agreement.

Is it necessary, then, to even consider the last figure, which is uncomfortably smaller than the others? It is necessary, indeed, for these numbers were obtained in a different way. They are the result of mathematical analyses, by M. King Hubbert, of the rates of growth and the beginnings of decline of discovery and production of oil and gas in the United States (Reference 5). The analyses are based on actual, known numbers rather than on predicted numbers. The changing rate of production is illustrated in Figure 21-7, which uses a logarithmic scale to emphasize the long period of persistently rapid growth before 1930 (a rate doubling every 8.4 years); then the abrupt decline due to the depression in 1930; the brief period of recovery; and the continuously decreasing rate of growth up to 1975. The distressing facts are that the period of decreasing rate of growth coincided with a period during which the market for oil and gas was expanding rapidly, more and more exploratory wells were being drilled, technology was improving each year, and geologic information was increasing at a prodigious rate. It is difficult to escape the conclusions that most underground oil and gas had been discovered and that the total resource was being diminished rapidly.

This interpretation is so important that we must be sure that the decline in production is not due to economic or political factors rather than to the amounts of fuel underground. As Hubbert suggested, we can look at data not affected by economics—the rate of discovery of fuel *per effort of search*. Figure 21-8 shows these data for oil. The numbers of feet of exploratory wells are shown at the bottom of the graph; they simply accumulate from left to right in units of hundreds of millions of feet. The dates in the upper part of the graph show how rapidly the drilling was accomplished. The bars indicate the amounts of oil discovered per foot of drilling for each of the accumulated units. Note that five times more exploratory drilling was done between 1960 and 1970 than between 1930 and 1940, yet approximately 28 billion barrels of oil were discovered during the earlier period compared with 13 billion barrels during the later one. The figure also shows why the 1972 U.S. Geological Survey estimates were so large; they were based on the *average* rate of discovery for the entire period (118 barrels per foot drilled) rather than on the rate in 1972 (30 barrels per foot drilled).

Hubbert first calculated and published his results at the end of 1961. He repeated his analysis in 1972, using the total data available, and his results were the same. We thus have 14 years of history against which to check his first prediction, and by and large the record bears it out. Why should his

460

TABLE 21-2

Oil and Gas Produced and Estimated to be Producible in the United States and Adjacent Continental Shelves, as of December 1974.

Source of estimate	Oil, in billions of barrels[1]			Gas, in trillions of cubic feet		
	Produced	Measured reserves	Undis-covered	Produced	Measured reserves	Undis-covered
U.S. Geological Survey (Ref. 2)	122	45	358	481	237	2,100
Nat'l. Petroleum Council (Ref. 3)	122	45	137	481	237	1,171
U.S. Geological Survey (Ref. 4)	122	45	127	481	237	686
M. King Hubbert (Ref. 5)	122	45	86	481	237	505

[1]Includes liquids condensed naturally from gas, which amount to about 13 percent of total liquids produced to date. A barrel is equal to 42 U.S. gallons.

method have been more accurately predictive than the studies made by industry and by the U.S. Geological Survey? One reason is that oil and gas are mobile fluids; as noted in Chapter 10 they typically leave rocks and flow to the surface. They are thus difficult to predict at depth. Another basic reason is that large reservoirs were found during the early stages of exploratory drilling, simply because of their size. The reservoirs remaining are mostly small and lie deep; many will never be found at a reasonable cost.

FIGURE 21-7

Production of oil in the United States, plotted on a logarithmic scale to show the long period of rapidly increasing growth (the dashed line is the average growth rate) and the changes since 1930. From M. King Hubbert (Reference 5), with data added for 1973–75.

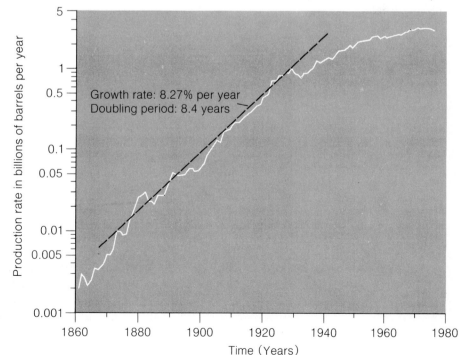

Growth rate: 8.27% per year
Doubling period: 8.4 years

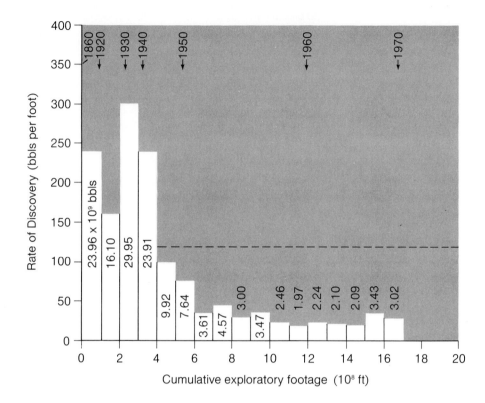

FIGURE 21-8
Progressive accumulation of feet of exploratory drilling (read the scale at the base of the diagram) relative to the rate of discovery of oil, in barrels discovered per foot of drilling. Each bar shows the average success of drilling for each 100 million (10^8) feet of exploratory holes drilled. The horizontal dashed line indicates the *average* success of drilling up to 1972, 118 barrels per foot. From M. King Hubbert (Reference 5).

A third reason for Hubbert's lower estimate is the essential difference in the methods of estimation. The geologists were estimating amounts of fuel they thought to be in the ground. Hubbert was looking at the trend of successes in finding those fuels. His estimate is not so much the amount of fuel in the ground as *how much industry will in fact recover.* That is also the estimate we need, and it indicates that other energy sources must be developed as soon as possible.

Alternative Mineral Fuels

Coal As we found in Chapter 14, coal occurs in distinct, persistent layers that can often be traced at the surface and can be measured at depth with a minimum of drill holes. Once formed coal stays in place. Since it is not destroyed by exposure at or near the surface, it can be obtained by low-cost surface mining. Adverse environmental affects, however, must be anticipated.

Identified, minable coal reserves in the United States are very large—approximately 1,570 billion tons as of 1975. Because the regions containing coal are large and the supply has not been pressed, there are also large tonnages yet to be discovered—probably at least 2,000 billion tons. In spite of these impressive amounts, however, only about 17 percent of the energy used in the United States in 1974 came from coal. Indeed, we exported about a billion dollars worth of coal that year, yet we imported large amounts of oil. This preference resulted partly from the "handiness" of oil and gas, already noted, and partly because oil and gas are cleaner burning. Air pollution is a major problem associated with burning coal and one that must be corrected.

A major share of this pollution results from the sulfur in coal, which is converted to sulfur oxides when the coal is burned. The coals located in the eastern part of the country, where the need for coal is greatest, typically contain 3 percent sulphur, which is three times the limit allowed by clean-air standards. The U.S. Bureau of Mines has developed a system of removing almost all the sulfur oxides from flue gases, and it is hoped that the system can be applied widely (Reference 6).

Low-sulfur coals are abundant in the western states, but transportation costs are too high to make them economically useful in the eastern states. One approach to this problem has been to use the western coals in large electrical generating plants near the mine sites and to transmit the electrical energy to regions that need it. Another possibility being considered is to convert the coal to gas or liquid fuel and then pipe it to eastern areas. Coal gasification and liquefaction, however, are at present too costly to be competitive with oil and natural gas and have pollution problems that must be worked out. One system under study would gasify coal layers by igniting them underground. Drill holes would carry the gas to the surface, where it could be cleaned as it emerged (Reference 6).

These innovations and improvements appear promising as of 1976 but have not been developed at full scale. We thus cannot be sure what proportion of our energy needs can be supplied by coal in the 1980s.

Oil shale and tar sands   *Oil shales* are fine-grained sedimentary rocks consisting mainly of carbonate minerals and clay and also containing abundant organic materials that were deposited with the sediments. The principal organic substance is *kerogen*, a complex mixture of organic compounds that can be converted to gaseous hydrocarbons by heating the rock. The gases condense as a liquid that can be refined into petroleum products.

Enormous quantities of oil shale occur in the Cenozoic basinal deposits described in Chapter 13, but the costs of mining and processing limit the usable parts of the deposits to a small fraction of the total—a body of shale 30 feet and more thick underlying 1,400 square miles of northwestern Colorado and adjoining parts of Utah. Conservative estimates indicate that this part of the deposits could produce about 54 billion barrels of oil, a reserve large enough to supply an important part of our future needs. A major problem, however, is that the massive mining and processing needed to obtain the fuel would probably be exceedingly damaging to waters flowing into the Colorado River as well as to the surface lands and the atmosphere. This "resource" is thus likely to be developed so slowly that it will not satisfy our needs for large quantities of fuel.

Tar sands (also called oil sands and bituminous sands) are porous formations that were filled with migrating petroleum liquids that became too viscous to be extracted except by mining and heating (or by heating in place, underground). The Athabaska sands of Alberta are the only large tar-sand resource in North America. They are estimated to contain 174 billion barrels of producible oil, but only a small part of the deposit is presently being mined. The growth of this industry is limited by the availability of capital and also by concern over the environmental effects of large-scale mining and production.

Energy from nuclear fission When atoms of ^{235}U (the isotope of uranium with an atomic weight of 235) are bombarded with neutrons, they *fission* (divide) into lighter atoms and additional neutrons. The neutrons then bombard neighboring ^{235}U atoms, starting a chain reaction that gives off large amounts of heat. The heat, when controlled, can be used to generate electricity. This simple fission is the basis of all nuclear power plants operating or currently being planned in the United States. The ^{235}U, however, is used up in the process (much like a burnable fuel), and it is a scarce substance. Large quantities of uranium ore have been discovered, but most contain less than 0.05 percent of uranium, of which only 0.7 percent is the fissionable isotope ^{235}U.

The remainder of the uranium is chiefly ^{238}U, which can be converted into fissionable fuel in what is called a *breeder reactor*, a device in which neutrons produced by fission of ^{235}U atoms bombard ^{238}U and convert it to an unstable isotope of plutonium, ^{239}Pu. The latter fissions into daughter products and neutrons much like ^{235}U and thus serves as additional fuel. The breeder reactor would solve the supply problems of the ordinary reactor were it not for the fact that the substances produced are exceedingly lethal. The widespread production and use of ^{239}Pu would also lead to the possibility that the metal might be stolen and used to construct nuclear weapons. Whether for these reasons or others, the development of breeder reactors in the United States is still under research and testing.

A major problem in using the normal fission reactor is that it produces radioactive substances that must be disposed of on a regular basis, and no place has yet been found to put the accumulating wastes safely. Some of the difficulties attending their disposal are: (1) they are exceedingly toxic; (2) they are in liquid form and thus can be entrained in groundwater; (3) they will remain dangerous for more than 250,000 years—a period long enough to disseminate even slow-moving substances widely; and (4) they may produce so much heat over the first several thousand years that they would melt their way out of containers buried in insulating materials such as rocks.

Another problem has risen because of concern that the reactors themselves may accidentally leak radioactive fluids. The Atomic Energy Commission has thus ordered increases in safety measures that have increased the costs of new reactors. This factor and the increasing costs of finding, mining, and refining uranium suggest that nuclear energy may become too costly to compete with other energy sources within 20 years. The supply of uranium ore, too, does not seem adequate to meet the needs of the program proposed for the 1980s (Reference 7). It thus appears that the "Atomic Age" will not bring us the reliable supplies of energy needed.

Nuclear fusion   Energy may also be obtained by *fusing* (joining) certain light elements at high temperatures, an action that produces large amounts of heat. The light element with the greatest potential for mass production is the hydrogen isotope called deuterium (D, in which the nucleus consists of a proton and a neutron). Controlled fusion would be the peaceful counterpart of the hydrogen bomb, in which the temperatures needed for fusion are supplied by detonating an atomic (^{235}U) bomb. A major problem in controlling fusion is devising a container that can withstand the temperatures, which are on the order of millions of degrees Celsius. Controlled fusion energy is thus at an early stage of research and may never become practical. It has a great appeal, however, because it produces little in the way of radioactive products and because the supply of deuterium in the oceans is very large.

Renewable Sources of Energy

Energy may be obtained from sources other than fuels. Certain natural processes transmit heat continuously to the earth's surface or have driving forces (kinetic energy) that can be used to generate electricity or do other kinds of work. These sources of energy are practically inexhaustible and are thus called renewable resources. Their immense potential is indicated in Figure 21-9, which shows the rate at which energy is contributed by each source (its power) as well as the general flow of energy among the sources.

Other than the waterpower obtained from river systems, these sources are developed only slightly compared to what they could be. Their underdevelopment is due partly to the need for new technology but far more to our having developed a fuels-dominated industry and home life. The increasing scarcity of mineral fuels and their polluting effects, however, suggest that we should consider converting to renewable sources of energy. Is this indeed possible?

Power from wind and moving water   Wind has been used to do work since ancient times, but variations in its strength and direction do not afford the steady supply of power needed for modern industry. Waterpower is based on the kinetic energy of water falling from higher to lower elevations. The falling water runs turbines and thereby generates electricity, but as Figure 21-6 shows, this is a minor source of energy. Although it is a clean and completely renewable source, the dams and reservoirs needed to develop it conflict with other uses of land and of rivers. Because of these problems, waterpower is developed about as much as it will ever be in the more industrialized regions of the world.

Another source of waterpower is the flow caused by tides. Power plants designed to use tidal flows in the Bay of Fundy, which has an exceptionally large tidal range (Figure 4-14), are under discussion between Canada and the United States. Tidal ranges are also over 12 feet and thus great enough to develop powerful tidal flows in the Cook Inlet, Alaska, and in some bays along the west coast of Canada. Energy generated from tidal sources could thus supply power to a few regions but will help little with needs elsewhere.

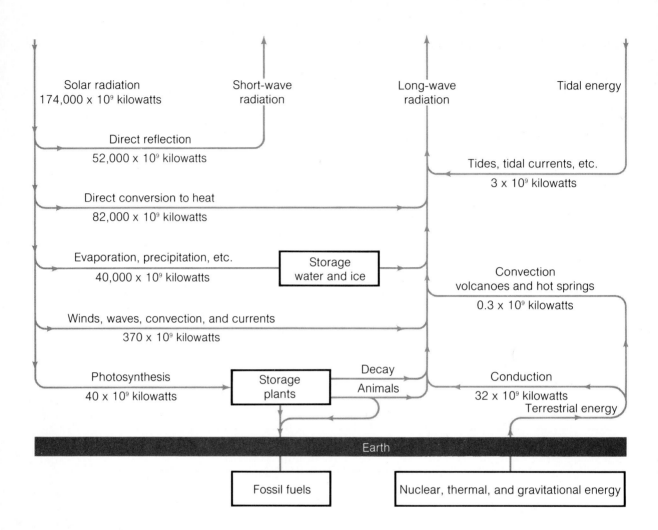

Solar radiation
174,000 x 10⁹ kilowatts

Short-wave radiation

Long-wave radiation

Tidal energy

Direct reflection
52,000 x 10⁹ kilowatts

Tides, tidal currents, etc.
3 x 10⁹ kilowatts

Direct conversion to heat
82,000 x 10⁹ kilowatts

Evaporation, precipitation, etc.
40,000 x 10⁹ kilowatts

Storage water and ice

Convection
volcanoes and hot springs
0.3 x 10⁹ kilowatts

Winds, waves, convection, and currents
370 x 10⁹ kilowatts

Photosynthesis
40 x 10⁹ kilowatts

Storage plants

Decay
Animals

Conduction
32 x 10⁹ kilowatts
Terrestrial energy

Earth

Fossil fuels

Nuclear, thermal, and gravitational energy

FIGURE 21-9

Flow sheet of energy for the earth, with rates of flow (energy per unit of time) shown for each part of the system. One kilowatt is equal to 1.34 horsepower or to 0.95 B.T.U. per second. From M. King Hubbert (Reference 5).

Geothermal energy   The resource called *geothermal energy* is the heat flowing from the earth's hot interior to its cool surface, illustrated diagrammatically in the lower right part of Figure 21-9. Part of this heat is inherited from the planet's accretion and early radioactivity, part comes from subsequent radioactivity, and part has been produced by frictional heating due to the deformation of rock bodies. Because rocks are poor conductors, the quantities flowing out at the surface are small compared to the amount of heat in rocks at depth. The heat available in the outer 6 miles of the earth, a depth within reach of drilling, has been calculated by Donald E. White to be approximately 3×10^{26} calories—17 million times the total energy consumed in the United States in 1970 (Reference 8).

   Most of this vast resource is too diffuse to be tapped economically now or in the foreseeable future, but we noted in Chapter 12 that pluton systems near the surface may supply large quantities of heat for hundreds or thousands of years. Another type of geothermal source consists of cracks or zones of broken rock along faults that reach deep into the earth and thus transmit heated water or steam to the surface. The ultimate source for this heat may be deeply buried plutons or more broadly heated metamorphic rocks. The distribution of hot springs shown in Figure 21-10 suggests that both plutons and faults are important, for the springs are by far most abundant in the region that was volcanically active and also broken by many faults in late Cenozoic time. This region is presently being prospected for potential geothermal sources, and the triangles in the figure indicate circulating systems that are hot enough to have potential for generating electricity (Reference 9). The gray band along the Gulf Coast is the location of a third kind of geothermal resource—water in sedimentary rocks heated by virtue of its great depth. The water is also confined under pressure (the load of the overlying rocks) so that it flows readily to the surface when tapped by drilling. These waters have temperatures of 140 to 170° and are exceedingly voluminous.

The principal use of these various hot waters and vapors will be to generate electricity. Only one system discovered so far (The Geysers, shown on the opening page of Chapter 12) produces dry steam that can be piped directly

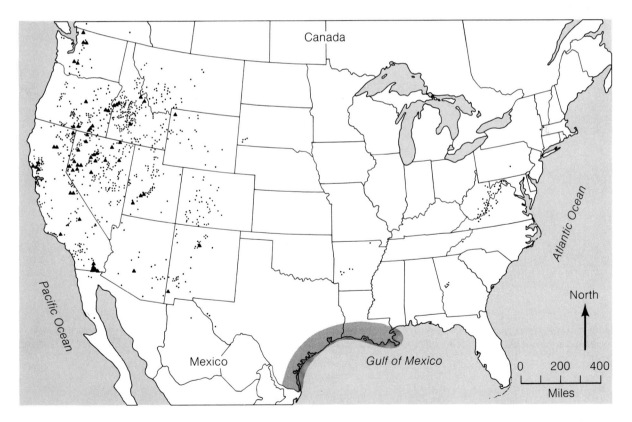

FIGURE 21-10
Distribution of thermal springs (dots) and circulating water systems in the temperature range of 90–150°C. (triangles). Thermal springs from Gerald A. Waring, "Thermal Springs of the United States and other countries of the world–a summary" (U:S. Geological Survey Professional Paper 492, 383 pp., 1965). Triangles from Reference 9.

from the ground into steam turbines. The other systems produce mixtures of steam and water from which the steam must be separated before being used. The hot water may be used to heat a liquid that boils at moderate temperatures (freon or isobutane) and is used to drive a turbine.

Because of their geographic occurrence, the systems discovered so far cannot provide energy for the eastern part of the United States, but they can afford comparatively inexpensive energy for local use. As the technology of heat exchange units improves, it is possible that deep waters such as those obtainable on the Gulf Coast can be used widely. For the next decade or so, however, geothermal sources are not likely to supply more than 10 percent of the total need.

Solar energy Daylight seems so benign compared to a nuclear reactor that we scarcely think of it as a source of energy to run an industrial society. Technological reviews have typically supported this notion, for they assert that solar energy is not for "now"—that we must await distant and somewhat mystical technological developments. In the 381-page book *U.S. Energy Outlook*, the National Petroleum Council included solar energy with "new energy forms" and dismissed it with three brief sentences (Reference 3, page 234). The federal government evidently held the same view at least until 1973, for its funding of research on solar energy was miniscule. The fact is, though, that sunlight has always warmed us, cooled us (by evaporation of moisture), grown our basic foods, produced oxygen by photosynthesis, circulated the air, pumped water from the saline ocean to all fresh water systems, and formed the mineral fuels! Moreover, people have already devised structures using solar energy to (1) heat water, (2) cook food, (3) heat and cool buildings, (4) obtain fresh water from salt water, (5) generate electricity, and (6) melt metals.

As Figure 21-9 shows, the basic supply of solar energy is larger by far than all other energy sources combined. Much of it, as the figure shows, is reflected to space, and on cloudy days the local supply is reduced greatly. Opponents of its use also cite its diffuseness; they claim that the large sum of energy is spread too thinly to be of practical value. In an average climatic situation, however, a roof measuring 25 by 50 feet, facing the sun, can collect enough solar energy to supply the household needs of an average family. The diffuseness of solar energy, in fact, turns out to be one of its basic strengths, for the use of many local, small units would greatly reduce the costs of building large generating plants, of finding and transporting mineral fuels to them, and of distributing power from the plants to all the users.

Proponents of solar energy suggest that solar units capable of powering major industrial systems would simply have to be larger. Mirrors might be used to gather energy and direct it on a confined liquid used in a turbine system, which would run machinery directly or generate electricity for uses nearby. Such a system would be environmentally clean. The electricity could also be used to generate hydrogen from water, and this fuel (which burns cleanly) could be used to run generators during cloudy periods and might take the place of mineral fuels in powering vehicles. If solar plants were located in areas with exceptionally large numbers of sunny days, as in the desert region described in Chapter 9, their output could be increased considerably. Another possibility is to install plants in warm oceans, where heat could be collected from the surface water. These marine units would produce electricity or hydrogen by the use of heat engines based on the temperature difference (around 33° F) between the warm surface water and the cold water at depths of 1,500 feet or so (Reference 10).

The development of large solar units is clearly for the future, although not necessarily the distant future. A 10,000 kilowatt plant designed for the Southwest should be tested by 1980. The possibility of home-sized units exists now. Barry Commoner has decribed studies showing that the integration of solar units with normal household sources is economical today (that is, cheaper than energy obtained *totally* from fuels) (Reference 11). The important relation, as he points out, is that the solar unit will become more and more economical as mineral fuels become scarcer and capital investments for new power plants become larger. Not only will homeowners who have invested in units be obtaining energy more cheaply, they will help in resolving all the major problems of the energy crisis. They will be conserving mineral fuels and will be reducing pollution as well as the special hazards from radioactivity. Most important, perhaps, they will be gaining independence and responsibility for running more of their own lives and effects.

Sources of Livelihood

We must finally consider other basic resources, ones that have generally been depleted toward "lower grade" and have been assigned "prices" even though they are invaluable. These are the water, air, soil, and plants that will remain essential for life regardless of our success in obtaining abundant energy and metals.

We should consider air first because its problems get handed along to the other resources. Air pollution is widespread and complex in many ways, but its principal cause is the burning of mineral fuels and the exposure of hydrocarbon-rich substances to sun and air. As mentioned for coals, the resulting oxides of sulfur are known to be serious pollutants, but there are a host of others that may damage the respiratory systems of plants and animals and create hazes that reduce the amounts of sunlight reaching the ground. Some organic substances, and all strongly radioactive particles, are known to cause cancers. Most of these pollutants could be intercepted by carefully designed systems on power plant chimneys—additions, however, that will cause increases in energy costs. Most existing smelters, too, need to be converted to meet clean-air standards—a costly procedure for the metals industry.

Water receives most airborne pollutants as well as massive liquid and solid wastes from industry and metropolitan areas, as noted in several earlier chapters. An additional pollutant that is increasing seriously is excess fertilizer in the waters that drain farmlands. The fertilizer promotes the rapid growth of algae, which first cloud the water and then, when they die, reduce

TABLE 21-3

Estimated Distribution of World Water Supply[1]

Occurrence	Volume in cubic miles	Percent of total water
Rivers (average)	300	.0001
Atmosphere	3,100	.001
Moisture above water table	16,000	.005
Saline lakes and inland seas	25,000	.008
Fresh-water lakes	30,000	.009
Groundwater to 13,000 ft. depth	2,000,000	.61
Ice sheets and glaciers	7,000,000	2.14
World Ocean	317,000,000	97.3
Total (rounded)	326,000,000	100.0

[1]From "Are we running out of water?" by Raymond L. Nace (U.S. Geological Survey Circular 536, 7 pp., 1967).

the dissolved oxygen content. Fuels also pollute water directly. Petroleum and derivative liquids form fine, pervasive films on the surface, and iron sulfide associated with coal reacts with oxygen and water to form sulfuric acid and therefrom a host of additional dissolved sulfur-bearing compounds and organic substances. If the surface mining of coal and oil shale proceeds at the scale that some have proposed, this type of pollution will be a serious problem.

The availability of water is also becoming problematical. Shortages are common in the western half of the country, and the supply in the eastern half will be inadequate in most regions by the year 2000. Water is used mainly in industry and agriculture, and its consumption will be increased markedly if we increase irrigation or convert to the widespread production of gas and liquid fuels from coal. Conflicts are now arising in the Southwest, where several large coal-conversion plants planned on Indian lands will take water needed for agriculture.

Desalinization of sea water could be a major source of additional water in coastal regions; perhaps solar energy could be used to accomplish this. The data in Table 21-3 suggest that the only other major source is ground-water, which has already been depleted seriously in some parts of the West. Moreover, much of the groundwater is so saline that it will have to be treated before use. A major contribution from marine biologists would be the development of edible plants that could be grown in saline or brackish water.

Soil conservation   Soils and plants are so closely related that depletions of one have always led to depletions of the other. The depletions are almost entirely due to agriculture and timber cutting, for although construction and surface mining destroy soil and vegetation they are highly localized. As described in earlier chapters, soil depletion is due basically to removal of vegetation, to exposure by tilling, and to chemical changes brought on by irrigation. We may add to these the use of chemical fertilizers, which deplete natural organic detritus and soil microflora and therefore change the soil physically.

A large share of these depletions might have been resolved by now had there been a continuation of the conservation programs and spirit started in the 1930s, but motivations and methods changed. Farming and forest cutting became increasingly large scale and mechanized after 1950, which required increases in capital investments and thus the need to produce more for less (or at least to attempt to do so). Federal policy in recent years has encouraged the industrialization of agriculture. However, large acreages cannot be treated efficiently with large machines without including local slopes or soil types that are unsuitable. These areas are then eroded or are depleted in other ways. The constant press of chemical fertilization and irrigation has led to additional depletions. In the Imperial Valley of California, for example, excessive irrigation led to soil problems resulting from the concentration of sodium ion, as described in Chapter 9. The damage was corrected by installing miles of drainpipe under fields, but the great cost of doing so induced further overirrigation, which led to an additional concentration of salt and the loss of much of that rich farmland.

Although the large amounts of fuels, chemicals, and heavy equipment needed for industrialized agriculture are major drains on other resources, it is often claimed that these outlays are necessary to produce the food we require. These claims are evidently far from true—at least judging from experiments with "minifarms," the extreme opposite of large-scale agriculture. These experiments tested an intensive agricultural method whereby small plots were hand-tilled to depths of several feet with organic materials (chiefly compost). Plants were grown so closely together that they shaded the entire surface and thereby conserved water and organic nutrients. The method was found to use one-half to one-sixteenth the average amount of water needed for other methods and to require little equipment and virtually no mineral fuels, chemical fertilizers, or insecticides. The remarkable result was that the plots produced an average of four times the food per unit area produced in the United States generally! One person, working a plot of only 2,500 square feet for 30 minutes a day through a six-month growing season, can thus obtain a year's full diet (Reference 12). Of tremendous importance is the fact that a special soil is not required; *the method is a soil-making one.* The experiments, for example, started with a difficult stony, clay-rich soil and changed it to rich, workable soil in 3 to 5 years.

If the simplicity of the method and the idea of people tilling soil by hand seem anachronistic, perhaps our views of the land and its productivity have become twisted. Indeed, our attitudes appear to be a major cause of some of our "problems" with natural resources. Wendell Berry, a writer who grew up close to farming in northwestern Kentucky, has looked closely at these attitudes and has seen a crucial interplay between two opposites. He has cast these two attitudes into the roles of "exploiter" and "nurturer," pointing out that we all encompass some of each. I quote part of his explanation (Reference 13):

> The exploiter is a specialist, an expert; the nurturer is not. The standard of the
> exploiter is efficiency; the standard of the nurturer is care. The exploiter's goal
> is money, profit; the nurturer's goal is health—his land's health, his own, his

family's, his community's, his country's. Whereas the exploiter asks of a piece of land only how much it can be made to produce, and how quickly it can be made to produce it, the nurturer asks a question that is much more complex and difficult: what is its carrying capacity? (That is: How much can be taken from it without diminishing it? What can it produce *dependably* for an indefinite time?) The exploiter wishes to earn as much as possible by as little work as possible; the nurturer expects, certainly, to have a decent living from his work, but his characteristic wish is to work *as well* as possible—he takes pride and pleasure in his work. The competence of the exploiter is in organization; that of the nurturer is in order—a human order, that is, that accomodates itself both to other order and to mystery.

In a human-dominated world, the ultimate ingredients of earth resources and earth systems are the understanding and the concern of the individual. Production, utilization, and conservation all stem essentially from them. This relation is especially crucial in nations with representative governments. Your increased awareness of the full earth system thus brings a responsibility. Are you willing to use your knowledge and to enlarge your part?

Summary

Our uses of earth resources must take into account their basic nature as well as their abundance and the results of their use. The following are some important relations:

1. Mineral deposits are highly localized concentrations of useful substances; they are the only places we can obtain these substances economically.
2. Our need for resources has increased as the population has grown and become more industrialized; the supply, on the other hand, continues to decrease, for the resources are essentially nonrenewable.
3. Ingenuity in mining and processing has kept the prices of some mineral commodities from increasing rapidly even though the number and concentration (grade) of the deposits is decreasing.
4. There have been recent large increases, however, in the prices of some important commodities; oil and gas are especially critical in this regard because they are our main sources of energy and are vital to obtaining all other commodities.
5. The quantities of domestic oil and gas that can be recovered economically are debatable; however, it seems likely that other energy sources will have to be developed on a large scale within the next decade or so.
6. Coal and uranium are abundant enough to supply the needed energy, but their extensive use still requires technological development and will create difficult, perhaps impossible, problems with pollution.
7. Of the renewable resources, the only one with universal potential is solar energy; it could be used effectively now through many small units, but

the development of large power plants will require a major effort over a period of many years.

8. A major reason to make this effort is the depletion of air, water, and life caused by pollutants from mineral fuels and radioactive materials; solar energy would be clean.

9. Regardless of which energy sources will be used in the future, our livelihood will depend on soils and plants, which must be improved and cultivated with greater care than we now exercise.

10. Our ultimate benefits from natural resources will depend on the understanding and concern of all persons using them.

REFERENCES CITED

1. J. David Lowell. Copper resources in 1970. *Mining Engineering*, vol. 22, pp. 67–73, April 1970.

2. P. K. Theobald, S. P. Schweinfurth, and D. C. Duncan. Energy resources of the United States. U.S. Geological Survey Circular 650, 27 pp., 1972.

3. National Petroleum Council. Committee on U.S. Energy Outlook. *U.S. energy outlook* (A report) 381 pp., 1972.

4. Betty M. Miller, *et al.* Geological estimates of undiscovered recoverable oil and gas resources in the United States. U.S. Geological Survey Circular 725, 78 pp., 1975.

5. M. King Hubbert. U.S. energy resources, a review as of 1972. Pp. 1–201 in a background paper printed for the use of the Committee on Interior and Insular Affairs, United States Senate, U.S. Government Printing Office, Washington, 1974.

6. Elburt F. Osborn. Coal and the present energy situation. *Science*, vol. 183, pp. 477–81, 8 February 1974.

7. Donald A. Brobst and Walden P. Pratt, eds. United States mineral resources. U.S. Geological Survey Professional Paper 820, 722 pp., 1973.

8. Donald E. White. Geothermal energy. U.S. Geological Survey Circular 519, 17 pp., 1965.

9. D. E. White and D. L. Williams, eds. Assessments of geothermal resources of the United States—1975. U.S. Geological Survey Circular 726, 155 pp., 1975.

10. Mark Swann. Sea thermal power. *Oceans*, vol. 9, no. 2, pp. 30–35, 1976.

11. Barry Commoner. *The poverty of power.* New York: Knopf. 1976.

12. Ecology Action of the Midpeninsula. Resource-conserving agricultural method promises high yields (1972–1975 research report summary). Ecology Action of the Midpeninsula, Palo Alto, Calif., 17 pp., January 12, 1976.

13. Wendel Berry. The unsettling of America (Part I). *The Nation*, vol. 222, no. 5, pp. 149–51, February 7, 1976. (The second part of Mr. Berry's essay appears in the February 14, 1976 issue, and an exchange with John R. Woods in the March 13, 1976 issue.)

ADDITIONAL IDEAS AND SOURCES

1. Brief descriptions of the occurrences, origin, and uses of the principal mineral commodities are given in *Our finite mineral resources* by Stephen E. Kesler (New York: McGraw-Hill, 120 pp., 1976). More thorough accounts of all the important commodities except for oil and gas are presented in U.S. Geological Survey Professional Paper 820 (Reference 7). Each of the sixty-eight parts were prepared by specialists.

2. A brief description of the worldwide occurrence of mineral deposits and the international economics based on them is given in *Minerals in world affairs* by Alexander Sutulov (Salt Lake City, Utah: The University of Utah Printing Services, 200 pp., 1972).

3. Three well-illustrated articles describing processes by which coal may be converted to oil and gas are "Oil and gas from coal" by Neal P. Cochran, "Clean power from dirty fuel" by Arthur M. Squires, and "The gasification of coal" by Harry Perry. All are in *Scientific American*, in the May 1976 issue (vol. 234), the October 1972 issue (vol. 227), and the March 1974 issue (vol. 230), respectively.

4. The intensive small-scale farming method mentioned near the end of the chapter has been developed widely but by comparatively few persons. The method is described thoroughly and in a very readable way by John Jeavons in *How to grow more vegetables* (*than you ever thought possible on less land than you can imagine*): (Palo Alto, Calif.: Ecology Action of the Midpeninsula, 82 pp., 1974). It is available from the publisher at 2225 El Camino Real, Palo Alto, Calif. 94306.

5. The book by Barry Commoner (Reference 11) includes important discussions of the efficiency of obtaining energy from the principal sources, as well as other interesting and provocative topics.

Evidence of internal order—the mineral fluorite (calcium fluoride) in a specimen from Rosiclaire, Illinois, that shows repeated development of the mineral's characteristic cubic crystals. Twice actual size.

Supplementary Chapter 1.
Minerals

Minerals

Constitution of Minerals
    Atoms and ions
    Minerals consisting of simple ions
    Minerals bonded by shared electrons
    The framework silicates called feldspars
    Silicates formed from chains and sheets
    Minerals with complex ions
    Mineral compositions
    The nature of minerals—a summary
Identification of minerals
    Cleavage
    Hardness
    Specific gravity (density)
    Crystal form
    Luster
    Color
    Streak
    Chemical and optical tests
List of minerals

Minerals are naturally occurring solids composed of atoms of one or more of the chemical elements. The specific properties of the different atoms lead directly to the properties of each mineral: (1) the order within its internal (crystalline) structure; (2) its intrinsic degree of cohesion or hardness; (3) its interactions with light and other kinds of radiant energy; (4) its density (weight per unit volume); and (5) its susceptibility to chemical changes.

The last property emphasizes the fact that no mineral is "forever." Each came into being because of specific conditions in some part of the earth, and each will change to other minerals as conditions require. Atoms, on the other hand, remain unchanged and are handed along from one mineral to the next. Minerals thereby provide a means of interpreting the changing conditions that make up the earth's history. These interpretations require an understanding of minerals as well as an ability to recognize them, and these are the chief topics of this chapter. Some mineral changes will be described briefly in this chapter, but most are presented in Chapters 8, 10, 12, 15, and 16 and in Supplementary Chapter 2.

Constitution of Minerals

Atoms and ions   Atoms are minute bodies—so small that mineral grains $\frac{1}{100}$ inch in diameter contain approximately 1×10^{22} (a billion trillion) of them. Each atom is nonetheless constructed of smaller particles held together by a highly complex (and not completely understood) energy system. At the center of each atom is a small but very dense *nucleus* consisting mainly of *protons* (particles carrying a positive electric charge) and *neutrons* (uncharged particles), along with various other particles. Around the nucleus orbit one or more *electrons*—lightweight particles that each carry one negative charge. Because atoms are electrically neutral, the number of electrons is equal to the number of protons in the nucleus. The number of protons is called the *atomic number*, and that number is unique for the atoms of each chemical element (Table S1-1).

The electrons are arranged around the nucleus in shells, in which the numbers of electrons increase outward (Figure S1-1A). The most important shell in most geological changes is the outer one, for it establishes the nature of most chemical interactions as well as the bonding that holds minerals together. In atoms with two or more electron shells, the outer shell is most stable when containing four pairs of electrons, each pair consisting of electrons that spin in opposite directions. Atoms that have six or seven electrons in their outer shells tend to achieve stability by gaining electrons to make eight, and those that have one or two tend to lose them to achieve a stable configuration (Figure S1-1B). These atoms thus acquire excess positive or negative charges and are called *ions*. As the figure indicates, ions are represented in chemical shorthand by the elemental symbol followed by pluses or minuses equal to the electrical charge the ion carries.

Minerals consisting of simple ions   Minerals such as halite (sodium chloride, NaCl) consist of ions, in this case Na^+ and Cl^-. These ions form a

TABLE S1-1

Atomic Numbers, Atomic Weights,
and Ionic Charges of Some of the Commoner Elements.

Name	Symbol	Atomic number	Atomic weight	Ionic charge
Aluminum	Al	13	26.98	(+3)[a]
Argon	Ar	18	39.94	0[b]
Calcium	Ca	20	40.08	+2
Carbon	C	6	12.01	(+4)
Chlorine	Cl	17	35.46	−1
Chromium	Cr	24	52.01	+2, +3
Copper	Cu	29	63.54	+1, +2
Fluorine	F	9	19.00	−1
Gold	Au	79	197.00	(+2)
Helium	He	2	4.00	0
Hydrogen	H	1	1.01	+1
Iron	Fe	26	55.85	+2, (+3)
Lead	Pb	82	207.21	+2
Magnesium	Mg	12	24.32	+2
Manganese	Mn	25	54.94	+2, +3
Molybdenum	Mo	42	95.95	+2
Nickel	Ni	28	58.69	+2
Nitrogen	N	7	14.01	(+5)
Oxygen	O	8	16.00	−2
Phosphorus	P	15	30.97	(+5)
Platinum	Pt	78	195.23	(+2, +4)
Potassium	K	19	39.10	+1
Rubidium	Rb	37	85.48	+1
Silicon	Si	14	28.09	(+4)
Silver	Ag	47	107.88	+1
Sodium	Na	11	23.00	+1
Strontium	Sr	38	87.63	+2
Sulfur	S	16	32.07	−2
Thorium	Th	90	232.12	(+4)
Tin	Sn	50	118.70	+2, +4
Titanium	Ti	22	47.90	(+4)
Tungsten	W	74	183.92	(+4)
Uranium	U	92	238.07	(+4)
Zinc	Zn	30	65.38	+2
Zirconium	Zr	40	91.22	(+4)

[a]Numbers in parentheses are for ions that do not ordinarily exist outside compounds.
[b]Does not form ions or compounds.

solution in water (for example, in the sea), but when the solution is evaporated to the point that the ions become crowded, they begin to join in grains of solid NaCl (halite). Because opposite electrical charges attract and similar ones repel, each Na^+ moves as close as possible to Cl^-s and stays as far away as possible from other Na^+s. The ions thus settle into the arrangement shown in Figure S1-2 (left), in which each ion is surrounded by six ions with the opposite charge. The sizes of the ions determine the simple cubic symmetry illustrated, and as a result halite grows in cubic or box-like crystals that also *cleave* (break in smooth planes) parallel to the three planes of the box form (Figure S1-2,

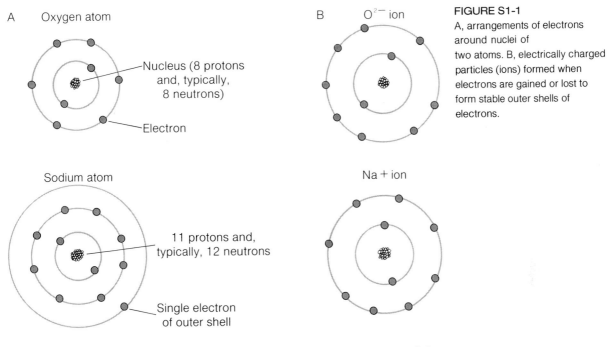

A   Oxygen atom

Nucleus (8 protons and, typically, 8 neutrons)

Electron

Sodium atom

11 protons and, typically, 12 neutrons

Single electron of outer shell

B O^{2-} ion

Na + ion

FIGURE S1-1

A, arrangements of electrons around nuclei of two atoms. B, electrically charged particles (ions) formed when electrons are gained or lost to form stable outer shells of electrons.

right). The cleavage results from the internal structure (arrangement) of the ions and is a property useful in identifying halite even in cases where the crystalline grains do not have box-like crystal shapes (because they often interfere with one another as they grow).

There is thus a difference between *crystalline substances*, which have internally ordered ions or atoms, and *crystals*, which are crystalline grains that have had the freedom or strength to grow geometric shapes.

Because each ion in halite carries only one electrical charge, the ionic bonds are not particularly strong and halite is easily scratched and quite

FIGURE S1-2

Left, arrangement of sodium and chloride ions in the mineral halite. Right, crystal of halite with its faces parallel to the cubic faces of the ion diagram, and with fragments having cleavage directions parallel to the faces of the crystal.

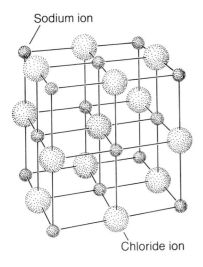

Sodium ion

Chloride ion

479

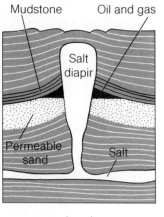

Mudstone   Oil and gas

Salt
diapir

Permeable
sand   Salt

0   1
|__|__|
Miles

Vertical section through a body of
salt that has intruded its way
upward through sedimentary
rocks, the flow taking place by
repeated gliding within the
crystals that compose the salt.

easily melted. Minerals with weak ionic bonds also tend to dissolve readily in
water because the water molecules, being polar, become arranged around each
ion freed from the crystal and thus keep it from recombining with the salt.
Halite is also quite ductile, gliding layer over layer parallel to the box faces of
Figure S1-2 and in the direction of their diagonals. As a result, halite rock
can flow slowly as a solid, and because it is a comparatively light mineral, large
bodies of it often rise through denser sedimentary rocks much as a drop of oil
rises through water. The resulting intrusions (*salt domes* or *diapirs*) are well
known, for they pierce and bend strata upward around them and thus create
traps for petroleum and natural gas (Figure S1-3). Sodium and chloride ions
thus have far-reaching geological effects.

Minerals bonded by shared electrons   Atoms such as those of silicon, carbon, aluminum, and phosphorus, which have 3, 4, or 5 electrons in
their outer shells, typically share electrons with other kinds of atoms rather
than lose or gain electrons to form ions. Silicon, the most abundant metallic
element in the outer part of the earth, shares its four outer electrons with four
oxygen atoms (Figure S1-4A). This type of bonding is called *covalent* or *shared-
electron-pair bonding* and is very strong because of the great stability it imparts
simultaneously to all the atoms. The grouping is especially stable because the
size of the silicon atom permits it to fit exactly in the small space at the center
of the four oxygen atoms, which are arranged symmetrically around it in the
shape shown in Figure S1-4B.

As the small diagram in Figure S1-4A shows, each oxygen needs one electron
to complete its four pairs. This is effected by joining a silicon atom to each

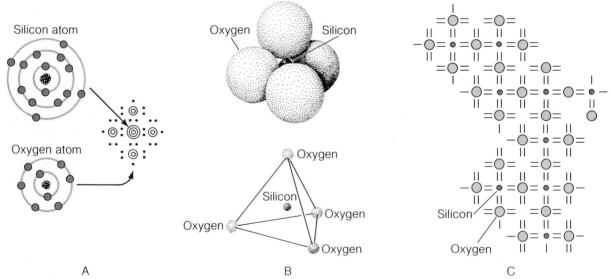

Silicon atom

Oxygen   Silicon

Oxygen atom

Oxygen

Silicon
                Oxygen
Oxygen
                Oxygen

Silicon

Oxygen

A   B   C

FIGURE S1-4

A, atoms of silicon and oxygen (left) and a diagram at reduced scale showing the sharing of eletrons in the outer shells of one silicon
and four oxygen atoms. B, small silicon fitting at the center of the four oxygen atoms, with the diagram below showing how the
oxygen atoms are positioned symmetrically around the silicon. C, diagram showing electron pairs in a continuously joined framework
of silicon and oxygen atoms.

TABLE S1-2

Abundances of the Main Elements in the Earth's Crust,
Estimated from Exposed Rocks.

Element	Percent by weight	Percent by volume
Oxygen	46.5	93.7
Silicon	28	0.9
Aluminum	8.5	0.5
Iron	5	0.4
Calcium	3.5	1.0
Sodium	3	1.3
Potassium	2.5	1.8
Magnesium	2	0.3
	99	99.9

SOURCE: Modified from Brian Mason. *Principles of Geochemistry*, New York, John Wiley & Sons, 1966.

Aluminum atom

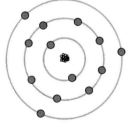

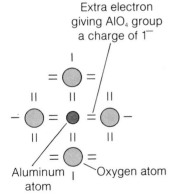

Extra electron giving AlO_4 group a charge of 1^-

Aluminum atom   Oxygen atom

FIGURE S1-5

Atom of aluminum (top) and reduced-scale diagram of AlO_4 group, stabilized by an extra electron that gives the group a charge of 1^-.

oxygen, then three more oxygens to each silicon, and so forth (Figure S1-4C). If you can imagine this framework being constructed of SiO_4 groups such as that of Figure S1-4B, perhaps you can see that the framework must extend into the page and out of it to form a three-dimensional structure. If you were to count the number of silicon atoms and the number of oxygen atoms in this large array, you would find that oxygen is twice as numerous as silicon; that is, this framework of minerals has the formula SiO_2.

SiO_2 (silica) occurs in several crystalline forms, the most common of which is quartz. This mineral has many properties based on the strength of its covalent bonds: it is hard; it is nearly insoluble in most fluids; it is inert to most chemical reagents; and it has a high melting point. Quartz lacks cleavage because of the three-dimensional arrangement of its bonds. The internal order of the mineral is evinced only when it has grown in fluids and is thus free to develop its characteristic six-sided crystals (Color Plate 14B).

The framework silicates called feldspars Minerals consisting largely of groups of bonded silicon and oxygen atoms are called *silicates*, and their great abundance is suggested by the preponderance of oxygen and silicon in the outer part of the earth (Table S1-2). In fact, most other elements listed in the table also occur mainly in silicate minerals, some as covalently bonded atoms and some as more loosely held ions. The most abundant minerals in the crust are the *feldspars*, which consist mainly of oxygen, silicon, and aluminum, in mixtures of three basic compounds: $KAl\,Si_3O_8$, $NaAlSi_3O_8$, and $CaAl_2Si_2O_8$. The most abundant feldspar, called plagioclase, consists of mixtures of the last two of the compounds; the other common feldspar, potassium feldspar, consists mainly of the first compound, with lesser amounts of the second.

The aluminum in the feldspars, like the silicon, is linked by covalent bonds to oxygen. However, aluminum atoms carry one less electron than do silicon atoms, so their presence requires an extra electron to produce the stable arrangement in all oxygen atoms (Figure S1-5). The electron introduces an

481

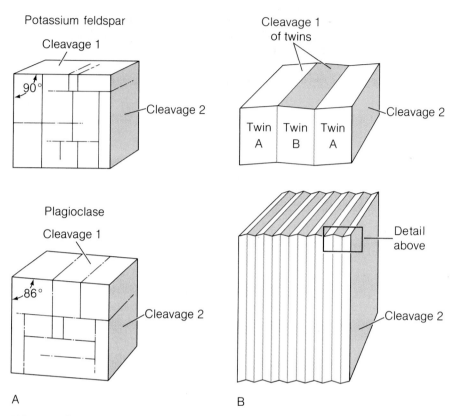

FIGURE S1-6

A, the two cleavage directions in potassium feldspar and plagioclase, intersecting at slightly different angles (the front of the diagrams is not a cleavage surface). B, multiple twinning in plagioclase, producing tabular segments with one cleavage inclined from one twin to the next and the other cleavage parallel.

extra negative charge that is compensated by the entry of a positive ion. Thus, for each single Al entering the framework a K^+ or Na^+ is added to form $KAlSi_3O_8$ or $NaAlSi_3O_8$, and where two aluminum atoms enter the framework, Ca^{++} is used to form $CaAl_2Si_2O_8$. The presence of the ions weakens the overall framework compared to that of quartz; as a result feldspars are somewhat softer than quartz and are more readily altered by chemical attack, an effect especially noticeable in soils. The feldspars also cleave along the two planar directions in which the ions are aligned, and this property makes these minerals quite readily distinguishable from quartz.

The cleavage directions also give a means of distinguishing the two kinds of feldspar, for the two planes intersect at nearly 90° in potassium feldspar and at about 85 ° in plagioclase (Figure S1-6A). Plagioclase can also be recognized because it typically develops tabular crystalline units that face in opposite directions—an arrangement called *multiple twinning* (Figure S1-6B). Note that one (the upper) cleavage surface is divided into two sets of planes that form narrow strips. These reflect light in sets of bright parallel lines (striations) that may be very thin but can usually be seen with a hand lens (magnifying

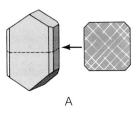

A

B

FIGURE S1-7

A well-formed crystal of pyroxene shown in parallel orientation to a crystal (B) partly dissolved by water percolating through a soil, showing spiny relics that are parallel to the silica chains. The cross section on the right shows the two cleavage directions within the crystal.

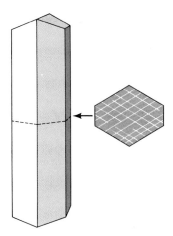

FIGURE S1-8

Crystal of amphibole, in which the chains of SiO_4 groups are oriented vertically. The cross section shows the two cleavage directions within the crystal.

glass). Twin striations are especially valuable in distinguishing the two feldspars where they form small grains in rocks such as granite and gabbro. In thin sections, the twin sets show clearly as parallel dark and light bands (Figure 18-12).

Silicates formed from chains and sheets In addition to the silicates linked by continuous three-dimensional frameworks of Si—O and Al—O bonds, some have strongly bonded units arranged in chains or sheets. *Pyroxene* is composed of chains of strongly bonded SiO_4 groups that have excess negative charges and are held together by the ions Mg^{++}, Fe^{++}, and Ca^{++}. The chains result in a prismatic crystal form and in two cleavage directions (Figure S1-7A). They also result in the minerals being attacked chemically parallel to the chains (Figure S1-7B).

Another important group of silicate minerals, the *amphiboles*, are based on double chains of SiO_4 groups, also bound together by positively charged ions. The double chains result in crystal faces and cleavage directions that lie at different angles (approximately 60° and 120°) from those of pyroxene (approximately 90°) and thus afford a means of distinguishing the two kinds of minerals (Figure S1-8).

The micas and other flaky silicate minerals (chlorite, clay minerals, talc) are constructed of sheets formed of joined SiO_4 groups. These carry excess negative charges and are held together in most species by positive ions (Figures 8-10 and 8-12). Again, the covalent bonds make the individual sheets strong; however, the ionic bonds holding them together are weak, and the minerals therefore are soft and can be cleaved readily into thin sheets (Figure S1-9).

Minerals with complex ions The silicate building units in the minerals garnet, olivine, zircon, and epidote are separate ions of covalently bonded silicon and oxygen. In most cases, the ions are single SiO_4 groups that have acquired four electrons to complete all outer shells of eight electrons and thereby carry four negative charges (Figure S1-10). These large ions are then joined with positive metal ions to complete each mineral. Olivine, for example, consists of Mg^{++} and Fe^{++}, ions of nearly the same size, in addition to SiO_4^{4-}, giving compositions ranging from Mg_2SiO_4 to Fe_2SiO_4.

These minerals are bonded by ionic forces like those in halite, but they are held together much more strongly because of the multiple electrical charges on the ions. All these silicates are hard and cleave only roughly along planes, if at all. The four positive charges on the zirconium ion (Zr^{4+}) make *zircon* ($ZrSiO_4$) one of the hardest and chemically least reactive of minerals. It is so slightly soluble in even the hottest rock melts that it crystallizes from them before all other silicate minerals—even though Zr^{4+} make up only a few parts in every 10,000 parts of melt. Garnet and olivine are less strongly bonded and as a result are somewhat soluble in soils, often being etched deeply by solution.

Other important minerals containing complex ions are the carbonates (calcite, aragonite, and dolomite) and the phosphate called apatite. Covalently bonded carbon and oxygen form the carbonate ion (CO_3^{--}), and covalently bonded phosphorus and oxygen form the phosphate ion (PO_4^{3-}). Calcium

483

FIGURE S1-9

A crystal of mica and thin flakes cleaved from it.

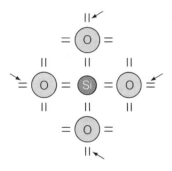

$SiO_4{}^{4-}$ ion

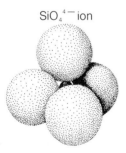

FIGURE S1-10

SiO$_4$ group stabilized as a separate unit by gaining four electrons (arrows). The resulting ion thus has a charge of 4−.

484

ion (Ca^{++}) is the principal metal ion in the minerals, although magnesium ion (Mg^{++}) is used to some extent in most carbonates and takes up half the metal-ion positions in dolomite, $CaMg(CO_3)_2$. These ionic minerals have properties between those of halite and those of the ionic silicates just described. The carbonates cleave in three directions, but they are harder, denser, less soluble, and less easily melted than halite. Apatite, as suggested by the three charges on the phosphate ion, cleaves only roughly and is harder and less soluble than the carbonates.

Mineral compositions About 2,200 minerals have been recognized and named as separate species. This number seems remarkably small considering the billions of possible compounds suggested by the number of elements (Table S1-1). The possibilities are reduced greatly by the limitations already noted: the numbers of electrons in the outer shells of atoms and the sizes of atoms and ions. That is, many combinations do not exist in nature because they are unstable. Another cause of the comparatively small number of minerals is that the proportions of elements in the outer part of the earth are such that most can be fitted into a few minerals. Feldspars, quartz, pyroxene, olivine, amphiboles, and micas make up all but a few percent of the earth's crust and outer mantle (to depths of 100 miles or so). Moreover, small numbers of unusual atoms can enter into certain positions in these minerals without disturbing them. The small number of strontium atoms, for example, may take the place of a scattering of Ca^{++} in plagioclase and pyroxene, so that a strontium silicate ($SrSiO_3$) does not form. This tendency for one element to substitute for another in silicate minerals is a major reason for the small number of minerals species.

Another simplifying effect due to these substitutions is that many minerals have a *range* of compositions. We have already noted that plagioclase compositions range from $NaAlSi_3O_8$ to $CaAl_2Si_2O_8$. In pyroxene, the ions Fe^{++}, Mg^{++}, and Ca^{++} may substitute for one another over a large range of compositions, so that most pyroxenes are mixtures of $FeSiO_3$, $MgSiO_3$, and $CaSiO_3$. Sometimes this relation is expressed by writing the general mineral composition (Fe, Mg, Ca)SiO_3. Amphiboles and micas have still greater ranges of composition due to substitutions. Only a moderate number of these variations have been named, and only a few will be introduced in this book. Actinolite and hornblende, for example, are similar amphiboles that have somewhat different compositions and form under different conditions.

Substitutions in minerals can take place only when the size and electrical charges on the ions are similar; otherwise the internal arrangement would be so altered that the mineral would be changed to another one. *The internal geometric (crystalline) arrangement of ions and atoms is thus the most basic aspect of a mineral species.*

A practical question is how can one recognize compositional variants in a species. One general rule is that varieties containing more iron (and often small amounts of titanium) are more deeply colored, commonly in brownish tints. The difference in color between biotite and muscovite is an example. Pyroxene, and amphibole, too, are faintly colored when free of iron and generally are deeper green or brown in varieties with iron, many being black or nearly so.

Garnet shows considerable variation in color dependent on composition (Color Plates 12B, 12H). Another property that changes with composition is density—iron-rich varieties are heavier. If a microscope is available, one can also determine that the *refractive index* of minerals (the degree to which they retard light passing through them) changes systematically with variations in composition.

The nature of minerals—a summary We can generalize the data and concepts discussed so far in some salient points:

1. Minerals are the earth's naturally occurring solid substances, and although a few are chemical elements, most are combinations of elements.
2. The elements may be present as atoms, as simple ions, or as complex ions consisting of several atoms. These differences are determined mainly by the number of electrons in the outer shell of each atom.
3. Atoms in minerals are joined mainly by sharing pairs of electrons and are thereby bonded strongly. Ions are joined by the electric charges they carry, and the strength of these bonds increases with the number of electric charges on the ions.
4. Some common minerals are simple ionic crystals (halite), and some are continous frameworks of shared-electron-pair bonds (quartz). Most, however, use both kinds of bonds to some degree.
5. The internal arrangement of atoms and ions and of the bonds holding them in place is the most basic aspect of a mineral species.
6. The compositions of minerals are set to a considerable degree by internal arrangement, but ions and atoms with similar sizes and similar electrical properties may substitute for one another over large compositional ranges.
7. A remarkably small number of minerals makes up all but a few percent of the outer part of the earth, suggesting a great degree of regularity in the ways that minerals form.

Identification of Minerals

The various physical properties mentioned in the foregoing section have been compared quantitatively, and they can be used in identifying minerals. Properties that can be estimated without involved equipment are described in the paragraphs that follow, and the specific values for each mineral are noted in the alphabetical list of minerals at the end of this chapter. Table S1-3 presents physical properties in a systematic way that can be used in identifying unknown samples. The unknown sample is classified first on the basis of cleavability, next on the basis of hardness, and finally on the basis of especially characteristic properties.

Cleavage As already noted, some minerals cleave along smooth planes parallel to one or more directions in their internal structures, and some do not. This property is intrinsic to a crystalline substance and therefore of great value

TABLE S1-3

Properties of Minerals Arranged for Identifying Unknown Samples.

A. Minerals having no cleavage (or in fine aggregates without cleavage)

HARDNESS

1	*talc* (in fine white to pale-colored waxy to somewhat shiny aggregates)
1–2	*sulfur* (yellow to brown; brittle; vitreous)
2–3	*gold* (yellow; metallic; ductile and malleable)
3–4	*serpentine* (pale green waxy aggregates; veins show fibrous structure)
3.5	*aragonite* (bubbles in dilute HCl; colorless or pale brown)
3.5–4	*chalcopyrite* (metallic; yellower than pyrite)
5	*apatite* (elongate crystals with vitreous to waxy luster; softer, waxy aggregates in sedimentary rocks)
5.5	*limonite* (yellow-brown to black ; powder (streak) colored yellow-brown)
5.5	*chromite* (black vitreous grains in peridotite)
5–6	*ilmenite* (dark gray to black, metallic grains)
5.5–6	*opal* (generally in pale colors; distinctly light in weight)
6	*hematite* (deep brownish red to metallic gray; streak brownish red)
6–6.5	*pyrite* (brassy colored and metallic; often in cubic or other equant crystals)
6	*magnetite* (metallic; magnetic; black; often in equant eight-sided crystals)
6.5	*olivine* (vitreous; pale yellow or green grains)
7	*quartz* (vitreous; hard; often in distinctive six-sided crystals)
7	*chalcedony* (waxy; translucent; often banded and colored)
7	*garnet* (tints of red or brown common; equant crystals; heavy)
7.5	*zircon* (often in four-sided elongate crystals; brown)

B. Minerals cleaving in one direction

HARDNESS

1	*talc* (in platy aggregates of crystals)
1–2	*graphite* (metallic; gray; marks paper)
2–2.5	*gypsum* (colorless; vitreous, often in rhombic crystals or silky aggregates)
2.5	*chlorite* (green; vitreous to waxy luster)
2.5	*muscovite* (pearly; cleaves in very thin elastic plates)
2.5	*biotite* (like muscovite but black or very dark green)
6	*hematite* (metallic; heavy; streak is brownish red)
6.5	*epidote* (vitreous; yellow-green; often in well-formed crystals)

C. Minerals cleaving in two or more directions

HARDNESS

2.5	*galena* (heavy; metallic; cleavages parallel to crystal faces, in three directions at right angles to one another)
2.5	*halite* (light; clear; salty taste)
3	*calcite* (bubbles in dilute HCl; distinctive crystal, cleavage forms)
3.5–4	*dolomite* (like calcite but heavier and bubbles in dilute HCl only if powdered first)
3.5–4	*sphalerite* (heavy; vitreous; yellow to brown or black)
4	*fluorite* (clear; cubic crystals with four cleavage directions; often colored)
5.5	*pyroxene* (black or green; two cleavage directions at approximately 90°)
5.5	*hornblende* (black to dark green or brown; two cleavages meeting at 60° and 120°)
5.5	*actinolite* (similar to hornblende but generally paler green and often in aggregates of small elongate crystals)
5.5	*glaucophane* (gray-blue; cleavages as in hornblende)
6	*potassium feldspar* (two cleavage directions meeting at 90°; often slightly pink and vitreous)
6	*plagioclase* (two cleavage directions meeting at somewhat more and less than 90°; often formed of multiple twins)

in identifying it and understanding its internal order. A small piece of a sample not showing obvious broken faces may be crushed and examined with a hand lens or binocular microscope. For grains as small as those in most rocks, a hand lens must generally be used to see the cleavage surfaces, which reflect light brightly because they are smooth planes (a cleavage surface on a grain of potassium feldspar forms a bright reflection in Color Plate 13G).

Hardness   This attribute is also basic to a crystalline substance and has been calibrated for minerals in a series of numbers from 1 to 10. The minerals and other substances listed below may be convenient as index materials, although any mineral of known hardness can be used. The hardness of an unknown mineral is determined by seeing which known minerals will scratch it and which will be scratched by it.

SCALE OF HARDNESS

1	talc	5.5	ordinary glass
2	gypsum	6	potassium feldspar
2.5	fingernail	7	quartz
3	calcite	8	topaz
3.5	copper penny	9	corundum
4	fluorite	10	diamond
5	apatite		

Specific gravity (density)   The weight of a mineral relative to the weight of an equal volume of water defines the mineral's *specific gravity*. Because water weighs close to 1 gram per cubic centimeter, the specific gravity of a mineral is the same as its density, except that the latter is expressed in grams per cubic centimeter. Perhaps because it is a number without units, specific gravity is customarily used by mineralogists. (It is generally abbreviated G.) The property is especially valuable if one has access to a laboratory where it can be determined accurately. Otherwise one must estimate specific gravity approximately by feeling the weight of a sample in the hand ("hefting" it). By practicing with known samples of about the same size (halite, quartz, dolomite, and fluorite give a useful range), one can develop a fairly sensitive feel for their comparative specific gravities.

Crystal form   This property would be very important were it not for the fact that most mineral grains have grown so that they interfere with one another and therefore show poorly developed crystal forms, if any. Well-formed crystals occasionally grow in cavities or early during the crystallization of a melt, and some minerals (such as garnet) grow with such strength that they establish crystal faces against other minerals. The color plates and other figures in this chapter illustrate the typical crystal forms of a number of common minerals.

Luster   The appearance of light reflected from a mineral's surface is called its *luster*. It is not an entirely reliable property because the surfaces of some minerals become tarnished and other minerals may be affected by deeper alterations. Plagioclase, for example, has a vitreous (glassy) luster when fresh but is commonly dull because of its being altered. An advantage of identifying minerals at the outcrop rather than in the laboratory is that one can often see occurrences of fresh as well as somewhat altered material and can therefore compensate for this effect (compare G and H in Color Plate 13).

Color Color is of great value in a first attempt at identification, but it is a variable property in many minerals, as already described in the section on mineral compositions. In addition, some normally colorless minerals are colored by small amounts of "foreign" elements or by the bombardment of nuclear particles and energy from radioactive substances. Quartz, for example, is colored lavender by traces of manganese and smoky brown by radioactive emanations (Color Plate 14B).

Streak The color of a small amount of powdered mineral, generally obtained by rubbing it against the unglazed part of a hard white tile, is useful in distinguishing some metallic sulfides and oxides (such as hematite, limonite, and magnetite).

Chemical and optical tests One of the few chemical tests that can be made easily and safely without unusual reagents is the application of a drop of dilute hydrochloric acid to minerals thought to be carbonates. Calcite will react with the acid to produce abundant bubbles of CO_2 in the drop, whereas dolomite will react only if powdered (such as along a scratch made with a knife).

The principal optical tests are made with polarized light, either by examining bits of minerals or by using thin sections. Cleavage fragments and crystal forms of small grains can thus be seen clearly, as in Figures 12-9 and 10, and the minerals can be observed to change colors in certain characteristic ways as they are rotated in the polarized light. Refractive indexes of minerals can be measured, and other effects of crystalline substances on light can be estimated from distinctive colors caused by interference of polarized light that has passed through the mineral. These various tests may be made with a *petrographic* (polarizing) *microscope* or with a simpler polariscope. Use of the latter has been described by John Sinkankas in *Mineralogy: a first course* (New York: D. Van Nostrand Co., 1966).

List of Minerals

(H = hardness;
G = specific gravity;
unless specified, luster ranges from vitreous to dull for each mineral)

Actinolite chiefly calcium-magnesium silicate, $Ca_2(Mg,Fe)_5Si_8O_{22}(OH)_2$; elongate crystals, separately or in fibrous masses (asbestos); cleavage in two directions (Figure S1-8); H = 5.5, G = 3.0; typically green (Color Plate 12A); formed by metamorphism at relatively low temperatures.

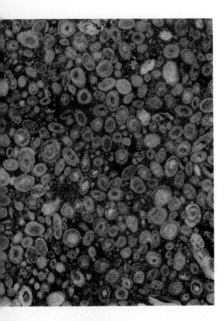

FIGURE S1-11
Pellets (*ooliths*) of fine grained earthy apatite, making up a rock called phosphorite, the principal source of fertilizer phosphate. Twice actual size.

FIGURE S1-12
Crystals of aragonite from Death Valley, California. Actual size.

Amphibole   a group of minerals including actinolite, hornblende, and glaucophane, each of which is described in this list.

Apatite   chiefly calcium fluophosphate, $Ca_5(PO_4)_3F$; as bone fragments and fine-grained aggregates in sedimentary rocks (Figure S1-11) and as small elongate six-sided crystals in other rocks (Color Plate 13B); cleavage poor or absent; $H = 5$; $G = 3.2$; colorless or various colors; sedimentary deposits are a valuable source for fertilizers.

Aragonite   calcium carbonate, $CaCO_3$; in needle-like crystals often forming fibrous masses (Figure S1-12); cleavage parallel to crystals but not apparent because of small size of most crystals; $H = 3.5$; $G = 2.9$; colorless or shades of brown; secreted in mollusk shells and formed by metamorphism of calcite at high pressures.

Biotite   variable alumino-silicate, commonly $K(Mg,Fe)_3AlSiO_{10}(OH)_2$; brightly reflecting flakes and tabular grains (Color Plate 12B), some being six-sided crystals; cleavage remarkable in one plane parallel to the flakes (Figure S1-9); $H = 2.5–3$; $G = 2.9$; black, dark brown, or dark green; widespread in granite, schist, and gneiss (Figure 10-16).

Calcite   calcium carbonate, $CaCO_3$; granular aggregates, locally in six-sided or rhombic crystals; cleaves readily in three planes parallel to rhombic crystal faces (Figure S1-13); $H = 3$; $G = 2.7$; typically colorless but may be tinted many colors; formed chiefly in sedimentary rocks (limestone) from shelly organic remains; a valuable raw material for cement and in the steel industry.

Chalcedony   silica, SiO_2; a variety of quartz in which the crystals are long and very thin, together forming layers (Color Plate 12C); colorless or in many colors; chiefly deposited in open cavities; banded varieties may be semiprecious stones (as agate and jasper).

Chalcopyrite   copper-iron sulfide, $CuFeS_2$; in irregular grains or in crystals with triangular faces; no cleavage; $H = 3.5–4$; $G = 4.2$; brass-yellow (and yellower than pyrite) though often with iridescent tarnish (Color Plate 12D); luster metallic unless tarnished; typically deposited by vapors associated with granite intrusions; principal source of copper.

Chlorite   complex Mg-Fe silicate; grains flaky but often so fine as to appear stony; cleavage in one plane parallel to flakes; $H = 2–2.5$; $G = 2.6–2.9$; typically green but sometimes almost black (Color Plate 12F); formed during metamorphism in the presence of abundant water.

Chromite   chiefly iron chromite, $FeCr_2O_4$; granular aggregates or single equant grains (Figure S1-14); no cleavage; $H = 5.5$; $G = 4.4$; black; luster metallic to pitchy; in peridotite; source of chromium.

Clay minerals   (see Chapter 8 for descriptions of the species kaolinite, illite, and expandable clay); formed by alterations near the earth's surface, in the presence of abundant water; uses various and valuable (Table 21-1).

FIGURE S1-13
Fragments of calcite broken from a large calcite crystal, showing the rhombic forms resulting from cleavage in three directions. One half actual size.

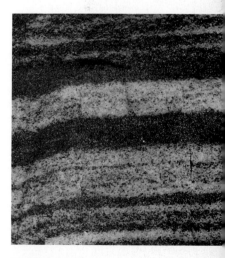

FIGURE S1-14
Layers of chromite grains (dark gray) and olivine grains (light gray) in a layered peridotite. Actual size.

489

Dolomite   calcium magnesium carbonate, $CaMg(CO_3)_2$; granular aggregates and rhombic crystals much like those of calcite (Color Plate 14H); cleavage in three planes as in calcite; H = 3–4; G = 2.9; white or tinted pale pink or brown; replaces calcite in sedimentary deposits and in low-temperature metamorphism; source of magnesium.

Epidote $Ca_2(Al,Fe)_3(OH)(SiO_4)_3$; granular aggregates or brightly reflecting crystals; cleavage in one plane but not always apparent; H = 6.5; G = 3.4; colored distinctively—pale yellow-green to deep brownish green (Color Plate 12F); formed by metamorphism at low to moderate temperature.

Feldspar   (see potassium-feldspar and plagioclase).

Fluorite   calcium fluoride, CaF_2; granular aggregates or cubic crystals (chapter opening page); cleavage in four planes (Color Plate 12G); H = 4; G = 3.2; colorless or often some shade of purple, yellow, or green; deposited by vapor in cavities; used as flux in smelting and as source of fluorine.

Galena   lead sulfide, PbS; irregular grains or cubic crystals (Color Plate 12E); cleavage in three planes parallel to cube faces; H = 2.5; G = 7.5; silvery (metallic) gray but often tarnished to dull dark gray; deposited by vapor or hot water in various rocks; source of lead.

Garnet   various compositions but most commonly $Fe_3Al_2(SiO_4)_3$ or $Ca_3Al_2(SiO_4)_3$; granular aggregates or twelve-sided crystals with rhombic faces; no cleavage; H = 7; G = 3.7 – 4.2; color variable but commonly some shade of red (Color Plate 12B and H); luster vitreous; chiefly in metamorphic rocks; used as an abrasive.

Glaucophane   silicate, $Na_2Mg_2Al_2Si_8O_{22}(OH)_2$ elongate crystals typical, some fibrous; cleavage as in Figure S1-8; H = 6; G = 3.1; distinctive blue-gray (Color Plate 13A); luster vitreous (or silky for fibrous forms); formed by metamorphism at high pressure and low temperature.

Gold   elemental gold, Au; grains, films, and small crystals in quartz or altered rocks, or rounded grains in stream-gravels (placer gold); no cleavage; grains distinctly malleable (spread easily by hammering); H = 2.5–3; G = 15–19 (varying with content of silver); distinctive bright metallic yellow; deposited by vapors associated with granite intrusions and in sedimentary deposits; a precious metal.

Graphite   elemental carbon, C; aggregates of flakes, some showing six-sided crystal shapes; cleavage distinct parallel to flakes; H = 1–2; G = 2.1; dark gray; luster metallic; in metamorphosed carbon-bearing rocks; a lubricant and refractory.

Gypsum   hydrated calcium sulfate, $CaSO_4 \cdot 2H_2O$; in fibrous aggregates and tabular crystals with rhombic outline (Figure S1-15); cleavage in one plane parallel to rhombic crystal faces; H = 2; G = 2.3; typically clear and colorless, or white; in sedimentary deposits; source of plaster, chemicals.

FIGURE S1-15
Common forms of gypsum, the lowest one cleaved into thin sheets. One half actual size.

Halite   sodium chlorite, $NaCl$; in granular aggregates or cubic crystals (Figure S1-2); cleavage in three planes parallel to cubic faces; $H = 2.5$; $G = 2.2$; typically colorless but may contain colored pigments; in sedimentary deposits and intrusions (Figure S1-3); source of salt, chemicals.

Hematite   ferric oxide, Fe_2O_3; locally in flaky crystals but typically in fine aggregates, rounded pellets, or colloform masses with prismatic grains (Color Plate 13C); typically no cleavage; $H = 6$; $G = 5.2$; dark red to black, but streak consistently brownish red; in sedimentary and metamorphic deposits; chief source of iron and steel.

Hornblende   a complex silicate, approximately $NaCa_2(Mg,Fe)_5AlSi_7O_{22}(OH)_2$; typically in moderately elongate crystals (Figure S1-16); cleavage as in Figure S1-8; $H = 5.5$; $G = 3.2$; generally green to black and dark brown; in a variety of igneous and metamorphic rocks.

Illite   (see page 182).

Ilmenite   ferrous titanate, $FeTiO_3$; in small grains and tabular crystals; no cleavage; $H = 5–6$; $G = 4.7$; black and metallic to pitchy in appearance; slightly magnetic; in igneous and metamorphic rocks and sands derived from them; source of titanium and titanium oxide.

Kaolinite   (see page 179).

Limonite   hydrated ferric oxides (a mixture of submicroscopic iron oxides and thus not a true mineral); brown earthy aggregates and nearly vitreous dark brown masses; no cleavage; $H = 4–5.5$; $G = 3–4$ (Color Plate 13D); formed near earth's surface in presence of water; a source of iron.

Magnetite   ferrous ferrite, $Fe^{II}Fe^{III}_2O_4$; equant grains, some being eight-sided crystals (Color Plate 13B); no cleavage; $H = 6$; $G = 5.1$; black (and streak black); metallic luster; strongly magnetic; widespread in igneous and metamorphic rocks and in sands derived from them; a source of iron and steel.

Mica   (see biotite, muscovite).

Muscovite   potassium-aluminum silicate, $KAl_3Si_3O_{10}(OH)_2$; in flakes or six-sided crystals; cleavage remarkable parallel to flakes; $H = 2–2.5$; $G = 2.8$; colorless to silvery brown, gray, or green; in metamorphic rocks and coarse granite (pegmatite); used as an electric insulator.

Olivine   magnesium-iron silicate, $(Mg,Fe)_2SiO_4$; equant grains and short prismatic crystals; no cleavage; $H = 6.5$; $G = 3.3$; pale green or yellow (in lavas often has an iridescent surface) (Color Plate 13E); chiefly in peridotite, gabbro, and basalt; some crystals are gemstones.

Opal   silicon oxide with water; silica is in the crystalline form called cristobalite, but the mixture is so fine that it looks like milky glass; no

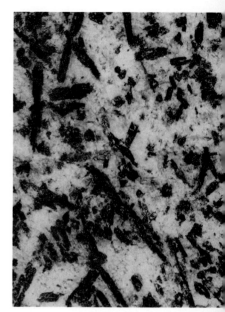

FIGURE S1-16
Hornblende crystals (black) in diorite, a rock that crystallized from a melt. Twice actual size.

cleavage; H = 5.5; G = 1.4; colorless or commonly tinted or opalescent in various colors (Color Plate 13F); as shells of diatoms (Figure 10-4) and deposited in veins; opalescent varieties are gemstones.

Orthoclase   (see potassium-feldspar).

Plagioclase   mixtures of $NaAlSi_3O_8$ and $CaAl_2Si_2O_8$; in roughly tabular crystals in igneous rocks (Color Plate 14G), otherwise equant grains; cleavages as in Figure S1-6; H = 6; G = 2.7; typically colorless to dark gray (Color Plate 13G and H); chief mineral in many igneous and metamorphic rocks and in sands derived from them.

Potassium-feldspar   chiefly $KAlSi_3O_8$; crystals tabular or elongate boxes (Color Plate 13B), but generally in irregular grains; cleavages as in Figure S1-6; H = 6; G = 2.6; commonly pale orange, pink, colorless, or gray (Color Plate 13G and H); in granite and related igneous rocks; used in porcelain and pottery.

Pyrite   iron sulfide, FeS_2; granular or in cubic or other distinctive crystal forms; no distinct cleavage; H = 6; G = 5; brassy yellow; luster metallic; (Color Plates 14A and 12D, lower left); in many kinds of rocks and ore deposits; a source of iron and sulfuric acid.

Pyroxene   mixtures of $MgSiO_3$, $FeSiO_3$ and $CaSiO_3$; roughly equant grains or short prismatic crystals (Figure S1-7); cleavage in two planes; H= 5.5; G = 3–3.9 (depending on iron content); pale green, brown, or black (Color Plate 14F and G); abundant in gabbro, basalt, and some metamorphic rocks formed at high temperature.

Quartz   silicon, SiO_2; irregular grains and distinctive six-sided crystals (Color Plate 14B); typically no cleavage; H = 7; G = 2.7; typically colorless but may be various colors; in many kinds of rocks and in large crystals deposited in cavities; used as a semiprecious stone and in electronics.

Serpentine   complex magnesium-rich silicate, chiefly $Mg_3Si_2O_5(OH)_4$; in fine waxy masses but locally fibrous (asbestos) (Color Plate 14C); no cleavage visible; H = 3–4; G = 2.5; typically light to dark green or black, often mottled in one specimen; formed chiefly by alteration of peridotite (olivine) in presence of water; a refractory (asbestos) and may be a semiprecious stone.

Sphalerite   zinc sulfide, ZnS; in crystals and granular aggregates; cleavages well developed in several planes intersecting at 60°; H = 3.5; G = 4; varies from pale yellow to reddish brown to black (with increasing content of iron) (Color Plate 12E); deposited by vapor or water in a variety of rocks; source of zinc.

PLATE 9

Above, part of the Wasatch Formation near Echo, Utah, showing red sandstone in somewhat wavy layers suggesting deposition in migrating stream channels. Below, thinly layered, light-toned rocks of the Green River Formation near Green River, Wyoming. The irregular brown bluffs above the regularly layered lake deposits are part of the Bridger Formation, deposited by streams.

PLATE 10
Right, conglomerate of approximately the same age as the Castlegate Sandstone. The largest cobble is 6 inches long. At the eastern edge of the Gunnison Plateau (labeled in Figure 14-2). Below, thick conglomerate layer filling a stream channel cut into the older, horizontal sandstone layers on which the hammer rests. Note the complete lack of layering within the conglomerate. Forty miles northwest of Price (labeled in Figure 14-2).

PLATE 11

Top, wall of a canyon eroded into the flank of Valles Volcano, New Mexico, exposing volcanic layers deposited from hot fragmental flows that were compacted and welded immediately after eruption. The layers are approximately 2 million years old and the red rocks beneath them are Cenozoic sedimentary rocks. Right, seacliff exposing faulted mixture (mélange) of chert (reddish, layered rocks), submarine lavas (greenish gray rocks in the lower right part of the view), and glaucophane schist (bluish gray rocks below the hammer). On the southern edge of the Santa Lucia Range, 50 miles south of Monterey, California. Photograph by Clarence A. Hall, Jr.

PLATE 12
Minerals and Rocks.

A. Crystals of actinolite (green, elongate) partly altered to reddish iron oxide in a schist. Actual size.

B. Biotite (black) and garnet (red) forming a schist. Actual size.

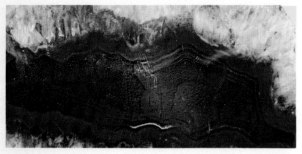

C. Part of a cavity filled partly by quartz (colorless) and then by banded orange chalcedony (agate). Actual size.

D. Chalcopyrite (yellow, locally tarnished reddish and bluish) and pyrite (grayish yellow) in altered granite. Actual size.

E. Crystals of galena (gray cubes) and sphalerite (small, reddish), in a lead-zinc ore. Actual size.

F. Fine-grained schist of chlorite (dark grayish green) with nodules of epidote (yellowish green). Actual size.

G. Cubic crystals of lavender fluorite and a crystal of yellow fluorite that shows the four cleavage directions.

H. Red garnet and colorless quartz making up the metamorphic rock called skarn (originally a limestone). Actual size.

PLATE 13
Minerals and Rocks.

A. Grains of glaucophane (blue), quartz (colorless), and garnet (small, dark) forming a schist. Thin section enlarged 15 times.

B. From left to right—crystals of magnetite (gray), zircon (dark brown), apatite (yellow), and potassium feldspar. Actual size.

C. Hematite, in colloform (bulbous) mass of radiating crystals and in flaky form with metallic luster. Half actual size.

D. Limonite in concentrically grown pellets (ooliths) forming a rock called ironstone. Actual size.

E. Grains of olivine (pale yellowish green) in a basalt. Actual size.

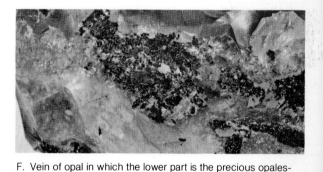

F. Vein of opal in which the lower part is the precious opalescent variety. Actual size.

G. Granite, showing potassium feldspar (pale orange), plagioclase (white), quartz (glassy), and biotite (black). Actual size.

H. Same granite as in G, but altered such that potassium feldspar is white, plagioclase yellow and pitted, and quartz gray.

PLATE 14
Minerals and Rocks.

A. Crystals of pyrite, showing typically striated crystal faces. Actual size.

B. Crystals of quartz, including the varieties amethyst (purple) and smoky quartz (grayish brown). Half actual size.

C. Serpentine, partly as fibrous asbestos that grew at right angles to the walls of a crack. Twice actual size.

D. Crystals of sulfur. Actual size.

E. Talc, forming large subparallel plates in a coarse schist that has been slightly wrinkled by buckling. Actual size.

F. Pyroxene (grayish green) in metamorphic marble. The white calcite has been stained tan on the exposed surface. Actual size.

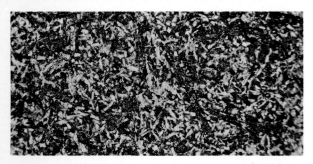

G. Platy crystals of plagioclase (white) surrounded by dark pyroxene in diabase (a fine-grained gabbro). Twice actual size.

H. Crystals of dolomite that grew on the walls of a cavity in limestone. Actual size.

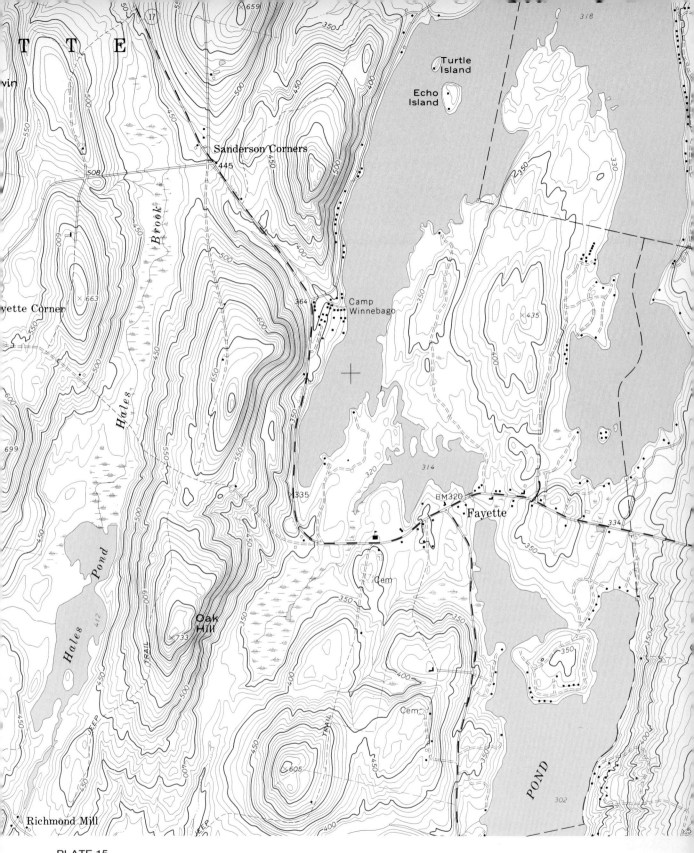

PLATE 15

Part of the Fayette quadrangle, Maine, of the U.S. Geological Survey. The scale is 1:24,000 (1 inch = 2,000 feet) and the contour interval is 10 feet.

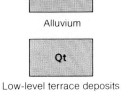

Qal

Alluvium

Qt

Low-level terrace deposits

QTf

High-level fluvial deposits

Occ

Calloway Creek Limestone

Og

Garrard Siltstone

Ocf

Clays Ferry Formation

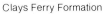

Ols
Olb

Olt

Olgc
cr
b

Lexington Limestone

Ot

Tyrone Limestone

North

| 0 | 1000 | 2000 |

Feet

Contour interval 20 feet

Ocn

Camp Nelson Limestone

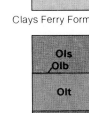

Oo

Oregon Limestone

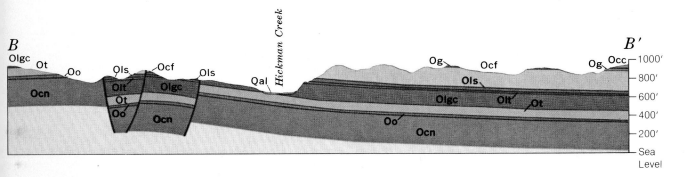

PLATE 16

Part of the geologic map and geologic sections of the Little Hickman quadrangle, Kentucky. Map GQ-792, by Don E. Wolcott, U.S. Geological Survey, 1969.

Sulfur elemental sulfur, S; granular aggregates and well-formed crystals (Color Plate 14D); no cleavage; H = 1.5–2.5; G = 2.1; yellow, orange, brown; in sedimentary and volcanic deposits; many industrial uses.

Talc magnesium silicate, $Mg_3Si_4O_{10}(OH)_2$; in fine waxy or silvery flaky aggregates (Color Plate 14E); cleavage in a plane parallel to flakes; H = 1; G = 2.7; colorless to pale green; the softness and flaky cleavage give the mineral a greasy feel; formed by metamorphism in presence of abundant water; used in soap, refractories.

Zircon zirconium silicate, $ZrSiO_4$; small, four-sided prismatic crystals (Color Plate 13B) or nearly equant grains; cleavage not usually visible; H = 7.5; G = 4.7; typically brown; widespread but scarce in granites and sand derived from them; source of zirconium, gemstones.

Granite (the rock with smaller grains) intruded by an irregular dike of pegmatite (the rock with larger grains). The minerals composing the two are much the same: mica (black), feldspar (white to pale gray), and quartz (medium gray). Indeed, the pegmatite probably formed from water-rich melt extruded from the granite when the latter was partly crystallized. One third actual size. Photograph by Ruperto Laniz and Perfecto Mari.

Supplementary Chapter 2.
Rocks

Rocks

Igneous Rocks
  Descriptions of igneous rocks
Sedimentary Rocks
  Descriptions of sedimentary rocks
Metamorphic Rocks
  Foliated rocks
  Granular rocks
  Rocks with mixed textures
  Grade of metamorphism

Rocks are made up of minerals in specific combinations and arrangements that tell much about the origin of each rock and therefore about the earth's history. The combinations and arrangements result from certain orderly processes that can be grouped into three broad categories. Each category is the basis of one of the three genetic groupings of rocks: (1) *igneous rocks*, which have formed from molten materials, either by the crystallization of minerals in a melt or by the chilling of a melt to glass; (2) *sedimentary rocks*, which are accumulations of fragmental or chemical sediments that have become consolidated during burial; and (3) *metamorphic rocks*, which form deep within the earth by heating, deformation, and chemical alterations of preexisting rocks. In a broad sense, the three groups form a continuum. The continuum starts with the eruption of new, molten materials to the outer part of the earth, then proceeds to the erosion of rocks to form fragments and ions that accumulate in sedimentary sequences, and finally ends by the deep burial and the heating of these various substances. We should note, too, that some rocks cannot be classified into one category and that they demonstrate the continuum directly. Igneous fragments exploded from volcanoes may fall in water and become incorporated directly into sediments. Sedimentary rocks pass by gradations into metamorphic rocks as they become buried deeply. Metamorphic rocks may be heated so strongly that they start to melt, thus grading into igneous rocks. The three broad groupings of rocks are thus interlinked in several ways.

To classify a rock, one must first examine the shapes, sizes, and arrangements of its grains. These constitute an aspect of the rock called its *texture*, analogous to the texture of cloth, which is due to the size, spacing, and weave of threads. The textures characteristic of igneous rocks result from the solidification of melts, especially from the growth of crystals during the early stages of cooling. The abundance of liquid at that stage makes it possible for minerals to grow as geometric crystals, which can be seen in lavas chilled suddenly to glass or to exceedingly fine-grained aggregates (Figure S2-1A). If, however, the melt crystallizes slowly and completely underground, the geometric crystals become surrounded by other grains that must accommodate to the remaining melt-filled spaces. Most grains thus end up with poorly developed crystal faces or are irregular (Figure S2-1B). In most igneous rocks, the early-formed grains include one or more of the minerals plagioclase, biotite, pyroxene, and hornblende. Quartz and potassium feldspar more commonly form at a later stage. The fairly consistent relation between the order of crystallization and the textural aspects of each mineral is thus one of the strongest indications of a rock's igneous origin.

The commonest texture of sedimentary rocks results from the fragmental nature of the grains and their tendency to lie with their longer dimensions more or less parallel to sediment layers (Figure S2-1C). Bits of shell, bone, or carbonized wood serve to categorize a rock as sedimentary. Thus, homogeneous, crystallized rocks (for example, many limestones) can be classified as sedimentary rocks at once because of the fossils in them.

The textures of metamorphic rocks reflect the fact that the grains have grown together in the solid state. Their mutual interference during growth

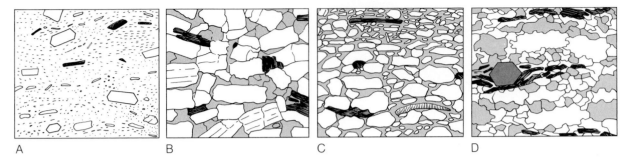

FIGURE S2-1

Characteristic textures of rocks. A, lava with well-formed crystals in a glassy or fine-grained matrix. B, rock from a pluton, with moderately formed crystals of feldspar (white) and dark minerals but with irregular grains of quartz (light gray) filling spaces between. C, sedimentary grains in a sandstone, surrounded by cement (gray). D, metamorphic rock, showing the irregular shapes of most grains (except for a well-formed crystal of garnet) and the tendency for grains to lie in parallel orientations; light gray grains are quartz, white ones feldspar.

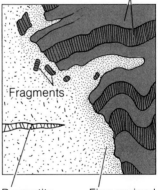

FIGURE S2-2

Some typical details along the border of a pluton (left) that has intruded into deformed (folded) rocks. The dike of coarse rock called pegmatite is thought to form from water-rich melt segregated from the rocks of the pluton just before they solidified completely.

results in irregular, interlocking shapes. Some common minerals, notably mica and amphibole, have enough crystallizing strength to grow in somewhat platy or prismatic crystal forms against quartz and feldspar, and a few minerals, for example, garnet, have the strength to form well-shaped crystals. The overall texture, then, is one of a graduated development of crystal forms, with feldspar grains irregular—thus differing from most igneous textures (Figure S2-1D). The figure illustrates another common aspect of metamorphic rocks—the tendency to develop a *foliation*, the parallel arrangement of platy and prismatic minerals. Foliation is typically due to the deformation of the rock during formation of the metamorphic minerals. Deformation also tends to contort sedimentary layers; thus, the metamorphic grains may lie oblique to deformed layers inherited from sedimentary rocks (Figure 15-11C).

Additional aspects of rock textures are described and illustrated in the remainder of this chapter, along with descriptions of the assemblages of minerals forming each rock. The descriptions refer mainly to small pieces of rock or to rocks as seen in small exposures because this is often all one has to work with. It should be noted, however, that identifications and interpretations are easier when one can examine exposures showing the interrelations among several rocks. As illustrated in Figure S2-2, for example, the igneous origin of rock bodies is undeniable when one can see the full effects of a melt: (1) its penetration into the surrounding rocks; (2) loose pieces of rock caught in it; (3) metamorphism of the surrounding rocks due to heat and vapors from the melt; (4) the finer grained margin caused by rapid cooling of the melt; and (5) small bodies consisting of the last minerals to crystallize from the melt, as in the photograph on the chapter opening page. Sedimentary rocks, too, are more readily identified and interpreted at exposures where their layering and fossil content are more apparent, and the distorted, recrystallized layers of metamorphic rocks are more understandable when seen in large exposures such as the right-hand part of Figure S2-2. Further examples of rock charateristics that can be seen at outcrops are described and illustrated in many other parts of the book.

498

Igneous Rocks

Igneous rocks are classified on the basis of composition and texture. In the tabulation in Figure S2-3, variations in composition are shown by the mineral proportions in the lower half of the diagram. The composition of any one rock is read along a vertical line by reference to the spacing of horizontal lines. For example, a rhyolite or granite represented by the line along the left edge of the diagram (an unusually quartz-rich variety) would consist of approximately 40 percent quartz, 50 percent potassium feldspar, 7 percent plagioclase, and a few percent biotite. Note the general variations in mineral proportions from left to right—the gradual decrease in quartz and potassium feldspar and the gradual increase in plagioclase. Note especially that the proportion of dark minerals (biotite, hornblende, and pyroxene) increases from left to right. The appearances of the rocks change accordingly and their densities (weights) do likewise (olivine as well as the dark minerals are denser than quartz and feldspars).

FIGURE S2-3

Classification of igneous rocks (rocks formed from melts), based partly on texture and partly on the proportions of the principal minerals.

Rock Textures	Rock Names				
Fragmental, finer grained than 0.2 inch	Rhyolite tuff	Dacite tuff	Andesite tuff	Basalt tuff	
Fragmental, coarser grained than 0.2 inch	Rhyolite breccia	Dacite breccia	Andesite breccia	Basalt breccia	No volcanic rocks in this range of composition
Glassy	Obsidian (solid) or pumice (frothy)				
Grains too small to be seen individually	Rhyolite	Dacite	Andesite	Basalt	
All grains large enough to be visible	Granite	Granodiorite	Quartz diorite	Gabbro	Peridotite

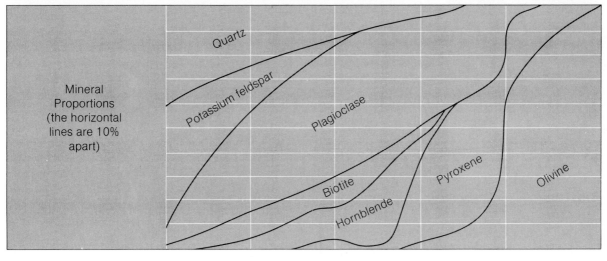

Mineral Proportions (the horizontal lines are 10% apart)

Quartz

Potassium feldspar

Plagioclase

Biotite

Hornblende

Pyroxene

Olivine

The textural differences among igneous rocks are even more obvious than the compositional variations and are a direct result of the rock's origin. The rocks listed in the upper four rows are typically erupted from volcanoes, and those in the lowest row crystallize underground. *Tuff* is a rock consisting of fragments of glass or minerals that are mainly smaller than 4 millimeters (0.2 inch) in diameter. Tuffs tend to be porous and light in weight because many fragments in them consist of chips and filaments of bubbly glass (Figure 12-7). Because the glass fragments are freshly erupted melt, they may be so hot when they accumulate that they compact into a solid rock called *welded tuff*—a rock that would look like a lava were it not for its relict fragments.

Melts that bubble and froth without exploding typically congeal into the frothy glass called *pumice*. Some viscuous lavas contain so little dissolved water that they neither explode nor froth but congeal into *obsidian*—a black or dark shiny glass that chips along curving fractures (Figure S2-4). Glasses with the composition of andesite or basalt are scarce because these lavas are so fluid that they crystallize easily; however, they chill to glass when erupted into water. Such glasses are duller than obsidian and are so unstable that they alter readily to green and yellowish masses of waxy material called *palagonite*, related to the clay minerals.

The largely crystalline but fine-grained rocks (rhyolite, dacite, and so on) occur mainly in lava flows or volcanic conduits and locally at the chilled margins of large intrusive bodies. Although most of their mineral grains are too small to be identified without a microscope, almost all carry at least a scattering of larger grains. Called *phenocrysts*, these grains are large enough to identify and thus provide a means of classifying the rocks (Figure S2-5).

The rocks in the lowest row of Figure S2-3 are those in which all mineral grains are large enough to be visible without a microscope. Those with grains smaller than 1 millimeter (0.05 inch) typically form small intrusive bodies or rapidly cooled margins of large bodies. They are sometimes named separately, for example, *fine-grained granite*; one of them, fine-grained gabbro, is often called *diabase*. Other interesting textural variants among coarse-grained rocks are those with a set of distinctly larger grains (*phenocrysts*) expressing the rapid or early growth of some minerals. These rocks are often referred to as being porphyritic, for example, *porphyritic granite* (Figure S2-6).

DESCRIPTIONS OF IGNEOUS ROCKS Additional details of rock varieties are included in the descriptions that follow in alphabetical order.

Andesite This abundant lava is characteristic of the arcuate chains of volcanoes rimming the Pacific Ocean. Tongues and layers of lava, tuff, and breccia are often mixed, although some volcanoes erupt one textural variety almost exclusively. Typical eruptions and recently erupted varieties of andesite are described in Chapter 12, and the genetic relations between volcanoes and larger earth features are discussed in Chapters 19 and 20.

Andesite typically has a medium to dark gray matrix of microscopic minerals and glass in which are set abundant phenocrysts of plagioclase and pyroxene, typically 0.5 to 3 millimeters (0.02 to 0.12 inch) in diameter. Scattered grains of olivine occur in some andesites, and some varieties carry abundant crystals of hornblende.

FIGURE S2-4
Rhyolite obsidian (the black glass) with a conchoidal (shell-like) fracture and groups of radiating crystals called spherulites (white). Actual size.

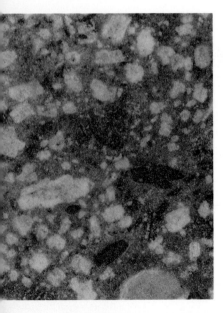

FIGURE S2-5
Dacite with abundant phenocrysts (large grains) of plagioclase and hornblende in a microscopically crystalline groundmass. Actual size.

The origin of andesite melt is a moot question. Its possible production from basalt melt by the settling of early-formed olivine is described in Chapter 12. Andesite may also be formed by the partial melting of basaltic rocks at great depth and by the mixing of basaltic and granitic melts.

Basalt This is the most widespread and abundant of volcanic rocks, forming much of the earth's crust under the oceans and appearing in vast lava fields and countless separate volcanoes on land (Chapters 12, 16, and 20). Basalt is the most fluid lava and therefore the most completely crystallized. The abundant tablets of plagioclase are generally just large enough to see with a hand lens (perhaps as flashes of light reflected from cleavage surfaces). Olivine commonly forms larger equant grains in the gray matrix (Color Plate 13E). The rock otherwise consists mainly of pyroxene and plagioclase, the latter sometimes forming large crystals (phenocrysts) (Figure 12-9). Because it is so liquid, the lava rarely retains gases to the point of becoming violently explosive, so that fragmental rocks are less abundant than in more viscous (silica-rich) melts. Where basalt is erupted underwater, however, it may chill and explode into mounds of glassy tuff, which are often associated with lavas made up of pillow-like masses (Figure 6-16).

Basaltic melt forms deep in the earth's mantle, as discussed in Chapter 16 and here under *peridotite.*

Dacite This comparatively silicic rock is abundant in many volcanic chains and is generally associated with andesite (Chapters 12 and 19). Except when chilled so rapidly that it forms obsidian, dacite is typically lighter in color than andesite and basalt and almost always carries abundant large grains of plagioclase in its fine matrix (Figure S2-5). Other typical grains are biotite and hornblende. Quartz typically crystallizes late and is thus present in the matrix rather than as phenocrysts. Because dacite melt is viscous, most dacite is exploded from volcanoes to form tuff and breccia; however, steep-sided domes (mounds) and thick lava flows are fairly common.

Dacite melt may form from basalt or andesite melt by the settling out of early-formed minerals, or it may form directly from the melting of rocks at great depth. Some dacite almost certainly forms by the mixing of basaltic and granitic melts.

Diorite A rather scarce granular rock consisting of plagioclase and hornblende or pyroxene, and differing from gabbro in containing fewer dark minerals (Figure S1-16). Forms small bodies often associated with quartz diorite.

Gabbro This rock forms when basaltic melt cools and crystallizes slowly beneath the surface. It forms most of the oceanic crust and a variety of plutons and parts of plutons in the continental crust. Most varieties are simple granular mixtures of plagioclase and pyroxene with small amounts of olivine (Figure S2-7). Hornblende may take the place of pyroxene and olivine in water-rich melts crystallizing at such great depth that the water is forced into the structures of growing minerals.

Because of the fluid nature of gabbro melt, crystals sink in it readily and accumulate in layers. Some layers are almost all pyroxene (a rock called

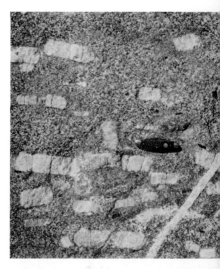

FIGURE S2-6
Porphyritic granite with well formed crystals of potassium feldspar aligned by flow. The largest crystal is three inches long.

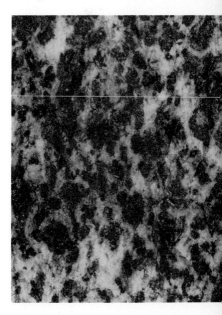

FIGURE S2-7
Gabbro, consisting of plagioclase (light gray) and pyroxene (dark gray). Actual size.

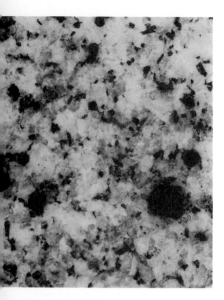

FIGURE S2-8
Granodiorite with scattered crystals of black hornblende (elongate, on the left) and biotite (equant, on the right). Actual size.

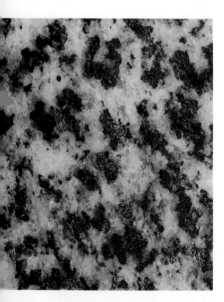

FIGURE S2-9
Quartz diorite, consisting of hornblende and biotite (dark gray), plagioclase, and quartz. Actual size.

pyroxenite), and some are almost all plagioclase (a rock called *anorthosite*). Accumulations of olivine may give rise to layers of peridotite. Fine-grained variants of gabbro (diabase) typically form small intrusions near the surface and are characterized by thin plates of plagioclase in larger pyroxene grains (Color Plate 14G).

Granite　This name is applied specifically as shown in Figure S2-3 but is also used generally for all coarse-grained igneous rocks consisting mainly of quartz and feldspar. The abundance of potassium feldspar can be judged in some rocks by the pinkish or orangish color of the mineral (Color Plate 13G). In some other cases it is distinctly grayer than the plagioclase and is often less altered (Color Plate 13H). The twin striations in plagioclase are helpful in estimating the amount of that feldspar (Supplementary Chapter 1).

Granite occurs widely in the continental crust, often as large, complex plutons (Chapter 12). Some granite melt has probably formed as the final residue of crystallization and crystal sinking in other melts, and some results from melting deep in the crust, perhaps in the mantle. In regions of rocks that were once deeply buried, granite bodies with hazy outlines appear to have formed by the metamorphic exchanges of substances without any large amount of melting (Figure 10-17). Some sedimentary rocks might be changed to granites with little change in composition. The origin of granites is thus most challenging!

Granodiorite　This less silicic "granite" is probably the most abundant igneous rock of large plutons, especially those grouped together in arcuate belts of mountains. Besides by its lesser amount of potassium feldspar, it can be distinguished from true granite by its greater amount of dark grains, especially of hornblende (Figure S2-8). It can probably originate in all the ways mentioned for granite.

Quartz diorite　This abundant rock of plutons is the coarsely crystallized equivalent of andesite lava. It can be distinguished from other quartz-bearing rocks by its abundant plagioclase and dark minerals, the latter often being hornblende (Figure S2-9). It is often associated with granodiorite and granite and probably forms in the same ways.

Rhyolite　This rock, the rapidly cooled equivalent of granite, is so viscous that it is erupted mainly in exploded (fragmental) varieties and as steep-sided domes and flows. The flows may consist of obsidian or of light gray to pink lava that carries large grains of quartz, feldspar, and biotite. The rock is commonly divided into thin layers resulting from the flow of the viscous melt. Similar layers in flows of obsidian are marked by sheets of spherical bodies called *spherulites*, kernels of radiating crystals that grow rapidly as the glass solidifies (Figure S2-4).

Peridotite　This rock results from the accumulation of olivine grains that crystallize at an early stage from gabbro melt. This origin is suggested by the occurrence of mineral layers (see gabbro) and by the fact that peridotite is not found as fine-grained rocks or as glass. Chromite, another mineral forming at an early stage in gabbro, may form separate layers in peridotite (Figure S1-14).

Most peridotites, however, are metamorphic rocks from the mantle, where they may have formed originally during the earth's early melting and differentiation. As described in Chapter 16, these rocks are characterized by irregular, interlocking grains formed by crystallization in the solid state (Figure 16-8). They are nearly devoid of feldspar, presumably because it has been melted from them and has been emplaced upward as basaltic melt. Thus, although the rocks are not igneous, they play an important part in igneous processes.

Sedimentary Rocks

Due to the history of the science, some sedimentary rocks are classified according to the sizes of their particles and the others are classified according to their mineral compositions. Table S2-1 shows the rock names based on grain sizes, the size groups being the same as those described in Chapter 1 (Figure 1-8). The grains in these rocks are mainly eroded particles—bits and pieces gained from soils and rocks. The most abundant minerals are typically quartz and clays (because most sediments are eroded soils); however, many other minerals may be eroded from rocks, some of the more common ones being mica and feldspars.

Table S2-2 shows the names of rocks based on the compositions of the principal mineral making up a given rock. Many of these rocks also started as fragmental (detrital) deposits; however, the particles in most cases consisted of shells and other hard parts of organisms—substances consisting of calcium carbonate, silica, or calcium phosphate (apatite). The detrital nature of these grains is commonly obscured and even obliterated when they are consolidated to form limestone, chert, or phosphorite. The resulting textures are granular and visibly crystalline rather than detrital, because the grains commonly recrystallize. Crystalline textures also result when calcite, dolomite, gypsum,

TABLE S2-1

Classification of Sediments and Sedimentary Rocks
Based on Sizes of Fragments (Detrital Grains).

Grain sizes		Sediment name	Rock name
mm	inches		
		Gravel	Conglomerate (breccia if grains angular)
2.0	0.1		
		Sand	Sandstone
0.06	0.002		
		Silt ⎫ ⎬ Mud Clay ⎭	Siltstone ⎫ Mudstone, ⎬ shale Claystone ⎭
0.002			

TABLE S2-2

Sedimentary Rock Names Based on Dominant Mineral Constituent.

Mineral	Rock	Mineral	Rock
Calcite or aragonite	Limestone	Halite	Salt
Dolomite	Dolomite or dolostone	Fine-grained apatite	Phosphorite
Opal, chalcedony, or fine quartz	Diatomite, chert	Silicates, carbonates, and oxides of iron	Ironstone
Gypsum	Gypsum		

or halite are deposited due to evaporation of sea water. A textural variety of some rocks in Table S2-2 results from chemical growth (accretion) of rounded pellets (*ooliths*) and nodules (Figure S1-11).

More than one mineral substance is abundant in many cases, and the rock is then given compounded names such as cherty limestone or dolomitic chert. Mixtures of the substances shown in the two tables also are common. Sand cemented by abundant dolomite is called dolomitic sandstone; limestone containing considerable sand is called sandy limestone, and so forth. The general rule is to base the rock name on the most abundant constituent.

The textures and names in Table S2-1 have great significance with regard to the processes by which the grains were transported and deposited. Especially characteristic of sedimentary rocks is the layering produced by long-term fluctuations in the energy or nature of the transporting systems. Coarse-grained sediments generally indicate highly energetic (swift, turbulent) parts of transporting systems, and fine-grained sediments indicate the less energetic parts. The abrasion (rounding) of grains is another significant textural feature, expressing the duration of transport. A particularly valuable attribute of texture is the degree to which the grains approach being one size, for some transporting agents sort and winnow mixtures of grains so that deposits are well sorted, and others do so to a lesser degree or not at all. The first eight chapters of the book describe modern sediments produced by a variety of eroding and transporting systems, and the descriptions that follow mention some of these specifically.

Sediments are consolidated into sedimentary rocks by processes described and illustrated in Chapter 10, which can be summarized briefly: (1) pressing together of grains, thus squeezing pore water out of them (compaction); (2) growth of new mineral grains around the original sediment grains (cementation); and (3) recrystallization of the original grains into new, interlocking groups of grains—an important process in clays and in the rocks in Table S2-2.

DESCRIPTIONS OF SEDIMENTARY ROCKS   The rock descriptions that follow are arranged alphabetically.

Chert   Chert is composed of opal or quartz grains that are so small that the rock has a vitreous or waxy appearance. A hard, chemically resistant rock, it is etched into prominence where it occurs in limestone and often forms resistant pebbles and sand grains in fragmental rocks. Layered cherts typically form from loose accumulations of shells made of silica—chiefly those of diatoms (Figure 10-4) and radiolarians (Figure 20-12). The initially loose aggregates gradually solidify after burial, some becoming hard but permeable rocks with the appearance of unglazed porcelain and some becoming so tightly cemented with added silica that they are tough vitreous cherts.

Claystone   This is a well-sorted fragmental rock consisting of clay minerals and of clay-size grains of quartz, mica, and other minerals. Clay gives the rock a waxy appearance and makes it feel smooth and slippery when wetted and rubbed. Claystones imply the erosion of well-developed soils and a transporting system that can sort very fine materials from coarser ones. Deposition may take place in deep, quiet water, in shallow water in bays, among the plants of marshy estuaries, or in swamps and lakes.

Conglomerate and Breccia   The coarsest fragmental rocks are called conglomerate when the fragments are at least somewhat rounded and breccia when they are angular. Most of these rocks are poorly sorted because of fine sediment that lodges among the larger fragments; however, some are far better sorted than others (compare Color Plate 7 with Color Plate 10 and Figure 15-22). They can form in a variety of situations but generally imply steep slopes and energetic currents. Because the fragments are large enough to be identified as rocks, they provide a valuable means of identifying source areas of sediments.

Diatomite   Loosely to moderately consolidated deposits of diatom shells are characteristically very light in weight, light in color, and with such large amounts of pore space that they are highly absorbent. See Figure 10-4 and the description of chert above.

Dolomite (or Dolostone)   The mineral dolomite, $CaMg(CO_3)_2$, is not normally precipitated from sea water, yet rocks consisting of it are quite common, especially in sedimentary sequences more than 100 million years old. The fact that many contain fossils that were once $CaCO_3$ shows that Mg^{++} ions have replaced Ca^{++} ions in the originally calcitic sediments. Many occurrences of limestone partially converted to dolomite support this idea. In lagoons where sea water is strongly evaporated and therefore unusually salty, the replacement occurs directly on the ocean floor. Some other dolomites are formed by replacement of Ca^{++} by Mg^{++} long after burial, perhaps because of the high salinity of some deep-seated groundwaters.

Gypsum   Where bodies of water evaporate rapidly, dissolved minerals may become so concentrated that they precipitate out in crystalline form. The first of these minerals is generally calcite, and the second is gypsum. Gypsum thus occurs as layers associated with limestone and fine-grained fragmental rocks, and it commonly is mixed with these substances. Many silts and clays deposited in saline waters contain separate crystals of gypsum (Figure S1-15).

Ironstone This name is used for a variety of sedimentary rocks consisting mainly of iron silicate, iron carbonate, and iron oxides. These rocks are characterized by their high density and by their color, which may be green, yellow, black, or red when the rocks are fresh but becomes rusty brown after the rocks are exposed. Bright green grains can be seen in some varieties, and some others consist of smoothly rounded grains that have accreted in concentric shells (Color Plate 13D). Ironstones generally form in shallow parts of the sea or in swamps and lakes; they are often associated with limestone and fine-grained sediments containing dark plant materials.

Limestone This widespread and variable rock owes its origin mainly to the crystallization of $CaCO_3$ in the hard parts of marine animals and certain of the algae. In reefs, these substances form solid rock masses due to the growth of the organisms, but most limestones represent accumulations of shells and fragments that are cemented and compacted after burial. Some, such as chalk, remain porous aggregates, and some recrystallize to coarse-grained rocks with only traces of their original organic particles. As described in the first part of Chapter 4, the organic particles may provide a means of identifying the environment of deposition. Limestones may also form by inorganic processes: by the erosion and deposition of grains of preexisting limestone or by the accretion of $CaCO_3$ on small round grains called pellets or ooliths (Figure 10-3). Slightly weathered and etched surfaces of limestones may show these various features more clearly than freshly broken surfaces.

Mudstone This rock results from the consolidation of mud, which is typically a poorly sorted mixture of silt and clay particles, often with abundant sand grains (Figure 10-15A). Mudstone typically breaks in angular chunks or spheroidal shells rather than in the flakes typical of shale. The grains of silt and sand may be hidden by the clay but give a gritty feel when rubbed against the teeth.

Phosphorite This rock results from marine accumulations of (1) fragments of bones and teeth; (2) certain phosphatic shells; (3) fecal pellets; and (4) inorganically precipitated apatite, which often accretes around grains where currents roll them enough to promote concentric growth (Figure S1-11). Phosphorite is generally associated with fine-grained limestone, shale, and chert rich in organic compounds. Phosphate evidently accumulates where organic detritus is abundant but under conditions prohibiting its use by other organisms, for example, where the content of dissolved oxygen is unusually low.

Salt When a body of sea water has been evaporated to about 10 percent of its original weight, halite begins to precipitate. Salt rock thus forms beds in sediments accumulated in strongly evaporating environments, such as in a shallow arm of the sea or in a saline lake. It is often associated with gypsum and fine-grained fragmental rocks.

Sandstone Sandstones form in a variety of environments. They are especially valuable rocks to study because their grains can be seen easily and can often be interpreted with respect to origin. Significant varieties are: (1) *arkose,* composed of feldspar, quartz, and mica and implying the rapid erosion of a terrain underlain by granite or gneiss; (2) *quartz sandstone,* composed of little besides quartz grains and thus indicating a source region with abundant sandstones or one so thoroughly weathered that silicate minerals other than quartz are reduced to clays and iron oxides (Figure 15-10 top); and (3) *graywacke,* a poorly sorted sandstone in which abundant clay and silt give the rock a dark color and in which the grains tend to be angular and to include a variety of mineral and rock fragments (Figure 20-14). Graywacke thus indicates rapid erosion and accumulation in layers that are not reworked and winnowed before being covered by another layer (as in the deposits made by debris flows and turbidity currents described in Chapter 6). The rock implies rapid uplift—mountain-making movements.

Examples of layer structures of sandstones and associated sediments formed by rivers are described in Chapters 3, 13, and 14; estuarine and shallow marine sands are described in Chapters 4, 5, 10, and 14; and wind-deposited sands are described in Chapters 4 and 8.

Shale Some sedimentary rocks consisting mainly of silt and clay split in thin sheets and flakes essentially parallel to layering (Figure 14-12). Such varieties are properly called shale. This name, however, is used by some geologists for all varieties of mudstone, siltstone, and claystone.

Siltstone Many rocks composed of silt are too fine grained to study without a microscope; however, the coarser silt sizes can be seen with a hand lens. A common characteristic of siltstones is that easily visible flakes of mica or broken shells tend to be sorted to these fine aggregates because of the large surface areas of the flaky grains.

Metamorphic Rocks

Metamorphism entails changes in preexisting rocks caused by heat, pressure, deformation, and the effects of chemically active fluids. These effects are almost always combined to some degree, as in these three common situations: (1) where rocks are deeply buried and thereby subjected to considerably increased temperatures and pressures (Chapter 10); (2) where rocks at any depth are intruded by bodies of melt that heat them, often deform them, and alter them with hot fluids (Chapter 12); and (3) where rocks are strongly deformed and are heated concurrently (Chapter 15).

In all cases, metamorphism causes the original mineral grains to crystallize to new grains. If crystallization takes place while the rock is being deformed, rotation of platy or elongate grains into more or less parallel orientations produces a foliated (flaky) texture. In contrast, rocks that are not deformed develop randomly oriented grains and therefore a granular texture. In both types of metamorphic rocks, some mineral grains may grow to larger sizes than others. Fluids in rocks may also promote growth of large grains, often producing sheets, rounded masses, or irregular groups of grains that differ in size from those in the rest of the rock. We have already noted, too, that some grains may grow in more geometrically perfect crystals than others.

FOLIATED ROCKS The first group of rocks that follow are those with foliated (platy or fibrous) textures, and they are presented in order of increasing grain size.

Slate The finest grained of the foliated rocks typically represents the first stage in the metamorphism of fine-grained sedimentary rocks and of some tuffs. The new metamorphic grains are too small to be visible, but the parallel growth of their flaky forms makes slate split readily into sheets or elongate slivers. Slate is harder and tougher than shale and commonly splits oblique to sedimentary layering. It is dull and is generally gray but may be green, red, or lavendar.

Phyllite Most grains in this foliated rock are too small to be seen without a microscope, but it splits in sheets that have a sheen due to the cleavage surfaces of minute mica and chlorite flakes or of amphibole needles. Some phyllites have scattered grains of distinctly larger size that give surfaces a spotted or slightly knobby appearance. Phyllite generally results from somewhat more advanced metamorphism than that producing slate and forms from the same kinds of rocks.

Schist Probably the most widespread metamorphic rock, schist consists of visible grains of mica, chlorite, or amphibole as well as more equant grains of quartz, feldspar, or other minerals. Because amphiboles are needle-like or prismatic, they form lineated, splintery schists (Color Plate 12A). Many schists have scattered grains of large size, garnet being especially common (Color Plate 12B, Figure 10-15E).

Greenschist This schist consists largely of chlorite, epidote, and actinolite (Color Plate 12F). It forms by the low-temperature metamorphism of basalt, andesite, or dolomitic shale.

Blueschist This schist contains the blue amphibole glaucophane and a variety of other minerals, chiefly chlorite, mica, garnet, quartz, and plagioclase (Color Plate 13A). It forms at low temperatures but high pressures from a variety of lavas and sedimentary rocks. The rock implies rapid subduction and deformation without much heating.

Gneiss This rock consists of small lenses, streaks, and layers of oriented flaky or prismatic minerals distributed through an abundant matrix of granular minerals (Figure 10-16). In most gneisses, the oriented groups consist chiefly of biotite, and the granular minerals are feldspars and quartz. These rocks form by the advanced metamorphism of fine-grained sedimentary rocks, sandstones, granite, and silica-rich lavas. Gneisses consisting of hornblende and plagioclase or of pyroxene and plagioclase result from the metamorphism of basalt, gabbro, andesite, and dolomitic shales. Garnet is a widespread mineral in gneisses.

GRANULAR ROCKS   The rocks that follow are those with granular textures.

Hornfels   Most of the grains in this rock are too small to be seen without a microscope, but their lack of parallel orientation is indicated by the rock's breaking with equal difficulty in all directions. Its toughness is due to the interlocking of the metamorphic grains. Most varieties of hornfels are dark gray, but some calcareous ones are pale tan or green. The broken surfaces have a minutely hackly, rough appearance.

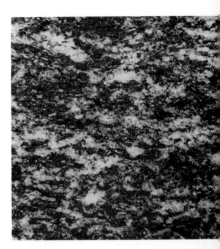

FIGURE S2-10
Amphibolite consisting mainly of hornblende (black) and plagioclase (white) and showing a distinct foliation. Actual size.

Mylonite   This is a special kind of granular rock in which the fine-grained parts are due to strong, rapid deformation of larger grains that then recrystallized into aggregates of very small ones (Figure 16-8). Some large grains appear to be rolled and smeared, for large rounded relics commonly lie in the fine, streaky matrix. The large relics are tyically feldspar, garnet, or pyroxene but never quartz, which is one of the first minerals to break down mechanically when rocks are deformed strongly. Mylonites form as sheets along deeply buried fault surfaces.

Granofels   This is a useful general name for visibly granular metamorphic rocks; however, many varieties are named more specifically on the basis of mineral content, as in the five rocks that are described next.

Amphibolite   This granular or weakly foliated rock consists of hornblende and plagioclase and typically forms from basalt, gabbro, or dolomitic shale (Figure S2-10).

Quartzite   This granular rock consists dominantly of quartz and is therefore a metamorphosed sandstone, siltstone, or chert (Figures 15-10 and 15-11A). Other minerals commonly present are micas, chlorite, calcite, or dolomite.

Marble   This granular or slightly foliated rock consists dominantly of calcite (if of dolomite, it is called a dolomite marble). Quartz, mica, and feldspars are commonly present. Abundant pyroxene or olivine (perhaps altered to sepentine) indicate that the original rock was a sandy or cherty dolomite and that the mineral dolomite reacted with silica during metamorphism to produce magnesium-rich silicates and calcite (Color Plate 14F).

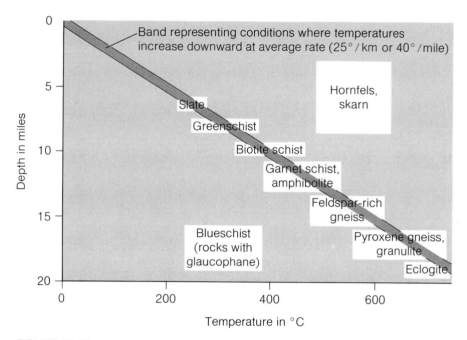

FIGURE S2-11

Diagram illustrating pressures (depth) and temperatures at which some common metamorphic rocks are thought to form. The band showing the increase of temperature with depth is for the average case, so that actual depths may vary considerably from one region to another.

Granulite Water-free minerals such as pyroxene, garnet, and feldspar characterize this granular to moderately foliated rock, thus indicating exceptionally high temperatures and deep burial.

Eclogite This granular rock has the composition of gabbro but consists of pyroxene and garnet *without feldspar*. Lack of feldspar is thought to reflect exceptionally high pressures and thus an origin in the mantle.

ROCKS WITH MIXED TEXTURES The rocks that follow have mixed textures, typically resulting from the action of fluids during metamorphism.

Skarn This unevenly granular rock forms from limestone containing various silicate materials and lying near an intrusive igneous body, typically granite or granodiorite. Silica, iron, and other components added by fluids from the intrusion cause the growth of garnet, pyroxene, or epidote and sometimes add metallic sulfides such as pyrite, chalcopyrite, and galena, in some cases making a valuable ore. The heterogeneous mixtures of minerals that result often include some well-shaped crystals (Color Plate 12H).

Migmatite This is a mixture of granite, granodiorite, or quartz diorite with schist or granular metamorphic rocks. The separate bodies are typically several inches thick, or less, and have a great variety of shapes, though many lie parallel to the foliation of the metamorphic host rock (Figure 10-17).

GRADE OF METAMORPHISM Most of the textural varieties of metamorphic rocks may form under a wide range of temperatures and pressures. The minerals present in the rocks help in determining these conditions more specifically because each set of minerals is most suitable under a limited range of temperatures and pressures. Chapter 10 presents this concept for metamorphosed mudstone. Textures also are helpful in estimating metamorphic grade. Typical temperatures and pressures suggested by the more common metamorphic rocks are shown in Figure S2-11.

Maps and Aerial Photographs

Map Scales and Orientation
Contours and Topographic Maps
    Drawing a topographic profile
    Topographic maps of the U.S. Geological Survey
    Interpreting landforms from topographic maps
Aerial Photographs
Geologic Maps
    Faults on maps
Mapping Rocks
    Measuring strike and dip of layers
    Constructing a cross section

Appendix A.
Maps and Aerial Photographs

In most earth studies it is exceedingly important to see the geographic distribution of certain kinds of information. Thicknesses of soils, amounts of rainfall, orientations of tributary streams, shapes of lakes, and dimensions of rock bodies are but a few examples. Topographic maps and aerial photographs show some of these things directly and can be used as a base on which to record other kinds of data.

Map Scales and Orientation

The great value of maps and aerial photographs lies in their being accurately scaled-down images of the earth's surface. Features on them are much smaller than the actual features on the ground, but their sizes relative to one another are the same and the directions between any two of them are the same. These relations on aerial photographs are produced by the camera lens, which is pointed vertically downward at the earth's surface. Like a small photograph of a person, the aerial photograph reproduces the features of the earth faithfully. Maps are accurately scaled by first surveying the positions of many points on the ground. The surveyed points are then used to control details added from aerial photographs or from more detailed surveys on the ground.

The final map, then, has the same *scale* throughout. The scale is shown in the margin of the map as a bar with increments of distance marked on it and usually also as a ratio between the size of a feature on the map and its size on the ground. On a map having the scale of 1:24,000, for example, 1 inch is equivalent to 24,000 inches or 2,000 feet on the ground.

In addition, maps indicate geographic orientation by a *north arrow* in their margins and by having their edges oriented north-south and east-west. We can thus use a protractor to measure accurately the orientation of any feature on the map. We can also go to the field and plot additional features on the map. As an example, the scratches made by a former glacier on exposed rock surfaces could be mapped by the following steps:

1. Measure the orientations of the scratches with a compass.
2. Locate the various exposures on the map.

3. Use a protractor and the orienting lines on the map to plot short lines representing the orientations of the scratches.

The map will then show the directions in which the glacier flowed across the area. It will also show the actual locations of the data, so that someone else may easily find and examine the scratches.

An important point is that maps and aerial photographs are things to be *used*. They have great value, and the purpose of this appendix is to describe them and to mention some of their uses.

Contours and Topographic Maps

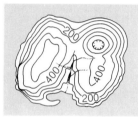

Contour interval = 100 feet

FIGURE A-1

Oblique aerial view of a small island and a map depicting its form by contours spaced at intervals of 100 feet.

The earth's vertical dimensions are not as obvious on maps and aerial photographs as are its horizontal dimensions. They can, however, be read accurately and even visualized by using maps called *topographic maps*. These maps have specific elevations, given in feet above sea level, printed at scattered localities. They also show the elevations of the ground surface elsewhere by lines called *contours*. A contour is a level line across the surface—like the line we might walk if we were to stay at one elevation and complete a loop around a hill. The contours shown on a map, however, are not just any level line; each is at a prescribed elevation above sea level and at the same *vertical* distance from the two adjacent contours. For example, the first contour inland from a coastline might be at an elevation of 40 feet above sea level; the next would then be at 80 feet, the next at 120 feet, and so on. The vertical distance between adjacent contours, called the *contour interval*, is selected from among the commonly used values of 5, 10, 20, 40, 100, and 200 feet on the basis of the scale, the purpose of the map, and the steepness of the slopes in the area. Some maps have a larger contour interval in mountainous areas and a smaller one in areas with subdued features.

To get some feel for how contours show topographic forms, examine Figure A-1. To sense each contour as level, try imagining that the island was once totally submerged and then sea level fell in increments equal to the contour interval. Each contour would thus have been a shoreline at one interval of emergence. Note how the spacing of contours shows that some slopes are gentle and some steep; try sensing the general change in surface shape around each ridge. Why do three contours become a single line at the vertical cliff? Note how the closed depression is indicated by contours with small lines (*hachures*) pointing downslope—a standard indication of closed depressions on topographic maps. The completeness of the map in some respects is remarkable. Follow, for example, the 100-foot contour from the ridge on the right into the valley. Note how it indicates a flat surface, in this case a floodplain, by the way it extends straight across the valley floor. Note, too, how it jogs far up the valley and back—expressing a trench cut by the stream into the floodplain.

In spite of all that is shown, however, perhaps you can see some limitations to contours. If you try to determine the height of the cliff, for example, you cannot be sure of the elevations at the base and top because contours "miss" all points between the prescribed contour elevations. The elevations of summits and hollows must thus be estimated from the contour nearest the summit or hollow.

Drawing a topographic profile To get a firmer feeling for contours and at the same time to explore one of their uses, we can construct a *topographic profile* of the island that will show its vertical dimensions more clearly. A profile shows the outline of a terrain as viewed horizontally, such as the view of land from a ship. It can be constructed by the following steps, illustrated in Figure A-2.

1. Prepare a sheet of paper by ruling a set of parallel lines equal to the number of contours on the island plus one for sea level, spacing them a contour interval apart at the scale of the map.
2. Draw a straight line across the part of the map you wish to see in profile.
3. Tape a strip of paper along this line and mark both shorelines precisely; then add short lines at each point where the line on the map crosses a contour, stream, summit, or other notable feature. Label each of these lines.
4. Remove the strip and lay it over the sheet of paper along the lowest of the ruled lines. Mark the positions of the shoreline points and draw lines from each across all the parallel lines, at right angles to them. Using these lines as guides to keep the strip in position, raise the strip to the second parallel line and mark points showing the positions of the first contour. Then raise the strip to the next line and mark the points for the second contour, and so forth.
5. When all the contour points, streams, and other features have been transferred to the lined sheet, connect them with a thin line, allowing for the extra height of ridge crests and the extra depth of valleys. This is the profile along the line drawn on the map.

FIGURE A-2

Constructing a profile through the island shown in Figure A-1, in steps described in the text.

If you try doing this along another line of your own selection, you may feel a certain imprecision in the method because of the distance between

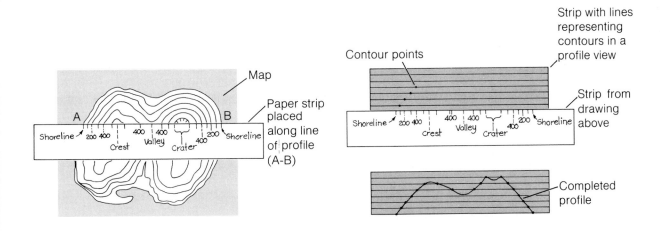

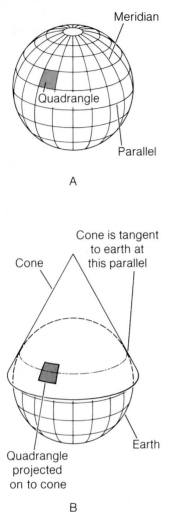

A

B

FIGURE A-3
A, a quadrangle on the spherical surface of the earth. B, quadrangle projected onto a cone that is tangent to the earth along the parallel passing through the quadrangle.

widely spaced contour lines and because of the approximate nature of the pencil work. Perhaps you will feel less apprehensive by learning that the contours are not all exactly in place on the map! The standards for topographic maps of the United States require that 90 percent of all mapped features be no more than 1/50th of an inch (0.5 millimeter) out of position and that 90 percent of all contours be less than $\frac{1}{2}$ contour interval off in elevation. Maps are thus somewhat approximate—as to some degree are all constructions and data based on measurements. We must thus work with reasonable care with profiles and other measurements from maps but must not assume that split-hair accuracy will make our constructions an absolutely true view of the earth's surface.

Topographic maps of the U.S. Geological Survey Topographic maps available for purchase by the public are published by the U.S. Geological Survey. These maps are often called *quadrangle maps* because they are bounded by four straight sides—two *meridians* (lines of longitude) and two *parallels* (lines of latitude) (Figure A-3A). Quadrangles thus form horizontal rows along lines of latitude and vertical sets along lines of longitude, so that the individual maps may be assembled readily into larger maps. The lines of latitude are parallel, but the meridians converge slightly toward the top (north end) of each map in the northern hemisphere (Figure A-3A). The maps are constructed as though projected outward from the earth's surface to a cone tangent to the earth at the latitude of the map (Figure A-3B). Because maps at each latitude are based on a separate cone, the projections are called polyconic (many-coned) projections. They constitute one of the most successful means of constructing flat maps from the spherical earth. Note that the convergence of meridians allows the scale to be the same at the top and the bottom of the map and on all maps in a wide region. At the same time, however, the directions toward north change slightly from the right to the left side of the map, though not enough to effect the uses we generally make of the maps.

Most quadrangles made today have scales of either 1:24,000 (1 inch = 2,000 feet) or 1:62,500 (1 inch = 1 mile). Each is named after some prominent feature on it. Color Plate 15 shows parts of a quadrangle in central Maine. The colors are standard for all topographic maps: brown for topography (contours and hachures); blue for all water features and glaciers; green for vegetation (omitted here because it tends to obscure the contours); and black, red, or lavender for features made by humans. Note that every fifth contour is heavier than the others and its elevation is labeled. These are called *index contours.* Many additional kinds of features are shown by standard symbols on other typographic maps, as illustrated in the pamphlet *Topographic maps,* free on request from the Map Information Office, NCIC-U.S. Geological Survey, 12201 Sunrise Valley Drive, Reston, Virginia 22092. Information on the availability and purchase of quadrangle maps may also be obtained from that office.

Interpreting landforms from topographic maps One can often determine the processes that have sculptured the land surface by examining the landforms shown on topographic maps. In Color Plate 15, for example, the

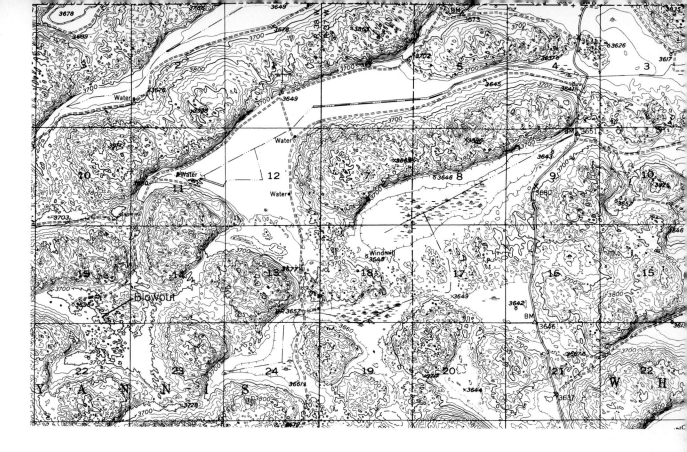

FIGURE A-4

Part of the Whitman quadrangle, central Nebraska, published in 1950 by the U.S. Geological Survey at a scale of 1:62,500 (1 inch = 1 mile).

elongate lakes and the parallel streamlined forms of the ridges and valleys express the approximately southward flow of the great ice sheet that eroded Maine during the last glaciation. The map thus gives a closeup view of erosional features such as those described in Chapter 7. By examining the map and the north arrow, what would you estimate to have been the exact direction of ice movement in this particular part of Maine?

Some very different topographic forms are shown in Figure A-4, which is part of a quadrangle map of the Nebraska Sand Hills, the gaint sand dunes described in Chapter 8. This map has a smaller scale than the one in Color Plate 15; each square (called a *section*) is a mile on a side. The contour interval is 20 feet, and north is toward the top of the map. The labeled contours are difficult to read, but you can determine the height of the dunes by using the specific elevations recorded here and there. You can also tell from the spacing of the contours that the southeast sides of the dunes are steeper than the other sides. What does this tell us about the wind direction during the period in which the dunes formed? Perhaps you can also see small elongate ridges and hollows oriented approximately northwest to southeast across many of the dunes. Note the flatness of the blowout areas between the dunes. An excellent example of a large blowout "frozen in action" is labeled in the lower left part of the map.

Examples of still other kinds of landforms that can be interpreted from topographic maps are shown in Figure A-5, part of a quadrangle that shows the south end of Death Valley, California. This area is shown in the photograph on the opening page of Chapter 9, and the features are like those described in the first part of the chapter. The bulging arcs of contours on the valley floor opposite each canyon mouth are alluvial fans; the smoothly curving

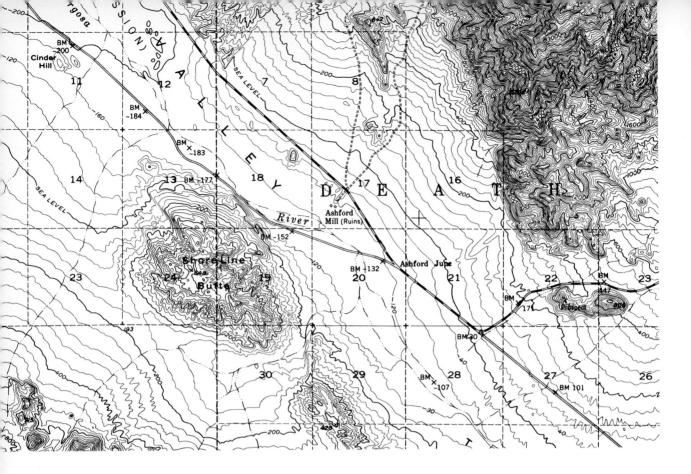

FIGURE A-5

Part of the Confidence Hills
Quadrangle, California, published
in 1950 by the U.S. Geological
Survey at a scale of 1:62,500
(1 inch = 1 mile).

groups of contours are alluvial fans presently active over most of their surfaces;
and the groups of irregular contours are fans being dissected along gullies. Note
that part of the valley is below sea level (the contours with a minus sign before
their elevation numbers). Note, too, that the rugged mountain peaks only a
few miles away are commonly 3,000 feet and more above sea level—striking
evidence of the rapid fault movements that formed the valley. If you compare
the map and photograph carefully, perhaps you can find relatively straight
lines along the edges of hills (near the center of the map) or mountains (along
the north edge) that are the surface expression of faults.

Aerial Photographs

The most common kind of aerial photograph and the one most used in earth
studies is called a *vertical photograph*—one taken from an airplane with a
camera pointed straight down at the earth's surface. These photographs have
many of the qualities of good maps and have endless additional details that
help in many studies: gully systems, waves, patterns of trees, widths of beaches
at specific tidal stages, and so on. Examples are shown in Figure 4-5, 5-3, and
the opening page of Chapter 7. Other photographs, taken with the camera
pointing down at some angle less than 90°, are *aerial obliques*. These views
are particularly valuable when taken with a wide-angle lens that includes the
full range from the horizon to the vertical (at the near edge of the photograph).
An example is Figure 7-1. Aerial obliques are unlike maps in that their scales
change greatly from foreground to distance; however, they are very useful

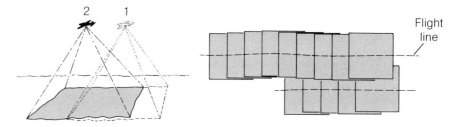

FIGURE A-6

Positions of aircraft when two successive photographs are taken, thus giving large overlaps between photographs, as shown below.

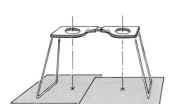

Photographs and stereoscope aligned parallel to flight line

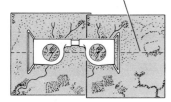

FIGURE A-7

A simple stereoscope shown in position to give a stereoscopic view of an area shown on each of two photographs.

because they show great tracts of country with a naturalness due to the three-dimensional appearance of the landforms.

Vertical photographs are taken at regular intervals from aircraft flown in parallel flight lines, and the lines are spaced to give overlapping strips of photographs (Figure A-6). Note that the photographs also overlap by 60 percent or more along each strip. Each pair of adjoining photographs therefore shows a sizable area photographed from two different positions, which makes it possible to see these areas as *stereoscopic images*—views in which vertical dimensions appear real. The principle is the same as for the stereoscopic image we see because each eye obtains a different view of an object. In fact, if one looks at one of two overlapping photographs with one eye, and the other photograph with the other, a stereoscopic view results. This viewing can be done with practice if the photographs are aligned exactly as they were taken and placed to the right and left of one another, with corresponding points about 2 inches apart. It is far easier, though, to see the image with a *stereoscope*, which has lenses that direct each eye, without effort, to the photo image beneath it (Figure A-7). The relief (appearance of the third dimension) is typically two to three times greater than normal, so that all landforms become distinct and even small features such as buildings and trees show noticeable height.

As dramatic and useful as these views may be, however, the photographs lack the contours that would make it possible to *measure* the vertical dimensions of landforms. Aerial photographs also lack north arrows from which to measure the orientations of linear features, as well as scale bars from which to measure distances accurately. One can overcome these disadvantages by using photographs in conjunction with a topographic map of the same area. A north arrow can be added to any photograph by finding a road or other straight feature on the ground or on a map, measuring its bearing (orientation relative to the north arrow), and using this bearing to determine north on the photograph. Similarly, the scale of the photograph can be determined by comparing the length of a feature on the photograph with the same feature scaled on the ground or on a map.

Aerial photographs can be obtained for most parts of the U.S. by (1) inquiring as to coverage of a specified area by writing to the Map Information Office, NCIC-U.S. Geological Survey, 12201 Sunrise Valley Drive, Reston, Virginia 22092; (2) purchasing a *photo index* (a photo copy, at reduced scale, of a layout of photographs such as the one shown in Figure A-6); and (3) selecting and ordering the photographs needed by referring to the index (which indicates photograph numbers, approximate scale, dates of the flights, and who has the photographs). The entire procedure is likely to take several weeks or longer.

Geologic Maps

Geologic maps show the distribution of the various rocks at the earth's surface and are thus a basic source of information for earth studies. Indeed, geologic maps typically include so much information that they are difficult to describe without referring to an example. The description that follows refers to part of the Little Hickman quadrangle, central Kentucky (Color Plate 16).

As you can see, the different rocks are indicated by colors that correspond to the formations named in the explanation on the right. A formation is a physically distinctive rock unit, one that can be recognized relative to adjoining formations (Chapter 13). All the formations in this quadrangle with the exception of one consist mainly of one kind of rock, as shown by the names. The exception, the Clays Ferry Formation, is a mixture of interlayered limestone, shale, and siltstone in roughly equal proportions. All the formations are described on another part of the published quadrangle sheet. On most other published maps they are described in the map explanation or in a bulletin or other document published with the map. The descriptions show how the geologist distinguished any two adjoining formations. These descriptions are especially important in the Little Hickman quadrangle because most of the formations are limestones.

A great value of the map is that we can carry it to the area and find the rock formations, because the formations and other features have been printed over the topographic map. At first this overprinting makes the map appear involved and confusing, but if you study the topographic lines first in a comparatively simple place, such as in the pink area on the right, you can get used to them quickly. Note the brown contour lines, numbered locally, which define groups of rounded hills. The blue lines of streams are especially helpful in seeing positions of valleys. Roads, buildings, and geographic names, which would be shown in black on a contour map, have been grayed so as not to be confused with the geologic lines and symbols, which are shown in black. Try examining the more complex area in the upper left part of the map. The sinuous blue band is Hickman Creek (named elsewhere on the map), and you will have to look closely in order not to confuse it with the blue areas of Tyrone Limestone. Note how the yellow band of alluvium (the floodplain deposits of the creek) beautifully defines the creek's sinuous valley. Look closely at the spacing of contours along both sides of the stream valley. Do you see that there are steep bluffs along the outer sides of most of the bends?

Faults on maps Perhaps the most noticeable things in the complex part of the map are the heavy black lines, which are faults. Notice that the colored area of each formation ends abruptly at each fault line. The abrupt endings express the offset of the rock formations by faulting. Indeed, the fault lines were probably located by mapping the distributions of the rocks and thereby finding the line along which each one suddenly ended.

Perhaps the faults will be easier to visualize if you now look at the cross section under the map, which is a constructed view of rocks beneath a topographic profile. You might think of the cross section as a deep vertical cut made along the line B-B' on the map. You are thus looking northeastward at the vertical side of this cut. If you compare the vertical section with the areas of formations along the line B-B' on the map, you will find each one exactly in place. The easiest way to check the positions is to lay the edge of a piece of paper along the line B-B' on the map, make pencil ticks at each end (at B and B'), and then mark each contact and fault similarly. In this way you can be sure that the three faults shown on the cross section are indeed those on the map.

The steep inclination (called the *dip*) of each fault is obvious on the cross section and is suggested on the map by the way the fault lines cross the terrain. No matter how steep the slopes or what their direction, the fault line remains nearly straight. Just the opposite is shown by the boundary lines (called *contacts*) between the formations. Notice that they run nearly parallel to the contours and must thus be level or nearly so—a relation shown clearly by the cross section. If you look closely at the top of the Tyrone Limestone (the line between the blue and brown areas) near the center of the map, you should be able to see that its elevation increases, for it crosses successively higher contours toward the northwest (left). Note that this relation is also expressed by the cross section. The down-faulted parts of the formations thus lie near the center of a low arch expressed by the inclination or dip of the formation layers.

Neither the map nor the cross section tells us when the formations were arched and faulted, but they do establish one important point about their age. If you follow the fault lines for a few inches, you can see that they are represented by dots in areas of alluvium and terrace deposits that are former floodplain deposits of the same creek. The dots mean that the deposits cover the fault and are not offset by it. Thus, no movement has taken place on the faults in recent geological times, and it is unlikely that these faults are active and pose a hazard.

The red lines shown on the map are called *structure contours*. Try to imagine the mapped area before erosion cut down nearly as deeply as it did to form the present landscape. Now imagine stripping away all the formations down to the top of the Tyrone Limestone. If you now drew contours on the surface, in feet above sea level, they would be the red lines on the map (Figure A-8). They are called structure contours because they express the structural shape of the rock formations. They show, for example, that the formations rise broadly from the southeast toward the faults. They show the same relations as the cross section does, only they show them over the entire map area. This information is of great value in predicting the movement of groundwater along permeable formations (Chapter 2) and in locating domes in the formations that might trap petroleum and natural gas (Chapter 10).

Mapping Rocks

Much of the U.S. has yet to be mapped in the geologic detail needed in many earth studies, so that sometime you might want to make your own geologic

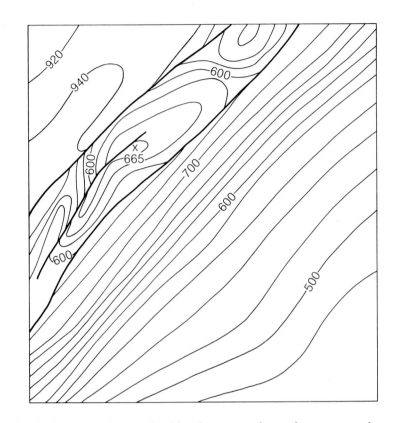

FIGURE A-8
Structure contours (thin lines) and faults
(heavy lines) traced from Color Plate 16
to show the general form of the contoured
surface (the top of the Tyrone Limestone,
before it was partly eroded away).

map. It is an intriguing procedure and evidently a natural one, for most people
get the knack after a few days' practice. The main reason probably is that the
rocks within a small area are always highly specific varieties; once one has
seen them in dozens of exposures—fresh and altered, wet and dry—one can
recognize them quickly anywhere. By mapping them one also gets an intimate
knowledge of their relations to soils, landforms, vegetation, and various
geologic features. Indeed, when you have completed a detailed map of a few
square miles, you will feel that no one else has ever gotten to know that
land better!

The materials needed are not involved or expensive: a topographic map or
aerial photograph on which to plot what you find; a folder to carry it in; a
notebook in which to record data and ideas; a hammer to break rocks—preferably with a pick end (a geologist's hammer) to pry and dig at exposures; a
hand lens to examine materials; some bags for samples; a scale (ruler) for
measuring distances on the map and small features at outcrops; a protractor to
measure and plot the inclination (dip) of layers; and a knapsack to carry the
lot. A compass and clinometer are also very helpful but not essential.

The general procedure is to examine exposed rock wherever you can find
it and to plot these places on your map, applying a colored dot or a letter
symbol to indicate what rock is at each place. Because one's aim is generally
to understand the geologic *history* of the area, one also describes and sketches
features that suggest the relative ages of rocks, how certain rocks formed, how
they were deformed, and possibly a number of other things (as suggested by

the studies described elsewhere in this book). If the rock formations you are tracing have already been described and named, read about them ahead of time and then see if they are indeed the same in your area (often they will be somewhat different). You can generally learn about published maps and bulletins from your state geologist or state geological survey.

You might first reconnoiter your area in order to see what rocks occur where. When you do this you can also get used to locating features on the map or photograph. Get in the practice of aligning the map so that it is oriented parallel to the surrounding terrain, for example, by turning it until a road shown on the map is parallel to the road on the ground. You can also align it with a compass. When it is oriented, locate nearby hills, streams, and buildings by glancing from terrain to map and back again. Note how the contours show the curving forms of hills and the breaks in their slopes. Practice locating your position closely enough so that you can mark it on the map with a pencil dot.

The main procedure now is to find rock outcrops. Start by looking along road cuts, streams, steep hillsides, and ridges. Locate the outcrops you find on the map, and mark them by a dot and some indication to identify the kind of rock (a color or letter symbol). When it becomes obvious that two kinds of rocks adjoin along a more or less definite line or course, you might concentrate on mapping this line systematically. This line will become a boundary or contact line like the fine black lines separating the colored areas in Color Plate 16, or it may turn out to be a fault. The actual boundary surfaces, which may be seen in only a few places, should be examined and described thoroughly in your notes.

Over large areas, however, the contact lines will have to be located approximately between outcrops of the two kinds of rock. Where outcrops are scarce, you can often use pieces of rock in the soil or the nature of the soil itself to see where the contact line must go. Burrowing animals often bring up fragments of the underlying rocks. As shown in Figure A-9, pieces of the lower of two formations can be used to locate a contact on hillslopes but not pieces of the upper formation. Play the sleuth! Walk a zig-zag course between the closest exposures of each rock formation and as you do so draw a line along the course of the hidden contact (Figure A-10). Once the line has begun to show a consistent direction you may be tempted to draw it ahead of your position, but you should realize that it may be offset by a fault or may suddenly curve along the form of a fold.

Measuring strike and dip of layers   It will be of great help to measure and plot orientations of rock layering in the formations, for the layering will be parallel to the contact surface if the formations lie on one another in a typical sedimentary sequence. The orientations of layering are measured and recorded as two lines: one showing the orientation of a horizontal line on the layers, called the *strike*, and one showing the inclination of the layers, called the *dip* (Figure A-11A). The strike line can be plotted by orienting the map carefully over the outcrop, sighting along a horizontal line on a layer surface, and drawing a line on the map parallel to this direction (Figure A-11B). You can also use a compass to measure the orientation of the line and a protractor to plot it on the map. To measure the dip of the layer, get in a position so that

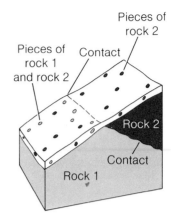

FIGURE A-9
Using pieces of float (fragments of rock in the soil) to locate the approximate position of the contact surface between two rock formations.

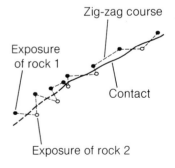

FIGURE A-10
Locating an unexposed contact approximately by walking a zig-zag course between exposures near the contact. The contact line is drawn solid where located closely and dashed where located approximately.

523

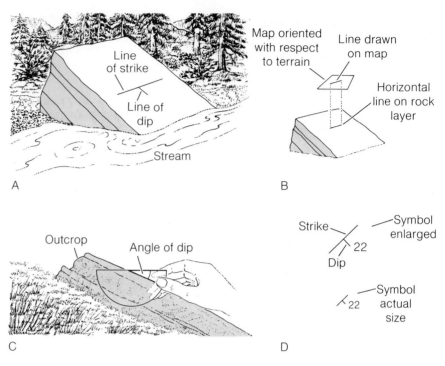

FIGURE A-11

A, exposure of layered rocks, with the directions of the strike (a horizontal line) and the dip (a line pointing down the slope) marked on a layer surface. B, by holding a map over the exposure and orienting it relative to the actual terrain, a line of strike and dip can be marked approximately on it. C, measuring the angle of dip (the inclination of the layers) with a transparent protractor. D, strike-and-dip symbol drawn on the map at the location of the exposure.

you are looking along a horizontal line (the strike) on a layer surface (Figure A-11C). As the figure shows, measure the angle of dip by holding the protractor at arm's length with its straight edge horizontal (parallel to the horizon). This measurement can be made more easily with a clinometer. The direction of dip is at right angles to the direction of strike and is plotted on the map by a short line on the downslope side of the strike line, with the angle of dip written at its end (Figure A-11D).

If strike and dip are plotted at scattered outcrops, spaced perhaps $\frac{1}{2}$ inch apart on your map, you will be able to see how formations trend across your area. Moreover, by comparing the orientations of the strike and dip symbols with the orientation of contact lines, you can interpret the nature of the contact surfaces. Figure A-12A shows the normal case in two layered formations, whereas Figures A-12B, C, and D show how symbols oblique to a contact indicate faults and Figure A-12E shows how they indicate an unconformity.

Constructing a cross section At any stage during the mapping you may want to see how the various features fit in the third dimension by constructing vertical cross sections such as those in Figure A-12. When the

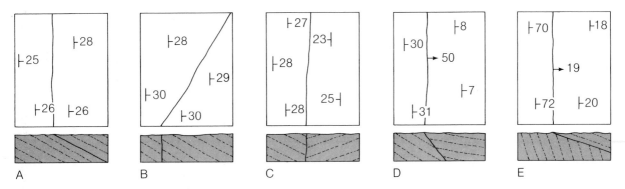

A B C D E

FIGURE A-12
Five maps and vertical cross sections showing how strike-and-dip symbols near a mapped contact line or fault can be used to interpret the surface underground. These possibilities must then be proven by finding an exposure of the surface itself so that its dip can be measured. A is thus a normal contact; B, C, and D are faults; and E is an unconformity.

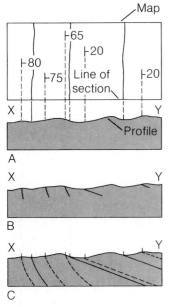

FIGURE A-13
A, projecting lines of strike from a map to a vertical section being constructed along the line X-Y. B, short lines plotted with a protractor to indicate the dip of layers. C, curving the lines so that each layer will have the same thickness along its dip.

mapping is completed (or seems so) you should draw cross sections across the entire area, such as those shown in Color Plate 16. The procedure is as follows:

1. Select a line on your map along which you want to see the configurations of the formations in vertical section. The most useful direction will be at right angles to the contacts, as the line B-B′ in Color Plate 16.
2. Construct a profile along this line by the method described in the first section of this appendix.
3. Using the same method, locate all points on the profile where the section line crosses contacts and faults.
4. Project nearby layer orientations to the line of section, by extending the strike lines of the strike-and-dip symbols (Figure A-13A).
5. Locate each of these points on the profile and use a protractor to construct the dip of layering, drawing a line about $\frac{1}{4}$ inch long beneath the surface (Figure A-13B).
6. If these lines are not nearly parallel, as in Figure A-11B, curve the layers beneath the surface so that the layered formations do not change in thickness (Figure A-13C).

From this cross section you can interpret some important events and features. The curving layers show that those formations were folded. It appears that the folded layers were then eroded and overlain by more layers and that all the formations were then tilted about 20° toward the east. There is also the possibility that the angular junction between the two sets of formations is a fault. The cross section thus causes you to return to the field to gather more information along this important line. Perhaps you will find crushed rock along it, suggesting it is a fault, or pebbles of the underlying rocks in the formation just above, suggesting it is an unconformity. In such ways a geologic history will evolve from your mapping.

Using Radioactive Isotopes to Determine Ages

The Radiocarbon (^{14}C) Method
The Potassium-argon Method
The Rubidium-strontium Method
The Uranium-lead Method

Appendix B.
Using Radioactive Isotopes to Determine Ages

The atoms composing each element have been classified into several varieties of somewhat different atomic weights. These varieties are called *isotopes*, and the weight differences are due to a difference of from one to four neutrons in the nuclei. As an example, nuclei of the most abundant isotope of carbon contain 6 neutrons, and nuclei of the next most abundant contain 7 neutrons. Because all isotopes of carbon contain 6 protons (the atomic number of the element), the atomic weights of the two isotopes are 12 and 13, respectively, and the isotopes are abbreviated ^{12}C and ^{13}C. The difference in weight causes the two isotopes to be slightly segregated in the ocean and the atmosphere, as described for isotopes of hydrogen and oxygen in the latter part of Chapter 7; however, the isotopes behave almost identically in chemical reactions.

Many isotopes, including the two just mentioned, are stable—that is, they remain unchanged. Certain other isotopes, however, are intrinsically unstable. The numbers of neutrons in their nuclei differ so much from an ideal "norm" that the nuclei break down (*decay*) to other nuclei. This change is said to be *radioactive* because it is accompanied by the emission of nuclear wave energy (*gamma radiation*) or of nuclear particles—*alpha particles* (consisting of two neutrons and two protons) or *beta particles* (nuclear electrons). As an example, the scarce carbon isotope with 8 neutrons in its nucleus (^{14}C) decays to nitrogen (^{14}N) by emitting a beta particle and thus increasing the nuclear charge from 6 to 7.

The rates of decay of many of the radioactive isotopes have been measured precisely and are essentially invariant, at least for all the physical conditions in the outer parts of the earth. Each radioactive isotope has thus been decaying at some specific rate since the formation of the rocks in which it occurs. Likewise, the products of decay have been accumulating at that rate. A convenient number with which to explore these changes quantitatively is the *half life* of the radioactive isotope—the time it takes for one half of the starting material to decay. During the first half life, the product of decay (the *daughter*) will increase from 0 to an amount equal to the remaining *parent* isotope (Figure B-1). During the next half-life period, the parent will again decrease by

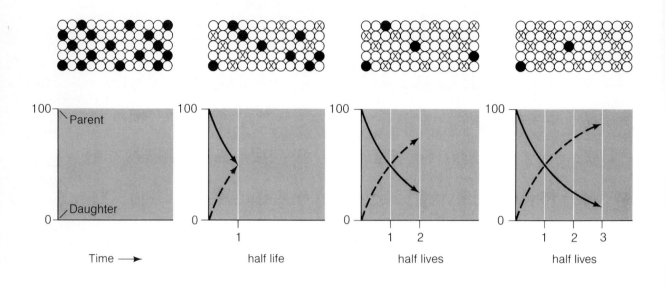

FIGURE B-1

Stages in the decay of a radioactive isotope (black) in a
mineral, with daughters labeled
X. The graphs chart the changing
proportions of parent and
daughter through three half lives.

one half, and the daughter will increase correspondingly, and so forth. The curves that can be constructed from the half-life quantities express the general decrease in the rate of production of the daughter due to the decreasing amount of parent. These are the basic curves for rock dating. If we can measure the amounts of parent and daughter in a suitable sample, we can use the rate changes expressed by the curves to calculate the age of the rock. Rocks dated essentially in this way have afforded some of the truly major advances in our knowledge of the earth and the solar system.

The methods used in measuring the parent and daughter and in calculating age vary from one isotope to another, depending on the nature of the substance and the history of the rock being dated. Some daughters, for example, are gases, which must be extracted by special methods and may be depleted or increased during their accumulation in the rock because of their mobility. The selection of a method also depends on the amounts of the isotopes in specific rocks. Another factor to consider is that some isotopes decay so rapidly that only young rocks can be dated by them, whereas other isotopes decay so slowly that they can be used only to date ancient rocks.

The methods described briefly here have been used in hundreds to thousands of cases and have given ages that are reproducible within useful limits. Many determined ages have proven to be consistent with geologic relations, as described in Chapter 13. All the methods, however, require involved, expensive equipment that must be operated by experienced scientists. Age determinations in commercial laboratories thus cost from several hundred dollars per determination on up. More complete and quantitative descriptions of the methods and the general theory of dating are given in *Ages of rocks, planets and stars* by Henry Faul (New York: McGraw-Hill, 109 pp., 1966), and a thorough and exceedingly well written description of one of the methods is *Potassium-argon dating* by G. Brent Dalrymple and Marvin A. Lanphere (San Francisco: W. H. Freeman, 258 pp., 1969).

The Radiocarbon (^{14}C) Method

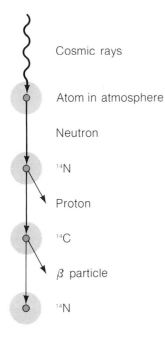

Cosmic rays

Atom in atmosphere

Neutron

^{14}N

Proton

^{14}C

β particle

^{14}N

FIGURE B-2

Chain of reactions in the formation and decay of radiocarbon atoms.

We have already noted the instability of the carbon isotope ^{14}C, also called radiocarbon. This isotope has a half life of approximately 5,570 years, which makes it useful in dating carbon-bearing materials younger than about 50,000 years. ^{14}C forms when atoms of ^{14}N acquire neutrons produced by cosmic ray bombardment of the upper atmosphere. As shown diagrammatically in Figure B-2, each ^{14}N activated by a neutron releases a proton and thus becomes ^{14}C. The ^{14}C, in turn, decays back to ^{14}N by emitting a beta particle. Because the production of ^{14}C and its decay back to ^{14}N go on at *approximately* constant rates, the ratio of ^{14}C to ^{12}C (the most abundant stable isotope) remains approximately constant in the atmosphere.

The ^{14}C in the atmosphere combines with oxygen to form carbon dioxide and thus is used in plants and animals during their growth. When a plant or animal dies, the ratio of ^{14}C/^{12}C in it is the same as that in the atmosphere, but the resistant parts (such as bones, shells, carbonized wood) that may then be preserved no longer exchange carbon with the atmosphere. Their ^{14}C decays gradually to ^{14}N, and their ratio of ^{14}C/^{12}C gradually diminishes. They can be dated by measuring their ^{14}C content by counting the number of beta particle emissions per unit of time and by comparing this quantity with the total mass of carbon. Age is then calculated by using the known atmospheric ratio already mentioned.

As noted, however, the atmospheric ratio remains only approximately constant, being affected by several changes that can be identified but have not been measured. One known change is an increase in the proportion of ^{12}C in the atmosphere due to the burning of coal and petroleum during the past 100 years. This burning has enriched the atmosphere in ^{12}C because the fuels used are too old to contain appreciable ^{14}C. The atmospheric explosion of nuclear bombs releases neutrons and thus leads to increases in atmospheric ^{14}C. Older variations in the ratio ^{14}C/^{12}C are probably caused by changes in the earth's magnetic field, which affects the cosmic rays entering the atmosphere. Cosmic radiation has evidently varied by about 10 percent during the past 5,000 years, as determined by radiocarbon dates on old trees that can be dated precisely by counting their growth rings.

These atmospheric variations, however, must have affected *all* plants and animals equally during any one period, so that ^{14}C dates can be used to match materials of equal age. The absolute ages are also exact enough to prove highly useful in their own right. They have revolutionized our understanding of the latter part of the ice ages (Chapter 7) and, especially, the course of human cultures.

The Potassium-argon Method

The isotope of potassium with an atomic weight of 40 (abbreviated ^{40}K) has a half life of 1,300 million years. This slow rate of decay would appear to limit the use of the isotope dating method to old rocks. However, potassium is so abundant in the common minerals mica and feldspar that the method can often be used to date rocks younger than a million years. Moreover, the abundance of rocks containing mica and feldspar make this the most frequently used method.

Nuclei of ^{40}K decay in two ways: (1) by emitting a beta particle and thereby changing to ^{40}Ca and (2) by capturing an electron and becoming ^{40}Ar (argon-40). The measurement of either daughter in a rock or mineral in which the content of ^{40}K had also been measured would thus be a basis for an age calculation. However, the ^{40}Ca cannot be used because it is an abundant isotope of calcium that may be introduced into rocks in other ways. Argon, on the other hand, is a chemically inert gas that can be separated and measured precisely. Argon from the air may enter the sample, but it can be identified by its typical isotopic composition and corrected for by calculation. The essentials of the method are as follows:

1. Analysis of the total amount of potassium in the rock or mineral.
2. Calculation of the amount of ^{40}K by using the known isotopic constitution of potassium.
3. Melting of the sample in a vacuum to extract and collect the argon.
4. Measurement of the proportion of ^{40}Ar in the argon sample, using a mass spectrometer.
5. Correction for atmospheric argon.
6. Calculation of the age from the amounts of ^{40}K and ^{40}Ar and from the known rate of decay.

The major limitation of the method derives from the fact that argon is a gas and may leak from the minerals in which it forms. The losses are generally large if a rock is heated to more than a few hundred degrees Celsius, so that any signs of metamorphism would make samples undesirable. Most potassium-bearing sedimentary minerals (for example, illite) also lose argon quite rapidly at low temperatures. Potassium-argon dating of sedimentary rocks must therefore depend on dating of associated igneous rocks that have been related geologically to the sedimentary rocks, as described in Chapter 13.

A great advantage of the method is that mica and feldspar separated from the same rock can be dated individually. If the dates are *concordant* (equal or nearly so) or if they differ consistently with each mineral's tendency to retain argon, the age of the rock can be considered reliable. Of special value is the fact that there are about 50 laboratories doing postassium-argon dating, so that important ages may also be checked for reproducibility.

The Rubidium-strontium Method

This method is based on an unstable isotope of rubidium ^{87}Rb, which decays by emitting a beta particle and thereby becomes a stable isotope of strontium, ^{87}Sr. Rubidium, however, is a rather scarce element, and the decay of ^{87}Rb is very slow (its half life is approximately 47 billion years). The method therefore gives only approximate ages of rocks younger than 50 million years. The most accurate ages are obtained from igneous rocks in single intrusions that have a considerable range of composition. The ratio of rubidium to strontium (Rb/Sr) is commonly much greater in silica-rich igneous rocks such as granite than it is in darker rocks such as gabbro, and the range in ratios provides a valuable means of determining an age. The method proceeds in these steps:

1. Collection of a half dozen or so samples, together showing a considerable range of rock composition.
2. Preparation of the samples by crushing and solution in acid, followed by separation of rubidium and strontium.
3. Analysis for amounts of rubidium and strontium.
4. Analysis of the isotopes with a mass spectrometer, including ^{86}Sr, which is a stable, common isotope of strontium that does not change in amount.
5. Plotting the ratios of the measured amounts of ^{87}Rb and ^{87}Sr to the stable isotope ^{86}Sr, such as in Figure B-3.

The age of the rock is then calculated from the slope of the line in Figure B-3, left. To see why this method should work, let us consider what has happened in the rocks since they crystallized from a melt. Figure B-3, right, shows the same kind of graph with solid dots representing the samples *as if they had been collected and analyzed just after they crystallized.* They all have the same $^{87}Sr/^{86}Sr$ ratio because they formed from a body of melt that came from the same source in the deep earth. (This is not an assumption—it has been proven by studies of recently formed igneous bodies.) As we have noted, however, the ratio $^{87}Rb/^{86}Sr$ differs among the different parts of the igneous body, as shown by the spread of the samples along the line in the figure. As time passes and ^{87}Rb decays to ^{87}Sr, the ratios of $^{87}Rb/^{86}Sr$ will decrease and the ratios of $^{87}Sr/^{86}Sr$ will increase, for the amount of ^{86}Sr remains the same. The changes are traced by the dashed arrows in Figure B-3, right. Note that

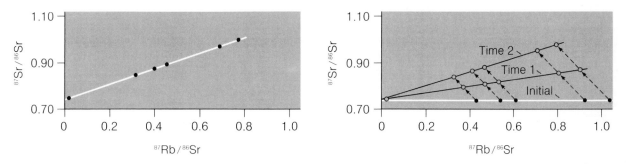

FIGURE B-3

Left, diagram comparing the ratio $^{87}Sr/^{86}Sr$ to the ratio $^{87}Rb/^{86}Sr$ in six samples from a single pluton. Using the known rate of decay of ^{87}Rb to ^{87}Sr, the slope of the line connecting the sample points can be used to calculate the age of the pluton. Right, diagram showing how all the samples would have had the same ratio of $^{87}Sr/^{86}Sr$ when the pluton first crystallized, so that each sample point would be changed systematically as the pluton aged. The solid line joining them will thus gradually steepen with time.

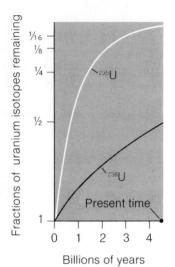

FIGURE B-4

Curves showing decay of ^{235}U and ^{238}U during the approximately 4.6 billion years since the earth formed.

the lengths of the arrows are proportional to the amounts of ^{87}Rb in each sample. Note, too, that their directions are the same, a relation due to the fact that the amount of ^{87}Sr produced in each sample is equal to the amount of ^{87}Rb that has decayed. Thus, an array of samples will describe a straight line *at any time*, and the line will become steeper as the rocks get older.

A special value of this method is that an accurate age is connoted when the sample points all lie close to a straight line. This relation implies that alterations have not moved rubidium and strontium within the system after the rocks first crystallized. Another value is that the age applies to an entire pluton rather than to only one rock from it.

The Uranium-lead Method

Two unstable isotopes of uranium, ^{235}U and ^{238}U, decay through a series of unstable daughter isotopes to stable isotopes of lead, ^{207}Pb and ^{206}Pb, respectively. The half life of ^{235}U is approximately 713 million years and that of ^{238}U is approximately 4,510 million years. Figure B-4 shows the two decay curves, either one of which can be used to calculate the age of a sample containing adequate uranium. The two curves also give a basis for calculating the gradual change in the ratio $^{207}Pb/^{206}Pb$, which provides another method of determining the age of a sample. Analysis of a sample for all four isotopes thus gives three separate determinations of the age of the mineral analyzed: (1) one based on the ratio $^{235}U/^{207}Pb$; (2) one based on the ratio $^{238}U/^{206}Pb$; and (3) one based on the ratio $^{207}Pb/^{206}Pb$. If the three ratios give essentially the same age, we can feel that the age is reliable. If the differences among the determinations are moderate, the age based on the ratio $^{207}Pb/^{206}Pb$ is generally favored because any alterations that remove lead are not likely to remove one isotope in preference to another. Indeed, lead is typically lost by minerals, so that the two uranium-lead ratios cannot be used to obtain an accurate age. However, a method has been devised to measure the lead losses and to use them in a graphical determination of the age (see the book by Henry Faul, already noted).

Up to a few decades ago few minerals contained enough uranium to be of use. Technological advances have now made it possible to use the mineral zircon, which generally contains small amounts of uranium. This advance is an important one, for zircon occurs widely in many rocks, being especially abundant in granites and related igneous rocks. A great advantage in using the uranium-lead method on zircon is that the mineral is typically unaffected by alterations that often spoil rocks for the potassium-argon and rubidium-strontium methods.

Index and Glossary

Including Units of Measure

A

Drill cores, 214, 216

Drilling from Glomar Challenger, 144, 413

Drumlins *hills of till (sediment deposited by a glacier) that are elongated and streamlined by flow of the overlying ice.* 145 (fig.), 157–59

Ductile substances *substances that can flow slowly but readily in the solid state.* 148, 330–32, 351, 360

Dune *a mound or hillock of sediment, most commonly sand, deposited by wind or moving water.*
  river dunes, 54–55, 314 (fig.)
  submarine dunes, 78 (fig.)
  wind dunes, 81–82, 173–75

Dust storms, 176

E

Earth, ellipsoidal shape, 382
  density of, 389
  deep mantle and core, 388–90
  lithosphere and asthenosphere, 362, 364, 381 (fig.), 383, 384
  magnetic field, 401–04
  early history of, 405
  oldest dated rocks of, 397
  seismicity of, 414, 424
  natural resources of, 453–74

Earthquakes, aftershocks, 245
  in Alaska, 1964, 233–35, 241–47
  in California, 7, 251
  causes and nature of, 236, 246
  controlling with water, 248–51
  damages related to water-charged sediments, 233
  in Denver, 1962–67, 248
  first-motion solutions for, 240, 245
  off Grand Banks, 139
  intensities of, 240
  magnitudes of, 238
  predictions of, 247, 253
  related to fluid pressures, 250
  related to subduction, 414, 424, 442

Earth system, 3

Earth tides, 424

Ebb tide *the seaward movement of water during the falling tide.* 87

Eclogite, 510

Eddies in flowing water, 13, 15, 37, 54

Eddies associated with wind dunes, 174 (fig.)

Elastic rebound theory *a name given to the slow setting and eventual rapid motion of rocks along a* fault, here called the stick-slip mechanism *of generating earthquakes.* 236

Electron *a minute, negatively charged particle that is the elemental energy unit (− 1) of negative electricity. Electrons are very light (each has about 1/1,000th the mass of a proton), but their energy space constitutes by far the greater part of the earth.*

Element *a substance made up of atoms having a specific number of protons in their nuclei.*

Elements, table of (with symbols), 478

Elements, abundances of common ones, 481 (table)

Ellesmere Island, glaciers, 147, 152 (fig.)

End moraine *an irregular ridge of sediment deposited at the terminus (end) of a glacier.* 148, 154, 159–60

Environments of sediment deposition deduced from rocks, continental, 288–92, 307–14, 341
  marine, 282–84, 310–12, 315, 440–42, 445

Eocene *a division of the Cenozoic Era.* 293 (fig.)

Eolian *pertaining to the wind.*

Ephemeral streams and lakes *water bodies that are dry except during rains or the rainy season.*

Epicenter *the point on the ground directly over the place where a fault generates an earthquake.* 238

Epidote, 490

Epoch *a period of geologic time formalized by a name and briefer than a Period (for example, the Pliocene, Miocene, etc. on p. 293, fig.).*

Era *the longest period of geologic time formalized by a proper name.* 295 (fig.)

Erosion *the loosening and carrying away (transport) of soil and rock particles.*
  in arid regions, 61–68, 198–202, 205
  by the freezing of water, 59
  by glaciers, 59, 148, 156–57
  of the Piedmont province, 25
  protection against by grasses, 25, 171, 176
  protection against by woodlands, 23–24, 60
  by impact of raindrops, 24
  by river systems, 32, 34–39, 61–68
  by tidal currents, 87–88
  due to urbanization, 43
  by waves, 106–108, 369–71
  by wind, 172–76, 400 (fig.)

ERTS images *photographs transmitted from Earth Resources Technology Satellite-1.* 5, 73, 80, 211

Estuaries *water bodies connected to the ocean (and thus affected by the tides) and also to a river or other fresh-water source on land (and thus varying in salinity from head to mouth).* 4, 73–96
  ancient deposits of, 309, 314

related to ore deposits, 271–73, 455–58
effects of resistance on mountain streams, 59
soils formed from, 176–79, 181–84, 185–86
Granodiorite, 271, 502
Granofels, 509
Granular *the textural aspect of a rock or sediment consisting of more or less equant grains (as most sands and granites).*
Granulite, 509
Graphite, 490
Grasses, annual root growth, 171
cause of deposition of loess, 176
evolution of, 298
soils formed by, 171–73, 187–89
Grassland soil, depletion of nutrients in, 189–90
distribution and importance, 171
erosion of, 176
origin of, 171–76, 187–89
Gravel bar (*See* Bar), 372
Gravitation (law of), 381
Gravity *the attractive force between the earth and any object in it, on it, or within this part of the universe; its cause is unknown.*
a cause of faulting, 336, 338, 339 (fig.)
a cause of folding, 332
over spreading rises, 423
results of precise measurements of, 382
variations near subduction zones, 417
Graywacke, 442, 507
Great Lakes region, continuing isostatic adjustment of, 382
glaciation in, 157, 162
uplifted former shorelines in, 375, 378 (fig.)
Great Salt Lake, Utah, 370
Greenland Ice Sheet, 150–55
general shape; nature of margins, 151, 153 (fig.)
isostatic depression caused by, 384
origin of end moraine, 154–55
precipitation and temperatures on, 151–54
rates of flow, 154
temperature history from, 162
transport and deposition of detritus, 154–55
Greenschist, 508
Ground moraine, 157
Groundwater *the more or less continuous body of water occupying cracks and pores in sediment and rocks.* 26–28, 470
compaction due to removal of, 220–23
effects of shallow groundwater on soils, 24, 183
effects on earthquake damage and faulting, 233–36, 248–51
flow through hot plutons, 272

Grouse Creek Mountains, Utah, deformed rocks, 328–343
Gulf Stream, 127–31
composition, 127, 130
meanders and closed loops, 128–29, 131 (fig.)
velocity and discharge, 127–28
Gully *a steep-walled valley of a small, generally ephemeral, stream; generally incised recently in easily eroded materials.* 25 (fig.), 305 (fig.)
Gypsum, 490
Gyre *a broad, horizontal circulation in the ocean, more or less circular in plan and of any duration.* 127 (fig.), 130 (fig.)

H

Hachures (on maps), 514
Halite, 477–79, 491
Half lives of radioactive isotopes, 527–28
Hardness of minerals, 487
Headcuts in arroyos, 63
Hectare = 10,000 square meters = 0.0039 square mile
Hematite, 491
High-angle faults, 337–39
Histograms of sediment sizes, 106
Holocene *the present period of geologic time, starting with the end of the last glaciation, approximately 10,000 years ago.*
Hornblende, 270, 491
Hornfels, 509
Hot (volcanic) fragmental flows, 258, 259
Hudson Bay, post glacial uplift, 367 (fig.), 382
Hudson (submarine) Canyon, 109 (fig.)
Humus *organic substances in soils, produced by decomposition of plant and animal remains.*
Hurricane Camille, 33
Hydrocarbon *any compound of hydrogen and carbon.*
Hydrology *the earth science treating the subject of water.*

I

Ice cap, 147 (fig.)
Ice lobe, 147, 160–161
Ice sheet *a glacier extending over thousands of square miles as a nearly flat mound of ice.* (*See* Greenland Ice Sheet, 150–55)
effects of former ice sheets, 156–61

P

A 7
B 8
C 9
D 0
E 1
F 2
G 3
H 4
I 5
J 6

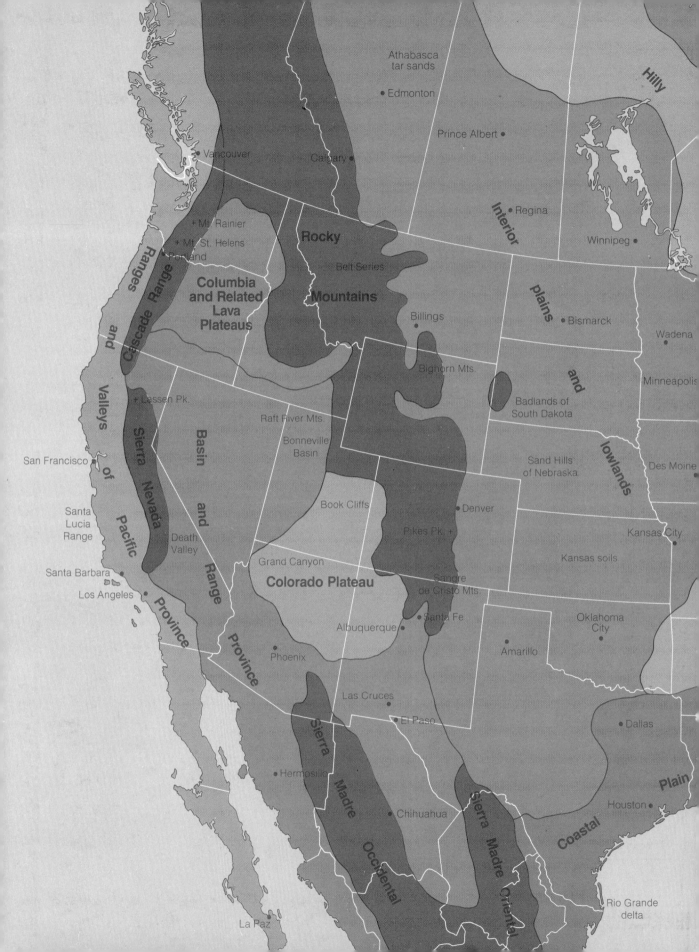